GLACIER HYDROLOGY AND HYDROCHEMISTRY

Edited by

M. SHARP
Department of Earth and Atmospheric Sciences, University of Alberta, Canada

K. S. RICHARDS
Department of Geography, University of Cambridge, Cambridge, UK

M. TRANTER
School of Geographical Sciences, University of Bristol, Bristol, UK

JOHN WILEY AND SONS
Chichester · New York · Weinheim · Brisbane · Singapore · Toronto

The papers in this volume were originally published in
Hydrological Processes – An International Journal
Volume 8 Issue 5 (465–480), **Volume 10 Issue 4** (479–508, 509–521, 523–539, 541–556, 557–578, 579–597, 599–613, 615–628, 629–648, 649–660), **Volume 10 Issue 10** (1411–1426), **Volume 12 Issue 1** (87–104, 105–134), **Volume 12 Issue 2** (191–219), **Volume 11 Issue 1** (59–78), **Volume 11 Issue 2** (211–224)

Other Wiley Editorial Offices

John Wiley & Sons, Inc., 605 Third Avenue,
New York, NY 10158-0012, USA

WILEY–VCH Verlag GmbH, Pappelallee 3,
D-69469 Weinheim, Germany

Jacaranda Wiley Ltd, 33 Park Road, Milton,
Queensland 4064, Australia

John Wiley & Sons (Asia) Pte Ltd, 2 Clementi Loop #02-01,
Jin Xing Distripark, Singapore 129809

John Wiley & Sons (Canada) Ltd, 22 Worcester Road,
Rexdale, Ontario M9W 1L1, Canada

Library of Congress Cataloging in Publication Data

Glacier hydrology and hydrochemistry / edited by M. Sharp, K.S.
   Richards, M. Tranter.
       p.   cm. — (Advances in hydrological processes)
   Includes bibliographical references and index.
   ISBN 0-471-98168-0 (alk. paper)
   1. Hydrology.   2. Glaciology.   3. Water chemistry.   4. Drainage.
I. Sharp, M. (Martin)   II. Richards, K. S.   III. Tranter, M.
(Martyn)   IV. Series.
GB2404.G54   1998
551.48—dc21   98-7026
                                               CIP

British Library Cataloguing in Publication Data

A catalogue record for this book is available from the British
Library

ISBN 0-471-98168-0

Typeset in 10/12 Times
Printed and bound in Great Britain by BPC Wheatons Ltd, Exeter

CONTENTS

Preface   vii

1   Introduction   1
M. Sharp, K. S. Richards and M. Tranter

2   Effect of Snow and Firn Hydrology on the Physical and Chemical Characteristics of Glacial Runoff   15
A. G. Fountain

3   Isotopic and Ionic Variations in Glacier River Water During Three Contrasting Ablation Seasons   29
W. H. Theakstone and N. T. Knudsen

4   The Hydrochemistry of Runoff from a 'Cold Based' Glacier in the High Arctic (Scott Turnerbreen, Svalbard)   47
R. Hodgkins, M. Tranter and J. A. Dowdeswell

5   Hydrochemistry as an Indicator of Subglacial Drainage System Structure: a Comparison of Alpine and Sub-Polar Environments   65
M. Tranter, G. H. Brown, A. J. Hodson and A. M. Gurnell

6   Impact of Subglacial Geothermal Activity on Meltwater Quality in the Jökulsá á Sólheimasandi System, Southern Iceland   81
D. M. Lawler, H. Björnsson and M. Dolan

7   Velocity–Discharge Relationships derived from Dye Tracer Experiments in Glacial Meltwaters: Implications for Subglacial Flow Conditions   103
P. W. Nienow, M. Sharp and I. C. Willis

8   Links Between Proglacial Stream Suspended Sediment Dynamics, Glacier Hydrology and Glacier Motion at Midtdalsbreen, Norway   119
I. C. Willis, K. S. Richards and M. J. Sharp

9   Impact of Post-mixing Chemical Reactions on the Major Ion Chemistry of Bulk Meltwaters Draining the Haut Glacier d'Arolla, Valais, Switzerland   139
G. H. Brown, M. J. Sharp, M. Tranter, A. M. Gurnell and P. W. Nienow

10   Experimental Investigations of the Weathering of Suspended Sediment by Alpine Glacial Meltwater   155
G. H. Brown, M. Tranter and M. J. Sharp

11   Statistical Evaluation of Glacier Boreholes as Indicators of Basal Drainage Systems   175

C. C. Smart

12   The Use of Borehole Video in Investigating the Hydrology of a Temperate Glacier   191
     L. Copland, J. Harbor, S. Gordon and M. Sharp

13   *In situ* Measurements of Basal Water Quality and Pressure as an Indicator of the Character of
     Subglacial Drainage Systems   205
     D. B. Stone and G. K. C. Clarke

14   Variability in the Chemical Composition of *in-situ* Subglacial Meltwaters   219
     M. Tranter, M. J. Sharp, G. H. Brown, I. C. Willis, B. P. Hubbard, M. K. Nielsen, C. C. Smart,
     S. Gordon, M. Tulley and H. R. Lamb

15   Seasonal Reorganization of Subglacial Drainage inferred from Measurements in Boreholes   239
     S. Gordon, M. Sharp, B. Hubbard, C. Smart, B. Ketterling and I. Willis

16   An Integrated Approach to Modelling Hydrology and Water Quality in Glacierized Catchments 269
     K. Richards, M. Sharp, N. Arnold, A. Gurnell, M. Clark, M. Tranter, P. Nienow, G. Brown,
     I. Willis and W. Lawson

17   Initial Results from a distributed, Physically based Model of Glacier Hydrology   299
     N. Arnold, K. Richards, I. Willis and M. Sharp

18   Towards a Hydrological Model for Computerized Ice-Sheet Simulations   329
     R. B. Alley

     Index   341

PREFACE

Traditionally, glacier hydrology has been a somewhat neglected branch of hydrology, and one which has had closer links with glaciology than with catchment hydrology. This history is reflected in the restricted range of approaches which have been employed in glacier hydrological studies and in the limited degree to which models of glacier hydrological systems have been developed. However, the situation is now changing, largely in response to a series of practical imperatives related to the uncertain response of glaciers and ice sheets to climate change, and to the implications of this response for regional water and power resources. Volumetrically, glaciers and ice sheets represent the most significant reservoir of fresh water on the surface of the Earth. The magnitude of this reservoir changes substantially on timescales of thousands of years in response to changes in global climate, impacting significantly on global sea level. Release of meltwaters from terrestrial ice masses during periods of deglaciation constitutes a major perturbation of continental hydrology, and may also induce large scale changes in ocean circulation. On shorter timescales, changes in glacier extent and the rate of glacier melting impact significantly on regional water resource availability and power generating capacity. Understanding the generation of runoff from glacier surfaces is therefore an important challenge for the science of global change. Meeting this challenge requires that we improve our understanding of the processes which convert glacier ice into liquid water, and especially of the processes which route this water from the site of production into extraglacial drainage systems. This in turn demands that we develop techniques to determine the structure of the internal drainage systems of ice masses, and methods to model their behaviour. Given the inaccessible and dangerous character of glacier interiors, this is no simple undertaking, and it is one which has hitherto been addressed largely with reference to small alpine glaciers. The task of determining whether similar processes also operated in large Pleistocene ice sheets is only just beginning to be confronted.

The papers in this volume, all of which have appeared in *Hydrological Processes*, deal with various aspects of these problems, although the principal focus is on deciphering the properties of glacier drainage systems in modern environments. The majority of papers are based on field studies, but a variety of approaches is reflected in the collection of papers. These include measurements of the hydrochemistry and sediment content of glacial meltwaters, dye tracing, and englacial and subglacial investigations conducted via boreholes created using the hot water drilling technique. In addition, several papers begin to approach the problem of modelling glacier drainage system behaviour for both modern and palaeo ice masses. We hope that the collection of papers both contributes to our understanding of hydrological processes in glacial environments and acts as a stimulus to further study, particularly in the areas of linking modelling with field investigations and of understanding the hydrology of large ice sheets.

Martin Sharp
Keith Richards
Martyn Tranter

1

INTRODUCTION

M. SHARP

Department of Earth and Atmospheric Sciences, University of Alberta, Canada

K. S. RICHARDS

Department of Geography, University of Cambridge, Cambridge, UK

AND

M. TRANTER

School of Geographical Sciences, University of Bristol, Bristol, UK

A major research theme within glaciology over the last 20 years has been the linkage between glacier hydrology and ice mass dynamics (Iken, 1981; Bindschadler, 1983; Iken et al., 1983; Kamb, 1987). This is now recognised as being fundamental in mediating the way in which glaciers and ice sheets respond to environmental change, and in determining the ways in which ice masses shape the landscape.

Where the base of a glacier or ice sheet is at the pressure melting point, water at the glacier bed can have a profound impact on rates and processes of glacier motion. This is true, regardless of whether the ice rests on a rigid substrate or on a layer of unconsolidated, permeable and potentially deformable sediment (Boulton and Jones, 1979; Clarke, 1987; Brown et al., 1987). In either case, the critical variable is believed to be the pressure of water within the subglacial drainage system. High water pressures promote decoupling between the glacier sole and both rigid and deformable beds (Iverson et al., 1995), and can reduce the strength and viscosity of subglacial sediments (Boulton and Hindmarsh, 1987). Both processes facilitate rapid basal motion.

The subglacial water pressure depends on, among other things, the meltwater flux through the drainage system, the gradients of ice pressure and bedrock slope which drive the water flow, the character and network properties of the drainage system as a whole, and the hydraulic geometry of individual drainage passageways (Röthlisberger, 1972; Kamb, 1987; Fowler, 1987; Hooke et al., 1990). The sign of the slope of the relationship between water pressure and water flux may also be determined by the drainage system geometry (Walder, 1986; Kamb, 1987; Walder and Fowler, 1994). The occurrence of fast glacier flow phenomena, such as ice streams (Bentley, 1987) and glacier surges (Raymond, 1987), may thus be explicable in terms of spatial and temporal variations in subglacial drainage system characteristics. For ice dynamicists, therefore, there is considerable interest in the nature of subglacial drainage systems, in the manner of their evolution through time, and in explanations of why they differ within and between glaciers.

Given this burgeoning interest, it seems appropriate to take stock of what has been achieved by studies of glacier drainage systems over the last two decades, to identify how research approaches have changed over this period, to review how changes in approach have led to the identification of new research questions and methodological challenges, and to discuss some of the critical themes emerging from current research. This collection of papers, all of which have previously been published in the journal *Hydrological Processes*, is designed to serve this purpose.

Despite the quite different specific research questions being addressed, many of the methodological challenges facing glacier hydrologists have already been faced by workers in hillslope and drainage basin hydrology. Traditionally, research in glacier hydrology has been published in glaciological and geophysical journals, rather than in the mainstream hydrological literature. As a result, glacier hydrology has remained somewhat isolated, has developed more slowly than other areas of hydrology, and has yet to take advantage of recent methodological advances in these areas. Collectively, the papers in this volume make a statement about the current state of glacier hydrology, which we hope will provide a focus for future work, and attract the interest of the broader hydrological community. We hope that the volume will stimulate increased interaction between glacier hydrologists and that broader community. Such an interaction should facilitate the development of new modelling strategies and assist in tackling the important issues of scaling between field observations and modelling at both glacier and ice sheet scales.

THE CHANGING NATURE OF RESEARCH IN GLACIER HYDROLOGY

Field studies in glacier hydrology

A major challenge facing pioneers of glacier hydrology was to determine the character of drainage systems located within and at the base of glaciers. Faced with the difficulty of making direct observations at the glacier bed, early workers relied heavily on indirect methods to make inferences about water sources and flow routing through glaciers. Prominent among these methods were dye tracing (Behrens et al., 1971; Lang et al., 1979; Theakstone and Knudsen, 1981; Collins, 1982; Burkimsher, 1983), the analysis of the solute and sediment concentration in glacial meltwaters (Collins, 1977, 1979; Raiswell and Thomas, 1984; Thomas and Raiswell, 1984; Gurnell and Fenn, 1985), and hydrograph shape analysis (Elliston, 1973).

All of these approaches are essentially "inverse" in character, in that system outputs are used as a basis for qualitative inferences about drainage system properties. Unfortunately, early attempts to apply these methods were hindered by the lack of well-defined theory to facilitate the interpretation of observations, and by the difficulty of obtaining independent evidence to corroborate interpretations. As the different techniques were rarely applied together, it was difficult to establish whether or not the conclusions drawn from them were mutually consistent. As a result, these early field studies had limited impact on the thinking of those concerned with incorporating a treatment of glacier hydrology into models of ice dynamics (e.g. Fowler, 1987; Alley, 1989).

More recently, several groups have sought to address some of these problems in an attempt to integrate field studies of glacier hydrology more effectively with theoretical development. Of particular interest in this respect is the research of Hooke and co-workers on Storglaciären, of Fountain at South Cascade Glacier, and of Clarke and his students at Trapridge Glacier. Our own work, initially at Midtdalsbreen in Norway and subsequently at Haut Glacier d'Arolla in Switzerland, has been conducted in a similar vein. These studies are characterised by a multi-dimensional approach to glacier hydrology, usually involving some combination of the field methods discussed above with direct investigations at the glacier bed and physical–mathematical modelling.

Investigations at the glacier bed have been facilitated by the ability to produce large numbers of boreholes rapidly using hot water drilling. Such subglacial investigations, which involve both *in situ* measurements of water pressure (Engelhardt, 1978; Hodge, 1979; Fountain, 1994; Smart, this volume) and water quality characteristics (Stone et al., 1993; Hubbard et al., 1995; Stone and Clarke, this volume; Tranter et al., this volume a; Gordon et al., this volume), and manipulation of the subglacial environment (Stone and Clarke, 1993; Waddington and Clarke, 1995; Iken et al., 1996; Kulessa and Hubbard, 1997), have generated new data which provide motivation for the development of more sophisticated theories of subglacial drainage (Stone, 1993; Hooke and Pohjola, 1994; Hubbard et al., 1995; Murray and Clarke, 1995; Clarke, 1996a). They also provide an independent means of testing the validity of inferences derived from the more traditional "inverse" methods (Lamb et al., 1995). The emphasis is thus increasingly on intensive case studies, demanding labour- and capital-intensive monitoring in which several interdependent indicators are measured simultaneously.

Whilst early work was largely empirical and inferential in character, modern research in glacier hydrology demands close integration and feedback among field research programmes, numerical modelling and theoretical development. Field studies provide new observations which guide the development of theory, which in turn provides a basis for the formulation of numerical models. Field studies also generate data which are essential for the initialisation and parameterisation of models, and which provide a basis for model testing and validation. This is not, however, a one way process. New theories can be used to predict how parameters such as meltwater chemical composition will reflect particular drainage configurations (Sharp, 1991; Tranter et al., 1993; Clarke, 1996a; Tranter et al., this volume b). Models which predict internal, system-state variables (such as basal water pressure and water throughflow times; Clarke, 1996b; Arnold et al., this volume) encourage new field approaches (such as hot-water borehole drilling and associated subglacial instrumentation) to allow the validation of models using spatially distributed internal data rather than simply lumped output data (Gordon et al., this volume). Thus the old "inverse" approaches are supplemented and enriched by newer "forward" approaches. In this respect, glacier hydrology mirrors conventional hydrology, where spatial distributions of the internal predictions of a model are increasingly emphasised. This is because the lumped output of a distributed model fails to do justice to the complexity of the model structure, and because good matches between observations and lumped outputs from distributed models can be achieved using different parameter combinations that imply completely different internal behaviours (e.g. Chappell and Ternan, 1992; Quinn and Beven, 1993).

Approaches to modelling glacier hydrology

At the same time as the first detailed field studies of glacier hydrology were being conducted, a number of theoreticians began to consider the form that subglacial drainage systems might take. Various possible drainage configurations, including water sheets and films (Weertman, 1972; Walder, 1982), linked cavities (Lliboutry, 1968; Walder, 1986; Kamb, 1987), channels incised into bedrock (Nye, 1973) and glacier ice (Röthlisberger, 1972; Shreve, 1972), and canals incised into deforming till (Walder and Fowler, 1994), have been proposed. Some workers have undertaken stability analyses to determine which configuration is most likely under particular circumstances (Kamb, 1987; Fowler, 1987; Walder and Fowler, 1994).

It is now widely recognised that drainage configurations may change over time as a consequence of changing glacier geometry or water flux (Raymond, 1987; Walder and Fowler, 1994; Fountain and Vaughn, 1995). However, there has been rather less appreciation that different types of drainage system may co-exist beneath a given ice mass. A number of field studies provide evidence for this (Humphrey et al., 1986; Raymond et al., 1995; Hubbard et al., 1995; Murray and Clarke, 1995; Richards et al., this volume; Gordon et al., this volume), and suggest a need to incorporate a treatment of interactions between system components in models of the full glacier hydrological system. This challenge has recently begun to be addressed (Clarke, 1996b; Alley, this volume), and is discussed further below.

For a long time, modelling of glacier hydrological systems lagged well behind the development of theory. Whilst some of the theories discussed above were used at an early stage as a basis for modelling glacier outburst floods (Nye, 1976; Spring and Hutter, 1981; Clarke, 1982), less sophisticated approaches were adopted for modelling more routine behaviour such as diurnal and seasonal runoff variations. Initially, such behaviour was investigated via the use of chemically based mixing models to separate meltwater discharge hydrographs into flow components allegedly attributable to runoff through different elements of the glacier drainage system (Collins, 1979; Collins and Young, 1981; Gurnell and Fenn, 1984; Tranter and Raiswell, 1991; Lecce, 1993). The results appeared to offer insight into the relative importance of different drainage system components and the nature of interactions between them. They were, however, limited by the arbitrary assumptions they made about drainage system structure and by the lack of realism of some of the assumptions which they made about meltwater chemistry (Sharp et al., 1995).

More sophisticated, but equally difficult to validate, were lumped catchment models, which treated glacier drainage systems as a series of serial or parallel linear reservoirs, each characterised by a distinct meltwater residence time (Baker et al., 1982). These forward models used sophisticated, often spatially distributed, energy balance approaches to compute meltwater inputs to the glacier drainage system (Escher-Vetter, 1985), but crude parameterisations to simulate flow routing through it. A major deficiency of such models was their inability to simulate temporal changes in drainage configuration and the geometry of individual drainage elements, and their effects on runoff processes. Furthermore, their performance could only be evaluated in terms of their ability to predict bulk runoff from a glacier. Thus, whilst they served a useful purpose as forecasting tools for water resources management, they shed little light on internal glacier hydrological processes.

The desire to explain aspects of glacier flow dynamics, such as temporal and spatial variations in rates of glacier motion, necessitates a different view of glacier hydrology. Such variations are probably attributable to changes in friction at the glacier bed and to changes in the stress field within the overlying ice, which arise because of the temporally and spatially restricted nature of such friction changes (Bahr and Rundle, 1996). It is highly likely that the causes of friction variations lie in events which take place within subglacial drainage systems, but such events are probably closely linked to events on the glacier surface (Fountain, this volume; Richards et al., this volume; Gordon et al., this volume). To understand where and when such events occur, it is necessary to have recourse to physically based, distributed models of glacier hydrology. Such models must simulate surface melt processes, changes in the distribution and hydraulic properties of surface snow and firn, vertical percolation and horizontal runoff through snow and firn (Fountain, this volume), and runoff over bare ice surfaces which change in geometry as melt progresses. They must also deal with the distributed input of runoff to the englacial drainage system, and simulate runoff through englacial and subglacial channels which grow or shrink over time due to wall melting, sediment erosion and creep closure by ice and subglacial sediments. Simulation of interactions between channel flow and more distributed flow in inter-channel areas is also necessary (Alley, this volume), as is allowance for the likelihood that channels are formed and destroyed on a seasonal basis (Richards et al., this volume; Arnold et al., this volume). Physically based models will also eventually have to deal with the role of glacier motion in opening and closing drainage pathways at the glacier bed.

Such models are challenging to construct, because of the uncertainty involved in specifying the character of subglacial and englacial drainage systems, and the location and hydraulic properties of major flow pathways. They are also demanding to run, because of the large amounts of data required to initialise and drive them. Demands for data increase still further if model performance is to be tested against spatially distributed patterns of internal system parameters (such as channel water pressure), rather than against lumped outputs (such as meltwater discharge). This will inevitably place severe limitations on the number of catchments to which such models can be applied. This emphasises the importance of developing specific research basins in contrasting hydrological environments in order that focused model development and testing can take place. Richards et al. (this volume) describe the task of assembling the database required to construct and run such a model, while Arnold et al. (this volume) present some initial model results. Whilst it is clear that model development is still at a very early stage, it is encouraging that model performance in runoff prediction is at least comparable to that of lumped models of glacier hydrology. The potential value of models of this sort is illustrated by their application in relating borehole observations of subglacial water pressure variations to model predictions of runoff variations into individual moulins (Gordon et al., this volume), and in simulating the relationships between water throughflow velocities and discharge on diurnal timescales (Arnold et al., this volume; Nienow et al., this volume).

The development of data-hungry distributed models in glacier hydrology is justified by the fact that the motivation for much current research is scientific rather than applied. Distributed models are unlikely to find widespread application as predictive tools in, for instance, the management of water resources for hydro-electric power generation. However, progressive simplification of complex models which have been tested in well-instrumented catchments may allow the development of more portable models, along with an understanding of how and why model performance degrades as model structure is simplified. This will provide a firmer basis for deciding on what sorts of models are appropriate for specific applied purposes, and increased insight into the limitations of their predictions.

MAJOR THEMES IN GLACIER HYDROLOGICAL RESEARCH

Although it is useful to identify changing approaches to modelling glacier hydrology and to be aware of the new role that field studies are playing in glacier hydrological investigations, it is also important to be aware of some of the major research problems that are emerging from these studies. The following section highlights a series of major research themes represented in the papers in this volume. Some of these relate to developments in the techniques used by glacier hydrologists, while others are concerned with gaps in knowledge which need to be filled for model development to proceed.

Borehole studies of glacier drainage systems

Models of the motion of temperate (or partially temperate) based glaciers suggest that rates of basal motion are strongly influenced by water pressure and/or water storage within subglacial drainage systems (Iken, 1981; Kamb et al., 1994). The relationship is unlikely to be straightforward, however, because subglacial water storage is probably a localised phenomenon, and because the magnitude and pattern of variation of water pressure will vary between different components of the subglacial drainage system. Water pressure fluctuations initiated in major drainage conduits may propagate into other components of the system with significant lagging and damping (Fountain, 1994; Hubbard et al., 1995; Alley, this volume). In addition to the direct effect which such events will have on glacier motion, there may be important indirect effects linked to adjustments within the glacier stress field which occur in response to localised changes in basal friction. Hence a part of the glacier which experiences increased rates of flow due to a local increase in water pressure may accelerate the flow of surrounding areas by pulling on sections of the glacier located upstream and marginal to it, and by pushing against downstream areas (Hooke et al., 1989).

To understand such phenomena, it is necessary to know something about the character of the subglacial drainage system. If it is channelised, water pressure events may be initiated in the vicinity of channels where water fluxes are concentrated. They will propagate rapidly downglacier, since hydraulic resistance is low in this direction, but will become damped and lagged as they propagate laterally against greater hydraulic resistance (Fountain, 1994; Hubbard et al., 1995; Alley, this volume). It is then important to understand where major channels are located (Sharp et al., 1993; Fountain and Vaughn, 1995), since this may be reflected in the form of the velocity field within the glacier (Harbor et al., 1997). If the drainage system is distributed in character, pressure events may be initiated synchronously over much wider areas, but downglacier propagation will generally be much less rapid than in a channelised system, due to the greater hydraulic resistance of distributed drainage systems (Kamb and Engelhardt, 1987; Stone and Clarke, this volume).

Boreholes create the possibility of determining directly the properties of the subglacial drainage system, but the process is not straightforward. Individual boreholes provide information about very localised areas of the glacier bed, and borehole arrays must be drilled to allow reconstruction of patterns in subglacial conditions from which the nature of the whole drainage system can be deduced (Murray and Clarke, 1995; Smart, this volume). Strangely, workers rarely provide an explanation of why boreholes were drilled in specific locations, suggesting that location is implicitly assumed not to be an important influence on the results obtained from borehole measurements. This assumption is clearly contradicted by results from high density drilling programmes, which show that major drainage channels are widely spaced and located towards glacier margins (at least in ablation areas), and that they affect water pressures over quite limited areas of the adjacent glacier bed (Hantz and Lliboutry, 1983; Fountain, 1994; Hubbard et al., 1995; Smart, this volume).

Questions arise concerning whether borehole arrays are drilled with a geometry and spacing which is well-matched to the configuration and natural length scales of subglacial drainage systems (Stone et al., 1994; Smart, this volume). Drilling programmes which use borehole arrays located near the glacier centreline and aligned parallel to glacier flow apparently have little chance of connecting to major channels. Consequently, inferences about subglacial drainage configuration which have been derived from such programmes on the implicit assumption that borehole water levels provide a direct measure of water pressure in subglacial channels must be treated with some scepticism. This applies, for instance, to the view that channels are broad and low in cross section, which is based on interpretation of results derived from such a drilling programme (Hooke et al., 1990).

The act of drilling a water-filled borehole may impose significant over-pressure on that part of the glacier bed to which the borehole connects. This over-pressure may trigger the formation of drainage channels which ultimately link up to the natural subglacial drainage system (Engelhardt, 1978). It is not clear whether measurements of the hydraulic properties of subglacial drainage systems derived from borehole experiments relate to the natural drainage system or to one created by drilling the borehole. Furthermore, open boreholes become part of the glacier drainage system and develop their own patterns of water supply, internal water circulation and drainage, all of which must be understood before measurements made at the base of boreholes can be interpreted in terms of subglacial conditions (Gordon et al., this volume). To date, little attention has been paid to whether drilling programmes conducted in different parts of the same glacier produce comparable results, or to whether results are reproducible between years (Smart, this volume). Further attention must be paid to all of these questions if we are to have confidence in borehole-based reconstructions of subglacial drainage conditions.

Meltwater quality as an indicator of subglacial drainage conditions

Where a more qualitative picture of flow routing and its changes over time is required, traditional inverse approaches, such as the analysis of meltwater quality variations, still have an important role to play. If this approach is to be successful, however, it is important to develop a good understanding of the processes by which meltwaters pick up solute and sediment in order that the most important controlling factors can be identified (Raiswell, 1984). It is also essential to understand how these controlling factors might vary between different types of drainage system, in order that likely water quality responses to specific hydrological conditions may be predicted (Sharp, 1991; Tranter et al., 1993). Conceptual models of the likely links between hydrological conditions, chemical weathering and sediment entrainment processes are now being developed and tested in glacierised basins in a range of climatic environments (Theakstone and Knudsen, this volume; Lawler et al., this volume; Tranter et al., this volume b; Hodgkins et al., this volume; Brown et al., this volume a), and attempts are being made to make these models more quantitative (Clarke, 1996a).

A number of critical themes are emerging from this work.

Atmospheric influences on meltwater chemistry

It is important to be able to separate atmospheric and crustal contributions to meltwater solute load if correct inferences about subglacial weathering processes are to be made. It is therefore necessary to understand the factors which control the relative importance of the two components (Theakstone and Knudsen, this volume), and to consider the impact that variations in the chemistry of meltwaters entering glaciers might have on rates and processes of chemical weathering (Fountain, this volume; Brown et al., this volume b). There is increasing evidence that snowmelt and icemelt may follow different flow pathways through glaciers, and atmospherically derived solute can act as an excellent tracer of snowmelt (Tranter et al., this volume a,b). Despite the problems of within-snowpack fractionation, analysis of the stable isotope composition of runoff may also help to discriminate between snow, ice and rainfall as sources of runoff (Behrens et al., 1971; Theakstone and Knudsen, this volume).

Lithological influences on meltwater chemistry

Whilst the subglacial hydrological environment exerts a major control on processes of solute acquisition through its influence on parameters such as meltwater residence time at the glacier bed and access to weatherable sediment and supplies of O_2/CO_2, bedrock lithology is also an important factor. Given relatively short water–rock contact times, weathering of trace components of bedrock with rapid dissolution kinetics (such as carbonate, evaporite and sulphide minerals) exerts a disproportionate influence on meltwater chemistry (Raiswell, 1984; Drever and Hurcomb, 1986). Variations in the relative abundance of carbonate and sulphide minerals in local bedrock and in the rock:water ratio in the weathering environment strongly influence the way in which meltwater chemistry evolves over time (Raiswell, 1984; Fairchild et al., 1994; Sharp, 1996). Understanding these effects is fundamental to linking water chemistry to drainage system properties and to water sources.

Suspended sediment as a solute source

Under some circumstances, there can be a close relationship between meltwater chemistry and the concentration of suspended sediment in meltwaters (Brown et al., this volume a). This implies that in-channel chemical weathering of suspended sediment can be an important process, and suggests a need to couple hydrological models capable of predicting meltwater sediment entrainment and transport with models of solute acquisition in order to maximise the information that can be gleaned from analyses of meltwater chemistry (Clarke, 1996a). Evidence that flushing of sediment from glaciers may be linked to glacier motion events (Humphrey et al., 1986; Willis et al., this volume) further emphasises the need for improved understanding of the feedbacks between glacier hydrology, basal motion, glacial erosion and suspended sediment dynamics.

Influence of glacier thermal regime on meltwater quality

Logistic constraints and interest in water resource-related aspects of glacier hydrology have meant that most field studies have been conducted on small valley glaciers in temperate alpine environments. In recent years, however, an increasing number of studies have been conducted in polar regions, where melt seasons are shorter, glacier ice is often cold (except at the surface during the melt season), and melt–runoff relationships are complicated by processes such as superimposed ice formation. Whilst subglacial drainage is largely absent from cold-based glaciers (Hodgkins et al., this volume), hydrochemical evidence suggests that it does occur in large, polythermal glaciers, in which basal ice is locally at the melting point (Tranter et al., this volume b; Skidmore, 1995; Wadham, 1997). The configuration of drainage systems within and below such glaciers is not well known, but it does appear that subglacial outflow is seasonally enhanced by inputs from surface melting.

Little is known about how surface melt penetrates large thicknesses of cold ice to reach the glacier bed, or about how subglacial waters breach marginal cold-based regions to exit the glacier. In some cases, outflow occurs throughout the year and results in the formation of solute-rich proglacial icings (Wadham, 1997), while in others, outflow ceases in winter and is re-initiated by major outburst floods during the summer melt season (Skidmore, 1995). In both cases, meltwater residence times at the glacier bed may be substantially greater than in temperate glaciers, and this may give runoff a distinctive chemical signature. Detecting this signature is often difficult, however, because small fluxes of subglacial meltwaters are rapidly mixed with and diluted by abundant supraglacial runoff.

Whilst it might be expected that runoff from cold-based glaciers would be dilute and dominated by atmospherically derived solute, this is not always the case (Hodgkins et al., this volume). Many currently cold-based glaciers may have had a different thermal regime during the Little Ice Age, when they were more extensive than today. Under such conditions, they may have eroded bedrock and deposited tills which are now exposed in ice-marginal and proglacial environments. Major ice-marginal streams are a common feature of polythermal and cold-based glaciers, and allow meltwaters to access and weather these sediments. Such streams also receive runoff from extraglacial areas which has been involved in weathering till and may have acquired a chemical composition very similar to that of subglacial drainage. In these cases, analyses of the chemistry of ice-marginal stream waters may lead to erroneous inferences concerning the extent of subglacial drainage.

As yet, virtually nothing is known about the chemistry of meltwaters draining glaciers which have been cold-based throughout their history and have produced little in the way of glacigenic sediments. Equally, little attention has been paid to the chemistry of waters produced by basal melting, or to the way in which such waters are removed from polar ice sheets in which they may be the only source of runoff. The discovery of large subglacial lakes beneath the Antarctic Ice Sheet (Kapitsa et al., 1996) raises a number of interesting questions about how these lakes are fed and drained, and about the chemical evolution of waters which may spend extremely long periods of time at the glacier bed.

Interpretation of bulk meltwater chemistry

Most glacier drainage systems contain a mix of supraglacial, englacial and subglacial components. Weathering rates and processes may vary substantially between these environments, producing waters with very different chemical signatures. However, these waters are commonly mixed prior to the point at which they are sampled for analysis. Unfortunately, mixing is rarely conservative due to the high reactivity of suspended sediment transported in meltwaters (Brown et al., this volume a,b), and there is no simple way of deducing the chemistry of the various source waters from that of the bulk runoff.

As yet there have been no attempts to simulate meltwater chemistry using coupled hydrological/geochemical models, though some attention is being paid to the development of appropriate models (Clarke, 1996a).

There have, however, been attempts to investigate the nature of subglacial weathering environments and their coupling to hydrological environments by *in situ* water sampling. This is a challenging task as boreholes provide the only obvious means of access to the glacier bed (Lamb et al., 1995; Stone and Clarke, this volume; Tranter et al., this volume a). Major problems arise in relation to developing a borehole sampling strategy which is appropriate given the unknown character of the drainage system, and in determining which components of the system have in fact been sampled. Uncertainty also exists as to the extent to which waters from different sources mix within the subglacial and borehole environments (Tranter et al., this volume a; Gordon et al., this volume), so that considerable effort must be expended on determining whether samples consist of true subglacial/englacial waters or whether they are products of weathering within the borehole environment. A major challenge lies in identifying the existence of reducing environments, such as might be expected to occur within water-saturated subglacial tills. The drilling of boreholes may result in the injection of oxygenated waters to the glacier bed and mixing of such waters with true subglacial waters may allow rapid obliteration of the chemical signatures of reducing conditions. Mixing of till pore waters with large volumes of well-oxygenated waters draining through subglacial channels also makes it difficult to identify contributions of such waters to bulk runoff.

The role of surface/subsurface linkages in the behaviour of glacier drainage systems

Glacier hydrologists have traditionally devoted much more energy to understanding the subglacial components of glacier drainage systems than they have to their supraglacial components. Studies of supraglacial environments have focused on melt modelling, analysis of water storage in the firn aquifer (Lang et al., 1977; Oerter and Moser, 1982; Fountain, 1989) and the morphology of supraglacial channels, but there has been little attempt to understand how the metamorphosis, thinning and removal of the supraglacial snowcover affects the pattern and rate of meltwater delivery to subglacial environments. Analyses of seasonal changes in the shape and timing of diurnal runoff hydrographs from glaciers provide clear evidence of the importance of this process (Fountain, this volume), while intensive dye tracing studies demonstrate a close link between the upglacier retreat of the snowline on the glacier surface and the timing of major changes in the rate at which water is transmitted through the subglacial drainage system (Nienow, 1993; Richards et al., this volume). Numerical modelling of runoff from the glacier surface into individual moulins indicates an abrupt change in the peakedness of diurnal hydrographs as supraglacial catchments become snow-free (Arnold et al., this volume). This may be sufficient to produce substantial reorganisation of subglacial drainage systems which are suddenly subjected to greatly increased peak daily water fluxes. Borehole studies provide clear evidence for the impact of the initiation of large diurnal discharge fluctuations on the pattern of water pressure variation at the glacier bed, and also on patterns of subglacial water flow (Gordon et al., this volume).

Even though there is persuasive evidence to suggest that important events within subglacial drainage systems fed by supraglacial meltwater are driven by events on the glacier surface, much remains to be learned about the precise nature of surface/subsurface coupling. The impact of supraglacial catchment morphology on the nature of runoff hydrographs into moulins has been little explored. The influence of the distribution of water input points on the character of the subglacial system and its water pressure behaviour is essentially unknown. Crevasses represent important input points. Their distribution is a function of ice flow dynamics, and it changes over time as crevasses are advected downglacier by ice flow and supraglacial streams are captured by the opening of new crevasses. Since ice flow is itself influenced by glacier hydrology, there are important feedbacks which need to be explored (Hooke, 1991). This will demand the development of whole-system models of glacier hydrology, models which may eventually need to be coupled to models of glacier dynamics.

Development of such models will require understanding of the properties of englacial drainage systems and of the way in which they link to subglacial systems. As yet, it is not clear how such understanding can be gained. Caving techniques allow access to only the largest passages (Holmlund, 1988), while studies of ice cores provide information about intergranular vein systems which probably do not transport large water fluxes (Raymond and Harrison, 1975; Nye and Frank, 1973). Borehole video photography reveals relatively large numbers of small englacial passages (Pohjola, 1994; Harper and Humphrey, 1995; Copland et al., this volume), and provides some insight into their distribution within a glacier, but these passages are probably not large enough to transport the volumes of water observed to drain into moulins. Techniques such as surface penetration radar (Hamran et al., 1996) have proved useful in identifying the general distribution of water within polythermal ice, but there is a need for new approaches to determine the three-dimensional geometry of larger englacial channels. This is important for establishing whether large englacial passages connect directly to major subglacial channels, as suggested by much dye tracing work, or whether such channels develop out of more distributed subglacial flow. It is also essential for determining whether and how surface-derived waters can penetrate large thicknesses of cold ice to allow surface-fed subglacial drainage in polythermal glaciers and ice sheets.

Interactions between the components of subglacial drainage systems

As outlined above, theoreticians have postulated the existence of a variety of subglacial drainage configurations. It is now evident that these are not mutually exclusive and that they mix in varying proportions to constitute real glacier drainage systems. The challenge of determining the geometry of such systems has already been discussed, but it is equally important to understand how they function as integrated wholes. Ultimately this will involve determining how and where water is delivered to the glacier bed, and how it is subsequently routed to the glacier terminus. This is unlikely to be a unidirectional process because the localised pattern of water delivery to the bed, the occurrence of large discharge fluctuations over short timescales and the variably pressurised character of subglacial channel flow combine to produce large and temporally variable hydraulic gradients. These can induce hydraulic damming of channel flow (Nienow et al., this volume; Arnold et al., this volume), and temporary transfer of waters from channels to other parts of the system (Hubbard et al., 1995; Gordon et al., this volume). Furthermore, large variations in channel pressure result in short-term changes in the distribution of support for the glacier overburden, which are probably associated with changes in the pattern and degree of separation between the glacier and its bed (Murray and Clarke, 1995; Gordon et al., this volume). These in turn influence the pattern of subglacial water flow through their effect on the distribution of water pressures across areas of the bed which are isolated from the direct influence of channels. Behaviour of this sort may produce elastic responses in the overlying ice, which need to be integrated over appropriate time and length scales to understand observed patterns of glacier motion (Bahr and Rundle, 1996; Fischer and Clarke, 1997).

Hydrological models for large ice sheets

Whilst most studies of glacier hydrology have focused on valley glaciers, there is an increasing need to understand the hydrology of large ice sheets (Alley, this volume). There is mounting evidence for fast flow within the former Laurentide and Fennoscandinavian ice sheets (Clark, 1994), and it has been suggested that this may be connected to North Atlantic "Heinrich Events" (Broecker et al., 1992; MacAyeal, 1993). Temporal changes in ice sheet hydrology may well be implicated in this behaviour (Fowler and Johnson, 1995), but most ice sheet models take no account of glacier hydrology. Those which do differ widely in how they represent it. Some assume that all subglacial drainage derives from basal melting, and that channelised water flow occurs only when the subglacial groundwater system is unable to transmit the whole of the imposed water flux (Boulton et al., 1995). Others take the evidence of esker sedimentology to suggest rapid variations in subglacial discharge which are unlikely to be driven by basal melting, and infer that surface-derived meltwaters were able to penetrate to the base of at least the marginal regions of ice sheets (Arnold and Sharp, 1992). Others believe that large volumes of water are stored beneath ice sheets (as in Vostok Lake, Antarctica; Kapitsa et al., 1996) and that these are periodically released in catastrophic outburst floods (Shaw, 1989; Shoemaker, 1992). To resolve these contrasting views, there is an urgent need for studies of the hydrology of sectors of modern ice sheets which might act as good analogues for Pleistocene ice sheets. These are most likely to be achieved in West Greenland, where there is substantial surface melting and evidence that surface waters do drain into the ice sheet (Braithwaite and Thomsen, 1989).

Even with field evidence to guide the construction of hydrological models for ice sheets, it will not be straightforward to translate modelling strategies developed for valley glaciers to the ice sheet scale. The grid size of most ice sheet models exceeds the size of most of the glaciers to which hydrological models have been applied, raising major questions about the way in which hydrological processes should be parameterised and about how field measurements of important parameters can be extrapolated to much larger spatial scales. Transient, seasonal and sub-seasonal phenomena appear to be important elements in the behaviour of contemporary glacier hydrological systems, and are certainly important in generating water pressure events which have a strong impact on glacier motion. They can only be effectively represented within hydrological models which operate on small timesteps and which can benefit from the availability of high temporal resolution data to drive them. There is no realistic prospect of coupling such models to ice sheet models because of the limitations of computing resources and the lack of appropriate data for the geological past. Steady-state hydrological models can be more readily coupled to ice sheet models, but it is questionable whether they can capture the processes which are most important for the operation of the system. There is therefore an urgent need for considerable ingenuity in tackling these problems of temporal and spatial scaling. Questions of how to validate models of the hydrology of past ice sheets have yet to be addressed in any serious manner, but a renewed interest in the geological products of ice sheet drainage systems is clearly called for.

CONCLUSION

This collection of papers reflects the changes in research approaches which have taken place in glacier hydrology over the last decade, and develops some of the most critical research themes which have emerged from this period of change. Investigations of glacier hydrology are characterised by rapidly evolving technologies, which allow direct observation within and at the bed of glaciers, by a growing methodological eclecticism, and by an increasing tendency for modelling and fieldwork to be conducted in parallel. In all these respects, glacier hydrology is moving down a path which has been trodden earlier by hillslope and drainage basin hydrology (such as the mutual testing of hypotheses with multivariate data, and the shift from lumped to semi-distributed modelling and associated internal testing).

Parallels between approaches to glacier and catchment hydrology have already been illustrated by the exploitation in glacier hydrology of the terrain models and geographical information systems that form the basis of semi-distributed rainfall–runoff models (Escher-Vetter, 1985; Sharp et al., 1993; Fountain and Vaughn, 1995; Arnold et al., 1996). In such models, input variables are specified to occur either in coherent spatial "patches" (Quinn and Beven, 1993), or as continuous surfaces dependent on topographic controls derived from a digital terrain model. As glacier hydrology continues to tread the path already navigated by hillslope and catchment hydrology, it will inevitably encounter many of the same problems and challenges that these sister disciplines have already begun to face. These include the incorporation of uncertainty into predictive models (Beven and Binley, 1992), the definition of relationships between the scales at which parameters can be measured and those appropriate for modelling (Beven, 1995), and the development of methods for achieving spatial sensitivity analyses (Lane et al., 1994). Glacier hydrology is already beginning to address the issue of interactions between channelised and distributed components of drainage systems, an issue which has parallels with the problem of treating interactions between matrix and macropore flow in structured soils on hillslopes. We hope that, by drawing the attention of the broader hydrological community to recent developments in glacier hydrology, we can stimulate an exchange of views and ideas between workers in these two areas which will be mutually enriching.

ACKNOWLEDGEMENTS

We thank Peter Nienow for his comments on an earlier draft of this paper. Martin Sharp acknowledges receipt of a Leverhulme Trust Linked Fellowship at the Institute for Advanced Studies, University of Bristol.

REFERENCES

Alley, R.B. 1989. 'Water pressure coupling of sliding and bed deformation. I. Water system', *J. Glaciol.*, **35**, 108–118.

Alley, R.B. This volume. 'Towards a hydrologic model for computerised ice sheet simulations'.

Arnold, N.S. and Sharp, M. 1992. 'Influence of glacier hydrology on the dynamics of a large Quaternary ice sheet', *J. Quat. Sci.*, **7**, 109–124.

Arnold, N.S., Willis, I.C., Sharp, M., Richards, K.S. and Lawson, W. 1996. 'A distributed surface energy balance model for a small valley glacier: I. Development and testing for the Haut Glacier d'Arolla, Valais, Switzerland', *J. Glaciol.*, **42**, 77–89.

Arnold, N.S., Willis, I.C., Richards, K.S. and Sharp, M. This volume. 'Initial results from a distributed, physically-based model of glacier hydrology'.

Bahr, D.B. and Rundle, J.B. 1996. 'Stick-slip statistical mechanics at the bed of a glacier', *Geophys. Res. Lett.*, **23**, 2073–2076.

Baker, D., Escher-Vetter, H., Moser, H., Oerter, H. and Reinworth, O. 1982. 'A glacier discharge model based on the results of field studies of energy balance, water storage and flow', *Int. Assoc. Hydrol. Sci. Publ.*, **138**, 103–112.

Behrens, H., Bergman, H., Moser, H., Rauert, W., Stichler, W., Ambach, W., Eisner, H. and Pess, K. 1971. 'A study of the discharge of alpine glaciers by means of environmental isotopes and dye tracers', *Z. Gletscherk. Glazialgeol.*, **7**, 79–102.

Bentley, C.R. 1987. 'Antarctic ice streams: a review', *J. Geophys. Res.*, **92(B9)**, 8843–8858.

Beven, K. 1995. 'Linkage parameters across scales: sub-grid parameterisations and scale dependent hydrology models', *Hydrol. Process.*, **9**, 507–525.

Beven, K. and Binley, A. 1992. 'The future of distributed models: model calibration and uncertainty prediction', *Hydrol. Process.*, **6**, 279–298.

Bindschadler, R.A. 1983. 'The importance of pressurised subglacial water in sliding and separation at the glacier bed', *J. Glaciol.*, **29**, 3–19.

Boulton, G.S. and Hindmarsh, R.C.A. 1987. 'Sediment deformation beneath glaciers: rheology and geological consequences', *J. Geophys. Res.*, **92(B9)**, 9059–9082.

Boulton, G.S. and Jones, A.S. 1979. 'Stability of temperate ice caps and ice sheets resting on beds of deformable sediment', *J. Glaciol.*, **24**, 29–43.

Boulton, G.S., Caban, P.E. and Van Gijssel, K. 1995. 'Groundwater flow beneath ice sheets: Part I – large scale patterns', *Quat. Sci. Rev.*, **14**, 545–562.

Braithwaite, R.J. and Thomsen, H.H. 1989. 'Simulation of run-off from the Greenland ice sheet for planning hydro-electric power, Ilulissat/Jakobshavn, West Greenland', *Ann. Glaciol.*, **13**, 12–15.

Broecker, W.S., Bond, G., McManus, J., Klas, M. and Clark, E. 1992. 'Origin of the Northern Atlantic's Heinrich Events', *Clim. Dyn.*, **6**, 265–273.

Brown, G.H., Sharp, M.J., Tranter, M., Gurnell, A.M. and Nienow, P.W. This volume a. 'Impact of post-mixing chemical reactions on the major ion chemistry of bulk meltwaters draining the Haut Glacier d'Arolla, Valais, Switzerland'.

Brown, G.H., Tranter, M., and Sharp, M. This volume b. 'Experimental investigations of the weathering of suspended sediment by Alpine glacial meltwater'.

Brown, N.E., Hallet, B. and Booth, D.B. 1987. 'Rapid soft-bed sliding of the Puget glacial lobe', *J. Geophys. Res.*, **92(B9)**, 8985–8998.

Burkimsher, M. 1983. 'Investigations of glacier hydrological systems using dye tracer techniques: investigations at Pasterzengletscher, Austria', *J. Glaciol.*, **29**, 403–416.

Chappell, N. and Ternan, L. 1992. 'Flow path dimensionality and hydrological modelling', *Hydrol. Process.*, **6**, 327–345.

Clark, P.U. 1994. 'Unstable behaviour of the Laurentide Ice Sheet over deforming sediment and its implications for climate change', *Quat. Res.*, **41**, 19–25.

Clarke, G.K.C. 1982. 'Glacier outburst floods from Hazard Lake, Yukon Territory, and the problem of flood magnitude prediction', *J. Glaciol.*, **28**, 3–21.

Clarke, G.K.C. 1987. 'Subglacial till: a physical framework for its properties and processes', *J. Geophys. Res.*, **92(B9)**, 9023–9036.

Clarke, G.K.C. 1996a. 'Lumped element model for subglacial transport of solute and sediment', *Ann. Glaciol.*, **22**, 152–159.

Clarke, G.K.C. 1996b. 'Lumped element analysis of subglacial hydraulic circuits', *J. Geophys. Res.*, **101(B8)**, 17 547–17 559.

Collins, D.N. 1977. 'Hydrology of an alpine glacier as indicated by the chemical composition of meltwater', *Z. Gletscherk. Glazialgeol.*, **13**, 219–238.

Collins, D.N. 1979. 'Quantitative determination of the subglacial hydrology of two alpine glaciers', *J. Glaciol.*, **23**, 347–362.

Collins, D.N. 1982. 'Flow routing of meltwater in an alpine glacier as indicated by dye tracer tests', *Beitr. Geol. Schweiz-Hydrol.*, **28**, 523–534.

Collins, D.N. and Young, G.J. 1981. 'Meltwater hydrology and hydrochemistry in snow and ice-covered mountain catchments', *Nordic Hydrol.*, **12**, 319–334.

Copland, L., Harbor, J., Gordon, S. and Sharp, M. This volume. 'The use of borehole video in investigating the hydrology of a temperate glacier'.

Drever, J.I. and Hurcomb, D.R. 1986. 'Neutralisation of atmospheric acidity by chemical weathering in an alpine drainage basin in the North Cascades Mountains', *Geology*, **14**, 221–224.

Elliston, G.R. 1973. 'Water movement through the Gornergletscher', *Int. Assoc. Hydrol. Sci. Publ.*, **95**, 79–84.

Engelhardt, H.F. 1978. 'Water in glaciers: observations and theory of the behaviour of water levels in boreholes', *Z. Gletscherk. Glazialgeol.*, **14**, 35–60.

Escher-Vetter, H. 1985. 'Energy balance calculations for the ablation period 1982 at Vernagtferner, Ötztal Alps', *Ann. Glaciol.*, **6**, 158–160.

Fairchild, I.J., Bradby, L., Sharp, M. and Tison, J-L. 1994. 'Hydrochemistry of carbonate terrains in alpine glacial settings', *Earth Surf. Process. Landforms*, **19**, 33–54.

Fischer, U.H. and Clarke, G.K.C. 1997. 'Stick-slip sliding behaviour at the base of a glacier', *Ann. Glaciol.*, **24**, 390–396.

Fountain, A. 1989. 'The storage of water in, and hydraulic characteristics of, the firn of South Cascade Glacier, Washington State, USA',

Ann. Glaciol., **13**, 69–75.

Fountain, A. 1994. 'Borehole water level variations and implications for the subglacial hydraulics of South Cascade Glacier, Washington State, USA', *J. Glaciol.,* **40**, 293–304.

Fountain, A. This volume. 'Effect of snow and firn hydrology on the physical and chemical characteristics of glacial runoff'.

Fountain, A. and Vaughn, B.H. 1995. 'Changing drainage patterns within South Cascade Glacier, Washington, USA, 1964–1992', *Int. Assoc. Hydrol. Sci. Publ.,* **228**, 379–386.

Fowler, A.C. 1987. 'Sliding with cavity formation', *J. Glaciol.,* **31**, 255–267.

Fowler, A.C. and Johnson, C. 1995. 'Hydraulic runaway: a mechanism for thermally regulated surges of ice sheets', *J. Glaciol.,* **41**, 554–561.

Gordon, S., Sharp, M., Hubbard, B., Smart, C.C., Ketterling, B. and Willis, I.C. This volume. 'Seasonal reorganisation of subglacial drainage inferred from measurements in boreholes'.

Gurnell, A.M. and Fenn, C.R. 1984. 'Flow separation, sediment source areas and suspended sediment transport in a proglacial stream', *Catena Suppl.,* **5**, 109–119.

Gurnell, A.M. and Fenn, C.R. 1985. 'Spatial and temporal variations in electrical conductivity in a pro-glacial stream system', *J. Glaciol.,* **31**, 108–114.

Hamran, S.E., Aarholt, E., Hagen, J-O. and Mo, P. 1996. 'Estimation of relative water content in a sub-polar glacier using surface penetration radar', *J. Glaciol.,* **42**, 533–537.

Hantz, D. and Lliboutry. L. 1983. 'Waterways, ice permeability at depth and water pressures at Glacier d'Argentière, French Alps'. *J. Glaciol.,* **29**, 227–239.

Harbor, J., Sharp, M., Copland, L., Hubbard, B., Nienow, P. and Mair, D. 1997. 'The influence of subglacial drainage conditions on the velocity distribution within a glacier cross-section', *Geology,* **25**, 739–742.

Harper, J.T. and Humphrey, N.F. 1995. 'Borehole video analysis of a temperate glacier's englacial and subglacial structure: implications for glacier flow models', *Geology,* **23**, 901–904.

Hodge, S.M. 1979. 'Direct measurement of basal water pressures: progress and problems', *J. Glaciol.,* **23**, 309–319.

Hodgkins, R. et al. This volume. 'The hydrochemistry of runoff from a cold-based glacier in the high arctic (Scott Turnerbreen, Svalbard)'.

Holmlund, P. 1988. 'Internal geometry and evolution of moulins, Storglaciären, Sweden', *J. Glaciol.,* **34**, 242–248.

Hooke, R.LeB. 1991. 'Positive feedbacks associated with erosion of glacial cirques and overdeepenings', *Geol. Soc. Amer. Bull.,* **103**, 1104–1108.

Hooke, R.LeB. and Pohjola, V.A. 1994. 'Hydrology of a segment of a glacier situated in an overdeepening, Storglaciären. Sweden', *J. Glaciol.,* **40**, 140–148.

Hooke, R.LeB., Calla, P., Holmlund, P., Nilsson, M., and Stroeven, A. 1989. 'A three year record of seasonal variations in surface velocity, Storglaciären, Sweden', *J. Glaciol.,* **35**, 235–247.

Hooke, R.LeB., Laumann, T. and Kohler, J. 1990. 'Subglacial water pressures and the shape of subglacial conduits', *J. Glaciol.,* **36**, 67–71.

Hubbard, B.P., Sharp, M., Willis, I.C., Nielsen, M.K. and Smart, C.C. 1995. 'Borehole water level variations and the structure of the subglacial hydrological system of Haut Glacier d'Arolla, Valais, Switzerland', *J. Glaciol.,* **41**, 572–583.

Humphrey, N., Raymond, C.F. and Harrison, W. 1986. 'Discharges of turbid water during mini-surges of Variegated Glacier, Alaska, USA', *J. Glaciol.,* **32**, 195–207.

Iken, A. 1981. 'The effect of subglacial water pressure on the sliding velocity of a glacier in an idealised numerical model', *J. Glaciol.,* **27**, 407–421.

Iken, A., Röthlisberger, H., Flotron, A. and Haeberli, W. 1983. 'The uplift of the Unteraargletscher at the beginning of the melt season – a consequence of water storage at the bed?', *J. Glaciol.,* **29**, 28–47.

Iken, A., Fabri, K. and Funk, M. 1996. 'Water storage and subglacial conditions inferred from borehole measurements on Gornergletscher, Valais, Switzerland', *J. Glaciol.,* **42**, 233–248.

Iverson, N.R., Hanson, B., Hooke, R.LeB and Jansson, P. 1995 'Flow mechanism of glaciers on soft beds', *Science,* **267**, 80–81.

Kamb, B. 1987. 'Glacier surge mechanism based on linked cavity configuration of the basal water conduit system', *J. Geophys. Res.,* **92(B9)**, 9083–9100.

Kamb, B. and Engelhardt, H.F. 1987. 'Waves of accelerated motion in a glacier approaching surge: the mini surges of Variegated Glacier, Alaska, USA', *J. Glaciol.,* **33**, 27–46.

Kamb, B., Engelhardt, H., Fahnestock, M.A., Humphrey, N., Meier, M., and Stone, D. 1994. 'Mechanical and hydrologic basis for the rapid motion of a large tidewater glacier. 2. Interpretation', *J. Geophys. Res.,* **99**, 15 231–15 244.

Kapitsa, A.P., Ridley, J.K., Robin, G. deQ., Siegert, M.J. and Zotikov, I.A. 1996. 'A large deep freshwater lake beneath the ice of central East Antarctica', *Nature,* **381**, 684–686.

Kulessa, B. and Hubbard, B. 1997. 'Interpretation of borehole impulse tests at Haut Glacier d'Arolla, Switzerland', *Ann. Glaciol.,* **24**, 397–402.

Lamb, H., Tranter, M., Brown, G.H., Gordon, S., Hubbard, B., Nielsen, M., Sharp, M., Smart, C.C. and Willis, I.C. 1995. 'The composition of meltwaters sampled from boreholes at the Haut Glacier d'Arolla, Switzerland', *Int. Assoc. Hydrol. Sci. Publ.,* **228**, 395–403.

Lane, S.N., Richards, K.S. and Chandler, J.H. 1994. 'Distributed sensitivity analysis in modelling environmental systems', *Proc. Roy. Soc. London Ser. A,* **447**, 49–63.

Lang, H., Schägler, B. and Davidson, G. 1977. 'Hydroglaciological investigations on the Ewigschneefeld – Grosser Aletschgletscher', *Z. Gletscherk. Glazialgeol.,* **12**, 109–124.

Lang, H., Leibundgut, Ch. and Festel, E. 1979. 'Results from tracer experiments on the water flow through the Aletschgletscher', *Z. Gletscherk. Glazialgeol.,* **15**, 209–218.

Lawler, D., Björnsson, H. and Dolan, M. This volume. Impact of subglacial geothermal activity on meltwater quality in the Jökulsá á Sólheimasandi system, southern Iceland.

Lecce, S. 1993. 'Flow separation and diurnal variability in the hydrology of Conness Glacier, Sierra Nevada, California, USA', *J. Glaciol.*, **39**, 216–222.

Lliboutry, L. 1968. 'General theory of subglacial cavitation and sliding of temperate glaciers', *J. Glaciol.*, **7**, 21–58.

MacAyeal, D.R. 1993. 'Binge/purge oscillations of the Laurentide Ice Sheet as a cause of the North Atlantic's Heinrich Events', *Palaeoceanography*, **8**, 775–784.

Murray, T. and Clarke, G.K.C. 1995. 'Black-box modelling of the subglacial water system', *J. Geophys. Res.*, **100(B7)**, 10 231–10 245.

Nienow, P.W. 1993. 'Dye tracer investigations of glacier hydrological systems', *unpublished PhD thesis*, University of Cambridge.

Nienow, P.W., Sharp, M. and Willis, I.C. This volume. 'Velocity–discharge relationships derived from dye tracer experiments in glacial meltwaters: implications for subglacial flow conditions'.

Nye, J.F. 1973. 'Water at the bed of a glacier', *Int. Assoc. Hydrol. Sci. Publ.*, **95**, 189–194.

Nye, J.F. 1976. 'Water flow in glaciers: jökulhlaups, tunnels and veins', *J. Glaciol.*, **17**, 181–207.

Nye, J.F. and Frank, F.C. 1973. 'Hydrology of the intergranular veins in a temperate glacier', *Int. Assoc. Hydrol. Sci. Publ.*, **95**, 157–161.

Oerter, H. and Moser, H. 1982. 'Water storage and drainage within the firn of a temperate glacier (Vernagtferner, Ötztal Alps, Austria)', *Int. Assoc. Hydrol. Sci. Publ.*, **138**, 71–81.

Pohjola, V.A. 1994. 'TV video observations of englacial voids in Storglaciären, Sweden', *J. Glaciol.*, **40**, 231–240.

Quinn, P. and Beven, K.J. 1993. 'Spatial and temporal predictions of soil moisture dynamics, runoff, variable source areas and evapotranspiration for Plynlimon, mid-Wales', *Hydrol. Process.*, **7**, 425–448.

Raiswell, R. 1984. 'Chemical models of solute acquisition in glacial meltwaters', *J. Glaciol.*, **30**, 49–57.

Raiswell, R. and Thomas, A. 1984. 'Solute acquisition in glacial meltwaters. I. Fjallsjökull (South-east Iceland): bulk meltwaters with closed system characteristics', *J. Glaciol.*, **30**, 35–43.

Raymond, C.F. 1987. 'How do glaciers surge: a review', *J. Geophys. Res.*, **92(B9)**, 9121–9134.

Raymond, C.F. and Harrison, W.D. 1975. 'Some observations on the behaviour of the liquid and gas phases in temperate glacier ice', *J. Glaciol.*, **14**, 213–234.

Raymond, C.F., Benedict, R.J., Harrison, W.D., Echelmeyer, K.A. and Sturm, M. 1995. 'Hydrological discharges and motion of Fels and Black Rapids Glaciers, Alaska, USA: implications for the structure of their drainage systems', *J. Glaciol.*, **41**, 290–304.

Richards, K.S., Sharp, M., Arnold, N., Gurnell, A., Clark, M., Tranter, M., Nienow, P., Brown, G., Willis, I. and Lawson, W. This volume, 'An integrated approach to modelling hydrology and water quality in glacierized catchments'.

Röthlisberger, H. 1972. 'Water pressure in intra- and subglacial channels', *J. Glaciol.*, **11**, 177–203.

Sharp, M. 1991. 'Hydrological inferences from meltwater quality data – the unfulfilled potential', in *Proc. Br. Hydrol. Soc. Third National Hydrology Symp.*, University of Southampton, 16–18 September 1991. Wallingford, Institute of Hydrology, 5.1–5.8.

Sharp, M. 1996. 'Weathering pathways in glacial environments – hydrological and lithological controls', in Bottrell, S.H. (Ed.) *Proceedings of the 4th International Symposium on the Geochemistry of the Earth's Surface*, Ilkley, UK, University of Leeds, 652–655.

Sharp, M., Richards, K., Willis, I., Arnold, N., Nienow, P., Lawson, W. and Tison, J-L. 1993. 'Geometry, bed topography and drainage system structure of the Haut Glacier d'Arolla. Switzerland', *Earth Surf. Process. Landforms*, **18**, 557–571.

Sharp, M., Brown, G.H., Tranter, M., Willis, I.C. and Hubbard, B. 1995. 'Comments on the use of chemically-based mixing models in glacier hydrology', *J. Glaciol.*, **41**, 241–246.

Shaw, J. 1989. 'Drumlins, Subglacial meltwater floods and ocean responses', *Geology*, **17**, 853–856.

Shoemaker, E.M. 1992. 'Subglacial floods and the origin of low relief ice sheet lobes', *J. Glaciol.*, **38**, 105–112.

Shreve, R.L. 1972. 'Movement of water in glaciers', *J. Glaciol.*, **11**, 205–214.

Skidmore, M.L. 1995. 'The hydrochemistry of a high Arctic glacier', *unpublished MSc thesis*, University of Alberta.

Smart, C.C. This volume. 'Statistical evaluation of glacier boreholes as indicators of basal drainage systems'.

Spring, U. and Hutter, K. 1981. 'Numerical studies of jökulhlaups', *Cold Reg. Sci. Tech.*, **4**, 221–244.

Stone, D.B. 1993. 'Characterisation of the basal hydraulic system of a surge-type glacier: Trapridge Glacier, 1989–92', *unpublished PhD thesis*, University of British Columbia.

Stone, D.B. and Clarke, G.K.C. 1993. 'Estimation of subglacial hydraulic properties from induced changes in basal water pressure: a theoretical framework for borehole response tests', *J. Glaciol.*, **39**, 327–340.

Stone, D.B. and Clarke, G.K.C. This volume. 'In situ measurements of basal water quality and pressure as an indicator of the character of subglacial drainage systems'.

Stone, D.B., Clarke, G.K.C. and Blake, E.W. 1993. 'Subglacial measurement of turbidity and electrical conductivity', *J. Glaciol.*, **39**, 415–420.

Stone, D.B., Meier, M.F., Lewis, K.J. and Harper, J.T. 1994. 'Drainage configuration and scales of variability in subglacial water system', *Eos (Trans. Amer. Geophys. Union)*, **75**, 22.

Theakstone, W.H. and Knudsen, N.T. 1981. 'Dye tracer tests of water movement at the glacier Austre Okstindbreen, Norway', *Norsk Geogr. Tidsskr.*, **35**, 21–28.

Theakstone, W.H. and Knudsen, N.T. This volume. 'Isotopic variations in glacier river water during three contrasting ablation seasons'.

Thomas, A.G. and Raiswell, R. 1984. 'Solute acquisition in glacial meltwaters. II. Argentière, French Alps: bulk meltwaters with open system characteristics', *J. Glaciol.*, **30**, 44–48.

Tranter, M. and Raiswell, R. 1991. 'The composition of the englacial and subglacial components in bulk meltwaters draining the Gornergletscher', *J. Glaciol.*, **37**, 59–66.

Tranter, M., Brown, G.H., Raiswell, R., Sharp, M. and Gurnell, A.M. 1993. 'A conceptual model of solute acquisition by alpine glacial meltwaters', *J. Glaciol.*, **39**, 573–581.

Tranter, M., Sharp, M.J., Brown, G.H., Willis, I.C., Hubbard, B.P., Nielsen, M.K., Smart, C.C., Gordon, S., Tulley, M. and Lamb, H.R. This volume a. 'Variability in the chemical composition of *in situ* subglacial meltwaters'.

Tranter, M., Brown, G.H., Hodson, A.J. and Gurnell, A.M. This volume b. 'Hydrochemistry as an indicator of subglacial drainage system structure: a comparison of alpine and sub-polar environments'.

Waddington, B.S. and Clarke, G.K.C. 1995. 'Hydraulic properties of subglacial sediment determined from mechanical response of water-filled boreholes', *J. Glaciol.*, **41**, 112–124.

14   M. SHARP, K. S. RICHARDS AND M. TRANTER

Wadham, J.L. 1997. 'The hydrochemistry of a High Arctic Polythermal-based Glacier: Finsterwalderbreen, Svalbard', *unpublished PhD thesis*, University of Bristol.
Walder, J.S. 1982. 'Stability of sheet flow of water beneath temperate glaciers and implications for glacier surging', *J. Glaciol.,* **28**, 273–293.
Walder, J.S. 1986. 'Hydraulics of subglacial cavities', *J. Glaciol.,* **32**, 439–445.
Walder, J.S. and Fowler, A. 1994. 'Channelized subglacial drainage over a deformable bed', *J. Glaciol.,* **40**, 3–15.
Weertman, J. 1972. 'General theory of water flow at the base of a glacier or ice sheet', *Rev. Geophys. Space Phys.,* **10**, 287–333.
Willis, I.C., Richards, K.S. and Sharp, M.J. This volume. 'Links between proglacial stream suspended sediment dynamics, glacier hydrology and glacier motion at Midtdalsbreen, Norway'.

2

EFFECT OF SNOW AND FIRN HYDROLOGY ON THE PHYSICAL AND CHEMICAL CHARACTERISTICS OF GLACIAL RUNOFF

ANDREW G. FOUNTAIN

US Geological Survey, PO Box 25046, MS-412, Denver, CO 80225, USA

ABSTRACT

Near-surface processes on glaciers, including water flow over bare ice and through seasonal snow and firn, have a significant effect on the speed, volume and chemistry of water flow through the glacier. The transient nature of the seasonal snow profoundly affects the water discharge and chemistry. Water flow through snow is fairly slow compared with flow over bare ice and a thinning snowpack on a glacier decreases the delay between peak meltwater input and peak stream discharge. Furthermore, early spring melt flushes the snowpack of solutes and by mid-summer the melt water flowing into the glacier is fairly clean by comparison. The firn, a relatively constant feature of glaciers, attenuates variations in water drainage into the glacier by temporarily storing water in saturated layer. Bare ice exerts opposite influences by accentuating variations in runoff by water flowing over the ice surface. The melt of firn and ice contributes relatively clean (solute-free) water to the glacier water system.

INTRODUCTION

Prediction of the rate and quantity of water flow through a glacier is complicated by the number of different hydrological processes that need to be considered. Surface processes include overland flow on bare ice and both unsaturated and saturated flow in snow and firn. Water in these surface layers generally drains into the body of the ice via crevasses. The englacial and subglacial drainage systems then route the water through and eventually out of the glacier. Englacial systems are largely unknown, whereas subglacial systems have been the subject of considerable attention (e.g. Weertman, 1972; Röthlisberger, 1972; Walder, 1986; Kamb, 1987). This paper addresses the physical processes of water flow though snow and firn on temperate glaciers and assesses their effects on runoff in glacial streams. In addition, snowmelt chemistry and chemical processes in the snow are reviewed with respect to their effect on stream waters flowing from glaciers. The effect of snow chemistry on basal waters and its influence on basal weathering is briefly reviewed.

PHYSICAL PROCESSES

Snow-free ice surfaces

In the bare ablation zone of a glacier, surface meltwater and rain flow across the ice surface to nearby moulins and crevasses where the water enters the body of the glacier (Stenborg, 1973). Water can accumulate along grain boundaries and within cracks in the surface ice. Little storage is expected in dynamic regions of the ablation zone because such ice is not long exposed at the glacier surface and subjected to solar radiation, thus the glacier ice exhibits narrow grain boundaries. As the surface ice is ablated, it is continually replenished as ice is advected from the glacier's interior by the emergence velocity (Meier and Tangborn, 1965; Paterson, 1981). In stagnant ice, however, such storage can be significant because the ice has been subjected to long periods of solar radiation that widens the grain boundaries, as often observed in melting lake ice (Ashton, 1980). Measurements by Larson (1977; 1978) show a water-table in the stagnant ice of Burroughs Glacier, Alaska (Figure 1). The maximum daily fluctuations of the water level are about 2 m and the hydraulic transmissivity of this active layer, derived from pump tests, is about $7 \cdot 7 \times 10^{-5}$ m^2 s^{-1}. However, it is unlikely that the storage volume is large, as indicated by the relatively quick response of the stream flow from Burroughs Glacier to diurnal variations in meltwater input (Larson, 1978). The quick response, typical for most glaciers, is indicative of the rapid flow of meltwater across the bare ice in the ablation zone and through the glacier's interior to the glacial streams (Fountain, 1992a).

Water flow in snow

When the snowpack is below freezing, rain and meltwater refreeze in the snowpack and release latent heat. This is the primary mechanism that warms a snowpack (Marsh and Woo, 1984; Conway and Benedict, 1994). The depth of water penetration depends on the snowpack temperature and structural characteristics

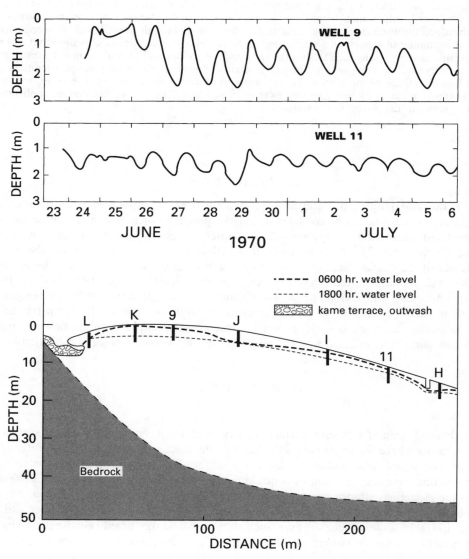

Figure 1. (a) Diurnal variations in the water-table of the stagnant ice zone of Burroughs Glacier, Alaska, USA. (b) Spatial water level variations. Modified from Larson (1977) with permission. Letters and numbers identify the different wells

and on the flux of water into the snow (Pfeffer *et al.*, 1990). Clearly, the refreezing of meltwater delays the penetration of water into and runoff from the glacier. In fact, for dry, cold snow, more than half of the surface melt is used to supply the irreducible water saturation (water retained by capillary forces) from the time the snowpack starts to melt until the time that the water reaches the base of the snowpack (Marsh and Woo, 1984).

For glaciers with large elevation ranges, the snowpack will warm earlier and stay warm longer at lower elevations than at higher elevations. If the elevation range of the glacier is large, snow may not melt in the upper reaches. Conversely, if the elevation range of a glacier is small, as for small alpine glaciers, there is little change in snow temperature. In either instance, the relation between snowpack temperature and elevation must be considered in assessing melt and runoff. Subsurface temperatures must also be understood given that meltwater may freeze in the lower layers of snow or in the firn. Surface ablation measurements may suggest a mass loss to runoff when in fact a redistribution of mass from the surface to the interior of the glacier has occurred (Trabant and Mayo, 1985). Neglecting the effects of snow and firn temperature can result in underestimates of the glacier mass balance (Trabant and Mayo, 1985) and overestimates of the predicted runoff.

The rate of flow through the snowpack is largely determined by the ice layers in the snow. Water stops flowing vertically when it encounters an ice layer and ponding above the ice layer occurs, which then saturates the snow. Water then flows laterally in the saturated snow to a gap in the ice layer before continuing downwards (Colbeck, 1973). Ice layers may disintegrate quickly, perhaps within hours when subjected to large meltwater fluxes (Gerdel, 1954, as interpreted by Male, 1980). The vertical flux of water in homogeneous snow is related to the intrinsic permeability of the snow and its effective saturation (Colbeck and Davidson, 1973)

$$u = \alpha k S^{*3} \tag{1}$$

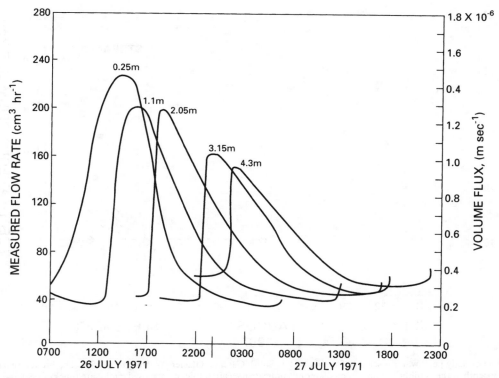

Figure 2. Diurnal meltwave at different depths below the snow surface. Note the steepening front of the wave. Modified from Colbeck and Davidson (1973) with permission

where u is the vertical liquid flux (volume of water flowing per unit area per unit time) in m s^{-1}, α is a constant equal to density multiplied by gravity divided by water viscosity, k is the intrinsic permeability and S^* is the fraction of pore volume containing moving water (effective saturation). It can be shown (Colbeck and Davidson, 1973) that the rate of vertical infiltration is related to the water flux

$$\frac{\mathrm{d}z}{\mathrm{d}t}\bigg|_u = 3\alpha^{1/3}k^{1/3}u^{2/3}\phi_e^{-1} \tag{2}$$

where $\mathrm{d}z/\mathrm{d}t$ is the rate of downward movement for flux u and ϕ_e is the effective porosity. This equation indicates that the rate of movement is dependent on the vertical flux and implies that larger meltwater fluxes will catch up with smaller fluxes and will form a wave characterized by a sharp front (Figure 2), like a shock front, followed by a slow recession. If Figure 2 is typical of a mature and homogeneous snowpack, such experiments indicate a wave speed of about $0.3\,\mathrm{m\,h^{-1}}$. This speed is roughly equivalent to the lateral flow speeds in the saturated zone of a ripe snowpack on a sloping glacier (Fountain, 1992b).

Once the water reaches the base of the snowpack, it either meets the surface of the underlying ice or percolates into the firn. Part of the water drains directly into crevasses, but this occurs under a small fraction of the total snow-covered area. If the water meets the ice surface, as in the case of an ice layer, it will saturate the snow to form a shallow water-table. The water then moves downslope and drains into the nearest crevasse. It is not uncommon to encounter slushy snow when walking in the ablation zone of a glacier in late spring. The water may saturate the full thickness of the snow, such as at the base of a steep slope. Under the right circumstances of water saturation and surface slope, the snowpack may fail, creating a slush avalanche (Onesti, 1987; Elder and Kattelmann, 1993). Also, ice crystals eroded from the ice surface or entrained from a nearby snowpack often form temporary ice jams in the small streams flowing over a glacier's surface (pers. obs.). Although both situations result in pulses of water entering crevasses or moulins, they are

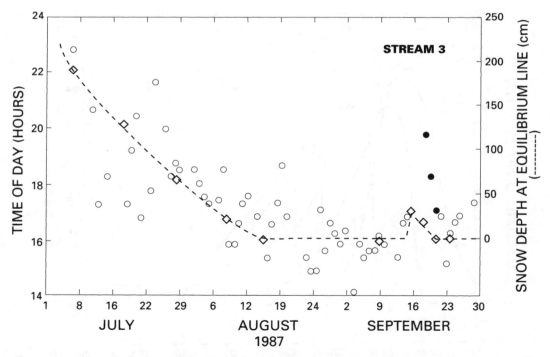

Figure 3. Time of daily peak water discharge for stream 3 at South Cascade Glacier, Washington, USA. The broken line is the interpolated snow depth at the equilibrium line; open diamonds indicate measured thickness; closed circles are times of peak daily discharge following a snowfall (adapted from Fountain, 1992b). Reproduced courtesy of the International Glaciological Society from the *Journal of Glaciology*, 1992, **38** (128), 191, figure 2

either rare in frequency or small in magnitude and probably do not appreciably affect either the rate at which water flows over the glacier or the discharge of water flowing from the glacier.

Snow-covered ice

The presence of snow on ice in the ablation zone significantly decreases the speed of water flow compared with that of flow over a bare ice surface. The delay is equal to the vertical transit time through the snowpack to the saturated zone plus the lateral transit time through the saturated zone to the nearest crevasse. This effect is shown in Figure 3. The time of day of peak flow in a proglacial stream decreases as the season progresses until mid-season, after which no further change is apparent (Fountain, 1992b). The earlier appearance of peak flow correlates with decreasing snow depth, measured at the glacier equilibrium line. Snow depth, in this instance, is a proxy measure for the elevation of the snowline on the glacier. Decreasing depth implies that the snowline is moving up-glacier and revealing more ice. As more bare ice is exposed, a greater amount of water moves quickly from the surface to the interior of the glacier. The trend of earlier peak flows terminates when the snowline reaches the firn line. Above the firn line, the firn itself stores water, as will be discussed later, precluding fast surface flow.

The effect of snowcover on the timing of peak flow was particularly well demonstrated by the 14 September snowfall (Figure 3), which covered the bare ice surface with a layer of snow 0·25 m deep. When the timing of peak flow could be reliably determined, four days later, the snow was 0·16 m deep and the peak was delayed by 3·75 h, compared with pre-snowfall values. As the snow depth decreased and the snowline retreated up-glacier, the time of peak flow appearance returned to pre-snowfall times. Both the seasonal change and short-term change caused by the late summer snowfall can be explained by vertical and lateral transit times through the snow (Fountain, 1992b).

The increasing amplitude of the diurnal variation in discharge from spring to mid-summer (Figure 4) can be partly explained by meteorological conditions and snowcover. In spring, compared with summer, the skies are cloudier, resulting in less solar insolation, the air temperatures are cooler and the albedo of the glacier is higher (Figure 5). These conditions produce less meltwater than in the summer and reduce the amplitude of diurnal variations in discharge. The amplitude is further attenuated by the presence of a snowpack over much of the ablation zone. The path length for vertical percolation increases with elevation on the

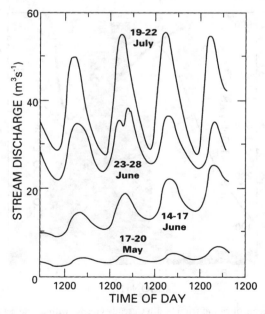

Figure 4. Daily variations in glacial stream discharge in the Matter-Vispa river basin, Switzerland. Modified from Elliston (1973) with permission

glacier because of the average increase in snow depth with elevation; thus the meltwater reaches the base of the snowpack at a different time. At the base, water flows laterally through the saturated layer and the path length depends on the distance between crevasses, which can vary significantly. If the path length, including the vertical and lateral components, were constant, then the diurnal variations of meltwater input would be delayed by some constant time lag. However, the path length is not constant because of variations in snow depth and in the separation distance between crevasses. The aggregate effect of these variations is to attenuate the amplitude of the meltwater wave entering the snowpack by the time it passes into the glacier via the crevasses. In contrast, when the ablation zone is snow-free, water is routed rapidly across the surface to the nearest crevasse and the delay between generation and penetration into the glacier is relatively small, which increases the amplitude of the diurnal variation of streamflow.

This explanation contrasts with the commonly held opinion (Elliston, 1973; Bezinge, 1981; Röthlisberger and Lang, 1987) that the decreasing delay results from the enlarging subglacial hydraulic system. Although some delay results from flow through any hydraulic system, the exact subglacial causes of the observed large delays have not been identified. Fountain (1992b) has argued that if the subglacial hydraulic system is envisioned as a pre-existing network of conduits, then we cannot explain the decreasing delay. Alternatively, the enlarging hydraulic system may be viewed as increasing its spatial coverage by connecting with subglacial regions previously isolated hydraulically during the winter. In this instance, the effect on the storage and delay of water flow is uncertain.

Under some circumstances, the thinning snowpack and increasing area of bare ice in the ablation zone favour development of large subglacial conduits. Hooke (1984) calculated a range of values of water discharge, glacier slope and thickness for which the rate of conduit enlargement exceeds that of closure. In this situation, conduits will most often be partly full at atmospheric pressure. To keep the ice walls from closing,

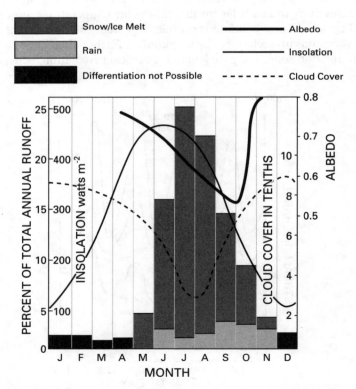

Figure 5. Seasonal variation in components of runoff from South Cascade Glacier with average glacier albedo, cloud cover and insolation (modified from Meier, 1969). Reprinted from *Journal of the American Water Works Association*, **61** (1) (Jan 1969), by permission. Copyright© 1969, American Water Works Association

the conduit needs to be pressurized from time to time. Calculations (Fountain, 1992a) show that for relatively thin ice (70 m) and a 10% slope, only a few hours of full-conduit flow at atmospheric pressure are required to compensate for several days of closure. These conditions often exist in the ablation zone of glaciers, particularly near the terminus where the ice is thin. For these situations, the damped variations of water flow in the glacier, caused by the presence of the snowpack, will minimally enlarge the conduit. Later in the season when the ablation zone is generally free of snow, larger discharge variations will enlarge the conduits to a greater extent than expected for large, but steady, discharges. The implication is that the position of the snowline may significantly influence the size of the conduits.

Firn

Once the water has percolated through the snow in the accumulation zone, it encounters the firn, the metamorphic transition between snow and glacier ice. Firn is porous and water percolates downwards through the unsaturated zone by presumably the same process as that for unsaturated snow. In temperate glaciers, percolating water forms a saturated layer at the base of the firn (Sharp, 1951; Schommer, 1977; Ambach *et al.*, 1978; Oerter and Moser, 1982; Fountain, 1989). Water is found from 0 to 40 m below the surface and drains to crevasses (Lang *et al.*, 1977; Schommer, 1977; Fountain, 1989). The water-table forms in the spring when water percolates into the firn and is depleted in the autumn when surface melt stops (Figure 6) and the firn water drains into crevasses. The maximum range in seasonal water levels at South Cascade Glacier, from which the saturated thickness is inferred, is 2 m (Fountain, 1989).

The presence of another porous medium below the snow layer indicates that the routing of water flow from the surface of the accumulation zone to the interior of the glacier is further delayed. For reasons similar to those discussed earlier — increasing snow and firn thickness with elevation and the variation in distance between crevasses — it is likely that spatially averaged drainage into the body of the glacier does not exhibit appreciable diurnal variations. This conclusion agrees with that of Humphrey *et al.* (1986). Furthermore, the volume of the slowly varying (baseflow) component of the streamflow can be explained by the volume of daily meltwater input into the accumulation zone (Fountain, 1992b).

Seasonal storage of water in the firn, which depends on the thickness of the saturated layer and the effective porosity of the firn, was estimated for South Cascade Glacier. The results indicated that about 11 cm of water averaged over the accumulation zone of the glacier, or $1.78 \times 10^5 \, \mathrm{m}^3$ (Fountain, 1989),

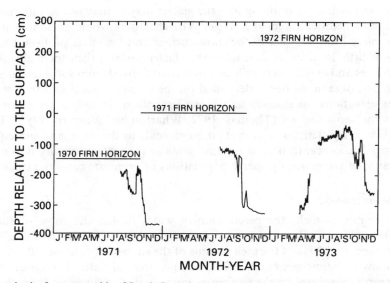

Figure 6. Seasonal variation in the firn water-table of South Cascade Glacier, Washington, USA (from Fountain, 1989). Reproduced courtesy of the International Glaciological Society from the *Annals of Glaciology*, 1989, **13**, 191, figure 2

were stored. This volume of water represents about 12% of the total spring storage measured at South Cascade Glacier by Tangborn *et al.* (1975). The drainage of the stored water contributes to the runoff from the glacier in the early winter months. Generally, the saturated layer completely drains by late November.

SNOW CHEMISTRY

Origin

The solute content of a snowpack, in part, reflects the chemistry of the atmosphere in which the snow formed and through which it falls. Ice crystals form by a number of different processes that involve either homogeneous or heterogeneous nucleation (Hobbs, 1974). The latter process requires a nucleus on which water vapour condenses. As the snowflake descends through the atmosphere it scavenges atmospheric contaminants, either by gas adsorption or particle adhesion (Junge, 1977). Adhering particles may include cloud droplets contaminated with acidic gases and solid particulates such as dusts or ashes (Magono *et al.*, 1979). Acidic gases, such as SO_2 and NO_2, can be absorbed by aerosol particles, rain drops and snow crystals and subsequently form acids such as sulphuric and nitric acids (Winkler, 1980). When the snowflake lands on a glacier, these contaminants are carried with it. In addition, when the snowpack forms, it traps local atmospheric gases within the pack. Gas exchange can continue in the pack as air is pumped through the snow by the topographically induced pressure variations as the wind blows over the surface (Clarke and others, 1987; Colbeck, 1989; Clarke and Waddington, 1991). Not only is gas exchanged, but the snow filters and traps airborne particulates (Cunningham and Waddington, in press). The effect of wind of pumping is a relatively near-surface phenomenon, so subsequent snow accumulation reduces or eliminates the air and particulate exchange between the atmosphere and interior of the snowpack.

The solute species expected in the snowpack depends on the trajectory of atmospheric flow and the location of a glacier relative to source regions. For example, snow in the Himalayas exhibits higher concentrations of sodium and chloride than the snow in the Karakoram, indicating a greater influence of monsoon moisture in the Himalayas (Wake *et al.*, 1990). Proximity to local pollution sources increases the snow acidity and related solutes such as nitrate and sulphate (Wagenbach and Münnich, 1988). Long-range transport of dust, such as Saharan dust to the Col du Dôme, French Alps, increases the concentration of aluminium and calcium (Maupetit, 1994). Therefore, the snow on any given glacier or group of glaciers may have a unique chemical signature depending on their location relative to natural and anthropogenic source regions, on the chemical characteristics of the source regions and on the patterns of local and long-range atmospheric transport. The chemical content of the meltwater draining into the glacier may or may not be different from that obtained from the dissolution of bedrock under the glacier. The chemistry would be similar if rockfalls and wind-borne particulates carry local bedrock to the snow surface and the effect of distant sources was relatively small. Conversely, if little local bedrock reaches the glacier surface, then the meltwater chemistry is controlled by other sources and may be very different from that derived from subglacial dissolution of bedrock.

During the summer, organic matter is deposited on the glacier. Insects and tree detritus are commonly observed at lower elevations on glaciers and living organisms, including ice worms (Goodman, 1971) and algae, inhabit the snow and ice (Thomas, 1972; Wharton and Vinyard, 1983). The by-products and decay of the organisms and detritus contribute organic acids to the snow and subsequently to the glacier and glacial streams. The concentration of organic acids in the water draining into and from a glacier is probably low because of the generally sparse populations of organisms and the small amount of detritus.

Processes of solute enrichment

Once the snow begins to melt, the percolation of water flushes the snow of solutes (Skartveit and Gjessing, 1979). However, the solute concentration in the meltwater flowing from the snow is not constant (Figure 7). Rather, the initial 30% of the melt volume of the snowpack contains 50–80% of the total solutes contained in the snow (Johannessen *et al.*, 1975). The preferential flush of solutes results from microscale processes that concentrate solutes on the surface of individual ice grains and from the macroscale vertical distribution of solutes in the snowpack (Bales *et al.*, 1989).

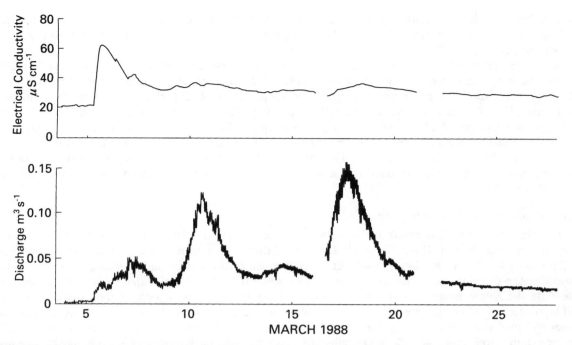

Figure 7. Electrical conductivity and water discharge in a snow-fed stream in Scotland during the first spring melt (modified from Morris and Thomas, 1985) Reproduced courtesy of the International Glaciological Society from the *Journal of Glaciology*, 1985, **31** (108), 191, figure 1

The microscale processes that concentrate the solutes are melt–freeze cycles (Bales *et al.*, 1989) and vapour transfer (Cragin *et al.*, 1993). These metamorphic changes to the crystal structure are concurrent with the chemical changes. When a snow grain begins to melt, it forms a water layer. The solute concentration in the water layer is equal to the sum of the solutes in the melting snow grain and on the grain surface. When the water refreezes, it rejects most of the solutes as the ice crystallizes, forming a rind of solutes on the outside of the grain. The solutes are excluded from the ice crystal lattice because of their inability to become incorporated into the ice lattice (Cobb and Gross, 1969; Gross *et al.*, 1987). The remaining trapped impurities are surrounded by ice rather than in the ice lattice itself. Repeated cycling of melting and freezing purifies the ice grain and increases the concentration of solutes on the grain surface.

Alternatively, solutes may concentrate on a grain surface through vapour transfer without an intervening liquid phase (Cragin *et al.* 1993). Vapour transfer results from temperature gradients produced by meteorological changes and by microscale vapour pressure differences. Temperature changes the vapour pressure such that surfaces with warmer temperatures have higher vapour pressures. Consequently, warmer ice grains lose mass to cooler ice grains. Microscale vapour pressures are also produced by the geometry of the ice grains themselves (Hobbs, 1974). A smaller grain, with a smaller radius of curvature, has a higher vapour pressure than a larger grain with a larger radius of curvature, thus smaller grains lose mass to larger grains. As an ice grain sublimates it leaves the solute behind, thus increasing the solute concentration on the grain surface. The solutes on the surface of the larger grain, on which the vapour is deposited, are maintained on the surface because they too are rejected from the growing ice front.

Processes that determine macroscale solute distribution include the initial solute content of the snow layers deposited during different storms and the redistribution of the solutes by meltwater flow. The initial concentration of solutes in the snowpack is determined by scavenging of the snowflakes, by the filtration of air moving through the snow and by dry deposition onto the snow surface, as previously discussed. The redistribution results from meltwater or rain acquiring the solutes as they wash over the snow grains. If the snow temperature is well below freezing the meltwater will refreeze, trapping the acquired solutes in a frozen layer. Such redistribution can produce increased solute concentrations near the base of the snow-pack and will enhance the effect of a concentrated flush of solutes after the snowpack warms to the freezing temperature (Colbeck, 1981).

Field investigations have indicated a preferential flushing of certain solute species, particularly sulphate and nitrate, in contrast with chloride (Davies *et al.*, 1982; Tsiouris *et al.*, 1985; Tranter *et al.*, 1986). Controlled laboratory experiments have shown that ice grains do not preferentially absorb any ion species from meltwater (Bales *et al.*, 1989; Cragin *et al.*, 1993). Rather, the apparent preferential flushing results from the original, non-homogeneous distribution of different solutes in the snowpack (Bales *et al.*, 1989). Cragin *et al.* (1993) showed that less soluble ions, such as sulphate, are more efficiently excluded during grain growth, by vapour transfer, than more soluble ions such as chloride. However, nitrate did not show preferential exclusion relative to chloride, indicating that the initial distribution of solutes in snow may be the primary cause for the preferential flushing of solutes.

Effect of concentrated solute flushing

The solute concentration entering the glacier is largely controlled by processes in the recent seasonal snow rather than the firn. Davies *et al.* (1982) showed that solutes from melting snow do not concentrate in the firn or ice, but rather are flushed out of the surface layers and presumably away from the glacier. However, the firn does have the effect, already described for diurnal meltwater variations, to delay and attenuate the solute pulse. In contrast, once the solutes leave the snowpack in the ablation zone they will be routed quickly to the glacial hydraulic system and hence to the glacial streams. Together, these pro-cesses will probably result in a strong initial pulse followed by several days of high solute concentration, perhaps similar to Figure 7 for snow. One effect of a concentrated flush of solutes is to change the chemistry of alpine lakes and streams, and the effect is particularly significant in basins that have a limited buffering capacity to neutralize acids (Melack *et al.*, 1985). These changes can also adversely affect aquatic biota (Hagen and Langeland, 1973).

The preferential flush of ions from the snow affects glacier erosion and inferences about the subglacial hydraulics based on stream chemistry. At the base of the glacier, the increased solute concentration increases the acidity and promotes chemical erosion (Reynolds and Johnson, 1972; Raiswell, 1984). The magnitude and significance of this increased erosion is uncertain. Two techniques used to infer the hydraulics of subglacial water flow — measurements of stream water chemistry (Raiswell, 1984; Raiswell and Thomas, 1984; Thomas and Raiswell, 1984) and of electrical conductivity in glacial stream water (Collins, 1978; 1979a; 1979b; Humphrey *et al.*, 1986; Fountain, 1992b) — make assumptions about the chemistry of the water enter-ing the glacier drainage system. For such investigations, particularly if taking place in the spring, ignoring the preferential flushing of solutes may invalidate interpretations gained from the studies. For example, Raiswell (1984) proposed using stream water chemistry to infer the hydraulic condition (pressurized or partly full) of subglacial passages with the assumption that atmospheric inputs of sulphate, via meltwater, are negligible and practically all the sulphate is derived from subglacial weathering. This assumption is significantly violated in the spring during the onset of melt when solutes are flushed from the snow. Ignoring the increased solute concentration, including sulphate, in the snowmelt would result in the misinterpretation of the hydraulic con-dition of the subglacial system. Similarly, the electrical conductivity of stream water has been used to separate the hydrograph of glacial streams into relatively clean englacial waters and comparatively solute-rich sub-glacial waters (Collins, 1979b; Vaughn, 1994). However, it is clear that such separations are useful only for those periods when the englacial water is largely devoid of solutes and cannot be applied during the early spring melt period.

CONCLUSIONS

Rapid changes in the seasonal snowpack greatly affect the flow-rate of water from glaciers. Interpretation of the variations in glacial runoff without considering the changing surface snow conditions can lead to mis-taken conclusions about englacial and subglacial conditions. The slow passage of water through snow strongly contrasts with the rapid overland flow on bare ice. Therefore, as the snowline retreats up the glacier and exposes larger areas of bare ice in the ablation zone, the meltwater flow increases in speed and volume. In response, glacial streams increase their diurnal variation in discharge and reach larger peak daily flows earlier in the day. The position of the snowline is thought to greatly affect the size of englacial and subglacial conduits.

The firn also affects the runoff of meltwater, but its effect is more subtle than that of the transient snow. The firn stores a significant fraction of the spring melt water. In the autumn, when surficial melt is reduced or absent, the firn drains water to the glacier. In addition, water flow through firn is significantly delayed and its variations are attenuated such that it probably provides much of the baseflow in glacial streams.

The chemistry of the water flowing into the glacier is controlled by the chemistry of the snowpack because the firn and ice have been leached of solutes. Most of the solutes are flushed from the snow early in the melt season because of microscale processes that concentrate the solutes on the exterior of the ice grains and macroscale processes that redistribute the solutes towards the base of the snowpack. The concentrated solutes increase chemical erosion at the base of the glacier, change the stream chemistry and may adversely affect the aquatic biota in the streams. Inferring subglacial processes during the early melt season based on interpretation of the electrical conductivity or geochemistry of the glacial streamflow must take into consideration the chemical processes in the snowpack.

REFERENCES

Ambach, W., Blumthaler, M., Eisner, H., Kirchlechner, P., Schneider, H., Behrens, H., Moser, H., Oerter, H., Rauert, W., and Bergmann, H. 1978. 'Untersuchungen der Wassertafel am Kesselwandferner, Ötztaler Alpen an einem 30 Meter tiefen Firnschacht', *Z. Gletscherk. Glazialgeol.*, **14**, 61–71.
Ashton, G. D. 1980. 'Freshwater ice growth, motion, and decay' in Colbeck, S. C. (Ed.), *Dynamics of Snow and Ice Masses.* Academic Press, New York. pp. 261–304.
Bales, R. C., Davis, R. E., and Stanley, D. A. 1989. 'Ion elution through shallow homogeneous snow', *Wat. Resour. Res.*, **25**, 1869–1877.
Bezinge, A. 1981. *Glacial Meltwater Streams, Hydrology and Sediment Transport: the Case of the Grande Dixence Hydroelectricity Scheme.* Birkhauser Verlag [in French] [see also translation in Gurnell, A. and Clark, M. (Eds), *Glacio-fluvial Sediment Transfer: an Alpine Perspective.* Wiley, New York. pp. 473–498].
Clarke, G. K. C. and Waddington, E. 1991. 'A three-dimensional theory of wind pumping', *J. Glaciol.*, **37**, 89–96.
Clarke, G. K. C., Fisher, D. A., and Waddington, E. D. 1987. 'Wind pumping: a potentially significant heat source in ice sheets', *Int. Symp. Physical Basis of Ice Sheet Modeling. IAHS Publ.*, **170**, 169–180.
Cobb, A. W. and Gross, G. W. 1969. 'Interfacial electrical effects observed during the freezing of dilute electrolytes in water', *J. Electrochem. Soc.*, **116**, 796–804.
Colbeck, S. C. 1973. 'Effects of stratigraphic layers on water flow through snow', *Cold Regions Res. Engin. Lab. Res. Rep.*, **311**, 33 pp.
Colbeck, S. C. 1981. 'A simulation of the enrichment of pollutants in snowcover runoff', *Wat. Resour. Res.*, **17**, 1383–1388.
Colbeck, S. C. 1989. 'Air movement in snow due to wind-pumping', *J. Glaciol.*, **35**, 209–213.
Colbeck, S. C. and Davidson, G. 1973. 'Water percolation through homogeneous snow', *Int. Symp. Role of Snow and Ice in Hydrology. IAHS Pub.*, **107**, 242–257.
Collins, D. N. 1978. 'Hydrology of an Alpine glacier as indicated by the chemical composition of meltwater', *Z. Gletscherk. Glazialgeol.*, **13**, 219–238.
Collins, D. N. 1979a. 'Hydrochemistry of meltwaters draining from an Alpine glacier', *Arctic Alpine Res.*, **11**, 307–324.
Collins, D. N. 1979b. 'Quantitative determination of the subglacial hydrology of two Alpine glaciers', *J. Glaciol.*, **23**, 347–362.
Conway, H. and Benedict, R. 1994. 'Infiltration of water into snow', *Wat. Resour. Res.*, **30**, 641–649.
Cragin, J. H., Hewitt, A. D., and Colbeck, S. C. 1993. 'Elution of ions from melting snow', *Cold Regions Res. Engin. Lab. Res. Rep.*, **93-8**, 13 pp.
Cunningham, J. and Waddington, E. D. 'Air flow and dry deposition of non-seasalt sulfate in polar firn: paleoclimatic implications', *J. Atmos. Environ.*, in press.
Davies, T. D., Vincent, C. E., and Brimblecombe, P. 1982. 'Preferential elution of strong acids from a Norwegian ice cap', *Nature*, **300**, 161–163.
Elder, K. and Kattelmann, R. 1993. 'A low-angle slushflow in the Kirgiz Range, Kirgizstan', *Permafrost Periglacial Process.*, **4**, 301–310.
Elliston, G. R. 1973. 'Water movement through the Gornergletscher', *Int. Symp. Hydrology of Glaciers. IAHS Publ.*, **95**, 79–84.
Fountain, A. G. 1989. 'The storage of water in, and hydraulic characteristics of, the firn of South Cascade Glacier, Washington State, U.S.A.', *Ann. Glaciol.*, **13**, 69–76.
Fountain, A. G. 1992a. 'Subglacial hydraulics of South Cascade Glacier, Washington, *PhD Dissertation*, University of Washington, 265 pp.
Fountain, A. G. 1992b. 'Subglacial water flow inferred from stream measurements at South Cascade Glacier, Washington, USA.', *J. Glaciol.*, **38**, 51–64.
Gerdel, R. W. 1954. 'The transmission of water through snow', *Trans. Am. Geophys. Union*, **35**, 475–485.
Goodman, D. 1971. 'Ecological investigations of ice worms on Casement Glacier, Southeast Alaska', *Ohio State Univ. Inst. Polar Studies Rep.*, **39**, 59 pp.

Gross, G. W., Gutjahr, A., and Caylor, K. 1987. 'Recent experimental work on solute redistribution at the ice/water interface, impli-
cations for electrical properties and interface processes', *J. Phys.*, **48**, 527–533.
Hagen, A. and Langeland, A. 1973. 'Polluted snow in southern Norway and the effect of the meltwater on freshwater and aquatic
organisms', *Environ. Pollut.*, **5**, 45–57.
Hobbs, P. V. 1974. *Ice Physics*. Clarendon Press, Oxford. 605 pp.
Hooke, R. LeB. 1984. 'On the role of mechanical energy in maintaining subglacial water conduits at atmospheric pressure', *J. Glaciol.*,
30, 180–187.
Humphrey, N., Raymond, C. and Harrison, W. 1986. 'Discharges of turbid water during mini-surges of Variegated Glacier, Alaska,
U.S.A.', *J. Glaciol.*, **32**, 195–207.
Johannessen, M., Dale, E. T., Gjessing, A., Henriksen, A., and Wright, R. F. 1975. 'Acid precipitation in Norway: the regional
distribution of contaminants in snow and chemical processes during snowmelt', *Int. Symp. Isotopes and Impurities in Snow and
Ice. IAHS Publ.*, **118**, 116–120.
Junge, C. E. 1977, 'Processes responsible for the trace content in precipitations', *Int. Symp. Isotopes and Impurities in Snow and Ice.
IAHS Publ.*, **118**, 63–77.
Kamb, B. 1987. 'Glacier surge mechanism based on linked cavity configuration of the basal water conduit system', *J. Geophys. Res.*,
92B, 9083–9100.
Lang, H., Schädler, B., and Davidson, G. 1977. 'Hydroglaciological investigations on the Ewigschneefeld-Gr. Aletschgletscher',
Z. Gletscherk. Glazialgeol., **12**, 109–124.
Larson, G. J. 1977. 'Internal drainage of stagnant ice: Burroughs Glacier, Southeast Alaska', *Inst. Polar Studies Ohio State Univ. Rep.*,
65, 33 pp.
Larson, G. J. 1978. 'Meltwater storage in a temperature glacier, Burroughs Glacier, Southeast Alaska', *Inst. Polar Studies Ohio State
Univ. Rep.*, **66**, 56 pp.
Magono, C., Endoh, T., Ueno, F., Kubota, F., and Itasaka, M. 1979. 'Direct observations of aerosols attached to falling snow crystals',
Tellus, **31**, 102–114.
Male, D. H. 1980. 'The seasonal snowcover' in Colbeck, S. C. (Ed.), *Dynamics of Snow and Ice Masses*. Academic Press, New York.
pp. 305–396.
Marsh, P. and Woo, M. 1984. 'Wetting front advance and freezing of meltwater within a snow cover 1. Observations in the Canadian
Arctic', *Wat. Resour. Res.*, **20**, 1853–1864.
Maupetit, F., Reynaud, L., Pourchet, M., Pinglot, J. F., and Delmas, R. J. 1994. 'Glaciological and glaciochemical activities of the
L.G.G.E. in the Mont Blanc area (French Alps)' in Haeberli, W. and Stauffer, B. (Eds), *Greenhouse Gases, Isotopes and Trace
Elements in Glaciers as Climatic Evidence of the Holocene. Rep. ESF/EPC Workshop, Zurich, 27–28 October 1992. Arbeitsheft
Nr. 14 of the Versuchsanstalt für Wasserbau, Hydrologie und Glaziologie der Eidgenössischen Technischen Hochschule, Zurich*,
48 pp.
Meier, M. F. 1969. 'Glaciers and water supply', *J. Am. Water Works Assoc.*, **61**, 8–12.
Meier, M. F. and Tangborn, W. V. 1965. 'Net budget and flow of South Cascade Glacier, Washington', *J. Glaciol.*, **5**, 547–566.
Melack, J. M., Stoddard, J. L., and Ochs, C. A. 1985. 'Major ion chemistry and sensitivity to acid precipitation of Sierra Nevada lakes',
Wat. Resour. Res., **21**, 27–32.
Morris, E. M. and Thomas, A. G. 1985. 'Preferential discharge of pollutants during snowmelt in Scotland', *J. Glaciol.*, **31**, 190–193.
Oerter, H. and Moser, H. 1982. 'Water storage and drainage within the firn of a temperate glacier Vernagtferner Oetztal Alps, Austria',
Int. Symp. Hydrological Aspects of Alpine and High-mountain Areas. IAHS Publ., **138**, 71–82.
Onesti, L. 1987. 'Slushflow release mechanism: a first approximation' in Salm, B. and Gruler, H. (Eds), *Avalanche Formation, Movement
and Effects. IAHS Publ.*, **162**, 331–336.
Paterson, W. S. B. 1981. *The Physics of Glaciers*. 2nd edn. Pergamon Press, New York. 380 pp.
Pfeffer, T., Illangasekare, T. H., and Meier, M. F. 1990. 'Analysis and modeling of melt-water refreezing in dry snow', *J. Glaciol.*, **36**,
238–246.
Raiswell, R. 1984. 'Chemical models of solute acquisition in glacial melt waters', *J. Glaciol.*, **30**, 49–56.
Raiswell, R. and Thomas A. G. 1984. 'Solute acquisition in glacial melt waters. I. Fjallsjökull (South-east Iceland): bulk melt waters
with closed-system characteristics', *J. Glaciol.*, **30**, 35–43.
Reynolds, R. C. and Johnson, N. M. 1972. 'Chemical weathering in the temperature glacial environment of the northern Cascade
Mountains', *Geochim. Cosmochim. Acta*, **36**, 537–545.
Röthlisberger, H. 1972. 'Water pressure in intra- and subglacial channels', *J. Glaciol.*, **11**, 177–203.
Röthlisberger, H. and Lang, H. 1987. 'Glacial hydrology' in Gurnell, A. and Clark, M. (Eds), *Glacio-fluvial Sediment Transfer: an
Alpine Perspective*. Wiley, New York. pp. 207–284.
Schommer, P. 1977. 'Wasserspiegelmessungen im Firn des Ewigschneefeldes, Schweizer Alpen', *Z. Gletscherk. Glazialgeol.*, **12**, 125–
141.
Sharp, R. P. 1951. 'Meltwater behavior in firn on upper Seward Glacier, St. Elias Mountains, Canada', *General Assembly of Brussels.
IAHS Publ.*, **32**, 246–253.
Skartveit, A. and Gjessing, Y. T. 1979. 'Chemical quality of snow and runoff during spring snowmelt', *Nordic Hydrol.*, **10**, 141–
154.
Stenborg, T. 1973. 'Some viewpoints on the internal drainage of glaciers', *Int. Symp. Hydrology of Glaciers. IASH Publ.*, **95**, 177–129.
Tangborn, W. V., Krimmel, R. M., and Meier, M. F. 1975. 'A comparison of glacier mass balance by glaciological, hydrological, and
mapping methods, South Cascade Glacier, Washington', *General Assembly of Moscow. IASH Publ.*, **104**, 185–196.
Thomas, W. H. 1972. 'Observations on snow algae in California', *J. Phycol.*, **8**, 1–9.
Thomas, A. G. and Raiswell, R. 1984. 'Solute acquisition in glacial melt waters. II. Argentière (French alps): bulk melt waters with
open-system characteristics', *J. Glaciol.*, **30**, 44–48.
Trabant, D. C. and Mayo, L. R. 1985. 'Estimation and effects of internal accumulation on five glaciers in Alaska', *Ann. Glaciol.*, **6**,
113–117.

Tranter, M., Brimblecomb, P., Davies, T. D., Vincent, C. E., Abrahams, P. W., and Blackwood, I. 1986. 'The composition of snowfall, snowpacks and meltwater in the Scottish highlands: evidence for preferential elution', *Atmos. Environ.*, **20**, 517–525.

Tsiouris, S., Vincent, C. E., Davies, T. D., and Brimblecombe, P. 1985. 'The elution of ions through field and laboratory snowpacks', *Ann. Glaciol.*, **7**, 196–201.

Vaughn, B. 1994. 'Stable isotopes as hydrologic tracers in South Cascade Glacier', *MS Thesis*, Department of Geology, University of Colorado, Boulder, 143 pp.

Wagenbach, D. and Münnich, K. O. 1988. 'The anthropogenic impact on snow chemistry at Colle Gnifetti, Swiss Alps', *Ann. Glaciol.*, **10**, 183–187.

Wake, C. P., Mayewski, P. A., and Spenser, M. J. 1990. 'A review of central Asian glaciochemical data', *Ann. Glaciol.*, **14**, 301–306.

Walder, J. S. 1986. 'Hydraulics of subglacial cavities', *J. Glaciol.*, **32**, 404–415.

Weertman, J. 1972. 'General theory of water flow at the base of a glacier or ice sheet', *Rev. Geophys. Space Phys.*, **10**, 287–333.

Wharton, R. A. and Vinyard, W. C. 1983. 'Distribution of snow and ice algae in western North America', *Madrono*, **30**, 201–209.

Winkler, P. 1980. 'Observations on acidity in continental and in marine atmospheric aerosols and in precipitation', *J. Geophys. Res.*, **85C**, 4481–4486.

3

ISOTOPIC AND IONIC VARIATIONS IN GLACIER RIVER WATER DURING THREE CONTRASTING ABLATION SEASONS

WILFRED H. THEAKSTONE

Department of Geography, University of Manchester, Manchester, UK

AND

N. TVIS KNUDSEN

Department of Earth Sciences, University of Aarhus, Denmark

ABSTRACT

The significance of the baseflow component of glacier river discharge in summer varies with geographical location, altitude, glacier geometry and glacier size. Baseflow is maintained by meltwater generated above the transient equilibrium line and by water released from temporary storage on, in or beneath the glacier. At the Norwegian glacier Austre Okstindbreen, where precipitation is generally high throughout the year and the summers are cool and wet, observations in three successive, but contrasting, years have shown that Na^+ ion concentrations in the glacier river water are influenced strongly by the amount of snowmelt. This itself depends on the preceding winter conditions, which determine the amount of accumulation, and on the current summer's weather. The efficiency of the glacier's drainage systems depends on the general progress of summer ablation. The speed with which the systems develop influences ion provision from subglacial sources. Ca^{2+} ion concentrations are largely determined by subglacial conditions. Oxygen isotope variations in glacier river water reflect the relative contributions made to total discharge by snow meltwater and other sources; the composition of the snow cover, which is a function of winter temperatures, has a strong influence. Ice meltwater has low isotopic variability, but the isotopic composition of rainfall varies markedly. A simple model of mixing of englacial and subglacial waters, each of a constant composition, cannot be applied to a high-latitude glacier of the size and altitudinal range of Austre Okstindbreen.

INTRODUCTION

Glacier river water is vital to the generation of electricity in Norway. Studies of the causes of variations of glacier river discharge therefore have practical importance. Seasonal variations of flow-routing of water through a glacier affect both subglacial sediment transfer and glacier movement; a better knowledge of the nature of the water which issues from glaciers should increase our understanding of those variations.

Glacier river discharge varies between years, between seasons and between days during the summer season (Collins, 1978; Theakstone and Knudsen, 1989). The variations are a response to changes in mass balance components (accumulation and ablation), to summer precipitation and to water storage and flow-routing. A baseflow component of river discharge is maintained during the summer by meltwater generated above the transient equilibrium line, and by water released from temporary storage on, in or beneath the glacier. Diurnal changes in discharge which are superimposed on the baseflow result principally from variations in ice melt below the transient equilibrium line: diurnal variations of melting at higher altitude are smoothed out as water percolates through snow and firn. The contribution of the baseflow component to total discharge varies with geographical location, altitude, glacier geometry and glacier size (Collins, 1978; Collins and Young, 1981; Theakstone and Knudsen, 1989; Brown and Tranter,

1990). Glacier river hydrographs are also influenced by irregular events, some related to liquid precipitation and others to the occasional collapse of ice into water courses within or beneath the glacier. In environments where precipitation is generally high throughout the year and the summers are cool and wet, such as at Norwegian glaciers close to the Arctic Circle, both liquid precipitation during the summer and snowmelt make large contributions to glacier river discharge.

It has been recognized for a long time that the chemical composition of the dissolved load of water draining from a glacier differs from that of precipitation and of water generated by the melting of snow or ice at the glacier surface (Rainwater and Guy, 1961; Keller and Reesmann, 1963; Slatt, 1972; Zeman and Slaymaker, 1975). More recent investigations have centred on the processes leading to enrichment of the water which penetrates into the glacier and have focused on the residence time of the water, the availability of freshly exposed material interacting with the water and the development of the internal drainage system within and underneath the glacier (Collins, 1979; Raiswell and Thomas, 1984; Thomas and Raiswell, 1984; Tranter et al., 1993). Parts of the internal drainage systems of glaciers are likely to close down in the winter, when little or no water enters them. During summer, as the transient equilibrium line moves up-glacier and the amount of surface ablation increases, the systems are likely to enlarge and to become more integrated. However, because the systems are inaccessible, their character and physical parameters have to be deduced from field measurements of water input and output and from theory (Röthlisberger and Lang, 1987).

The dissolved load of glacier rivers commonly displays systematic patterns of variation with time, together with some superimposed random deviations. As the summer progresses, the dissolved load may change in response to alterations of the glaciers' drainage systems or to changes in chemical reactions within them (Tranter et al., 1993). The impurities in the water, together with its isotopic composition, provide evidence about variations of water sources, flow routing and processes of solute acquisition. Most studies of glacier hydrochemistry have focused on electrical conductivity as a surrogate for total dissolved solids, or on variations of specific cations or anions, and there have been relatively few attempts to investigate water sources and flow routing on the basis of the stable isotopic composition of snow, ice and river water in glacierized catchments; such an approach is reported in this paper.

Most glacier hydrological investigations have been conducted in Alpine environments, either on steep, small, high-lying glaciers or on larger glaciers with a major tongue descending to lower altitudes. Many of these glaciers receive rather little accumulation and have experienced several years with a negative mass balance as a result of high values of summer ablation. In such circumstances, a major proportion of the meltwater draining from them originates from icemelt: snowmelt is important only during the early part of the ablation period (Gurnell et al., 1992). In contrast, most glaciers in the Okstindan area, Norway (66°N, 14°07′E), including Austre Okstindbreen (Figure 1), which is the largest, have experienced above-normal precipitation and low ablation during recent summers. The mass balance of Austre Okstindbreen has been positive in most years (Knudsen, 1993) and snowmelt has made an important contribution to river discharge almost until the end of the summer.

Austre Okstindbreen covers 14·01 km^2. It is bigger, and has a more maritime climate, than many glaciers at which hydrochemical studies have been made. Sampling of river water discharging from the glacier in three successive, but contrasting, summers has shown that ionic concentrations in the water are influenced by the amount of snow which accumulates during the winter, and by the melting conditions in the summer. The depth of the accumulated snow is measured in late winter and late summer, as part of a standard mass balance programme (Knudsen, 1993), by probing at intervals of 100 m along 14 km of profiles, and by direct observation in a number of pits excavated to the previous summer's surface.

The oxygen isotopes in the snow which covers Austre Okstindbreen in winter are influenced by the temperature during snowfall. Thus the ratio of the heavy and light isotopes of oxygen in water which results from melting of the snow varies considerably. In contrast, the variability of the relative concentrations of these isotopes in water generated by the melting of ice is slight. Therefore, the accumulated snow and the course of its melting determine the stable isotopic composition of the river water, and conditions in the glacier's accumulation area are a principal control of both the quantity and the quality of the water which issues from the glacier in summer.

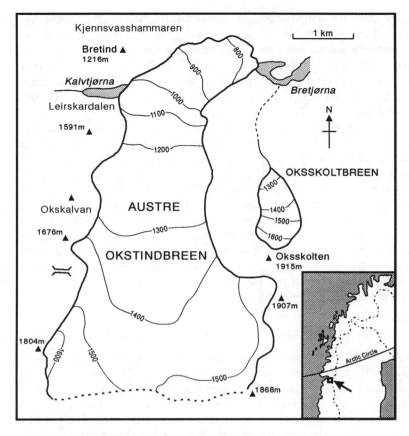

Figure 1. Location map of the glacier Austre Okstindbreen

HYDROLOGY OF AUSTRE OKSTINDBREEN

Glacio-hydrological studies at Austre Okstindbreen were started in 1976 as part of the Okstindan Glacier Project, undertaken jointly by personnel from the universities of Aarhus, Denmark and Manchester, UK. The glacier is located at around $66°01'N$ $14°07'E$, close to the border between Norway and Sweden (Figure 1). Its accumulation zone, most of which is between 1300 and 1500 m, covers about $11 \cdot 4 \, km^2$; the highest parts are more than 1700 m above sea level (Knudsen, 1993). The glacier ends in the Oksfjelldalen valley, where it has been retreating through a proglacial lake (740 m) for more than 20 years. Much of the glacier front has retreated from the lake during the last four years.

There is no detectable outflow of water from Austre Okstindbreen in winter. River discharge usually starts in May and does not continue beyond October. At the start of the melt season, flooding of the valley floor may occur upstream of temporary dams formed by snow and ice which still cover lakes and the river in Oksfjelldalen. In most years, water appears at the front of Austre Okstindbreen within a few days of the start of snowmelt, generally at one or two discrete positions.

Between 1976 and 1986, two rivers, one larger than the other, issued from the glacier front in summer (Theakstone and Knudsen, 1989), but in 1987 and more recent years, outflow was restricted to a single river. The position at which the principal river left the glacier did not change during the course of any summer except that of 1985, when the major outlet changed its location during a storm in July (Karlsen, 1991). In 1987, 1988 and 1989, the river had a short subaerial course before entering the northern side of the proglacial lake, where it rapidly deposited a coarse-grained delta. In 1990 it left the glacier from a more central position. In most summers until 1989, the discharge of the glacier river was affected by the drainage of an ice-dammed lake, Kalvtjørna (Knudsen and Theakstone, 1988).

Observations at the start of the melt season at Austre Okstindbreen have shown that water draining from the snow which covers the lower part of the glacier flows on top of a superimposed ice layer 1–5 cm thick. It is only when the superimposed ice has melted that meltwater begins to penetrate into the glacier through crevasses (Stenborg, 1968). The supply of water, which is considered to play a part in the development of the glacier's internal drainage system, then depends on the amounts of melting and liquid precipitation. Dye tracer tests suggest that the efficiency of the drainage system is determined by the general progress of summer ablation (Theakstone and Knudsen, 1981).

In several recent years, the equilibrium line at Austre Okstindbreen has been no higher than 1250 m, and about 70% of the glacier has been covered by the remains of the previous winter's snow at the end of the summer; on several occasions, heavy summer snowfalls have covered the glacier down to about 1000 m for several days. In these years, the ablation of ice below the equilibrium line has been low and the glacier's internal drainage systems, which cannot be observed directly, are likely to have been poorly developed.

Above the equilibrium line, meltwater drains diffusely through the snow until it is intercepted by ice layers, many of which are seen to disappear during the summer. The meltwater then passes down to the firn–ice interface. Both direct observations during summers with high ablation and extensive snow depth soundings conducted as part of an annual mass balance study have shown that the accumulation area is heavily crevassed. Thus the water reaching the ice surface must enter the glacier rather quickly. It then may penetrate to the bottom, or to a position close to the bottom, either above the icefall that starts at about 1200 m, or within the icefall.

RIVER DISCHARGE AT AUSTRE OKSTINDBREEN

River discharge at Austre Okstindbreen has been recorded every summer since the early 1980s. Throughout the period, a gauging station has been maintained at the outlet of the proglacial lake. Observations have shown that ice emerging from the glacier river portal is carried rapidly to the lake outlet by the strong current. For much of the 1980s, the water level was recorded within 30 m of the point at which the river first emerged from the glacier, and this dual recording revealed that the lake has little damping effect on the discharge measurements at the gauging station. Water which has been temporarily stored within the glacier contributes to river discharge at Austre Okstindbreen, but its importance tends to decline through the summer. Water balance calculations for the catchments in the four years 1981–1984 indicated that 37–47% of the river discharge during periods of 26–29 days in July and early August was supplied by stored water (Svendsen, unpublished data). In recent years, the contribution, determined for slightly shorter periods, has been smaller (Table III). Between 1981 and 1987, despite variations of snow ablation, ice ablation, precipitation and release of water from storage, the mean daily river discharge at Austre Okstindbreen in July was remarkably uniform (Table I).

Glacier river discharge at a particular time results from periodic events (controlled by varying temperature and radiation), single events (the results of precipitation), trend and noise. Using values of air temperature and

Table I. Mean daily river discharge, Austre Okstindbreen

Period	Mean daily discharge ($\times 10^{-6}$ m^3)
4 July–4 August 1981	0·72
5 July–26 August 1982	0·74
8 July–31 July 1983	0·74
5 July–31 July 1984	0·72
6 July–3 August 1985	0·77
5 July–2 August 1986	0·83
5 July–31 July 1987	0·65
7 July–10 September 1988	1·10
11 July–17 September 1989	0·59
5 July–13 September 1990	0·80

Table II. Mean specific winter, summer and net mass balance (metres water-equivalent) at Austre Okstindbreen

Year	Winter	Summer	Net
1987–1988	1·5	3·4	−1·9
1988–1989	3·7	2·2	+1·5
1989–1990	3·0	2·7	+0·3
1978–1992	2·2	2·5	−0·3
1986–1992	2·5	2·3	+0·2

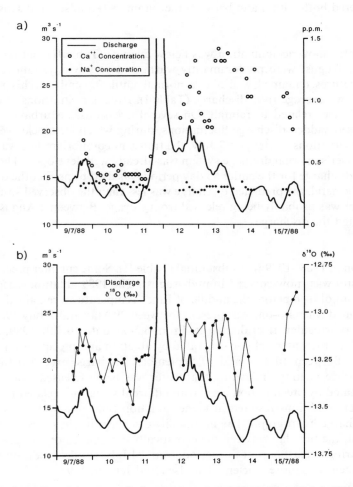

Figure 2. (a) During the period of sampling of river water discharging from Austre Okstindbreen in July 1988, discharge was influenced strongly by rainfall. Much of the previous winter's snow already had melted and Na^+ concentrations in the water were low. Because the glacier's internal drainage systems were well developed and water had access to a substantial part of the bed, Ca^{2+} concentrations were higher than usual. They were particularly high after heavy rain fell on 11 July. (b) Oxygen isotope variations were in phase with discharge during the first part of the 1988 sampling period. After the heavy rainfall of 11 July, both discharge and $\delta^{18}O$ values were more variable than before it

Table III. Water storage at Austre Okstindbreen

Period	Stored water (%)
10 July–3 August 1988	23
10 July–3 August 1989	10
10 July–30 July 1990	26

net radiation measured at 20-minute intervals, and precipitation determined on a daily basis, Jensen (unpublished data) computed temporal variations in river discharge from Austre Okstindbreen for the summers of 1988, 1989 and 1990. He analysed the frequencies and magnitudes of events using a Fast Fourier Transform procedure. Computed discharge variations were compared with the recorded discharge. Cross-correlation functions between meteorological factors and discharge indicated that the response of discharge to a given meteorological change varied with time. The delay between water inputs at the surface and water release in the glacier river differed both within and between the summers (Knudsen and Jensen, 1992).

1987–1988

In 1987–1988, winter snow accumulation was below average (Table II) and the ablation season started early. July and early August were predominantly warm and very windy; rain fell frequently during the summer and temperatures remained well above normal until September. This resulted in exceptionally high ablation and above-average river discharge (Table I). Irregular variations of discharge were common (Figure 2), many of them related to rainfall. On several occasions, disturbance of the glacier's internal drainage system caused sudden discharge fluctuations, during which ice blocks were carried from beneath the glacier. When observations started, on 7 July, the transient equilibrium line was at a high level and it is probable that the glacier's internal drainage system was already well developed. The regression equation for air temperature and discharge for the entire 65 day period was used to predict the discharge on a daily basis (Knudsen, 1989). In general, the predicted discharge was less than the observed value for the period 7 July–2 August, when water was probably being released from storage. Between 3 August and 10 September, the equation overestimated the discharge.

1988–1989

Snow accumulation in 1988–1989 was substantial (Table II). Some summer precipitation fell as snow and most of the glacier still was snow-covered in mid-August. The 1989 ablation season started late and little water left Austre Okstindbreen before the middle of June. Temperatures remained low until the last part of July. Kalvtjørna drained when no-one was present, between 29 May and 7 July 1989, probably on 28 June, a day of high, but very irregular, river discharge. From 6 June until 10 July 1989, discharge of water from the glacier was less than the water volume generated by ablation and precipitation: about $4\cdot8 \times 10^6$ m^3 was stored in the glacier. Between 10 July and 17 September, discharge exceeded the combined supplies of precipitation and surface melting, indicating that water was being released from storage. Before 20 July, discharge was dominated by precipitation events, and diurnal variations related to ablation were rare (Figure 3). Later, warmer conditions resulted in higher ablation rates, but wet weather still had a dominant influence on the discharge hydrograph. The mean daily discharge in 1989 was very low (Table I). Between 10 July and 3 August, the time lag between the climatically determined discharge and the actual discharge was 2–4 days, but during the last part of the period, from 24 July, the delay was only 1–1·5 days, suggesting that the drainage system was more efficient (Knudsen and Jensen, 1992).

1989–1990

Winter accumulation in 1989–1990 was well above average (Table II), but the 1990 ablation season began early and summer ablation was slightly above average. Once again, the importance of the glacier as a reservoir influencing discharge was evident, with changes of discharge lagging behind those of the

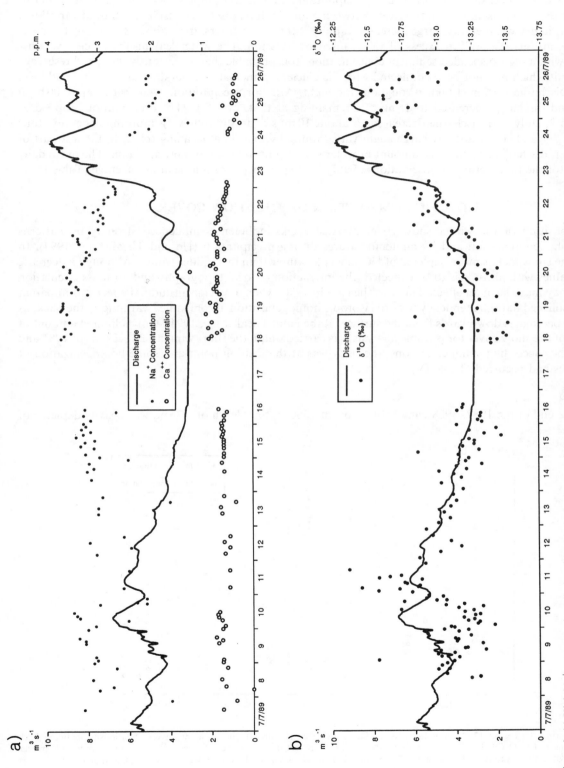

Figure 3. (a) High Na$^+$ concentrations during the period of sampling of river water at Austre Okstindbreen in July 1989 reflected the dominant influence of snow meltwater on discharge. Sampling was interrupted for two periods (16–18 July and 23–24 July). A period of low ablation in the middle of the month resulted in very low discharge, with little variation. Subsequently, as ablation rates increased, Na$^+$ concentrations varied in phase with discharge. Ca^{2+} concentrations were low throughout the period of sampling. (b) During a period of low ablation in mid-July 1989, snow meltwater, released from storage or delayed in its passage through the glacier, made a major contribution to river discharge. This resulted in low δ^{18}O values. From 20 July, as ablation and discharge increased, the increasing contribution of ice meltwater was reflected in higher δ^{18}O values

weather on several occasions. Before 5 July, precipitation and ablation supplied $5 \cdot 0 \times 10^6 \, m^3$ more water to the glacier than was discharged in the river. Between 5 and 18 July, discharge displayed diurnal variations. On 20 July, in dry weather, discharge increased rapidly within a few hours, from about 8 to about $11 \, m^3 \, s^{-1}$ (Figure 5), an increase well in excess of that caused by ablation. The water leaving the glacier was discoloured, with a high suspended sediment concentration. It is probable that a previously untapped reservoir, or a conduit which had not been connected with the glacier's principal internal drainage system, had been opened. Discharge remained high throughout the night. A major precipitation event began at 10.00 h on 21 July and discharge increased to $15 \, m^3 \, s^{-1}$, remaining at that level for 24 h. Precipitation stopped at 21.00 h on 22 July and discharge decreased to about $10 \, m^3 \, s^{-1}$ on the following morning. The next four days were dry and sunny and discharge again varied diurnally, but on a declining trend. In the last part of the summer, discharge exceeded the amount of water supplied by ablation and precipitation. The mean daily discharge for the total period of observations (5 July–13 September) was a little above average (Table I).

ISOTOPIC COMPOSITION OF THE SNOW COVER

The composition of the winter snow cover at Austre Okstindbreen is influenced strongly by patterns of atmospheric circulation and by air temperatures during precipitation (He and Theakstone, 1994). In general, the snowpack is more depleted of ^{18}O in a cold winter than in a milder winter. Although a decrease of $\delta^{18}O$ values with altitude is to be expected, the interaction of local topography and air mass circulation patterns may mask such an effect. This was the case in 1990, when a reverse altitudinal effect was apparent (He, unpublished data). A tendency towards isotopic homogenization accompanies warming of the pack, as meltwater percolates downwards from the surface (Raben and Theakstone, 1994). Only a limited amount of sampling of the snow cover for isotopic analysis was carried out in the short winter field seasons of 1987 and 1988. In the succeeding summer seasons, the pack was at the melting point and some homogenization of $\delta^{18}O$ values had occurred (Table IV).

1987–1988

Near the end of the 1987–1988 winter, the snow at 1360 m, to a depth of 1·25 m, was heavily depleted of

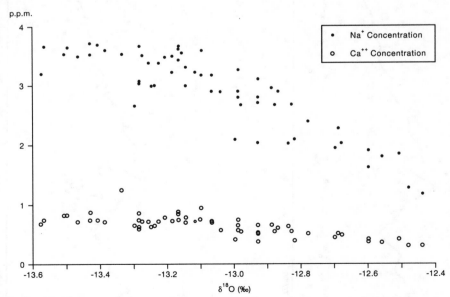

Figure 4. During the period 17–26 July 1989, as discharge increased (Figure 3), there was an inverse relationship between Na^+ concentrations and $\delta^{18}O$ values. $\delta^{18}O$ values above about $-12 \cdot 8$‰ reflect the dominance of icemelt; lower values represent a greater contribution of snowmelt to river discharge. Ca^{2+} concentrations show a weaker relationship with $\delta^{18}O$ values, as much of the Ca^{2+} is acquired subglacially

Table IV. $\delta^{18}O$ values of accumulated snow at Austre Okstindbreen

Date	Altitude (m)	Sampled depth (m)	No. of samples	Mean $\delta^{18}O$ value (‰)	Minimum $\delta^{18}O$ value (‰)	Maximum $\delta^{18}O$ value (‰)
22 May 1988	1360	1·25	22	−15·50	−21·84	−10·70
23 May 1988	840	1·70	6	−13·16	−15·78	−10·77
19 July 1988	1450	2·20	38	−11·19	−12·86	−9·96
31 May 1989	1240	2·25	5	−12·07	−12·56	−11·18
31 May 1989	1000	1·80	8	−10·21	−12·00	−9·30
21 July 1989	1450	6·30	43	−11·38	−14·67	−10·08
24 July 1989	1300	1·92	14	−11·03	−11·95	−10·26
7 May 1990	1470	2·40	23	−11·97	−17·20	−8·20
15 July 1990	1470	2·10	37	−10·73	−11·97	−9·59
20 July 1990	1370	4·10	41	−11·64	−12·76	−10·30
24 July 1990	1230	2·80	27	−12·51	−13·56	−10·93
16 July 1990	980	1·90	36	−12·87	−15·84	−10·79
26 July 1990	980	1·10	11	−11·25	−12·46	−10·34

^{18}O (Table IV). Only the top 0·5 m of the pack was at the melting point. At 840 m, however, the 1·7 m deep pack which covered the glacier ice was isothermal at the melting point and was less depleted in the heavy isotope. The early onset of melting and the warm conditions of the 1988 ablation season resulted in a net loss of material from most of the glacier's accumulation zone, and firn was widely exposed in mid-July; the ^{18}O content of snow samples collected from a pit at 1450 m, above the transient equilibrium line, was much higher than that recorded at the two lower sites in May (Table IV).

1988–1989

The 1988–1989 winter was much warmer than normal, especially after December. In late May 1989, the snowpack at 1240 m and 1000 m was richer in ^{18}O than at the end of previous winters, but an altitude effect (increasing depletion of the heavy isotope with increasing altitude) was apparent (Table IV). On 21 July, 6·3 m of snow remained at 1450 m; the thickness of the pack had limited the degree of isotopic homogenization (Table IV). The uppermost metre of the underlying firn, however, was isotopically homogeneous mean $\delta^{18}O$ value −11·38‰). On 24 July, the 1·92 m thick snow cover at 1300 m was isotopically heavier than that at 1450 m.

1989–1990

In May 1990, substantial variations of $\delta^{18}O$ values in the snowpack at 1470 m reflected varied air temperatures during winter episodes of precipitation; there was no evidence of disruption of the isotopic stratigraphy by percolating meltwater, or of isotopic homogenization. By July, ^{18}O enrichment had occurred (Table IV). At 1370 m, on 20 July, the previous summer's surface was at a depth of 4.1 m; variations of $\delta^{18}O$ values in the underlying firn (20 samples, mean −11·04‰, minimum −11·67‰, maximum −10·45‰) were slight and the lowermost metre of the sampled material was almost homogeneous. At 1230 m, most of the 2·8 m of snow which covered glacier ice on 24 July was even more depleted of ^{18}O than was the pack at 1370 m, and the trend of decreasing $\delta^{18}O$ values with decreasing altitude continued down to 980 m (Table IV). However, melting and compaction which reduced the thickness of the pack by 0·8 m in 10 days was accompanied by ^{18}O enrichment (Table IV).

HYDROCHEMICAL OBSERVATIONS AT AUSTRE OKSTINDBREEN

Sampling and analytical methodology

Analyses of about 3000 samples of glacier river water and 1700 samples of snow collected at Austre Okstindbreen during the last decade have shown that both the river's dissolved load and the isotopic composition of the river water vary, both from year to year and during the course of the summer, and that snow at and near the surface of Austre Okstindbreen is much less depleted of the heavy isotope of oxygen, ^{18}O, than is the river water which discharges from the glacier (Theakstone, 1988; 1991; Theakstone and Kundsen, 1989; Knudsen, 1990).

Snowpack samples were collected in a continuous column from freshly exposed pit walls using a 1000 cm^3 stainless-steel tube. They were transferred directly to polyethylene bags, which were then sealed. Samples were allowed to melt before being transferred to polyethylene bottles which had been rinsed with hydrochloric acid and rinsed several times with distilled water. Some of the sample water was used to rinse the bottle before the sample was bottled. Samples were kept in a snow pit at about 0°C until they were transported to the laboratory, where they were kept in a cold room at 4°C before analysis.

River water samples were collected where the river issued from the glacier using a 24-bottle Manning S-4050A automatic liquid sampler with a 2-h sampling interval. The intake was well away from the river banks, at a depth varying between 0·3 m and 0·5 m, where the water depth was 1–2 m. Samples were transferred to polyethylene bottles and filtered at the field station through a 0·45 μm Whatman filter, the filter paper having been cleaned with about 25 ml of the sample. Thereafter, the samples were treated and stored in the same manner as the snowpack samples.

Ca^{2+}, Mg^{2+}, Na^+ and K^+ were determined by flame atomic absorption spectrometry on a Perkin-Elmer 5100 PC using an autosampler. It was necessary to dilute some samples to obtain results within the linear range of the calibration graph. The calibration was checked for each 10 samples. If a drift of >2% was observed, a new calibration was performed and the samples were analysed again. Three values were determined for each sample. The determination of divalent cations was undertaken with added 1 : 250 strontium chloride to prevent interference. The accuracy of the determinations generally was within ±5%.

Results

1985–1987. In 1985 and 1986, Na^+ concentrations were lower than in previous years (Theakstone and Knudsen, 1989), but those of Ca^{2+}, K^+ and Mg^{2+} were high (Table V). In 1987, all cation concentrations were low. In 1985, diurnal variations of $\delta^{18}O$ values of glacier river water were superimposed on a declining trend until the change of drainage conditions which accompanied the mid-July storm; after the event, values were lower than before it. In 1986, recorded $\delta^{18}O$ values were unusually high. In 1987, when drainage of the glacier-dammed lake, Kalvtjørna, terminated before the basin was completely empty, they were lower than usual (Theakstone and Knudsen, 1989).

1987–1988. In 1988, when river sampling was started on 9 July, very little snow remained on the glacier below 1400 m. Much of the Na^+ which had been present in the snowpack at the end of winter must have been removed as the snow melted. During the afternoons of 9, 11 and 12 July, there were heavy showers of rain. About 30 mm fell on 11 July, causing a rapid increase in the discharge, which disrupted operation of the automatic water sampler for some hours. Sampling ended in the middle of July when the station was destroyed by another sudden flood. Although some Na^+ from higher lying snow continued to reach the glacier river throughout the seven days of sampling, the mean Na^+ concentration in the water during the

Table V. Mean cation concentrations (ppm) and $\delta^{18}O$ values (‰) in river water, Austre Okstindbreen (standard deviation in parentheses)

	Na^+	Ca^{2+}	Mg^{2+}	K^+	$\delta^{18}O$
1983	1·26 (0·25)	0·35 (0·06)	0·25 (0·06)	0·31 (0·05)	−13·01 (0·36)
1984	0·67 (0·21)	0·23 (0·08)	0·10 (0·09)	0·23 (0·46)	−13·25 (0·26)
1985	0·53 (0·18)	3·29 (1·51)	0·33 (0·13)	0·58 (0·35)	−13·36 (0·33)
1986	0·67 (0·24)	5·65 (5·84)	0·50 (0·33)	1·12 (0·59)	−12·72 (0·23)
1987	0·80 (0·23)	0·56 (0·22)	0·18 (0·08)	0·23 (0·10)	−13·80 (0·26)
1988	0·34 (0·03)	0·79 (0·35)	0·17 (0·17)	0·17 (0·10)	−13·18 (0·12)
1989	3·14 (0·35)	0·67 (0·11)	0·49 (0·07)	0·35 (0·07)	−13·05 (0·26)
1990	1·73 (0·25)	0·66 (0·24)	0·28 (0·04)	0·27 (0·04)	−13·58 (0·15)

period was very low (Table V); a slight decrease with time was evident (Figure 2a). Concentrations of Ca^{2+}, Mg^{2+} and K^+ were slightly higher than in most previous summers: similarly high values had been recorded only in 1985 and 1986 (Theakstone and Knudsen, 1989; Knudsen, 1990). This, together with the 1–2 day lag between temperature and discharge variations, indicates that water was stored for some time in contact with sources of these ions. Apparently, water was able to reach the bed, where it acquired more Ca^{2+} from subglacial sources; the glacier's drainage systems may have been in contact with the bed in many places. Ca^{2+} concentrations were particularly high after the 11 July event (Figure 2a).

In general, $\delta^{18}O$ values of glacier river water during the short 1988 sampling period were similar to those of several earlier summers (Table V). The weather conditions did not favour ablation-related diurnal variations of $\delta^{18}O$ values, but variations between 9 and 11 July tended to be in phase with the discharge (Figure 2b). Variations were a little more pronounced after the heavy rainfall on 11 July than during the preceding period (Figure 2b).

1988–1989. In 1989, water samples were collected during a generally cold period of 20 days (7–26 July). Sampling was interrupted for two periods (16–18 July and 23–24 July). Much of the glacier remained snow-covered into August and the glacier's drainage systems were probably not well developed until late in the

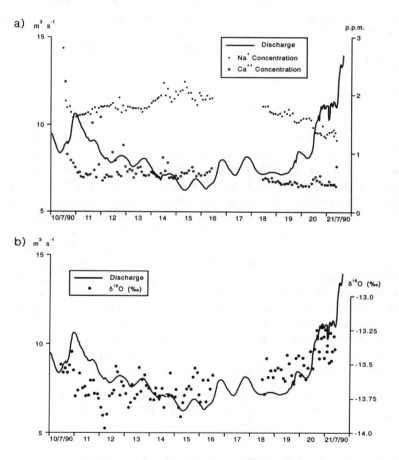

Figure 5. (a) As river discharge decreased during the first half of the 1990 sampling period at Austre Okstindbreen, the increasing contribution of old snow to river discharge was reflected in increasing Na^+ concentrations. After sampling was re-started on 18 July, ice meltwater played a larger part, and Na^+ concentrations decreased. Water released from the glacier on 20 July, a rain-free day, had slightly elevated Ca^{2+} concentrations; the continuing decrease in Na^+ concentrations indicated that the water was dominated by icemelt, rather than by snowmelt. (b) $\delta^{18}O$ values of river water discharging from Austre Okstindbreen were exceptionally low during the first half of the 1990 sampling period, reflecting the importance of snow meltwater with a low heavy isotope content. Much of the water released from the glacier during the flood event of 20–21 July may have been ice meltwater, previously stored in or below the glacier tongue

Table VI. Range and mean $\delta^{18}O$ values (‰) of sources of glacier river water sampled at Austre Okstindbreen, 1980–1990

Source	Minimum	Mean	Maximum	Median	No. of samples
Glacier ice	−12·78	−11·61	−10·79	−11·57	39
Ice meltwater	−12·72	−11·94	−10·49	−11·78	31
Snow (May)	−21·84	−12·79	−7·23	−12·14	112
Snow (July)	−15·60	−11·54	−8·66	−11·38	250
Snow meltwater	−13·75	−12·10	−10·46	−12·08	16
Firn	−12·05	−10·91	−10·30	−10·73	35
Rain water	−15·53	−10·52	−6·49	−10·16	38

summer. Between 10 and 20 July, rain fell frequently, but there was little ablation. The river discharge was low and water balance calculations indicated that it was maintained largely by water released from storage. Because icemelt did not dilute the stored water, ionic concentrations were high (Table V). The very high, but variable, Na^+ concentrations (Figure 3a) reflected the dominance of snowmelt. Fluctuations of Ca^{2+} concentrations, which were slight, occurred mainly on days with precipitation. Between 20 and 22 July, as ablation increased, Na^+ concentrations varied with discharge, although the rising trend of discharge was accompanied by a declining trend of Na^+ concentrations (Figure 3a).

$\delta^{18}O$ values rose to a high level on 11 July 1989 and then declined through the period of low ablation (Figure 3b). From 20 July, with increasing ablation, $\delta^{18}O$ values increased, without displaying clear diurnal variations. In mid-July, when much of the glacier was still covered by snow, most of the water discharging in the glacier river was meltwater from old snow. Discharge generally was low (Figure 3b). Later in the month the influence of old snow on the composition of the river water was reduced as larger contributions from icemelt and rain resulted in higher river discharge. As snow ages, it becomes richer in the heavy isotope of oxygen (Stichler, 1987; Theakstone, 1991) and it loses ions by leaching (Davies et al., 1987; Raben and Theakstone, 1994). The water which drains from the snow into the glacier is therefore depleted in the heavy isotope, i.e. it has low $\delta^{18}O$ values, and enriched in the ions which are leached from the snow. This results in an inverse relationship between $\delta^{18}O$ values and Na^+ concentrations (Figure 4). Between 18 and 23 July, as the river discharge increased, $\delta^{18}O$ values increased and Na^+ concentrations decreased; diurnal variations were superimposed on the trends. Samples of rainfall collected in all recent years at Austre Okstindbreen have shown that the isotopic composition of rain water varies considerably, both between events and within a single event (Theakstone, 1991). It is not surprising, therefore, that rain had no clear, specific influence on the stable isotopic composition of the river water.

1989–1990. In 1990, rainfall events were slight: most precipitation fell in showery weather, rather than in heavy storms, but increases in discharge caused by rainfall were larger than those resulting from daily ablation variations. Although Na^+ concentrations were lower than those recorded in the previous summer, in general they were relatively high (Table V) and highest at times of lowest discharge (Figure 5a). With the exception of some high values recorded on days with precipitation, Ca^{2+} concentrations varied little, although they tended to be slightly higher when discharge was low (Figure 5a).

During a period of low discharge in mid-July, when the contribution of the baseflow component to

Table VII. Mean ionic concentrations (ppm) in snow, ice and rain at Austre Okstindbreen (*n* = number of samples)

	Na^+	Ca^{2+}	Mg^{2+}	K^+	*n*
Snow (May)	1·24	0·04	0·13	0·04	420
Snow (July)	0·23	0·12	0·04	0·05	31
Glacier ice	0·10	0·04	0·02	0·00*	5
Rain water	0·15	0·05	0·04	0·07	3

* Below the detection limit

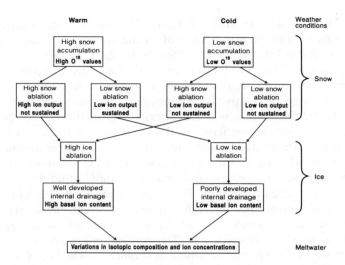

Figure 6. Schematic representation of the possible chemical evolution of glacier meltwater through the summer period. High accumulation is characteristic of warm winters and low accumulation of those which are cold. $\delta^{18}O$ values are higher in snow which accumulates at relatively high temperatures than in that which accumulates in cold winters. Subsequently, the ablation rate determines the chemical characteristics of the snow meltwater and supraglacial ice meltwater: an early start to the ablation season and a high rate of snow ablation result in rapid depletion of ions in the pack. A low rate of ablation of snow after a winter of high accumulation results in a sustained period of elution of ions from the pack. When the snowpack at the end of winter is thin, depletion of ions is rapid, even with a low ablation rate. The development of internal drainage systems depends on the amount of water which penetrates into the glacier. This, in turn, controls the acquisition of subglacial ions. Substantial ablation of ice may follow a winter of high snow accumulation if the ablation season starts early and the rate of snow ablation is high. After a cold winter, ice ablation is likely to be low if the ablation season starts late and the rate of ablation of snow is low. Both the isotopic composition of the water leaving the glacier and the ionic concentrations within it depend on the pre-existing weather conditions and vary with the ablation rates of snow and ice

discharge was large (Figure 5), $\delta^{18}O$ values of the river water were the lowest recorded in the three summers 1988–1990. A substantial proportion of the baseflow is supplied by meltwater from snow. During the 20–21 July event, $\delta^{18}O$ values were high (Figure 5b), suggesting that the water responsible for the increase in river flow may have been ice meltwater, which is enriched in the heavy isotope (Theakstone, 1988).

DISCUSSION

Theakstone (1988) tabulated the range and median $\delta^{18}O$ values of sources of glacier river water collected at Austre Okstindbreen between 1980 and 1987. Further sampling between 1988 and 1990 (Theakstone, 1991) has confirmed that winter snow, rain and glacier ice differ in their stable isotopic composition (Table VI). Glacier ice has a $\delta^{18}O$ value of about $-11\cdot6\text{‰}$. Supraglacial stream water collected in the ablation area is dominated by ice meltwater and so has a very similar value, as there is no isotope fractionation during melting; some snow meltwater may contribute to the supraglacial streams. Meltwater formed from snow is more depleted in the heavy isotope than that formed from ice, with $\delta^{18}O$ values centred on $-12\cdot1\text{‰}$. The composition of winter snow varies from year to year. Most of the snow samples collected in May have been below the melting point and their wide range of $\delta^{18}O$ values reflects the variability of air temperatures during snowfall in winter. In general, winter snow is more depleted of ^{18}O than is glacier ice. As the temperature of the snow increases to the melting point, water movement through the pack results in changes of the initial isotopic composition. Both the onset of melting and its progress vary from year to year, but in July most of the residual snow is at the melting point and has become enriched in ^{18}O, with a tendency towards homogenization; clearly, the water passing out of the snowpack must be depleted of the heavy isotope. Water which has passed through the snowpack percolates more slowly through the denser firn below, and isotopic exchange during densification results in a high ^{18}O content in firn, which is isotopically homogeneous, with a low range of $\delta^{18}O$ values. The isotopic composition of rainfall is highly variable.

The mean concentrations of Na^+, Ca^{2+}, Mg^{2+} and K^+ in samples of snow, ice and rain are shown in

Table VII. Most of the snow sampled in May was below the melting point, whereas that sampled in July was melting: the loss of ions from the pack as a result of the melting and percolation of water is evident. The higher values of Ca^{2+} recorded in July can be accounted for by the deposition of dust onto the snow surface in summer. Glacier ice is pure, with very low levels of any of the four ions. The composition of rainfall is likely to be variable, depending in part on the trajectory of the air masses from which it falls; the small number of recorded samples may not be representative of the input to the glacier as a whole during the summer.

The winter snow at Austre Okstindbreen is characterized by relatively high Na^+ ion concentrations. Because the amount of snow which accumulates on the glacier differs from year to year, and the melting of the snow depends on summer weather conditions, the Na^+ ion concentrations in glacier river water vary from one year to another (Figure 6). Ca^{2+} ion concentrations in the snow are much lower than those of Na^+. The almost total absence of supraglacial debris at Austre Okstindbreen suggests that most of the Ca^{2+} in the glacier river has a subglacial origin. The same applies for K^+ and Mg^{2+}.

Glacier river chemistry at Austre Okstindbreen displayed different characteristics in three successive years. In 1988, the ablation season started early, after a winter with below-average snowfall, and leaching of ions from the snowpack started at an early date. This resulted in low concentrations of Na^+ ions in the glacier river water at the start of the period of sampling (Figure 2). The high Ca^{2+} concentrations at that time suggest that water which had acquired Ca^{2+} subglacially was being released from the glacier. Marked diurnal variations of $\delta^{18}O$ values of the river water during the first two days of sampling resulted from variations of the amount of melting of glacier ice. Although ablation was above average during the warm summer, frequent falls of rain made a considerable contribution to the generally high river discharge and generally resulted in increased $\delta^{18}O$ values of the river water.

In 1989, ablation started late, after a winter with the highest snow accumulation measured during the last 20 years, and the summer was cool and wet. River discharge was low and the influence of ice ablation was limited: the extensive snow cover on the glacier in late summer meant that discharge was dominated by snowmelt, a situation reflected in very high Na^+ concentrations through the period of measurement (Figure 3). Rainfall influenced the glacier river hydrograph but, between 20 and 22 July, which was rain-free, the snowmelt-dominated baseflow component was diluted by icemelt during the day, and diurnal variations of discharge, Na^+ concentrations and $\delta^{18}O$ values were recorded.

In 1990, ablation started early, after a winter in which accumulation was higher than the mean. Despite high ablation rates, river discharge was only slightly above average and, for much of the summer, water apparently went into storage within the glacier. High Na^+ concentrations and low $\delta^{18}O$ values (Figure 5) resulted from the major contribution of snowmelt to discharge.

Collins (1978) introduced a simple two-component mixing model of glacier river chemistry, in which the bulk meltwater emerging from the glacier is assumed to consist of an englacial component and a subglacial component. Much of the former is supplied by water from the melting ice surface below the transient equilibrium line, whereas the latter has a relatively prolonged contact with basal materials, thereby acquiring a substantial solute load. The two components are considered to mix in conduits close to the point at which the bulk meltwater emerges from the glacier, and the inverse relation between meltwater discharge and conductivity has been interpreted as the result of dilution of a rather uniform baseflow component by meltwater formed by ice ablation on the lower part of the glacier, which moves quickly to the glacier margin. The conductivity is thought to be determined by the proportions of chemically pure water from icemelt and of water from the accumulation area which has been chemically enriched during transport (Collins, 1978). However, Tranter and Raiswell (1991) suggested that the composition of the subglacial component depends on mixing processes, such as might occur if water of different composition is transferred between subglacial reservoirs at specific times during the diurnal discharge cycle. Clearly, if separation of the hydrograph into englacial and subglacial components is based on the composition of the subglacial component remaining constant, errors will be introduced.

At Austre Okstindbreen, as in most temperate areas with high annual precipitation, wet deposition is more important than dry deposition (Conklin, 1991). Most of the ions entering the glacier are provided by the winter snow cover, which contains a record of the atmospheric chemistry during the formation of snow crystals and the subsequent accumulation of the snowpack (Figure 6). After a winter with high accumulation, abundant ions are potentially available to meltwater, both as particles within individual snow crystals and absorbed on the outer side of grains. Before melting starts, the ions are rather immobile, and the stratigraphic variations are retained within the snowpack. When ablation starts, ions are leached at a rate which exceeds that of the actual melting: the chemical load is removed from the pack as an ionic pulse (Johannessen and Henricksen, 1978; Raben and Theakstone, 1994). In 1991, most of the ions had been leached from the snowpack at Austre Okstindbreen when less than 30% of the snow had melted (Raben, unpublished data). The first meltwater to leave the pack is rich in those ions which were present in the snow. The isotopic composition of this water reflects the influence of winter temperatures on the snow; all snow is more depleted of ^{18}O than is glacier ice, and snow which has accumulated in a cold winter is more depleted than that which has accumulated in a warm winter.

The first contribution to glacier river discharge is made by snow meltwater from the glacier tongue, and its $\delta^{18}O$ value is lower after a cold winter than after a winter which is less cold. Initially, ionic concentrations in the river water are likely to be high because of the dominance of snowmelt. Later, the inputs of ions and isotopes from the snow to the glacier's drainage systems depend on the rate of surface melting and the thickness and stratigraphic variations of the pack. If there is an early start to the ablation season after a winter in which accumulation was low and there were few melting and refreezing episodes, the supply of ions from the snow will be exhausted early in the summer and isotopic homogeneity will be achieved quickly (Figure 6). Conversely, during an ablation season which starts late after a winter of high accumulation, both light isotopes and ions will be available from the snow cover well into the summer; the presence of ice layers in the pack may delay their depletion still further.

During summer rainfall events, ions are provided both by the rain itself and by water which acquires them as it runs over rock or sediment surfaces and penetrates beneath the glacier. Falls of rain which are particularly heavy (Figure 2) may result in uplift of the glacier from its bed, allowing increased contact between water and bed material, and so causing an increased acquisition of ions there. Water which passes slowly through the glacier, including that which is stored for some time at the bed, may react relatively rapidly with bed materials, including both rock flour and coarser sediments, thereby maximizing solute acquisition (Tranter et al., 1993).

The proportion of river water which has passed quickly through supraglacial stream channels and englacial ice-walled channels, and has had little opportunity to acquire a high solute load, is larger during summers with high ablation rates than in those which are characterized by cold weather. Water generated by ablation of the tongue reaches the glacier river more rapidly than that which results from snowmelt above the transient equilibrium line, and its $\delta^{18}O$ value is high, reflecting the dominance of glacier ice in its composition (Table VI). The opportunities for such water to make a large contribution to river discharge are higher after a winter of low accumulation than after a winter in which the snow cover is thick. An early onset of melting, causing the snow cover to disappear from the tongue before the summer is far advanced, increases the opportunity for melting ice to make a major contribution to glacier river discharge.

At the onset of surface melting, conduits at and close to the bed of the glacier have small diameters, because of closure in winter, when water inputs are low or non-existent. A great amount of water is not needed to fill the conduits. As the summer progresses, the conduits enlarge. The development of the glacier's internal drainage systems during the summer depends on the amount of ablation and its variation with time, and on the amount of liquid precipitation. As the latter usually constitutes a minor portion of the discharge through the summer, ablation is the most important determinant of drainage system development. In general, development is slower in summers with low ablation than in those during which the transient equilibrium line moves quickly up-glacier and melting supplies abundant water.

CONCLUSIONS

Austre Okstindbreen is larger, and has a more maritime regime, than many glaciers at which hydrochemical studies have been made. The content of individual ions in river water emerging from the glacier in summer depends on the amount of snow present at the beginning of the ablation period and on the melting conditions. There is a higher loading of Na^+ at the end of a winter in which snow accumulation has been above average.

During the early part of the ablation season, the river water is dominated by snow meltwater from the lower part of the glacier and the adjacent parts of the catchment. Later in the summer, as the transient equilibrium line moves up-glacier, icemelt increases, but the amount of snow meltwater supplied from the glacier's accumulation area also increases. Towards the end of the summer, ionic concentrations in the snow which remains at the glacier are low; consequently, there is a trend of decreasing Na^+ concentrations in the glacier river water through the summer. If the ablation season starts early, and the snow cover is relatively thin, Na^+ concentrations in the glacier river towards the end of summer are likely to be particularly low. However, some Na^+ does remain at the glacier as long as snow survives, and some is supplied by summer rainfall.

The summer's weather plays a significant part in determining the amount of water entering the glacier and, hence, the speed with which its internal drainage systems develop. This influences the opportunities for subglacial acquisition of ions, notably Ca^{2+}. If the drainage system is poorly developed, as in 1989, the variability of Ca^{2+} in glacier river water is low. Most meltwater is provided by snowmelt, resulting in high Na^+ values in the river water; the $\delta^{18}O$ values depend on those of the snow. If it is well developed and rainfall events occur, as in 1988, large fluctuations of Ca^{2+} occur, whereas the Na^+ values remain near-constant, as Na^+ concentrations in old snow and rain are similar. $\delta^{18}O$ values are determined by ice meltwater and rainfall. In 1990, the Na^+ concentrations in the river water were intermediate between those of 1988 and 1989, as the contribution of snowmelt to discharge decreased and that of icemelt increased through the summer.

Water released from storage contributes to river discharge. Because storage at the glacier bed differs from one summer to another, absolute values of ionic concentrations in the river also differ. The stored water may include variable proportions of ice meltwater and snow meltwater, and its isotopic composition may vary markedly from one time to another. Rainfall affects the chemical composition of the glacier river, in part through its effect on previously stored water. The $\delta^{18}O$ value of rain water depends both on the transport path of the water vapour between its oceanic source and the glacier and on the air temperature at the time of precipitation. At Austre Okstindbreen, $\delta^{18}O$ values of river water are generally higher during, or immediately after, precipitation events than in the period which precedes them. The icemelt component of glacier river discharge has low oxygen isotope variability. Thus the isotopic composition of the river water is influenced strongly by that of the snow cover, which itself is a function of winter temperatures. When discharge is low and flow is maintained principally by melting snow from high altitudes, $\delta^{18}O$ values of the river water tend to be low.

The observations at Austre Okstindbreen in three contrasting summers indicate that glacier hydrochemistry may vary between years at a single glacier. The complexity of the influence of different accumulation and ablation conditions on the development of the glacier's internal drainage systems means that a simple model of mixing of englacial and subglacial waters, each of a constant composition, cannot be applied to a glacier of the size and altitudinal range of Austre Okstindbreen, where the development of the drainage system between years is highly variable.

ACKNOWLEDGEMENTS

We thank Professor Jens Tyge Møller for his long-term contribution to the work of the Okstindan Glacier Project, He Yuanqing, Frank Jacobsen and Lars Jensen for field assistance, and Nick Scarle for cartographic work. The programme at Okstindan has been supported by funding from the Natural Environment Research Council (Research Grant GR3/7253) and Statkraft, Norway.

REFERENCES

Brown, G. H. and Tranter, M. 1990. 'Hydrograph and chemical separation of bulk meltwaters draining the Upper Arolla glacier, Vallais, Switzerland', *Int. Assoc. Hydrol. Sci. Publ.*, **193**, 429–437.
Collins, D. N. 1978. 'Hydrology of an Alpine glacier as indicated by the chemical composition of meltwater', *Z. Gletscherk. Glazialgeol.*, **13**, 219–238.
Collins, D. N. 1979. 'Hydrochemistry of meltwaters draining from an Alpine glacier', *Arc. Alpine Res.*, **11**, 307–324.
Collins, D. N. and Young, G. J. 1981. 'Meltwater hydrology and hydrochemistry in snow- and ice-covered mountain catchments', *Nordic Hydrol.*, **12**, 319–334.

Conklin, M. H. 1991. 'Discussion of "Dry deposition to snowpacks" ' in Davies, T. D., Tranter, M., and Jones, H. G. (Eds), *Seasonal Snowpacks. Processes of Compositional Change*. pp. 67–70. Springer-Verlag, Berlin.

Davies, T. D., Brimblecombe, P., Tranter, M., Tsiouris, S., Vincent, C. E., Abrahams, P. and Blackwood, I. L. 1987. 'The removal of soluble ions from melting snowpacks' in Jones, H. G. and Orville-Thomas, W. J. (Eds), *Seasonal Snowcovers: Physics, Chemistry, Hydrology*. pp. 337–392. Springer-Verlag, Berlin.

Gurnell, A. M., Clark, M. J., and Hill, C. T. 1992. 'Analysis and interpretation of patterns within and between hydroclimatological time series in an alpine glacier basin', *Earth Surf. Process. Landforms*, **17**, 821–839.

He, Y. and Theakstone, W. H. 1994. 'Climatic influence on the composition of the snow cover at the glacier Austre Okstindbreen, Okstindan, Norway, 1989 and 1990', *Ann. Glaciol.*, **19**, 1–6.

Johannessen, M. and Henricksen, A. 1978. 'Chemistry of snow meltwater: changes in concentration during melting', *Wat. Resour. Res.*, **14**, 615–619.

Karlsen, E. 1991. 'Variations in grain-size distribution of suspended sediment in a glacial meltwater stream, Austre Okstindbreen, Norway', *J. Glaciol.*, **37**, 113–119.

Keller, W. D. and Reesman, A. L. 1963. 'Glacial milks and their laboratory-simulated counterparts', *Geol. Soc. Am. Bull.*, **74**, 61–76.

Knudsen, N. T. 1989. 'Water discharge from Austre Okstindbreen, Norway', *Okstindan Glacier Project Rep.*, **89.4**, 16 pp.

Knudsen, N. T. 1990. 'Dissolved load in glacier river water, Austre Okstindbreen, Okstindan, Norway', *Okstindan Glacier Project Rep.*, **90.2**, 20 pp.

Knudsen, N. T. 1993. 'Mass balance, meltwater discharge and ice velocity at Austre Okstindbreen, Nordland, Norway 1991–92', *Okstindan Glacier Project Rep.*, **93.4**, 20 pp.

Knudsen, N. T. and Jensen, L. 1992. 'Time series analysis of radiation, temperature and discharge at Austre Okstindbreen, Nordland, Norway', *Nordisk Hydrol. Konf. 1992 (NHP-rapport nr. 30)*, 718–727.

Knudsen, N. T. and Theakstone, W. H. 1988. 'Drainage of the Austre Okstindbreen ice-dammed lake, Okstindan, Norway', *J. Glaciol.*, **34**, 87–94.

Raben, P. and Theakstone, W. H. 1994. 'Isotopic and ionic changes in a snow cover at different altitudes: observations at the glacier Austre Okstindbreen in 1991', *Ann. Glaciol.*, **19**, 85–91.

Rainwater, F. H. and Guy, H. P. 1961. 'Some observations on the hydrochemistry and sedimentation of the Chamberlin Glacier area, Alaska', *US Geol. Surv. Prof. Pap.*, **414-C**.

Raiswell, R. and Thomas, A. G. 1984. 'Solute acquisition in glacial melt waters. I. Fjallsjøkull (south-east Iceland): bulk meltwaters with closed-system characteristics', *J. Glaciol.*, **30**, 35–43.

Röthlisberger, H. and Lang, H. 1987. 'Glacial hydrology' in Gurnell, A. M. and Clark, M. J. (Eds), *Glacio-fluvial Sediment Transfer*. pp. 207–284. John Wiley & Sons Ltd.

Slatt, R. M. 1972. 'Geochemistry of meltwater streams from nine Alaskan glaciers', *Geol. Soc. Am. Bull.*, **83**, 1125–1132.

Stenborg, T. 1968. 'Glacier drainage connected with ice structures', *Geogr. Ann.*, **50A**, 25–53.

Stichler, W. 1987. 'Snowcover and snowmelt processes studied by means of environmental isotopes' in Jones, H. G. and Orville-Thomas, W. J. (Eds), *Seasonal Snowcovers: Physics, Chemistry, Hydrology*. pp. 673–726. Springer-Verlag, Berlin.

Theakstone, W. H. 1988. 'Temporal variations of isotopic composition of glacier-river water during summer: observations at Austre Okstindbreen, Okstindan, Norway', *J. Glaciol.*, **34**, 309–317.

Theakstone, W. H. 1991. 'Stable isotope investigations at Austre Okstindbreen, 1980–1990', *Okstindan Glacier Project Rep.*, **91.3**, 46 pp.

Theakstone, W. H. and Knudsen, N. T. 1981. 'Dye tracer tests of water movement at the glacier Austre Okstindbreen, Norway', *Norsk Geogr. Tiddskr.*, **35**, 21–28.

Theakstone, W. H. and Knudsen, N. T. 1989. 'Temporal changes of glacier hydrological systems indicated by isotopic and related observations at Austre Okstindbreen, Okstindan, Norway, 1976–87', *Ann. Glaciol.*, **13**, 252–256.

Thomas, A. G. and Raiswell, R. 1984. 'Solute acquisition in glacial melt waters. II. Argentière (French Alps): bulk meltwaters with open-system characteristics', *J. Glaciol.*, **30**, 44–48.

Tranter, M. and Raiswell, R. 1991. 'The composition of the englacial and subglacial component in bulk meltwaters draining the Gornergletscher, Switzerland', *J. Glaciol.*, **37**, 59–66.

Tranter, M., Brown, G., Raiswell, R., Sharp, M., and Gurnell, A. 1993. 'A conceptual model of solute acquisition by Alpine glacial meltwaters', *J. Glaciol.*, **39**, 573–581.

Zeman, L. J. and Slaymaker, H. O. 1975. 'Hydrochemical analysis to discriminate variable runoff source areas in an alpine basin', *Arc. Alpine Res.*, **7**, 341–351.

4

THE HYDROCHEMISTRY OF RUNOFF FROM A 'COLD-BASED' GLACIER IN THE HIGH ARCTIC (SCOTT TURNERBREEN, SVALBARD)

R. HODGKINS,[1] M. TRANTER[2] AND J. A. DOWDESWELL[3]

[1]Department of Geography, Royal Holloway, University of London, Egham, Surrey TW20 0EX, UK
[2]Department of Geography, University of Bristol, University Road, Bristol BS8 1SS, UK
[3]Institute of Earth Studies, University of Wales, Aberystwyth, Ceredigion SY23 3DB, UK

ABSTRACT

There are still relatively few hydrochemical studies of glacial runoff and meltwater routing from the high latitudes, where non-temperate glacier ice is frequently encountered. Representative samples of glacier melt-water were obtained from Scott Turnerbreen, a 'cold-based' glacier at 78° N in the Norwegian high Arctic archipelago of Svalbard, during the 1993 melt season and analysed for major ion chemistry. Laboratory dissolution experiments were also conducted, using suspended sediment from the runoff. Significant concentrations of crustal weathering derived SO_4^{2-} are present in the runoff, which is characterized by high ratios of $SO_4^{2-} : (SO_4^{2-} + HCO_3^-)$ and high $p(CO_2)$. Meltwater is not routed subglacially, but flows to the glacier terminus through subaerial, ice marginal channels, and partly flows through a proglacial icing, containing highly concentrated interstitial waters, immediately afront the terminus. The hydrochemistry of the runoff is controlled by: (1) seasonal variations in the input of solutes from snow- and icemelt; (2) proglacial solute acquisition from the icing; and (3) subaerial chemical weathering within saturated, ice-cored lateral moraine adjoining drainage channels at the glacier margins, sediment and concentrated pore water from which is entrained by flowing meltwater. Diurnal variations in solute concentration arise from the net effects of variable sediment pore water entrainment and dilution in the ice marginal streams. Explanation of the hydrochemistry of Scott Turnerbreen requires only one major subaerial flow path, the ice marginal channel system, in which seasonally varying inputs of concentrated snowmelt and dilute icemelt are modified by seepage or entrainment of concentrated pore waters from sediment in lateral moraine, and by concentrated interstitial waters from the proglacial icing, supplied by leaching, slow drainage at grain intersections or simple melting of the icing itself. The ice marginal channels are analogous neither to dilute supra/englacial nor to concentrated subglacial flow components.

INTRODUCTION

There are still relatively few hydrochemical studies of runoff and meltwater routing from high latitudes, where non-temperate glacier ice and 'cold-based' glaciers are frequently encountered (e.g. Hagen *et al.*, 1993). The current glacial hydrochemical literature concentrates largely on studies of temperate or 'warm-based', mid-latitude glaciers (e.g. Collins, 1979; Raiswell and Thomas, 1984; Fountain, 1992; Brown *et al.*, 1994), which have very different meltwater flow paths to non-temperate, or 'cold-based', glaciers, from which subglacial drainage is believed to be largely absent. Instead, meltwater is understood to be routed predominantly subaerially. The penetration of water along intergranular boundaries is largely absent in non-temperate ice, and large-scale permeability (crevasses, moulins) is limited by the relatively slow rate of ice deformation and low rates of glacier activity (consequent on low mass balance gradients), and by the formation of superimposed ice, i.e. the refreezing of meltwater (Hodgkins, in press). Intuitively, 'cold-based'

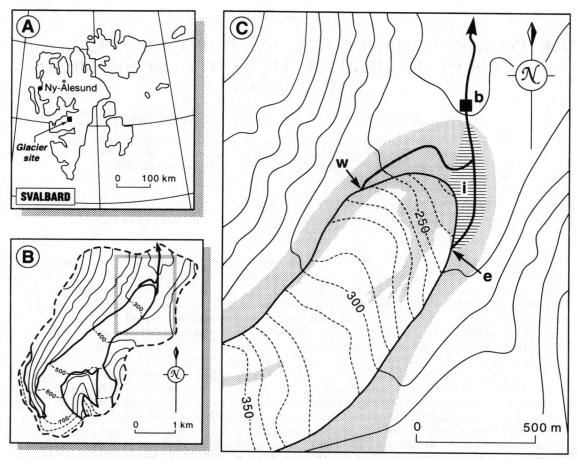

Figure 1. Location of the study glacier and of drainage features discussed in the text. (A) Location of the catchment within the Svalbard archipelago. (B) The Scott Turnerbreen catchment. Contours are in m a.s.l. with an interval of 100 m. The shaded area is enlarged in part C. (C) Lower Scott Turnerbreen. Stippling indicates glacial moraine. Contours are in m a.s.l. with an interval of 50 m off-glacier, 10 m on-glacier. 'e' and 'w', respectively, mark the sampling locations for the east and west ice marginal meltwater streams (where the streams debouch from the glacier margins), 'i' denotes the proglacial icing and 'b' is the location of bulk meltwater sampling and discharge and suspended sediment monitoring

glaciers should give rise to relatively dilute meltwaters, since access comminuted subglacial sediment is restricted, and residence times are short for surface runoff. Temporal variability in chemical composition (other than that arising from the input of snowmelt: Tranter *et al.*, 1996) should also be limited, because there is no mixing of meltwaters from supra/englacial and subglacial sources. To date, there is no detailed account of the hydrochemistry of 'cold-based' glaciers in the mainstream literature. Hence this paper investigates the hydrochemistry of runoff from such a glacier, Scott Turnerbreen, in the Norwegian high Arctic archipelago of Svalbard (76–81°N, 10–35°E), with the aim of linking variations in runoff chemistry to changes in meltwater and solute sources. A principal concern is whether the observed hydrochemistry can be explained by a drainage system that does not contain a subglacial component.

SITE DESCRIPTION

Scott Turnerbreen is a 3·3 km^2 glacier, of altitude range 230–680 m a.s.l., located in central Spitsbergen (78° 06′N, 15° 57′E), occupying a catchment of total area 12.8 km^2 (Figure 1). The catchment geology is

(a)

(b)

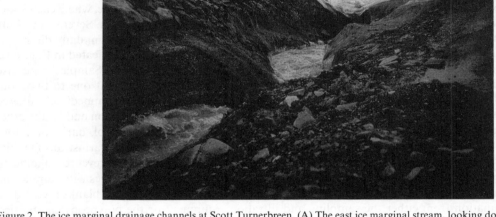

Figure 2. The ice marginal drainage channels at Scott Turnerbreen. (A) The east ice marginal stream, looking down-glacier at *ca.* 310 m a.s.l. (B) The west ice marginal stream, looking down-glacier at *ca.* 450 m a.s.l. In both cases, note the subaerial flow, steep ice or ice-cored channel sides and abundant sediment supplies

dominated by gently dipping Tertiary sandstones, shales and siltstones (Major and Nagy, 1972). Basal temperatures measured in boreholes (Hodgkins, 1994) were $-4.1°C$ (surface altitude 305 m a.s.l., 42 m ice depth) and $-3.3°C$ (surface altitude 595 m a.s.l., 54 m ice depth). The observed temperature gradients indicate that pressure melting requires at least 95 m ice thickness. It is therefore unlikely that the glacier bed is temperate even at its deepest point, 76 m.

Meltwater drainage from the glacier is dominated by two ice marginal channels, the east and west streams (Figure 1c; Figure 2), of which the east carries *ca.* 80% of the total discharge. The course of these channels is through ice-cored moraine and stagnant ice at the glacier margins. They are essentially subaerial in character, although in places they flow beneath the glacier margins. Both streams debouch from the glacier margins in the proglacial region, up-valley of a prominent end-moraine from which the glacier has retreated since the 1930s. The east stream (Figure 2a) issues over a proglacial icing (*aufeis* or *naled*), which accumulates over winter in the basin between the glacier terminus and the end-moraine (Figure 1c) and is discussed in more detail below. The west stream (Figure 2b) flows across an area of moraine before reaching its confluence with the east stream inside the end-moraine. The proglacial stream below this confluence is called the bulk meltwater (Figure 1c). This directly breaches the end-moraine before flowing down the main valley flood-plain, across which it braids extensively.

The steep head and side walls of the catchment hold *ca.* 20% of the annual snowfall at the onset of snow-melt. Vegetation is very scarce on the slopes, which consist primarily of frost-shattered debris and moraine. The lowermost sections of the slopes adjoining the glacier consist of ice-cored moraine on which an active layer of *ca.* 0·1 m depth develops during the melt season. Water-saturated sediment is delivered to the ice marginal channel banks by subaerial mass wasting (Figure 2). Some of the sediment enters the stream flow directly. However, most is probably mobilized during rising discharge, as the increased water volume enlarges the channels' wetted perimeters, and bank sediment becomes entrained by flowing water (Hodgkins, 1996).

METHODOLOGY

Sampling programme and procedure

The collection, processing and analysis of samples is described fully in Hodgkins *et al.* (in press). This paper is specifically concerned with meltwater samples collected from the east and west ice marginal streams and from the bulk stream.

Seventeen samples were collected from both the east and the west streams, between 22 June and 10 July 1993. These were obtained from open channels at the fixed locations indicated in Figure 1c, where each issues from the glacier margin. One sample from each was collected daily, in the early afternoon. Seventy-two bulk meltwater samples were collected, at times as close as possible to minimum and maximum daily discharge: 1000 and 1700 hours, respectively, on average. These were obtained from the location indicated in Figure 1c, between 14 June and 10 July 1993 (from the onset of proglacial surface drainage). Samples were also collected during six intensive sessions, in which five to seven samples were obtained at one to two hour intervals. Samples were collected in a prerinsed, 500 ml polypropylene bottle, then immediately filtered through 0·45 μm cellulose nitrate membranes. Two 60 ml aliquots of filtrate (one for cation and Si, the other for anion and alkalinity analyses) were stored, air-free, in polypropylene bottles in cool, dark conditions until return to the UK. A third 60 ml aliquot was used for pH measurement at Norsk Polarinstitutt facilities at Svalbard airport no later than 14 days after collection; air-free storage should have prevented significant change in sample pH. Care was taken to prerinse the filtration apparatus and storage bottles with sample and filtrate as appropriate, to minimize contamination. Analysis of field protocol and storage blanks revealed no significant contamination from either the sampling and filtration equipment or the polypropylene bottles. Samples were also obtained from the snowpack, from glacier ice and from the proglacial icing; their collection and treatment are described in Hodgkins *et al.* (in press).

The stage of the proglacial stream was measured every 5 minutes, with a Druck PDCR830 pressure transducer, in a stable stretch of unbraided proglacial bulk stream *ca.* 250 m downstream of the confluence of the east and west streams (Figure 1c), and logged as hourly means over the period 15 June–11 July 1993. Discrete discharge measurements were undertaken at the same location by the relative dilution method (Water Research Association, 1970), and used to obtain a rating relationship for converting stage to discharge. The typical error of this relationship is ±8%. Suspended-sediment concentration (SSC) was measured concurrently by discrete automatic sampling in the same location, with an Epic Products 1011 vacuum sampler, collecting 250 ml stream water samples. As the stream was highly turbulent, point sampling should have yielded representative estimates of SSC (Gurnell *et al.*, 1992). The sampling interval varied between 1 and 3 h, according to the frequency with which samples could be processed. Samples were filtered in a perspex pressure filtration apparatus using preweighed Whatman grade 40 paper with an initial penetration pore size of 8 μm, and were returned to the UK for drying and reweighing. The effective pore size of the paper is rapidly reduced by clogging during filtering, so it is anticipated that most of the sub-8 μm fraction is actually retained (Gurnell *et al.*, 1992). This is important in the context of the laboratory dissolution experiments (described below), as this fraction is likely to be the most geochemically reactive (the surface area, and hence quantity of reactive surface sites, increases with decreasing sediment grain size: Brown *et al.*, 1996).

Laboratory analyses

The pH was determined with an Orion 290a portable pH meter with a Ross combination electrode, calibrated with Orion low ionic strength buffers of pH 4·10 and 6·97. The precision of the measurements is ± 0.2 pH units. The concentrations of major base cations (Na^+, K^+, Mg^{2+}, Ca^{2+}) and strong acid anions (Cl^-, NO_3^-, SO_4^{2-}) were determined on a Dionex 4000i ion chromatograph. The precision of the analysis varies from $\pm 3\%$ at concentrations in excess of 50 μeq l^{-1} to $\pm 100\%$ at 1 μeq l^{-1}. The concentration of Si was determined by flow injection analysis, using a Tecator FIAstar 5010 system with FIAstar 5023 spectrophotometer, V100 injector and type III Chemifold. The precision of the analysis is ± 2 μmol l^{-1}. Alkalinity (HCO_3^-) was determined by titrating 25 ml of filtered sample to an end-point pH of 4·5, using 0·01 M HCl (standardized with 0·01 M Na_2CO_3), detected with BDH 4·5 mixed indicator. HCO_3^- is overestimated at low concentrations.

Charge balance errors, *CBE*, were calculated from

$$CBE = \frac{(\Sigma^+ - \Sigma^-)}{(\Sigma^+ + \Sigma^-)} \tag{1}$$

expressed as a percentage, where Σ^+ is the sum of the measured positively charged equivalents, and Σ^- is the sum of the measured negatively charged equivalents. Mean *CBE*s for snow, east and west streams, icing interstitial water and bulk runoff samples were 0·0, −8·8, −6·9, −4·0 and −2·3%, respectively.

Laboratory dissolution experiments

Laboratory dissolution experiments were carried out on suspended sediment sampled from the bulk runoff throughout the sampling season. Suspended sediment was brushed off several of the filter papers used for SSC determinations to provide a composite sample of *ca.* 60 g. The composite sample was homogenized by stirring with a plastic spatula. The methodology adopted for the experiments is similar to that of Brown *et al.* (1994). Suspended sediment (10·0 g) was introduced into 1·0 l of deionized water in a plastic beaker, open to the prevailing atmosphere. The solution–sediment mixture was continuously agitated by a magnetic stirrer and plastic-coated stirring bar for a period of 24 h. All apparatus was contained in a refrigerator in which temperatures ranged from 0 to 2°C. Sufficient time was allowed before the experiment commenced for the apparatus to achieve thermal equilibrium. Periodically, 0·02 l of mixture was withdrawn with a prerinsed plastic syringe and immediately passed through a 0·45 μm filter membrane. The filtrate was analysed for Ca^{2+}, as above. In all, five replicate experiments were performed.

RESULTS

Results are summarized in Table I and selected time-series are presented in Figures 3–6. Essential features of the data are detailed below.

Si variation

The kinetics of silicate/aluminosilicate and quartz dissolution are slow (Raiswell, 1984; Dove; 1994). Hence, relatively high concentrations of Si in meltwater may indicate routing through a drainage structure with a long residence time (Lamb *et al.*, 1995). Time series of the variation of Si in bulk meltwater and in the

Table I. Hydrochemical characteristics of bulk stream ($n = 56$), east ($n = 17$) and west ($n = 17$) ice marginal stream meltwaters, during the period 22 June–10 July 1993, and interstitial water in the icing ($n = 40$) during the period 4 May–30 June 1993. The median is given, with the range in parentheses

Species/variable	Bulk (proglacial) meltwater	East ice marginal meltwater	West ice marginal meltwater	Icing interstitial water
Σ_{ions}, µeq l^{-1}	1100 (62–2600)	970 (750–3800)	530 (440–2200)	5100 (100–17 000)
Si, µmol l^{-1}	3·0 (2·3–4·3)	2·9 (1·9–4·1)	1·9 (1·7–2·4)	14 (2·3–63)
Na^+, µeq l^{-1}	210 (110–740)	160 (110–1500)	99 (74–490)	1600 (260–5400)
K^+, µeq l^{-1}	9·4 (5·1–19)	6·5 (5·7–23)	5·6 (4·7–11)	18 (4–59)
Mg^{2+}, µeq l^{-1}	150 (99–290)	140 (88–390)	72 (58–250)	270 (65–1100)
Ca^{2+}, µ l^{-1}	180 (120–300)	180 (130–420)	95 (76–310)	560 (130–2900)
HCO_3^-, µeq l^{-1}	170 (110–260)	200 (170–230)	200 (140–260)	1800 (350–4600)
*SO_4^{2-}, µeq l^{-1}	130 (96–200)	140 (120–300)	32 (18–49)	830 (0–3200)
*SO_4^{2-}/total SO_4^{2-}, %	78 (57–95)	82 (53–93)	67 (21–93)	99 (89–99)
SEF	0·42 (0·33–061)	0·43 (0·36–0·51)	0·16 (0·08–0·19)	0·34 (0·28–0·41)
$\log_{10} p(CO_2)$	−3·2 (−3·5 to −2·8)	−3·1 (−3·3 to −3·0)	−3·2 (−3·3 to −3·0)	−3·2 (−3·5 to −2·7)

east and west ice marginal streams are given in Figure 3b. The median concentration of Si in bulk meltwater was 3·1 µmol l^{-1} (the median is used as the measure of central tendency, as many of the sample data sets are highly skewed), with a minimum of 2·3 and a maximum of 8·5 µmol l^{-1}. The median concentrations of Si in east and west ice marginal stream meltwaters were 3·1 (range 1·5–3·9) and 2·3 (range 1·5–2·3) µmol l^{-1}, respectively. These concentrations compare with medians of 0·8 (range 0·0–1·5) µmol l^{-1} for melting snow, and 0·0 (range 0·0–4·7) µmol l^{-1} for glacier ice (Hodgkins et al., in press). Hence meltwater appears to possess elevated concentrations of Si with respect to the main meltwater sources. The temporal pattern of Si concentrations (Figure 3b) was one of a quasi-exponential decline from a maximum of 8·5 µmol l^{-1} to ca. 3 µmol l^{-1} for the period 14–22 June. The Si concentration of east ice marginal stream meltwater samples was consistently greater than that of west stream samples, but the concentration of bulk stream samples was often higher than both of these, e.g. 22–24 June and 5–9 July.

SO_4^{2-} variation

The total SO_4^{2-} concentration in the meltwater is partitioned into snowpack (i.e. that derived from snowfall scavenging of sea salt and other atmospheric aerosols) and non-snowpack (i.e. that derived from chemical weathering of crustal material) sources. It is assumed that all Cl^- in meltwater is derived from the snowpack, and that SO_4^{2-} is eluted from the snowpack with the same $SO_4^{2-} : Cl^-$ ratio as dry snow (0·126; Hodgkins et al., in press). The non-snowpack SO_4^{2-}, *SO_4^{2-}, is defined as

$$*SO_4^{2-} = SO_4^{2-} - (0.126Cl^-) \tag{2}$$

where SO_4^{2-} is the total concentration of SO_4^{2-} (µeq l^{-1}) and Cl^- the total concentration of Cl^- (µeq l^{-1}) in meltwater. Time-series of the concentration of *SO_4^{2-} in bulk meltwater and the ice marginal streams are presented in Figure 4b. The median concentration of *SO_4^{2-} in bulk meltwater was 140 µeq l^{-1} (range 96–510 µeq l^{-1}). The median concentration of *SO_4^{2-} was 140 µeq l^{-1} (range 120–300 µeq l^{-1}) and 32 µeq l^{-1} (range 18–49 µeq l^{-1}) in east and west ice marginal stream meltwaters, respectively. These concentrations compare with median concentrations of total SO_4^{2-} of 11 and 13 µeq l^{-1} in melting snow and glacier ice, respectively (Hodgkins et al., in press). The source of *SO_4^{2-} is ultimately crustal weathering, and most likely sulfide oxidation, given the predominance of shales and siltstones in the catchment bedrock.

The temporal pattern of bulk meltwater *SO_4^{2-} concentrations is one of a steady decline from a maximum of 510 µeq l^{-1} at the start of the monitoring period to 360 µeq l^{-1} on 20 June. There was a sudden fall to 180 µeq l^{-1} on 21 June, coinciding with the onset of the discharge of snowmelt concentrated by elution (Hodgkins et al., in press), which appeared to dilute *SO_4^{2-} concentrations until 26 June, when the

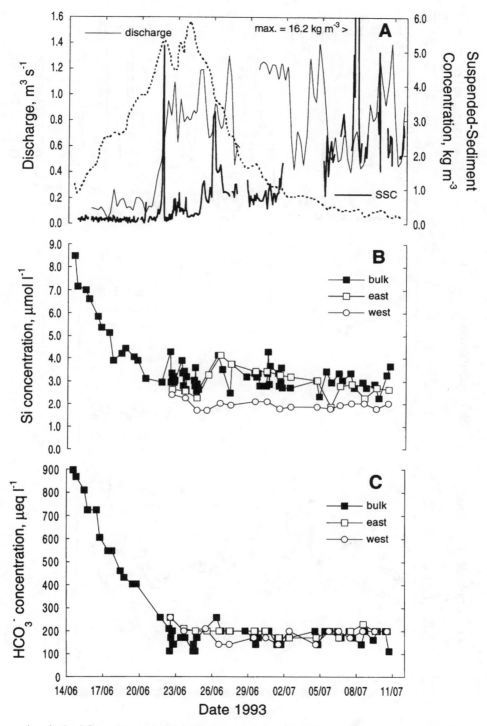

Figure 3. Time-series obtained from the proglacial bulk stream and from the east and west ice marginal streams. (A) Meltwater discharge and suspended sediment concentration (proglacial bulk stream). The dashed line is the non-dimensionalized, snowpack-derived, solute concentration in the bulk meltwater. (B) Si concentrations. (C) HCO_3^- concentration

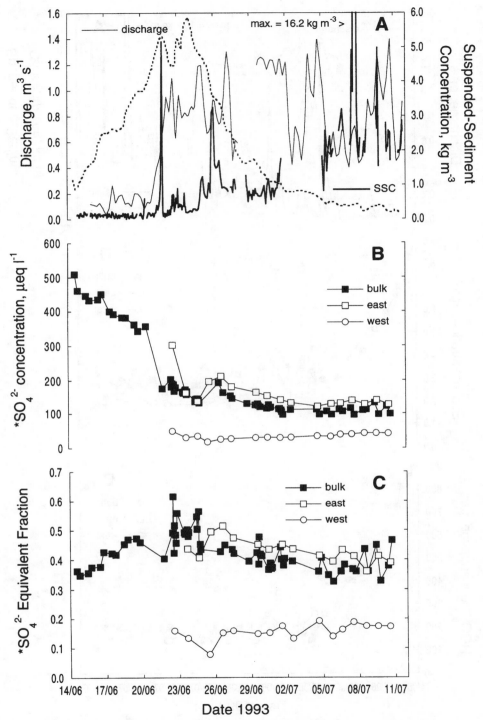

Figure 4. Time-series obtained from the proglacial bulk stream and from the east and west ice marginal streams. (A) Meltwater discharge and suspended sediment concentration (proglacial bulk stream). The dashed line is the non-dimensionalized snowpack-derived, solute concentration in the bulk meltwater. (B) $^{*}SO_4^{2-}$ concentrations. (C) $^{*}SO_4^{2-}$ equivalent fraction, SEF

concentration rose to 190 μeq l^{-1}. Thereafter, it declined steadily until 1 July, when the concentration varied around 100 μeq l^{-1} (Figure 4b). East ice marginal stream $*SO_4^{2-}$ concentrations were typically *ca.* 10 μeq l^{-1} higher than those of the bulk meltwater, but their temporal pattern was similar. West ice marginal stream $*SO_4^{2-}$ concentrations were *ca.* 100 μeq l^{-1} lower than in the bulk meltwater throughout the monitoring period, and after an initial slight fall, rose very gradually for the remainder of the period, to 42 μeq l^{-1} by the end.

$*SO_4^{2-}$ formed *ca.* 80% of the total SO_4^{2-} load of both the bulk and east ice marginal stream meltwaters, but only *ca.* 67% of that in the west ice marginal stream meltwater. The proportion of $*SO_4^{2-}$ in the total SO_4^{2-} of bulk meltwater was *ca.* 95% at the start and end of the monitoring period, but decreased to *ca.* 60% around 23 June, corresponding to the peak discharge of concentrated snowmelt. The proportions and temporal pattern are very similar in the east ice marginal stream, but the minimum proportion of $*SO_4^{2-}$ in the total SO_4^{2-} of the west ice marginal stream was only *ca.* 20% during the snowmelt concentration peak. Tranter *et al.* (1996) suggest that the preferential elution of SO_4^{2-} relative to Cl^- cannot be responsible for this pattern, since it would lead to an overestimation of $*SO_4^{2-}$ when Cl^- concentrations are high, in the early season, and an underestimation when Cl^- concentrations are low, in the late season. The observed pattern is attributed to the elution and depletion of snowpack SO_4^{2-} with respect to a more constant source of $*SO_4^{2-}$ from crustal weathering. It is noteworthy that, as for Si, there were high $*SO_4^{2-}$ concentrations in the earliest part of the monitoring period (preceding the snowmelt concentration peak) when SSC was very low (Figure 4a).

$*SO_4^{2-}$: HCO_3^- equivalent ratio variation

The ratio $*SO_4^{2-}$: ($*SO_4^{2-}$ + HCO_3^-), sometimes called the sulfate mass fraction (Tranter *et al.*, 1993), but here called sulfate equivalent fraction (SEF, where concentrations are in equivalents per unit volume), can be used as an index of the proportion of chemical weathering promoted by sulfide oxidation and dissociation of carbonic acid. It is likely that sulfide oxidation and carbonate dissolution are coupled [Equation (3)], since these reactions are more rapid than silicate dissolution.

$$4FeS_2(s) + 16CaCO_3(s) + 15O_2(aq) + 14H_2O(aq) \rightleftharpoons 16Ca^{2+}(aq) + 16HCO_3^-(aq) + 8SO_4^{2-}(aq) + 4Fe(OH)_3(s)$$

pyrite   calcite   ferric oxyhydroxides

$$(3)$$

Inspection of Equation 3 reveals that coupling of sulfide oxidation and carbonate dissolution gives rise to an SEF of 0·50. Dissolution of silicates [Equation (4)] and carbonates may also be promoted by carbonic acid (derived from atmospheric CO_2, CO_2 in bubbles in ice, or microbial oxidation of organic carbon). These types of carbonation reactions decrease the SEF.

$$CaAl_2Si_2O_8(s) + 2CO_2(aq) + 2H_2O(aq) \rightleftharpoons Ca^{2+}(aq) + 2HCO_3^-(aq) + H_2Al_2Si_2O_8(s)$$

anorthite (Ca feldspar)   weathered feldspar surfaces (4)

The variation of the SEF in the bulk and ice marginal stream meltwaters is shown in Figure 4c. The SEF varied between 0·33 and 0·62 in the bulk meltwater, with a median of 0·42. The SEF of the two ice marginal streams differed, with a mean value of 0·43 in the east stream (range 0·36–0·51), but only 0·16 in the west stream (range 0·08–0·19). There was no statistically significant relationship between discharge and SEF. There was no strong trend in the variation of the SEF over the monitoring period, except for a slight increase over the first six days, and a brief sharp increase during the period of concentrated snowmelt discharge, 22–25 June (Hodgkins *et al.*, in press). Thereafter the SEF declined very slightly before varying around a value of *ca.* 0·40 from 4 July. Figures 3c and 4b show that the concentrations of HCO_3^- and $*SO_4^{2-}$ declined over the monitoring period. The more rapid decline of HCO_3^- concentrations relative to $*SO_4^{2-}$ in the earliest days of the monitoring period accounts for the slight increase in the SEF over the same interval.

p(*CO₂*) *variation*

The relative rates of proton supply and consumption by chemical weathering impart a characteristic $p(CO_2)$ signature to meltwater, indicative of the prevailing weathering regime (Tranter *et al.*, 1993). Relatively high $p(CO_2)$ conditions indicate that proton supply is more rapid than proton depletion by chemical weathering. $p(CO_2)$ is calculated from Equation (5),

$$\log_{10}p(CO_2) = \log_{10}[HCO_3^-] - pH + pK_{CO_2} + pK_1 \tag{5}$$

where HCO_3^- concentrations are in units of mol l^{-1}, $pK_{CO_2} = 1{\cdot}12$ and $pK_1 = 6{\cdot}58$ at 0°C (Ford and Williams, 1989). The length of the $p(CO_2)$ time series is constrained by the availability of pH measurements (Figure 5b), and is consequently shorter than the ionic concentration time-series in Figures 2 and 3.

The variation of $p(CO_2)$ during the monitoring period is presented in Figure 5c. The partial pressure of CO_2 in the atmosphere is $10^{-3{\cdot}5}$ atm, hence values of $\log_{10}p(CO_2)$ (g) of $-3{\cdot}5$ denote equilibrium with atmospheric $p(CO_2)$. Values greater and less than $-3{\cdot}5$ denote high and low $p(CO_2)$ conditions, respectively. It is clear from Figure 5c that the majority of meltwater samples exhibited high $p(CO_2)$ characteristics. The bulk meltwater was therefore supersaturated with CO_2. Re-equilibration takes place relatively slowly, because of the shallow concentration gradient between the gaseous and aqueous phases, with timescales of at least 30 minutes (Raiswell and Thomas, 1984). Bulk meltwater $p(CO_2)$ therefore probably reflects ice marginal stream values, particularly that of the east stream (since it has the greater discharge). There is no statistically significant relationship between $\log_{10}p(CO_2)$ and SEF of the bulk meltwater.

Diurnal solute variation

There is little evidence for systematic diurnal variation in the concentrations of cations and anions on any of the intensively sampled days (22–24 June and 29 June–1 July). However, neither were there any regular diurnal variations in discharge during these periods. Nevertheless, there were large variations in the overall concentration and relative proportions of ions between days, indicative of seasonal variations (Hodgkins *et al.*, in press). Diurnal variation in the concentration of most solute species can be detected from approximately three weeks into the melt season (e.g. Figures 3b, 3c and 4b). From 4 July, the second daily (maximum discharge) bulk meltwater sample was less concentrated in all ions, except for K$^+$ (NO$_3^-$ is equivocal), than the first (minimum discharge) in 36 out of 42 paired cases, with all differences in excess of error limits. This was also a period of more regular diurnal variations in discharge (Figure 6). A χ^2 test confirms that these variations are non-random with a probability exceeding 99·9%. Dilution at maximum discharge is therefore regarded as a genuine effect. The mean measured diurnal concentration amplitudes were: 20 µeq l^{-1}, Na$^+$; 13 µeq l^{-1}, Mg^{2+}; 16 µeq l^{-1}, Ca^{2+}; 51 µeq l^{-1}, HCO$_3^-$; 20 µeq l^{-1}, SO$_4^{2-}$; 14 µeq l^{-1}, Cl$^-$ (these are not necessarily the maximum amplitudes as only two samples were collected each day).

Laboratory dissolution experiments

The change in Ca^{2+} concentration over a time interval of between one minute and one day (Figure 7) can be described by the expression

$$\log_{10}[Ca^{2+}] = 0{\cdot}65 + 0{\cdot}26 \log_{10}t \tag{6}$$

where [Ca^{2+}] is the concentration of Ca^{2+} in µeq l^{-1} and t is the duration of the experiment in s. This equation explains some 99% of the variance exhibited by the eight data points, each of which is the mean of the values obtained from the five replicate experiments.

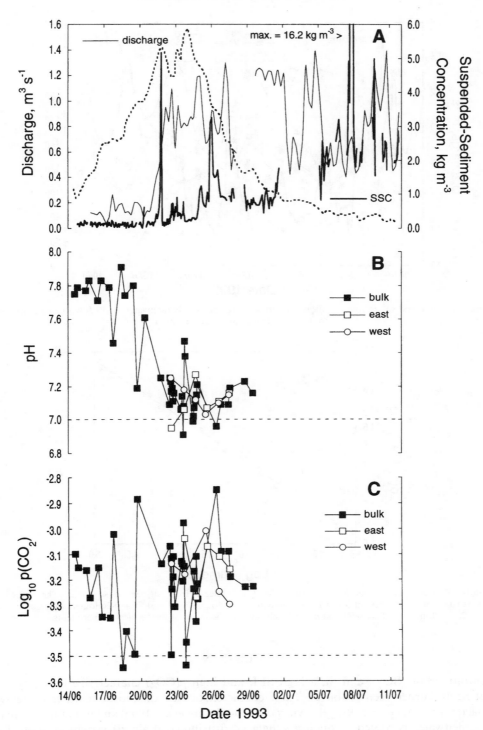

Figure 5. Time-series obtained from the proglacial bulk stream and from the east and west ice marginal streams. (A) Meltwater discharge and suspended sediment concentration (Proglacial bulk stream). The dashed line is the non-dimensionalized, snowpack-derived, solute concentration in the bulk meltwater. (B) pH. (C) Partial pressure of carbon dioxide, $p(CO_2)$

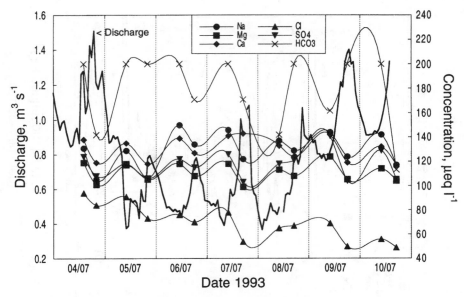

Figure 6. Diurnal discharge and solute concentration variations in proglacial bulk meltwater, 4–10 July: the concentration curves are produced by cubic spline interpolation

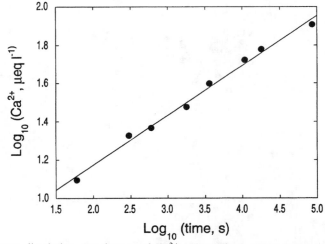

Figure 7. Results of laboratory dissolution experiments of Ca^{2+} release from suspended sediment. The suspended sediment concentration was 10 kg m^{-3}. The best fit linear regression equation is $\log_{10}[Ca^{2+}] = 0.65 + 0.26 \log_{10}t$, with Ca^{+} in µeq l^{-1} and t, time, in s. $R^2 = 0.99$ and $n = 8$: each point is the mean of five replicate experiments

DISCUSSION

High concentrations of *SO$_4^{2-}$ and Si, a high SEF and high $p(CO_2)$ characterize the early season bulk meltwater at Scott Turnerbreen (Table I). These characteristics are diagnostic features of delayed flow at temperate glaciers (Tranter *et al*., 1993; Brown *et al*., 1994; Lamb *et al*., 1995) and might therefore be taken to indicate that meltwater is routed through a subglacial distributed drainage system at Scott Turnerbreen (cf. Hodgkins *et al*., 1995). Furthermore, the progressive decline in concentrations of *SO$_4^{2-}$ (Figure 4b) and Si (Figure 3b) would then be consistent with the development and up-glacier expansion of channelized subglacial drainage, yielding an increasing proportion of dilute icemelt in the bulk meltwater. These chemical variations would therefore tend to suggest that the drainage system structure of Scott Turnerbreen could be described by a two-component subglacial model. This cannot be the case, given that borehole measurements indicate that the glacier bed is below the pressure-melting temperature and that moulins are absent (Hodgkins, 1994). However, there are subtle differences between the hydrochemistry of runoff at Scott Turnerbreen and that described in the temperate ('warm-based'), mid-latitude study glacier of Tranter *et al.* (1993), Brown *et al.* (1994) and Lamb *et al.* (1995): Haut Glacier d'Arolla. The SEF is relatively constant, it does not vary inversely with discharge and is not associated with $p(CO_2)$ (cf. Tranter *et al.*, 1993). Furthermore, the Si concentration of the runoff is much lower, 2·3–8·5 μmol l^{-1} (Table I), compared with 10–31 μmol l^{-1} at Haut Glacier d'Arolla (Lamb *et al.*, 1995). These characteristics suggest that a different hydrochemical model is necessary for this 'cold-based' glacier.

The icing as a source of solute

The proglacial icing accumulates over winter in the basin between the glacier terminus and the end-moraine (Figure 1c). It is formed by the refreezing of meltwater stored within the glacier at the end of the melt season, which continues to issue from the glacier during winter. Meltwater forms ponds at certain points within the ice marginal channels towards the end of the melt season: a net meltwater storage of $1·7 \times 10^5$ m^3 was estimated between 4 and 18 August 1992, and frozen meltwater was observed within the east ice marginal channel in May 1993, about a month before the start of the melt season (Hodgkins, 1994). The mean winter discharge, estimated from the water equivalent icing volume and the duration of the winter season, is *ca.* 0·01 m^3 s^{-1}. The ice marginal channel floors are incised into saturated heterogeneous sediment. Assuming Darcian flow through a channel floor sediment matrix, a hydraulic conductivity of $6·9 \times 10^{-3}$ m s^{-1} is estimated, yielding winter meltwater residence times of up to *ca.* 140 h, compared with *ca.* 1 h during the melt season (Hodgkins, 1994). It is not proposed to deal with the characteristics and formation of the icing in detail here, as this requires a discussion of the glacier drainage system which is beyond the scope and purpose of this paper.

The very high solute concentrations observed in samples from the icing can be attributed to the following features of winter meltwater drainage: (1) the late season meltwater that enters winter storage is initially dilute and hence particularly chemically aggressive; (2) residence times of meltwater in contact with sediment are very long during the winter season, allowing even the slowest chemical reactions, such as the dissolution of silicates, to add solute to the meltwater; (3) chemical weathering is promoted by the presence within the catchment sediments of sulfide minerals with relatively rapid dissolution kinetics, since oxidation of sulfides yields protons for weathering; (4) there is a high ratio of the surface area of sediment available for weathering to the meltwater volume; and (5) solute rejection by freezing meltwater not only concentrates the residual meltwater, but also increases CO_2 (aq) (as this is not readily incorporated into the ice lattice), which dissociates to provide protons that promote further weathering.

Interstitial water from the proglacial icing is highly concentrated in all ions, except for Cl^- (Table I). The median total solute concentration of interstitial water in the icing is 5200 μeq l^{-1}, including 940 μeq l^{-1} SO_4^{2-} and 1800 μeq l^{-1} HCO_3^-, with 14 μmol l^{-1} Si (compared with 55, 11 and 0 μeq l^{-1} and 0·8 μmol l^{-1}, respectively, in melting snow, and 100, 13 and 0 μeq l^{-1} and 0·0 μmol l^{-1} in glacier ice). Given that both the east and bulk streams flow through the icing during the melt season (Figure 1c), there is clearly potential for solute acquisition. Strong evidence that this occurs is provided by the elevated concentration of Si in the bulk stream (sampled downstream of the icing), relative to both ice marginal streams (sampled upstream of the icing). In-channel chemical weathering of suspended sediment is ruled out as a possible source of the additional Si, since the transit time between the east and bulk stream sampling sites is estimated to be of the order of 10 minutes only, which is insufficient to yield significant extra Si by dissolution (Raiswell, 1984; Dove, 1994). Likewise, Na^+ is the most dominant cation in the icing (Table I). East stream samples, which have not flowed through the icing, contain median concentrations of 160 μeq l^{-1} Na^+ and 180 μeq l^{-1} Ca^{2+}. Bulk stream samples, which have flowed through the icing, also contain a median concentration of 180 μeq l^{-1}

Ca^{2+}, but the median Na^+ concentration is increased to 210 µeq l^{-1} (Table I). The chemical weathering of suspended sediment may yield some additional Na^+, but the lack of increase in the concentration of Ca^{2+} suggests that some other process is responsible, such as the acquisition of solute from the icing.

The icing is the most likely source of the additional solute to the bulk runoff. Acquisition of solute from the icing must involve the flow or mixing of the concentrated interstitial water that is present in the icing into the meltwater streams that flow through the icing during the melt season. Whether this mixing is facilitated by leaching, slow drainage at grain intersections or simple melting of the icing itself is unknown. The contribution of solutes from the icing to the bulk runoff can be quantified approximately as follows. The median sum of dissolved ions in the bulk runoff for the period 22 June–10 July is 1100 µeq l^{-1}, whereas the corresponding values for the east and west ice marginal channels are 970 and 530 µeq l^{-1}, respectively (Table I). It is assumed that the relative contributions of the east and west ice marginal streams to the bulk runoff are 80 and 20% respectively, that conservative mixing occurs and that no other water or solutes contribute to the bulk runoff, other than from the interstitial waters of the icing. The composition of the interstitial waters is approximated by the median sum of dissolved ions for the period 4 May–30 June, namely 5100 µeq l^{-1} (Table I). A unit of bulk runoff therefore contains four parts east ice marginal stream, one part west ice marginal stream and x parts of icing interstitial water. Conservation of the equivalents of dissolved ions requires that x is 0·21, and that the bulk runoff therefore contains *ca.* 4% icing interstitial water. The solute flux contributed by the icing interstitial water is therefore *ca.* 19% of the bulk runoff flux. Clearly, the magnitude of this flux will vary during the melt season, particularly as the icing becomes depleted in solute.

Chemical weathering in sub-aerial environments

Meltwaters of the east and west ice marginal streams contain significant solute concentrations, in marked contrast to the dilute supraglacial streams of alpine glaciers (Gurnell and Fenn, 1984), The possibilities for subaerial solute acquisition at Scott Turnerbreen are outlined below.

The catchment lithology offers considerable potential for chemical weathering, as it consists of intensely frost-shattered Tertiary sediments with a high organic and sulfide content (Major and Nagy, 1972). The east and west streams flow in ice marginal locations, in contact with abundant moraine deposited during glacier retreat since the 1930s (Hodgkins, 1994). Accordingly, each stream is turbid: the mean SSC during the 1993 monitoring period was 1·0 kg m^{-3}. Furthermore, fresh sediment is input to the streams along their entire length, and the availability of sediment for fluvial transport increases during the melt season (Hodgkins, 1996). It should be emphasized that this sediment supply may be a vestige of Little Ice Age conditions, when this glacier was substantially thicker than at present and probably partially warm-based (Hodgkins, 1994). Had Scott Turnerbreen always been cold-based, or had it not retreated significantly in recent time, this sediment supply might now be absent. Meltwater in the ice marginal channels is largely in contact with the atmosphere, allowing the diffusion of CO_2 and O_2 into solution. The combination of a high and ongoing supply of potentially reactive sediment and free access to gases that promote chemical weathering [Equations (2) and (3)] potentially explains the relatively high concentrations of solute found in these subaerial meltwaters. Rapid and significant sulfide oxidation would have to occur in the stream channels to account for the high $p(CO_2)$ and high SEF of the east ice marginal stream. The relatively short duration of rock–water contact could explain the relatively low concentrations of Si in the ice marginal streams.

An alternative possibility is that chemical weathering occurs in the active layer of saturated, ice cored lateral moraine that transports sediment to the ice marginal channels through subaerial mass wasting. There are high rock:water ratios in this environment, and potentially higher concentrations of trace reactive minerals, such as carbonates and sulfides, than in suspended sediment in the ice marginal streams, since a proportion will be exhausted *en route* to the ice marginal channels. Steady seepage of concentrated pore water from the lateral moraine into the ice marginal streams could occur, pore water could be entrained along with sediment during periods of rising stream stage, or delivered directly to the channel through mass wasting (the latter probably taking the form of frequent events of small magnitude). High $p(CO_2)$ and high SEF could be generated owing to the sulfidic nature of the sediment.

The following calculations suggest that significant chemical weathering is more likely to occur within the lateral moraine, rather than within the ice marginal channels themselves. The Manning equation can be used to define the mean flow velocity, v (m s^{-1}), in an open channel,

$$v = R^{0.67} S^{0.5} n^{-1} \qquad (7)$$

where R is the hydraulic radius (channel cross-sectional area/wetted perimeter, m), S is the channel gradient (m m^{-1}), and n is a constant dimensionless roughness coefficient (Richards, 1982). The length and altitude range of the east ice marginal channel give a gradient, $S = 300/3500 = 0.09$ m m^{-1}. A value of 0.1 is adopted for n. The east ice marginal channel width is 2.0–3.0 m. Given the steep sides of the channel, composed of ice or ice-cored moraine, it is considered that the channel cross-sectional shape is reasonably approximated by a rectangle, and that channel width is more likely to be conserved than channel depth during rising stream stage. For the purposes of these calculations, channel width was therefore fixed in 0.1 m steps in the range 2.0–3.0 m, and the corresponding channel depth allowed to vary such that the observed discharge was matched.

The mean diurnal discharge amplitude in July was *ca.* 0.40 m^3 s^{-1}, the range being *ca.* 0.50–0.90 m^3 s^{-1} (Figure 3a). At low discharge (0.50 m^3 s^{-1}), $R = 0.20$–0.17 m (channel cross–sectional area 0.49–0.56 m^2) gives $v = 1.0 - 0.90$ m s^{-1} and a meltwater residence time of 58–65 minutes. At high discharge (0.90 m^3 s^{-1}), $R = 0.27$–0.23 m (channel cross-sectional area 0.73–0.81 m^2) gives $v = 1.2$–1.1 m s^{-1} and a meltwater residence time of 47–52 minutes. Results for a range of channel geometries are given in Table II. These times can be substituted into Equation (6) to derive an estimate of the Ca^{2+} released by 10 kg m^{-3} of suspended sediment in the ice marginal stream at high and low discharge. The results are 38 and 36 µeq l^{-1},

Table II. Low (0.50 m^3 s^{-1}) and high (0.90 m^3 s^{-1}) discharge meltwater residence times for the east ice marginal channel, calculated from Equation (7), and corresponding Ca^{2+} concentrations arising from the weathering of suspended sediment, calculated from Equation (6). A is the channel cross-sectional area, R the hydraulic radius and v the mean flow velocity

Channel width (m)	Channel depth (m)	A (m^2)	R (m)	v (m s^{-1})	Discharge (m^3 s^{-1})	Residence time (min)	Ca^{2+} conc. (µeq l^{-1})
2.0	0.25	0.49	0.20	1.0	0.50	58	38
2.1	0.24	0.50	0.19	1.0	0.50	58	38
2.2	0.23	0.51	0.19	0.99	0.50	59	38
2.3	0.22	0.51	0.19	0.97	0.50	60	38
2.4	0.22	0.52	0.18	0.96	0.50	61	38
2.5	0.21	0.53	0.18	0.95	0.50	61	38
2.6	0.21	0.53	0.18	0.94	0.50	62	38
2.7	0.20	0.54	0.17	0.93	0.50	63	38
2.8	0.20	0.55	0.17	0.92	0.50	63	38
2.9	0.19	0.55	0.17	0.19	0.50	64	38
3.0	0.19	0.56	0.17	0.90	0.50	65	38
2.0	0.36	0.73	0.27	1.2	0.90	47	36
2.1	0.35	0.74	0.26	1.2	0.90	48	36
2.2	0.34	0.74	0.26	1.2	0.90	48	36
2.3	0.33	0.75	0.25	1.2	0.90	49	36
2.4	0.32	0.76	0.25	1.2	0.90	49	36
2.5	0.31	0.77	0.25	1.2	0.90	50	36
2.6	0.30	0.77	0.24	1.2	0.90	50	36
2.7	0.29	0.78	0.24	1.1	0.90	51	36
2.8	0.28	0.79	0.24	1.1	0.90	51	36
2.9	0.28	0.80	0.23	1.1	0.90	52	36
3.0	0.27	0.81	0.23	1.1	0.90	52	36

respectively (Table II). These values, derived from a solution with a rock: water ratio one order of magnitude greater than the mean value of SSC in the bulk meltwater, are at least a factor of three lower than Ca^{2+} concentrations found in the east ice marginal stream, and at least a factor of two lower than those found in the west ice marginal stream (Table I). Despite the limitations of the simple dissolution experiments and the measured rate law as realistic analogues of chemical weathering in ice marginal streams, these simple calculations suggest that the dominant subaerial chemical weathering environment is more likely to be the lateral moraine adjoining the ice marginal channels than within the streams.

These calculations cannot, however, account for the observed diurnal variation in Ca^{2+} concentration of 16 µeq l^{-1} in the Scott Turnerbreen bulk runoff during July. The calculated residence time difference of 6–18 minutes yields a concentration difference of no more than 3 µeq l^{-1} from Equation (6). Diurnal variation in meltwater residence time in the ice marginal channels does not therefore appear to exert a significant kinetic control on meltwater chemistry. Given the foregoing discussion, it is likely that the observed diurnal variation in meltwater chemistry arises from the varying effects of sediment pore water entrainment and dilution in the ice marginal channels. At low discharge, both entrainment and dilution are relatively low; at high discharge, both entrainment and dilution are relatively high. Greater entrainment would tend to increase meltwater concentration, whereas greater dilution would tend to decrease concentration. Overall, the dilution effect appears to dominate, with more dilute runoff at high discharge.

CONCLUSIONS

The chemistry of the bulk meltwater at Scott Turnerbreen, a 'cold-based' glacier in the Norwegian high Arctic archipelago of Svalbard, is controlled by: (1) seasonal variations in the input of solutes from snowmelt and icemelt (Hodgkins *et al.*, in press); (2) proglacial solute acquisition from an icing early in the melt season; and (3) subaerial chemical weathering within saturated, ice-cored lateral moraine adjoining drainage channels at the glacier margins, sediment and concentrated pore water from which is entrained by flowing meltwater. The abundant sediment supply, which is important in explaining hydrochemistry at Scott Turnerbreen, may be a vestige of Little Ice Age conditions, when the glacier ice was thicker and warmer than at present. The difference in meltwater residence times at low and high discharges does not account for observed diurnal variations in solute concentration. It is probable that these variations arise from the net effects of variable sediment pore water entrainment and dilution in the ice marginal streams. This account of the hydrochemistry of Scott Turnerbreen requires only one major subaerial flow path, the ice marginal channel system, in which seasonally varying inputs of concentrated snowmelt and dilute icemelt are modified by seepage or entrainment of concentrated pore waters from sediment in lateral moraine, and by concentrated interstitial waters from a proglacial icing, supplied by leaching, slow drainage at grain intersections or simple melting of the icing itself. The ice marginal channels are therefore analogous neither to dilute supra/ englacial nor to concentrated subglacial flow components.

ACKNOWLEDGEMENTS

This work was supported by NERC grant GR9/946 (to M.T. and J.A.D.) and NERC studentship GT4/91/ AAPS/11 (to R.H.). Ms M. Zeeman performed the laboratory dissolution experiments, and Mrs J. Mills was responsible for laboratory analysis. The paper was greatly improved by the robust and provocative reviews of two anonymous referees, and the editorial comments of M. J. Sharp.

REFERENCES

Brown, G. H., Sharp, M. J., Tranter, M., Gurnell, A. M., and Nienow, P. W. 1994. 'The impact of post-mixing chemical reactions on the major ion chemistry of bulk meltwaters draining the Haut Glacier d'Arolla, Valais, Switzerland', *Hydrol. Process.*, **8**, 465–480.
Brown, G. H., Tranter, M., and Sharp, M. J. 1996. 'Experimental investigations of the weathering of suspended sediment by Alpine glacial meltwater', *Hydrol. Process.*, **10**, 579–597.

Collins D. N. 1979. 'Hydrochemistry of meltwaters draining from an Alpine glacier', *Arct. Alp. Res.*, **11**, 307–324.

Dove, P. M. 1994. 'The dissolution kinetics of quartz in sodium chloride solutions at 25° to 300°C', *Am. J. Sci.*, **294**, 665–712.

Ford, D. C. and Williams, P. 1989. *Karst Geomorphology and Hydrology*. Unwin Hyman, London. 601 pp.

Fountain, A. G. 1992. 'Subglacial water flow inferred from stream measurements at South Cascade Glacier, Washington, U.S.A.', *J. Glaciol.*, **38**, 51–64.

Gurnell, A. M. and Fenn, C. R. 1984. 'Flow separation, sediment source areas and suspended sediment transport in a pro-glacial stream', in Schick, A. P. (Ed.) *Channel Processes: Water, Sediment, Catchment Controls, Catena* suppl. **5**, Elsevier, Amsterdam. pp. 109–119.

Gurnell, A. M., Clark, M. J., Hill, C. T., and Greenhalgh, J. (1992). 'Reliability and representativeness of a suspended sediment monitoring programme for a remote alpine proglacial river', *IAHS Publ.*, **213**, 191–200.

Hagen, J. O., Liestøl, O., Roland, E., and Jørgensen, T. 1993. *Glacier Atlas of Svalbard and Jan Mayen*. Norsk Polarinstitutt Meddelelser, Norsk Polarinstitutt, Oslo. p. 129.

Hodgkins, R. 1994. 'The seasonal evolution of meltwater discharge, quality and routing at a High-Arctic glacier', *Unpublished PhD Thesis*, University of Cambridge, UK.

Hodgkins, R. 1996. 'Seasonal trends in suspended-sediment transport at an Arctic glacier, and their implications for drainage system structure', *Ann. Glaciol.*, **22**, 147–151.

Hodgkins, R. in press. 'Glacier hydrology in Svalbard, Norwegian High Arctic', *Quat. Sci. Rev.* (Spec. Iss.).

Hodgkins, R., Tranter, M., and Dowdeswell, J. A. 1995. 'The interpretation of hydrochemical evidence for meltwater routing at a High-Arctic glacier', *IAHS Publ.*, **228**, 387–394.

Hodgkins, R., Tranter, M., and Dowdeswell, J. A. in press. 'Solute provence, transport and denudation in a High-Arctic glacierised catchment', *Hydrol. Process.*

Lamb, H. R., Tranter, M., Brown, G. H., Hubbard, B. P., Sharp, M. J., Gordon, S., Smart, C. C., Willis, I. C., and Nielson, M. K. 1995. 'The composition of subglacial meltwaters sampled from boreholes at the Haut Glacier d'Arolla, Switzerland', *IAHS, Publ.*, **228**, 395–403.

Major, H. and Nagy, J. 1972. 'Geology of the Adventdalen Map Area', *Norsk Polarinstitutt Skrifter 138*. Norsk Polarinstitutt, Oslo.

Raiswell, R. 1984. 'Chemical models of solute acquisition in glacial melt waters', *J. Glaciol.*, **30**, 49–57.

Raiswell, R. and Thomas, A. G. 1984. 'Solute acquisition in glacial meltwaters. I. Fjallsjökull (south-east Iceland): bulk meltwaters with closed-system characteristics', *J. Glaciol.*, **30**, 35–43.

Richards, K. S. 1982. *Rivers, Form and Process in Alluvial Channels*. Methuen, London. pp. 361.

Tranter, M., Brown, G. H., Raiswell, R., Sharp, M. J., and Gurnell, A. M. 1993. 'A conceptual model of solute acquisition by Alpine glacial meltwaters', *J. Glaciol.*, **39**, 573–581.

Tranter, M., Brown, G. H., Hodson, A. J., and Gurnell, A. M. 1996. 'Hydrochemistry as an indicator of subglacial drainage system structure: a comparison of alpine and sub polar environments', *Hydrol. Process.*, **10**, 541–556.

Water Research Association, 1970. 'River flow measurement by dilution gauging', *Water Res. Assn. Tech. Paper TP74*. Department of the Environment, Reading.

5

HYDROCHEMISTRY AS AN INDICATOR OF SUBGLACIAL DRAINAGE SYSTEM STRUCTURE: A COMPARISON OF ALPINE AND SUB-POLAR ENVIRONMENTS

MARTYN TRANTER

Department of Geography, University of Bristol, Bristol BS8 1SS, UK

GILES H. BROWN

Centre for Glaciology, Institute of Earth Sciences, University College of Wales, Aberystwyth SY23 3DB, UK

ANDREW J. HODSON

Department of Geography, University of Southampton, Southampton SO9 5NH, UK

AND

ANGELA M. GURNELL

School of Geography, University of Birmingham, Birmingham B15 2TT, UK

ABSTRACT

The anion compositions (SO_4^{2-}, HCO_3^- and Cl^-) of runoff from the Haut Glacier d'Arolla, Switzerland and Austre Brøggerbreen, Svalbard are compared to assess whether or not variations in water chemistry with discharge are consistent with current understanding of the subglacial drainage structure of warm- and polythermal-based glaciers. These glacial catchments have very different bedrocks and the subglacial drainage structures are also believed to be different, yet the range of anion concentrations show considerable overlap for SO_4^{2-} and HCO_3^-. Concentrations of Cl^- are higher at Austre Brøggerbreen because of the maritime location of the glacier. Correcting SO_4^{2-} for the snowpack component reveals that the variation in non-snowpack SO_4^{2-} with discharge and with HCO_3^- is similar to that observed at the Haut Glacier d'Arolla. Hence, if we assume that the provenance of the non-snowpack SO_4^{2-} is the same in both glacial drainage systems, a distributed drainage system also contributes to runoff at Austre Brøggerbreen. We have no independent means of testing the assumption at present. The lower concentrations of non-snowpack SO_4^{2-} at Austre Brøggerbreen may suggest that a smaller proportion of runoff originates from a distributed drainage system than at the Haut Glacier d'Arolla.

INTRODUCTION

The chemical composition of glacial runoff has been used to model or constrain interpretations of the structure of the subglacial drainage system of glaciers in the Alps and on Svalbard. Hydrochemical research in the Alps was pioneered by Collins (1978; 1979) and has helped to define a number of aspects of the hydroglacial system (Lemmens and Roger, 1978; Thomas and Raiswell, 1984; Souchez and Lemmens, 1987; Sharp, 1991; Tranter and Raiswell, 1991; Brown and Tranter, 1990; Tranter *et al.*, 1993a; 1993b). By contrast, hydrochemical studies of glacial runoff on Svalbard are restricted (Pulina, 1984; Kostrzewski *et al.*, 1989; Vatne *et al.*, 1992). A central assumption which governs the use of water chemistry to model subglacial drainage systems is that a unique chemical signature is imparted to the water body as it flows from the source to the glacier terminus, and that this chemical signature is diagnostic of either the water source (snow- or icemelt) or the reservoir through which the water has flown. Variations in water chemistry may thus be used to examine the storage and release of water from subglacial reservoirs (Collins, 1979). Detailed dye tracing and borehole studies have resulted in a re-evaluation of the different types of subglacial drainage system (Nienow *et al.*, Unpublished data; Sharp *et al.*, 1993b), and it is now less clear that runoff chemistry can be used to reconstruct the finer detail of complex subglacial drainage systems. However, runoff chemistry is believed to give information about the main characteristics of the subglacial drainage system (Tranter *et al.*, 1993a).

This paper illustrates the strengths and weaknesses of meltwater chemistry in reconstructions of subglacial drainage systems by examining the runoff from a warm-based Alpine glacier, the Haut Glacier d'Arolla (Switzerland) and a polythermal sub-Polar glacier, Austre Brøggerbreen (Svalbard), which are believed to have very different subglacial drainage characteristics. Factors governing solute acquisition by glacial meltwaters are outlined first, followed by a brief review of what is known about subglacial drainage structures at the Haut Glacier d'Arolla and Austre Brøggerbreen. Thereafter, the paper examines whether or not variations in the anion content of runoff are consistent with current hydroglacial models.

Solute acquisition by glacial meltwaters

Glacial runoff acquires solute from two primary sources, namely the atmosphere and the chemical weathering of rocks and related materials such as glacial flour, dust in snowpacks and moraines. The former source provides sea salt, acidic nitrate and sulphate aerosols, temporarily stored in snowpack and ice, and gases such as CO_2 and O_2. The latter source gives rise to the base cations, Ca^{2+}, Mg^{2+}, Na^+ and K^+, and increases the concentrations of dissolved anions such as HCO_3^- and SO_4^{2-} (Tranter *et al.*, 1993a). Ions from these primary sources may precipitate out of solution to form secondary sources, which include subglacial precipitates such as carbonates (Hallet, 1976). Most ions found in meltwaters, apart from the early snowmelt, are derived from the chemical weathering of minerals (Souchez and Lemmens, 1987). The magnitude and rate of solute acquisition from chemical weathering is usually dependent on four first-order variables, namely the reactive mineralogy, the reactive surface area to water ratio, the duration of rock–water contact and the availability of other species involved in the chemical weathering reactions, such as protons and oxygen (Raiswell, 1984; Tranter *et al.*, 1993a; Brown *et al.*, this issue). Geochemically reactive minerals, in order of decreasing reactivity, include evaporites (e.g. halite and gypsum), carbonates (e.g. dolomite and calcite), sulphides (e.g. pyrite) and aluminosilicates (e.g. feldspars), and it is common to find that evaporites, carbonates and sulphides have a dominant control on water chemistry, despite being present in only trace amounts in the bedrock (Holland, 1978). Increasing the reactive surface area to water ratio increases the amount of solute in the solution per unit time, and increasing the rock–water contact time results in increasing solute concentrations, although the increase with time is often non-linear (e.g. Brown *et al.*, this issue). Simple dissolution reactions, such as the dissolution of halite or sea salt, do not involve the consumption of other dissolved species. However, the dissolution of rock flour normally involves surface exchange, carbonation and oxidation reactions, which consume protons, dissolved CO_2 and O_2 respectively (Tranter *et al.*, 1993a). Hence the presence of these species may limit the chemical weathering of glacial flour and control the magnitude of solute acquisition by glacial runoff.

How these master variables map onto the hydroglacial drainage system has been discussed by Brown *et al.* (this issue). Clearly the reactive mineralogy is a function of the bedrock and debris contained within the catchment and may be modified either by sorting during transport or exhausted by chemical weathering processes. The reactive surface area to water ratio is a function of the suspended sediment concentration and particle size distribution, the presence or absence of microparticles (Tranter, Unpublished data) and the size distribution and concentration of debris (for example, unfrozen subglacial debris) through which water may be flowing. The rock to water contact time is a function of the residence time of water in a particular flowpath or reservoir and the availability of limiting species, such as protons, O_2 and CO_2, depends on factors such as the proximity to a gas source (such as the atmosphere or gas bubbles in ice) and turbulence. Subglacial hydrology has an important influence on these factors. Hence we anticipate that glaciers with different bedrocks and different subglacial hydrologies will exhibit different water chemistries. The following sections review what is known of the subglacial hydrology of Alpine and sub-Polar glaciers and the bedrock of the study areas.

Subglacial hydrology at Haut Glacier d'Arolla and beneath sub-Polar glaciers

Ice at the sole of warm-based glaciers is at the pressure melting point and at least a thin film of water resides at the ice–bedrock interface. By contrast, ice at the sole of cold-based glaciers is largely frozen to the bedrock (Sugden and John, 1976), but water may still exist at the bed at ice–bedrock temperatures below freezing if the rate of water flow is greater than the rate of freezing (Sharp, pers. comm.). Many glaciers on Svalbard are polythermal (Dowdeswell, 1986), being frozen to the bedrock near the snout and warm-based under the accumulation area. The position of the cold-temperate ice transition is dependent on ice thickness, climate and flow-rate (Blatter and Hutter, 1991; Blatter, 1987). The subglacial hydrology of warm-based Alpine glaciers is reasonably well established (see Figure 1), but the subglacial drainage system of polythermal-based sub-Polar glaciers on Svalbard is poorly understood (see Figure 1). This lack of understanding remains, despite there being much interest in the nature of the drainage system (Hagen *et al.*, 1991; Vatne *et al.*, 1992), which is believed to influence the frequency and magnitude of surge activity (Clarke *et al.*, 1984; Kamb *et al.*, 1985; Kamb, 1987; Dowdeswell *et al.*, 1991).

Haut Glacier d'Arolla. Diurnal variations in the electrical conductivity and discharge of Alpine glacial runoff formed the basis of the hypothesis that there are two components of glacial runoff, one dilute and the other concentrated (Collins, 1978). It was assumed that the concentrated waters originated from water

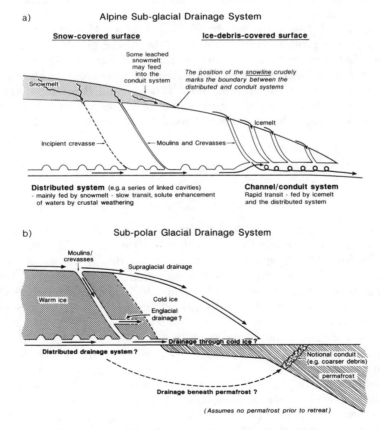

Figure 1. Schematic representation of the subglacial drainage systems beneath Alpine (after Nienow *et al.*, unpublished data) and sub-Polar glaciers (after Hodson, unpublished data)

flowing along subglacial drainage paths near to or at the bed and that the dilute waters arose from englacial drainage routes. This hypothesis has been refined and extended by dye tracing studies (Nienow *et al.*, Unpublished data) and by attempts to identify tracer ions which better define component sources or flowpaths (Tranter and Raiswell, 1991; Brown and Tranter, 1990; Tranter *et al.*, 1993a; 1993b; 1994). It is likely that the subglacial drainage structure beneath warm-based Alpine glaciers has two elements (see Figure 1). There is an inefficient *distributed drainage system* underlying snow-covered ice, whereas an efficient *channelized drainage system* (flanked by a distributed drainage system) underlies snow-free ice nearer the snout. The channelized system grows up-glacier at the expense of the distributed system as the snowline retreats (Nienow *et al.*, Unpublished data). Waters that have passed through the distributed system before contributing to bulk runoff are called *delayed flow*, whereas waters that have only passed through the channelized system are called *quickflow*.

Sub-Polar glaciers. The subglacial drainage system underlying sub-Polar polythermal-based glaciers is only poorly known. One possible model of the structure is shown in Figure 1. A significant proportion of icemelt in the vicinity of the snout drains directly off the surface (Sugden and John, 1976) or may be englacially routed without contact with the bed (Hagen *et al.*, 1991). However, melt may also be routed to the warm-based bed (Boulton, 1972; Hagen *et al.*, 1991; Vatne *et al.*, 1992). It is unclear whether or not water from the warm-based area can flow through the cold-based region near the snout. Some drainage may occur through englacial channels above the bed (Sugden and John, 1976; Hagen *et al.*, 1991), but it is believed that most of this englacial flow originates directly from surface melting rather than from water that re-emerges from subglacial drainage (Hagen *et al.*, 1991). However, there may be a restricted drainage system running through the cold ice along the bed (Dowdeswell and Drewry, 1989; Clarke and Blake, 1991; Vatne *et al.*, 1992) or within unfrozen/permeable subglacial till or sediment (Liestol, 1975; Kamb *et al.*, 1985; Clarke, 1987; Lauritzen, 1991).

A fundamental difference between the subglacial hydrology of Haut Glacier d'Arolla and Austre Brøggerbreen is that a greater proportion of runoff from the Alpine glacier is believed to originate from meltwaters routed along the bed in a distributed drainage system. All other factors being equal, this should result in the Alpine runoff being more concentrated in solutes derived from subglacial chemical weathering in the distributed drainage system, namely HCO_3^- and SO_4^{2-} (Tranter and Raiswell, 1991). In this paper, we examine the anion content of runoff as the relative amounts of HCO_3^- and SO_4^{2-} may give information on the relative proportion runoff that has been derived from the distributed system (Tranter *et al.*, 1993a).

STUDY AREAS

The Haut Glacier d'Arolla is located in the Val d'Hérens, Switzerland (see Figure 2) and has an area of $\sim 6 \cdot 3 \, \text{km}^2$. The main ice tongue descends to the snout at 2560 m and is fed from a compound basin of maximum elevation 3838 m. The glacier has been in retreat during recent years (by 720 m during 1967–1989). The glacier is warm-based and has a maximum ice thickness of 180 m (Sharp *et al.*, 1993a). The catchment is underlain by amphibolite and schists and gneisses of the Arolla series. The bedrock contains trace amounts of geochemically reactive minerals, such as carbonates and sulphides (Brown, Unpublished data).

Austre Brøggerbreen is located about 3 km to the south-west of Ny-Ålesund, north-west Svalbard (see Figure 3). It has an area of $\sim 11 \, \text{km}^2$ (Hodson, Unpublished data) and drains into Kongsfjord via the Bayelva River. The headwalls of the watershed rise to 720 m and the snout lies at ~ 50 m. The bedrock consists of Devonian, Carboniferous and Tertiary sedimentary rocks, which include limestone, dolomite, shale and red sandstones (Hodson, Unpublished data). The glacier has retreated considerably over the last 100 years. The total ice volume may have decreased by 50%, and the present annual mean rate of retreat is $\sim 0 \cdot 4$ m/yr. The maximum ice thickness is currently ~ 130 m (Liestol, 1988). The glacier base is polythermal. Temperature measurements from two boreholes

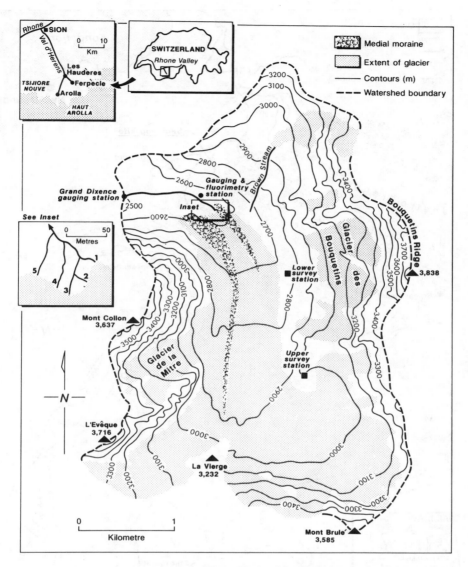

Figure 2. Location of Haut Glacier d'Arolla, Switzerland

suggest that the glacier is warm-based beneath ice thicknesses of >70–80 m and cold-based beneath shallower ice (Hagen *et al.*, 1991).

METHODOLOGY

At the Haut Glacier d'Arolla, discharge estimates were derived from continuous stage records at a rectangular weir located in the Grande Dixence S.A. meltwater intake structure, 1 km from the glacier snout (see Figure 2). The error in discharge measurement is typically ±3% (Brown, Unpublished data). Sampling was undertaken in both 1989 and 1990 (Julian days 152–240 in both seasons). On most days, two samples were collected at times approximating maximum (17.00 h) and minimum discharge (10.00 h).

At Austre Brøggerbreen, stage was continuously measured by recording pressure transducer variations

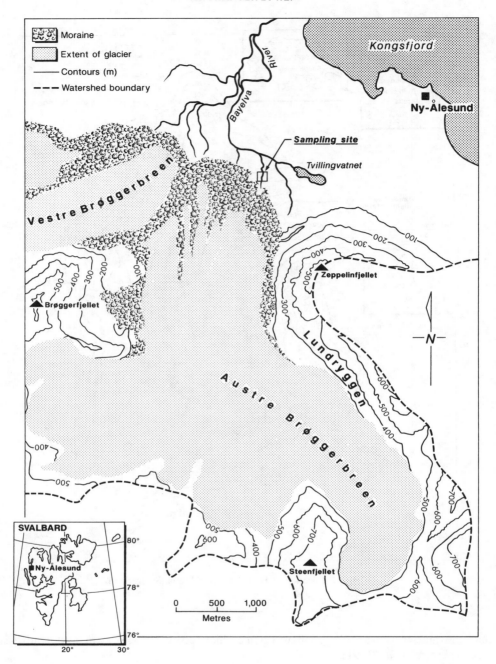

Figure 3. Location of Austre Brøggerbreen, Svalbard

about 200 m from the outflow of the main lateral stream on the eastern side of the glacier (see Figure 3). Discharge was estimated from current meter determinations and discharge rating curves. Errors associated with these measurements are ±10% (Hodson, Unpublished data). Water sampling was conducted during the 1991 and 1992 field seasons (Julian days 179–224 and 162–218, respectively). Meltwater samples were collected, where possible, on alternate days close to diurnal maximum and minimum discharge, depending on the weather conditions and logistical constraints.

Table I. Range of selected parameters measured at each glacier throughout the sampling season. Units of discharge are m^3/s, and for other dissolved species, μeq/l.

	Discharge	SO$_4^{2-}$	non-snowpack SO$_4^{2-}$	HCO$_3^-$	Cl$^-$
HGA 1989	0·4–6·5	30–240	NA	180–370	0·03–3·3
HGA 1990	0·4–7·2	20–170	NA	200–460	0·9–6·7
AB 1991	1·0–2·3	21–110	15–71	230–520	39–540
AB 1992	0·5–4·5	10–140	5·7–75	145–390	21–550

HGA, Haut Glacier d'Arolla; AB, Austre Brøggerbreen; and NA, not applicable

Sample treatment

All samples were filtered through 0·45 μm cellulose nitrate membranes immediately after collection. Filtered samples were stored in pre-cleaned polyethylene bottles. Full details of sampling treatment and storage may be found in Brown (Unpublished data) and Hodson (Unpublished data).

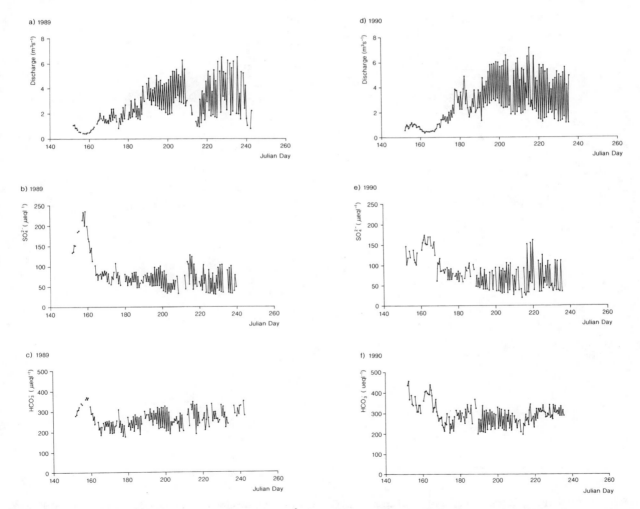

Figure 4. Haut Glacier d'Arolla: variations in discharge, SO$_4^{2-}$ and HCO$_3^-$ concentrations in runoff during the 1989 and 1990 sampling seasons

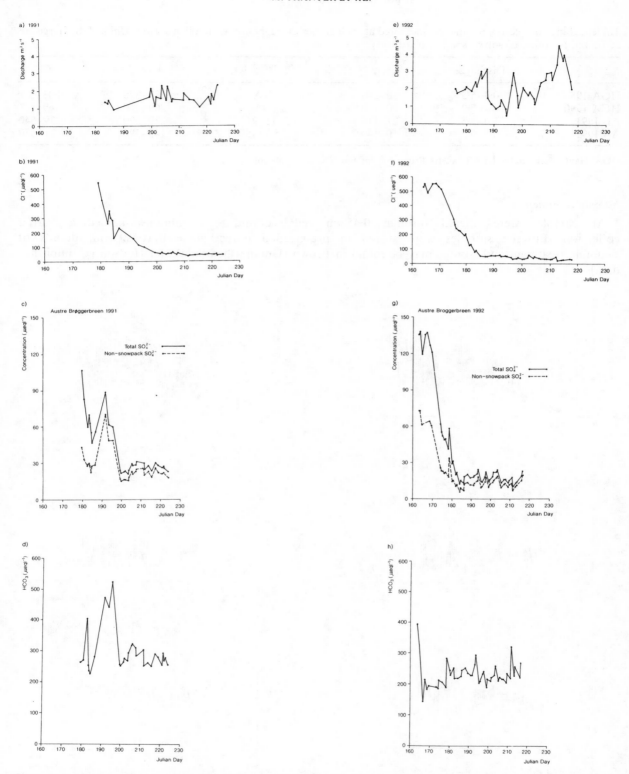

Figure 5. Austre Brøggerbreen: variations in discharge, Cl⁻, SO₄²⁻ and HCO₃⁻ concentrations in runoff during the 1991 and 1992 sampling seasons

Analysis

SO_4^{2-} and Cl^- were determined by ion chromatography on a Dionex 4000i ion chromatograph using either an AS4A analytical column (Haut Glacier d'Arolla samples) or a Fast Anion-1 column (Austre Brøggerbreen samples) and a sample loop of $50\,\mu l$. The precision and accuracy of the measurements is typically $\pm 3\%$. HCO_3^- was determined by titration with $1\,mM$ HCl to an endpoint of pH 4·5 using BDH 4·5 indicator. Precision was typically $\pm 5\%$ for concentrations $>100\,\mu eq/l$. Full details of analysis can be found in Brown (Unpublished data) and Hodson (Unpublished data). Values for Cl^- at Haut Glacier d'Arolla are presented for illustrative purposes only, as they are near the quantification limit for the analytical conditions.

RESULTS

We may reasonably anticipate that there is greater chemical weathering potential in the Austre Brøggerbreen basin because of the greater proportion of carbonate in the bedrock. In addition, the sulphide content of shale is usually of the order of a few per cent by weight (Holland, 1978), so that there is high potential for sulphide oxidation to provide SO_4^{2-} and protons for additional chemical weathering. By contrast, the bedrock of Haut Glacier d'Arolla is less reactive. The most geochemically reactive materials are the trace amounts of carbonate and sulphide, and the ferromagnesian minerals of the amphibolite. The greater reactivity of the bedrock at Austre Brøggerbreen may partially offset the reduction in the potential for chemical weathering given a more limited distributed drainage system. Hence it is unclear from theoretical considerations whether or not runoff at Austre Brøggerbreen should be more dilute or more concentrated than at the Haut Glacier d'Arolla. However, we anticipate that ions derived predominantly in the distributed system should be in greater evidence at Haut Glacier d'Arolla.

Results for each glacier are summarized in Table I. There is considerable overlap in the range of concentrations encountered at each site for ions other than Cl^-. Higher discharges and concentrations of SO_4^{2-} are recorded at Haut Glacier d'Arolla, whereas higher Cl^- concentrations are recorded at Austre Brøggerbreen. Generally, HCO_3^- concentrations are comparable between glaciers. The temporal variations in the parameters are briefly described in the following sections.

Haut Glacier d'Arolla

Figure 4 depicts the variation in discharge for the 1989 and 1990 sampling seasons, with a range of $0·4$–$7·2\,m^3/s$. Both discharge ($<2\,m^3/s$) and the diurnal amplitude ($\leq 0·6\,m^3/s$) are lowest during June (Julian days 152–176) and both parameters increase as the ablation season progresses. High discharge and increasingly peaked hydrographs of high diurnal amplitude ($>3·0\,m^3/s$) occur during July and August (Julian days 190–240). Short periods of flow recession are the result of snowfall (e.g. Julian days 212–217, 1989).

The lower concentrations of Cl^- at Haut Glacier d'Arolla compared with those of Austre Brøggerbreen (see Table I) derive from the continental location of the former compared with the maritime location of the latter. Both HCO_3^- and SO_4^{2-} concentrations are highest during the early season (up to 460 and $240\,\mu eq/l$, respectively). Thereafter concentrations decrease and show an inverse diurnal association with discharge (see Figure 4). The relative decrease in HCO_3^- is less than for SO_4^{2-}.

Austre Brøggerbreen

The climatic controls on runoff variations have been reported by Gurnell *et al.* (in press). The range in recorded discharge is from $\sim 0·5$ to $4·5\,m^3/s$ (see Figure 5). Discharge does not vary systematically, but does generally increase during the sampling period. Diurnal variations in discharge are small compared with those recorded in the Alps and are of the order of $0·5$–$1\,m^3/s$.

Figure 5 shows the variation in HCO_3^-, SO_4^{2-} and Cl^- during both sampling periods. Concentrations of Cl^- are highest ($500\,\mu eq/l$) at the start of the sampling season (Julian days 164–168), coinciding with the onset of the thaw and drainage of snowmelt from the glacier surface. It is likely that more concentrated snowmelt drained the glacier surface before sampling started and that maximum concentrations were not recorded. Concentrations decrease exponentially and level out at $\sim$50 μeq/l after Julian day 184. The 1991 data are similar to those of 1992 (see Figure 5), despite marked differences in discharge between seasons. There is a break in the exponential decline (Julian days 182–184), coinciding with a period of high rainfall superimposed on a period of recession flow (Hodson, Unpublished data).

The concentration of HCO_3^- is variable during both sampling seasons (140–520 μeq/l). Higher values are usually found at the beginning of the sampling season, but maximum concentrations were recorded during a period of flow recession in 1991. The main source of HCO_3^- in solution is likely to be from carbonation reactions (Reynolds and Johnson, 1972; Tranter et al., 1993a) and occurs in supraglacial channels as well as in subglacial environments. Small diurnal variations were recorded and are typically of the order of 20–40 μeq/l.

There are two sources of SO_4^{2-} in the bulk runoff, namely the snowpack and the oxidation of sulphides in comminuted bedrock, moraine and rock flour. The snowpack sulphate is derived largely from sea salt aerosol and the average ratio of Cl^- to SO_4^{2-} in the snowpack is 0·118 (Hodson, Unpublished data), compared with 0·103 for seawater (Holland, 1978). We have calculated the non-snowpack SO_4^{2-} (i.e. that derived from chemical weathering), $^{non\text{-}snowpack}SO_4^{2-}$, by assuming that all Cl^- in runoff is derived from the snowpack. We assume that there is no change in the ratio of SO_4^{2-} to Cl^- during snowmelt, hence

$$^{non\text{-}snowpack}SO_4^{2-} = {}^{total}SO_4^{2-} - (0\cdot118Cl^-) \tag{1}$$

where $^{total}SO_4^{2-}$ denotes the SO_4^{2-} concentration measured in the runoff. Both $^{total}SO_4^{2-}$ and $^{non\text{-}snowpack}SO_4^{2-}$ concentrations are shown in Figure 5. The range of concentrations measured is 10–140 and 6–75 μeq/l, respectively. Concentrations of both species decrease during both sampling periods, but $^{total}SO_4^{2-}$ concentrations decrease faster, so that $^{non\text{-}snowpack}SO_4^{2-}$ is $\sim$50% of $^{total}SO_4^{2-}$ at the start of sampling period and $\sim$80% at the end. This is not a consequence of the preferential elution of SO_4^{2-} with respect to Cl^- from the snowpack (Tranter, 1989), as this leads to an overestimation of $^{non\text{-}snowpack}SO_4^{2-}$ when Cl^- concentrations are high during the early season and an underestimation when Cl^- concentrations are low during the later season. Hence the pattern observed would be accentuated by preferential elution. Rather, the increasing proportion of $^{non\text{-}snowpack}SO_4^{2-}$ reflects the leaching and non-replacement of the snowpack SO_4^{2-} source compared with the relatively durable source of $^{non\text{-}snowpack}SO_4^{2-}$ from chemical weathering. Diurnal variations in $^{total}SO_4^{2-}$ arise largely from variations in $^{non\text{-}snowpack}SO_4^{2-}$, which are of the order of 5 μequiv./l, and it is $^{non\text{-}snowpack}SO_4^{2-}$ concentrations, rather than snowpack SO_4^{2-}, that increase during a period of flow recession during 1991 (Julian days 186–192).

DISCUSSION

Runoff chemistry and subglacial drainage structure at the Haut Glacier d'Arolla

Mass balance calculations show that the snowpack provides $\sim$1–2% of the SO_4^{2-} load of runoff. We have not corrected the $^{total}SO_4^{2-}$ concentration for $^{snowpack}SO_4^{2-}$ because there is only a small Cl^- concentration with which to normalize the SO_4^{2-} data. Hence we assume that $^{total}SO_4^{2-} \approx {}^{non\text{-}snowpack}SO_4^{2-}$.

The growth of the channelized system at the expense of the distributed drainage system beneath the Haut Glacier d'Arolla (Nienow et al., Unpublished data) produces a systematic variation in the chemistry of the bulk meltwaters. It is believed that carbonation reactions are dominant in the channelized system and that sulphide oxidation and carbonate dissolution are important chemical weathering reactions in the distributed system (Tranter et al., 1993a). Laboratory dissolution experiments confirm that carbonation reactions dominate interactions between dilute solutions and glacial flour that has passed through the hydroglacial system and therefore characterize weathering reactions in the channelized system (Brown et al., 1994; this issue). Water samples obtained from boreholes to the glacier bed contain high SO_4^{2-} concentrations and high $SO_4^{2-}/(SO_4^{2-} + HCO_3^-)$ ratios and are believed to be representative of distributed system drainage (Sharp et al., 1993b).

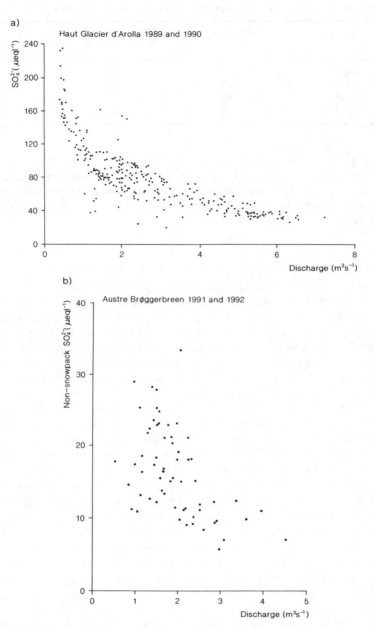

Figure 6. Variation in sulphate and $^{\text{non-snowpack}}SO_4^{2-}$ concentration with discharge at Haut Glacier d'Arolla and Austre Bróggerbreen, respectively, for both sampling seasons

Figure 6 shows that both HCO_3^- and SO_4^{2-} have negative associations with discharge. The linear regression equations for the best-fit lines are presented in Table II, giving a crude representation of the rate of decrease in both species per unit discharge (because Figure 2 shows that the association with discharge is non-linear). The $SO_4^{2-}/(SO_4^{2-} + HCO_3^-)$ ratio is inversely associated with discharge also (see Figure 7). Waters draining from the distributed system into the channelized system impart a high $SO_4^{2-}/(SO_4^{2-} + HCO_3^-)$ ratio to the bulk runoff, whereas carbonation reactions in the channelized system drive the ratio downwards. The proportional contribution of delayed flow to bulk runoff decreases as

Table II. Linear regression of the anions with discharge for both sampling seasons at Haut Glacier d'Arolla

Sampling season	Species	Gradient	Intercept	Correlation coefficient	n
1989	HCO_3^-	−12	300	−0·445	148
1990	HCO_3^-	−17	340	−0·575	151
1989	SO_4^{2-}	−19	130	−0·758	142
1990	SO_4^{2-}	−18	130	−0·818	136

the channelized system grows up-glacier throughout July and August and the relative magnitude of carbonation reactions increases (Brown *et al.*, 1994). Hence as discharge increases during July and August, the $SO_4^{2-}/(SO_4^{2-} + HCO_3^-)$ ratio of bulk runoff decreases. The results shown in Figure 7 are consistent with the two-component subglacial drainage model deduced from dye tracing returns (Nienow *et al.*, Unpublished data).

Runoff chemistry and subglacial drainage structure at Austre Brøggerbreen

The main source of Cl^- to Austre Brøggerbreen runoff is snowmelt (Hodson, Unpublished data) and the pattern of runoff concentrations (see Figure 4) is similar to the exponential decrease of laboratory snowmelt experiments, where the first meltwaters are more concentrated than the original solute content of the snow by factors typically in the range of 3–10 (Johannessen and Henriksen, 1978; Tranter, 1989). There is little evidence of increased Cl^- concentrations during periods of recession flow (Tranter *et al.*, 1994), suggesting that most snowmelt is rapidly routed to the glacier terminus, in agreement with previously published work.

Figure 5 shows that both HCO_3^- and $^{non-snowpack}SO_4^{2-}$ increase in concentration during the 1991 recession flow (Julian days 184–194). These increased concentrations can arise from changes in two factors. Firstly, the residence time of turbid water in supraglacial and englacial channels increases during flow recession. This will give rise to an increase in HCO_3^- if there is no significant reduction in the rock to water ratio and also to SO_4^{2-} if there is reactive sulphide in the suspended sediment. Secondly, delayed flow may make a greater contribution to runoff during the flow recession. Field observations do not suggest the influence of a third possible factor, increased suspended sediment concentrations.

We infer that processes of solute acquisition may not be directly comparable in Alpine and sub-Polar hydroglacial systems, as the supraglacial channels of sub-Polar glaciers may contain relatively high concentrations of suspended sediment (Hodson, Unpublished data; Hodgkins *et al.*, Unpublished data). Sulphide oxidation and carbonate dissolution may therefore occur in these channels, depending on the mineralogy and the severity of the chemical weathering that the suspended sediment underwent before transport through the supraglacial channels. Most of the suspended sediment appears to originate from mud slumps and lateral moraine at Austre Brøggerbreen (Hodson, Unpublished data). Hence it is possible that these sediments liberate SO_4^{2-} on transport if there has been relatively little leaching of the sediment or if freeze–thaw cycles have exposed fresh reaction sites. Dissolution experiments have yet to be performed on these sediments and their geochemical reactivity remains unassessed.

Table III shows that there is, at most, a very weak association between Cl^-, HCO_3^- and discharge for both sampling seasons, although Cl^- exhibits a good negative correlation during 1992. High values of Cl^- from the early sampling season were not included in the calculation of the linear regression, as these were clearly outliers from the remainder of the data set. The most significant association with discharge is for SO_4^{2-} and $^{non-snowpack}SO_4^{2-}$. The poor association of HCO_3^- with discharge is probably a function of the fact that carbonation can take place in both supraglacial, englacial and subglacial environments, and so is controlled by factors such as suspended sediment concentration, access to a CO_2 source and rock–water contact time (Brown *et al.*, this issue). The better inverse association between $^{non-snowpack}SO_4^{2-}$ and discharge suggests that (a) there are fewer sources of this ion compared with HCO_3^-, (b) SO_4^{2-} release

a)

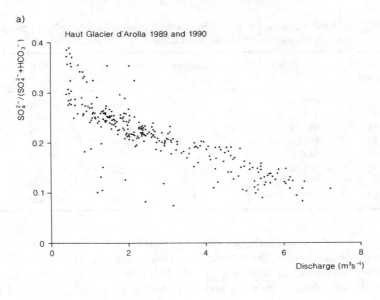

b)

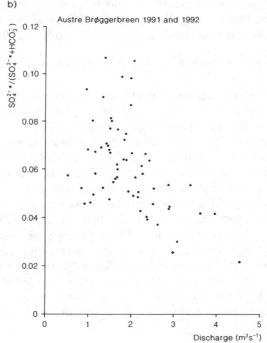

Figure 7. Variations in the mass fraction of sulphate (defined as $SO_4^{2-}/(SO_4^{2-} + HCO_3^-)$, where units are eq/l) in runoff from the Haut Glacier d'Arolla and Austre Brøggerbreen during both sampling seasons. The * denotes $^{\text{non-snowpack}}SO_4^{2-}$

during chemical weathering is less sensitive to changes in residence time and water to rock ratio and/or (c) $^{\text{non-snowpack}}SO_4^{2-}$ is largely confined to delayed flow, which makes a smaller proportional contribution to runoff as discharge increases. Figure 6 shows the inverse association of $^{\text{non-snowpack}}SO_4^{2-}$ with discharge and Figure 7 shows that the ratio of $^{\text{non-snowpack}}SO_4^{2-}/(^{\text{non-snowpack}}SO_4^{2-} + HCO_3^-)$ is inversely associated with discharge. Figures 6 and 7 show that runoff from Haut Glacier d'Arolla also exhibits similar variability, attributed to the existence of a distributed drainage system (see earlier). Our conclusion is therefore that if the provenance of SO_4^{2-} and HCO_3^- is similar at both Austre Brøggerbreen and the Haut Glacier

Table III. Linear regression of the anions with discharge for both sampling seasons at Austre Brøggerbreen. Early season high Cl^- values are not included

Sampling season	Species	Gradient	Intercept	Correlation coefficient	n
1991	Cl^-	+3·5	44	+0·166	18
1992	Cl^-	−7·1	55	−0·605	34
1991	HCO_3^-	2·2	270	0·031	22
1992	HCO_3^-	−0·29	230	−0·100	38
1991	SO_4^{2-}	−3	32	−0·346	18
1992	SO_4^{2-}	−3	24	−0·749	34
1991	non-snowpack SO_4^{2-}	−5·8	31	−0·505	22
1992	non-snowpack SO_4^{2-}	−2·5	19	−0·422	38

d'Arolla, then there is a distributed drainage system beneath Austre Brøggerbreen and delayed flow is a component of runoff from Austre Brøggerbreen. This assertion is supported by the increase in non-snowpack SO_4^{2-} during the period of recession flow (Julian days 186–192, 1991). Hence, there must be a connection between the distributed drainage system underlying the thicker ice and the glacier terminus, which possibly transmits delayed flow through the cold basal ice beneath the thinner ice (Dowdeswell and Drewry, 1989; Clarke and Blake, 1991; Vatne *et al.*, 1992) or within unfrozen/permeable subglacial till or sediment (Liestol, 1975; Kamb *et al.*, 1985; Clarke, 1987; Lauritzen, 1991).

Comparison of HCO_3^- and SO_4^{2-} concentrations, and their variation with discharge, at the Haut Glacier d'Arolla and Austre Brøggerbreen

The range of HCO_3^- concentrations encountered at both sites is similar (see Table I). Direct comparison of the SO_4^{2-} concentration of the glaciers is misleading, as a significant proportion of the SO_4^{2-} at Austre Brøggerbreen is derived from the snowpack. Comparison of SO_4^{2-} at Haut Glacier d'Arolla with non-snowpack SO_4^{2-} at Austre Brøggerbreen is more meaningful as both quantities are derived from chemical weathering. Table I shows that concentrations are higher at Haut Glacier d'Arolla, despite there being only trace amounts of SO_4^{2-} in the bedrock. This suggests that runoff at Haut Glacier d'Arolla contains a greater proportion of delayed flow than at Austre Brøggerbreen.

Comparison of the rate of change in the concentration of the anions per unit change in discharge may give some indication of dilution processes. At the Haut Glacier d'Arolla, Table II shows there is a strong negative association of concentration with discharge. This represents the dilution of delayed flow by quick-flow. At Austre Brøggerbreen there is a smaller change in concentration per unit discharge (cf. Tables II and III), which might indicate that (a) there is less difference between the composition of supraglacial and subglacial waters than at Haut Glacier d'Arolla or (b) there is little subglacial water to dilute. Given the potential reactivity of the bedrock, it is more likely that the latter is the case for SO_4^{2-} and given the relatively high turbidity of supraglacial meltwaters at Austre Brøggerbreen, the former is also true for HCO_3^-.

CONCLUSIONS

This paper has illustrated the utility of water chemistry variations as a predictor of the overall structure of a subglacial drainage system beneath a glacier whose subglacial drainage system was hitherto largely undetermined. Comparison of the concentrations of HCO_3^- and SO_4^{2-} (derived from chemical weathering) in runoff at the Haut Glacier d'Arolla and the Austre Brøggerbreen, and their variation with discharge suggests that a distributed drainage system underlies both glaciers and that there is a component of delayed flow in runoff from both glaciers. A central, but unsubstantiated, assumption is that the provenance of SO_4^{2-} is similar in both glacial drainage systems. It appears that delayed flow makes a bigger contribution to runoff at Haut Glacier d'Arolla than at Austre Brøggerbreen.

ACKNOWLEDGEMENTS

This work was supported by NERC via grants nos. GR3/7004 and GR3/8114, Fellowship No. GT5/F/AAPS/3 and Studentship Nos. GT4/88/AAPS/56 and GT4/90/AAPS/63. The assistance of C. T. Hill, M. J. Clark and J. O. Hagen during the course of this work was greatly appreciated. The manuscript was greatly improved following constructive comments by M. J. Sharp, K. S. Richards and two anonymous reviewers. Neil Young's "Love to burn" provided solace during revision of the text.

REFERENCES

Blatter, H. 1987. 'On the thermal regime of an Arctic valley glacier: a study of White Glacier, Axel Heiberg island, N.W.T., Canada', *J. Glaciol.*, **33**, 200–211.

Blatter, H. and Hutter, K. 1991. 'Polythermal conditions in Arctic glaciers', *J. Glaciol.*, **37**, 261–269.

Boulton, G. S. 1972. 'The role of thermal regime in glacial sedimentation' in Price, R. J. and Sugden, D. E. (Eds), *Polar Geomorphology. Inst. Br. Geogr. Spec. Publ.*, **4**, 1–19.

Brown, G. H. and Tranter, M. 1990. 'Hydrograph and chemograph separation of bulk meltwaters draining the Upper Arolla glacier, Valais, Switzerland', *IAHS Publ.*, **193**, 429–437.

Brown, G. H., Sharp, M. J., Tranter, M., Gurnell, A. M., and Nienow, P. N. 1994. 'The impact of post-mixing chemical reactions on the major ion chemistry of bulk meltwaters draining the Haut Glacier d'Arolla, Valais, Switzerland', *Hydrol. Process.*, **8**, 465–480.

Brown, G.H., Tranter, M. and Sharp, M. J. 1996. 'Experimental investigations of the weathering of suspended sediment by alpine glacial meltwater', *Hydrol. Process.*, **10**, 579–597.

Clarke, G. K. C. 1987. 'Subglacial till: a physical framework for its properties and processes', *J. Geophys. Res.*, **92** (B9), 9023–9036.

Clarke, G. K. C. and Blake, E. W. 1991. 'Geometric and thermal evolution of a surge-type glacier in its quiescent state: Trapridge Glacier, Yukon Territory, Canada, 1969–89', *J. Glaciol.*, **37**, 158–169.

Clarke, G. K. C., Collins, S. G., and Thompson, D. E. 1984. 'Flow, thermal structure and subglacial conditions of a surge-type glacier', *Can. J. Earth Sci.*, **21**, 232–240.

Collinis, D. N. 1978. 'Hydrology of an alpine glacier as indicated by the chemical composition of meltwater', *Z. Gletscherk. Glazialgeol.*, **13**, 219–238.

Collins, D. N. 1979. 'Quantitative determination of the subglacial hydrology of two alpine glaciers', *J. Glaciol.*, **23**, 347–361.

Dowdeswell, J. A. 1986. 'Drainage-basin characteristics of Nordaustlandet ice caps, Svalbard', *J. Glaciol.*, **32**, 31–38.

Dowedswell, J. A. and Drewry, D. J. 1989. 'The dynamics of Austfonna, Nordauslandet, Svalbard: surface velocities, mass balance, and subglacial melt water', *Ann. Glaciol.*, **12**, 37–45.

Dowdeswell, J. A., Hamilton, G. S., and Hagen, J. O. 1991. 'The duration of the active phase of surge-type glaciers: contrasts between Svalbard and other regions', *J. Glaciol.*, **37**, 388–400.

Gurnell, A. M., Hodgson, A., Clark, M. J., Bogen, J., Hagen, J. O., and Tranter, M. 1994. 'Water and sediment discharge from glacier basins: and arctic and alpine comparison', *IAHS Publ.*, **224**, 325–334.

Hagen, J. O., Korsen, O. M., and Vatne, G. 1991. 'Drainage pattern in a subpolar glacier: Broggerbreen, Svalbard' in Gjessing, Y., Hagen, J. O., Hassel, K. A., Sand, K., and Wold, B. (Eds), *Arctic Hydrology. Present and Future Tasks*. Norwegian Committee for Hydrology, Oslo. pp. 121–131.

Hallet, B. 1976. 'Deposits formed by subglacial precipitation of $CaCO_3$', *Geol. Soc. Am. Bull.*, **87**, 1003–1015.

Holland, H. D. 1978. *The Chemistry of Atmospheres and Oceans*, Wiley Interscience, New York.

Johannessen, M. and Henriksen, A. 1978. Chemistry of snow meltwater: changes in concentration during melting', *Wat. Resour. Res.*, **14**, 615–619.

Kamb, W. B. 1987. 'Glacier surge mechanism based on linked cavity configuration of the basal water conduit system', *J. Geophys. Res.*, **92** (B9), 9083–9100.

Kamb, W. B., Raymond, C. F., Harrison, W. D., Engelhardt, H., Echelmayer, K. A., Humphrey, N., Brugman, M. M., and Pfeffer, T. 1985. Glacier surge mechanism: 1982–83 surge of Varigated Glacier, Alaska, *Science*, **227**, 469–479.

Kostrzewski, A., Kaniecki, A., Kapusunsk, J., Klimczak, R., Stach, A., and Zwolinski, Z. 1989. 'The dynamics and rate of denudation of glaciated and non-glaciated catchments, central Spitsbergen', *Polish Polar Res.*, **10**, 317–367.

Lauritzen, S. E. 1991. 'Groundwater in cold climates: interaction between glacier and karst aquifers' in Gjessing, Y., Hagen, J. O., Hassel, K. A., Sand, K., and Wold, B. (Eds), *Arctic Hydrology. Present and Future Tasks*. Norwegian Committee for Hydrology, Oslo. pp. 139–144.

Liestol, O. 1975. 'Pingos, springs and permafrost in Spitsbergen', *Norsk Polarinst. Arbok*, **1975**, 7–29.

Liestol, O. 1988. 'The glaciers in the Kongsfjorden area, Spitsbergen', *Norsk Geogr. Tidsskr.*, **42**, 231–238.

Pulina, M. 1984. 'The effects of cryochemical processes in the glaciers and permafrost in Spitsbergen', *Polish Polar Res.*, **5**, 137–163.

Raiswell, R. 1984. 'Chemical models of solute acquisition by glacial melt waters', *J. Glaciol.*, **104**, 49–57.

Reynolds, R. C. and Johnson, N. M. 1972. 'Chemical weathering in the temperate glacial environment of the Northern Cascade Mountains', *Geochim. Cosmochim. Acta*, **36**, 537–554.

Sharp, M. J. 1991. 'Hydrological influences from meltwater quality data: the unfulfilled potential', *Proceedings of the 3rd British Hydrological Symposium*, Southampton, pp. 5.1–5.8.

Sharp, M. J., Richards, K., Willis, I., Arnold, N., Nienow, P., Lawson, W., and Tison, J. L. 1993a. 'Geometry, bed topography and drainage system structure of the Haut Glacier d'Arolla, Switzerland', *Earth Surf. Process. Landforms*, **18**, 557–571.

Sharp, M. J., Willis, I. C., Hubbard, B., Nielsen, M., Brown, G. H., Tranter, M., and Smart, C. C. 1993b. Water storage, drainage evolution and water quality in Alpine glacial environments', *Interim Rep. NERC Grant No. GR3/8114*.

Souchez, R. A. and Lemmens, M. M. 1987. 'Solutes' in Gurnell, A. M. and Clark, M. J. (Eds), *Glacio-fluvial Sediment Transfer*. Wiley, Chichester. pp. 285–303.

Sugden, D. E. and John, B. S. 1976. *Glaciers and Landscape. A Geomorphological Approach*. Edward Arnold, London.

Thomas, A. G. and Raiswell, R. 1984. 'Solute acquisition in glacial meltwaters. II. Argentiere (French Alps): bulk melt waters with open system characteristics', *J. Glaciol.*, **104**, 44–48.

Tranter, M. 1989. 'Controls on the composition of snowmelt' in Davies, T. D., Tranter, M., and Jones, H. G. (Eds), *Seasonal Snowpacks. Processes of Compositional Change. NATO ASI Series*, **G28**, 241–271.

Tranter, M. and Raiswell, R. 1991. 'The composition of the englacial and subglacial components in bulk meltwaters draining the Gornergletscher', *J. Glaciol.*, **125**, 59–66.

Tranter, M., Brown, G. H., Raiswell, R., Sharp, M. J., and Gurnell, A. M. 1993a. 'A conceptual model of solute acquisition by Alpine glacial meltwaters, *J. Glaciol.*, **39**, 573–581.

Tranter, M., Brown, G. H., and Sharp, M. J. 1993b. 'The use of sulphate as a tracer of delayed flow in Alpine glacier runoff', *IAHS Publ.*, **213**, 89–98.

Tranter, M., Brown, G. H., Hodson, A., Gurnell, A. M., and Sharp, M. J. 1994. Nitrate variations in glacial runoff from Alpine and sub-Arctic glaciers, *IAHS Publ.*, **223**, 299–311.

Vatne, G., Etzelmuller, B., Odegard, R., and Sollid, J. L. 1992. 'Glaciofluvial sediment transfer of a subpolar glacier, Erikbreen, Svalbard', *Stuttgarter Geograph. Studien*, **117**, S. 253–266.

6

IMPACT OF SUBGLACIAL GEOTHERMAL ACTIVITY ON MELTWATER QUALITY IN THE JÖKULSÁ Á SÓLHEIMASANDI SYSTEM, SOUTHERN ICELAND

D. M. LAWLER

School of Geography, The University of Birmingham, Edgbaston, Birmingham B15 2TT, UK
(Email: D.M.Lawler@bham.ac.uk)

H. BJÖRNSSON

Science Institute, University of Iceland, Dunhaga 5, 107 Reykjavík, Iceland

AND

M. DOLAN

Quality Department, North West Water, Dawson House, Great Sankey, Warrington WA5 3LW, UK

ABSTRACT

The influence of subglacial geothermal activity on the hydrochemistry of the Jökulsá á Sólheimasandi glacial meltwater river, south Iceland, is discussed. A radio echosounding and Global Positioning System survey of south-west Myrdalsjökull, the parent ice-cap of the valley glacier Sólheimajökull, establishes the geometry and position of a subglacial caldera. A cauldron in the ice-cap surface at the basin head is also defined, signifying one location of geothermally driven ablation processes. Background H_2S concentrations for the Jökulsá meltwaters in summer 1989 show that leakage of geothermal fluids into the glacial drainage network takes place throughout the melt season. Chemical geothermometry (Na^+/K^+ ratio) applied to the bulk meltwaters tentatively suggests that the subglacial geothermal area is a high-temperature field with a reservoir temperature of $\approx 289-304°C$. A major event of enhanced geothermal fluid injection was also detected. Against a background of an apparently warming geothermal reservoir, the event began on Julian day 205 (24 July) with a burst of subglacial seismic activity. Meltwater hydrochemical perturbations followed on day 209 and peaked on day 213, finally leading to a sudden and significant increase in flow on day 214. The hydrochemical excursions were characterized by strong peaks in meltwater H_2S, SO_4^{2-} and total carbonate concentrations, transient decreases in pH, small increases in Ca^{2+} and Mg^{2+} and sustained increases in electrical conductivity. The event may relate to temporary invigoration of the subglacial convective hydrothermal circulation, seismic disturbance of patterns of groundwater flow and geothermal fluid recruitment to the subglacial drainage network, or a cyclic 'sweeping out' of the geothermal zone by the annual wave of descending groundwater. Time lags between seismic events and meltwater electrical conductivity responses suggest mean and maximum intraglacial throughflow velocities of $0·032-0·132\,m\,s^{-1}$, respectively, consistent with a distributed drainage system beneath Sólheimajökull. Because increases in flow follow hydrochemical perturbations, the potential exists to use meltwater hydrochemistry to forecast geothermally driven flood events in such environments.

INTRODUCTION

Research problem

Meltwater hydrochemistry studies have proved extremely valuable in shedding light on hydrological pathways, water balances and storages, solute acquisition patterns, chemical weathering efficacy and solute yields in glacierized catchments (e.g. Collins, 1977; Raiswell, 1984; Raiswell and Thomas, 1984; Thomas and Raiswell, 1984; Gurnell and Fenn, 1985; Fenn, 1987; Souchez and Lemmens, 1987; Tranter and Raiswell, 1991; Fairchild *et al.*, 1994). However, a number of reservations have been voiced (e.g. Sharp,

1991; Brown *et al.*, 1994) and several research lacunae remain. Firstly, much of this research has focused on alpine regions and comparatively few studies from the very different environments of the subarctic have emerged. Secondly, most previous studies have concentrated on relatively small catchments. Thirdly, despite the large number of glacierized catchments which are geothermally influenced (e.g. in Iceland and the Pacific rim), their hydrochemistry is poorly understood. Often characterized by substantial subglacial geothermal heat fluxes capable of melting ice at the glacier sole, knowledge of the driving processes and their consequences in these environments is important for understanding subglacial water production and pressures, glacier sliding velocities, seasonal and long-term development of the glacial drainage network, certain types of jökulhlaup and the chemical evolution of bulk meltwaters.

The present work was inspired by the study of Sigvaldason (1963), who found geothermal products in the meltwaters of the Jökulsá á Sólheimasandi glacial river in south Iceland. His work raised the exciting possibility of linking subglacial geothermal activity with meltwater quality and subsequent discharge increases — thereby providing a potential hydrochemical basis for forecasting geothermally or volcanically driven jökulhlaups. Several of such connections have since been suggested in Iceland. For example, one or two days before a jökulhlaup was observed in the Skaftá meltwater river (north-west Vatnajökull), a sulphurous smell was noted locally (Björnsson, 1977: 77). Comparable features have been observed for the Skeidará river (south-west Vatnajökull), which carries geothermal products from the Grímsvötn subglacial lake (Björnsson, 1988: 88–89; Steinthórsson and Oskarsson, 1983: 77). Fenn and Ashwell (1985) recorded in the Kverkjökull stream in northern Vatnajökull a strong increase in meltwater temperature and electrical conductivity accompanied by a sulphurous smell, and suspected some geothermal influence. Meltwater and geothermal fluid chemical analyses were presented by Gíslason and Eugster (1987a; 1987b) to help infer system characteristics for sites in northern Iceland.

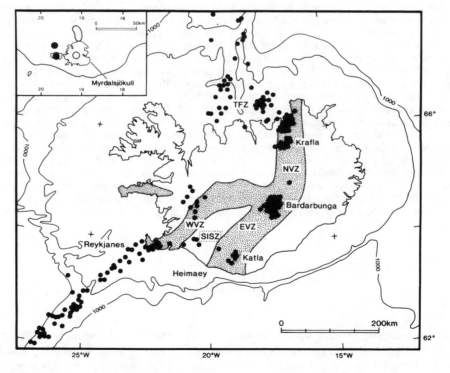

Figure 1. Vulcanicity and earthquake epicentres in Iceland (from Einarsson, 1991). NVZ, Northern Volcanic Zone; WVZ, Western Volcanic Zone; and EVZ, Eastern Volcanic Zone, which incorporates the Katla volcanic system. The WVZ and EVZ are bridged in the south by the South Iceland Seismic Zone (SISZ). The inset shows the Myrdalsjökull ice-cap and localized magnetic anomalies identified by Jonsson *et al.*, 1991. The one in the middle of Myrdalsjökull is a strongly negative magnetic anomaly which corresponds to the Katla caldera

Aims and objectives

The aim of this paper is to evaluate the impact of subglacial geothermal activity on the meltwater quality of the Jökulsá á Sólheimasandi (abbreviated to Jökulsá here) glacial river in southern Iceland. Specifically, the objectives are: (1) to define the catchment boundary conditions, including ice surface and subglacial topography, to indicate ice thicknesses and the position of subglacial geothermal areas and drainage divides; (2) to detect episodic subglacial melt season geothermal activity through a programme of continuous monitoring of meteorological conditions, meltwater discharge and electrical conductivity; (3) to make inferences about the nature of such subglacial geothermal activity using a more detailed bulk meltwater hydrochemical analysis; and (4) to estimate intraglacial throughflow velocities from an analysis of time lags between subglacial seismic events and subsequent meltwater hydrochemical perturbations.

STUDY AREA

Jökulsá á Sólheimasandi catchment

Iceland, with its accessible subarctic glacierized catchments and vigorous seismic, volcanic and geothermal activity, forms an ideal study area for this research. Vulcanicity in Iceland is related to its position on the Mid-Atlantic ridge and a distinctive inverted Y-shaped distribution of volcanic activity is evident (Figure 1). The Jökulsá glacierized catchment in southern Iceland (Figure 2) was chosen specifically because: (i) the basin is believed to include at least one geothermal area; (ii) Sigvaldason (1963) found geothermal indicators in the bulk meltwaters here; (iii) a fairly dense seismometric network exists in the area; (iv) the basin is reasonably large; (v) there is an Orkustofnun (Icelandic National Energy Authority) gauging and water sampling station (BGS; Figure 2) 4 km from the snout of the basin's valley

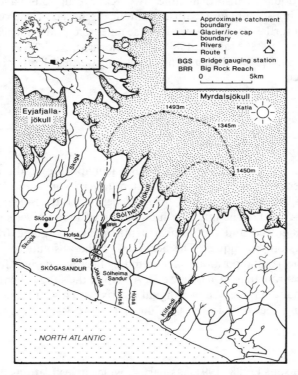

Figure 2. Jökulsá á Sólheimasandi catchment in southern Iceland, including south-west Myrdalsjökull, the valley glacier Sólheimajökull and the extraglacial mountains and sandur. Note the positions of the centre of the Katla subglacial volcanic zone, BRR water quality monitoring site and BGS gauging station

glacier, Sólheimajökull; and (vi) there is minimal human interference in the basin, so that water quality fluctuations should strongly reflect natural processes.

The Jökulsá basin consists of two main parts (Figure 2), namely the Myrdalsjökull parent ice-cap and the 8 km long advancing Sólheimajökull glacier which flows out from between two ice domes in a valley 1–2 km wide (Björnsson, 1979; Lawler, 1994a; 1994b). The drainage area at the bridge gauging station (BGS) is approximately 110 km^2, of which 78 km^2 (71%) is glacierized (Figure 2). Hydroclastic and acid volcanic rocks dominate the solid geology of the basin (Carswell, 1983). The extraglacial mountains (Figure 2) are composed of erodible palagonite tuffs and volcanic breccias (Sæmundsson, 1979; Tómasson, 1990). Soils are generally very thin on the terrace surfaces and hillslopes. The land is sparsely vegetated, largely with grasses and mosses. In this wettest part of Iceland, the annual precipitation (1931–1960) increases from ≈1600 mm at BGS (altitude 60 m) to over 4000 mm near the highest boundary of the catchment (1493 m). Mean January and July air temperatures are 0°C and 10–12°C, respectively [Eythorsson and Sigtryggsson, 1971 (cited in Björnsson (1979))]. Bankfull discharge at BGS is estimated to be 100 m^3 s^{-1} and peak runoff occurs on average in late July (Lawler, 1991; Lawler *et al.*, 1992).

Katla volcanic system

The Katla volcanic system underlying Myrdalsjökull (Figures 1 and 2) is one of the two most active volcanoes in Iceland (Björnsson, 1975). Two associated areas of seismic activity have been identified by Einarsson and Björnsson (1979): one in the south-east of Myrdalsjökull and one in the south-west, not too distant from Sólheimajökull (Figure 2). Katla is mapped as a high-temperature area in Fridleifsson (1979), which Thorarinsson (1967: 197) considers to be fed through a fissure system by a deep-seated layer of basalt magma. A subglacial caldera has been identified here from the presence of a strong negative magnetic anomaly surrounded by a positive field under Myrdalsjökull (Figure 1; Jonsson *et al.*, 1991). The Jökulsá á Sólheimasandi meltstream takes the local name *Fulilaekur* ('foul-smelling river'), which relates to its sulphurous odour. In historical times two Katla eruptions have occurred in the southern Myrdalsjökull caldera causing jökulhlaups with maximum discharge of the order of 10^5 m^3 s^{-1} to drain down Sólheimajökull, and a further 18 are known from sedimentological evidence to have drained eastwards (Thorarinsson, 1975; Jónsson, 1982; Maizels, 1989, 1991; Björnsson, 1992). Eruptions occur approximately every 40–60 years and there are presently concerns about an 'overdue' Katla eruption, which has provided extra impetus to this research.

METHODS

Definition of subglacial topography

The bedrock topography of south-west Myrdalsjökull was mapped by radio echosounding in May 1990 (Figure 3). The radio echosounder is a mono-pulse system (Sverrisson *et al.*, 1980; Björnsson, 1988) with the following specifications: duration of pulses, 0·2 μs; repetition rate, 1 kHz; antenna length, 30 m; and receivers bandpass filter, 2–5 MHz. The intensity modulation of the receiver's signal was recorded photographically, subsequently digitized and the vertical depth of ice computed from the digitized sounding records using general inversion techniques. The accuracy of the absolute ice thickness measurements is of the order of ±15 m. The horizontal position of the sounding lines was derived from Global Positioning System (GPS) surveys and is accurate to ±10 m (Figure 3). A map of the subglacial topography was constructed by interpolation between sounding lines using a digital matrix with a grid spacing of 200 × 200 m. Because the spacing between sounding lines is typically 500–1000 m (Figure 3), features smaller than 1 km across are not represented.

Field monitoring programme

For four summer melt seasons (1988–1991), as part of wider investigations into glaciofluvial solute and sediment dynamics here (e.g. Dolan, unpublished data; Lawler, 1991, 1994a; 1994b; Lawler *et al.*, 1992), we established a meltwater quality and quantity measurement programme. Discussion here concentrates on

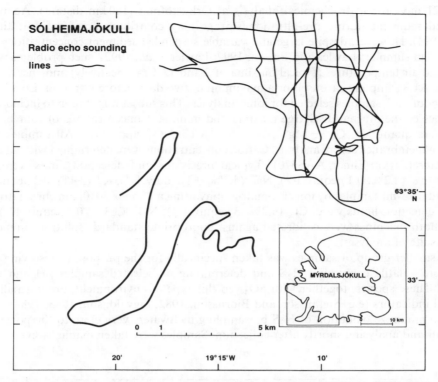

Figure 3. Sounding lines used in the radio echo survey of the upper basin of Sólheimajökull in May 1990

field season 1989 (5 June–1 September). The water quality monitoring station, 'Big Rock Reach' (BRR), was positioned on the left bank of the main Jökulsá meltwater channel, as near as logistically possible to the glacier portal (approximately 1 km downstream; Figure 2). Monitored variables included river stage, temperature, turbidity and electrical conductivity (EC). Information was recorded on Grant Instruments SQ8 eight-bit dataloggers scanning at frequencies $\leqslant 10$ minutes. Stage was later converted to discharge at BGS using time-specific rating curves based on salt dilution techniques, float gauging, current metering and channel geometry soundings (e.g. see Lawler *et al.*, 1992). Electrical conductivity was recorded in-stream using a Kent Instruments CM3 conductivity probe connected to a Walden Precision Apparatus CM25 conductivity meter. Electrical conductivity values are reported here in $\mu S\,cm^{-1}$ and are uncorrected for temperature changes, as is standard procedure (e.g. Collins, 1977; Fenn, 1987) for glacial meltwaters of low and near-constant temperature such as here (summer diurnal variation typically 0·5–2·5°C; Dolan, unpublished data). A Didcot Instruments automatic weather station (AWS) was used to monitor the meteorological variables relevant to glacial runoff production (solar and net radiation receipts; wind velocity and direction; wet and dry bulb and screen air temperatures; precipitation intensity). The AWS Campbell Scientific CR10 logger scanned at 10-second intervals and stored hourly means. The AWS was located on Skógasandur approximately 3 km to the south-west of the glacier snout at an altitude of $\approx 85\,m$ a.s.l.

Hydrochemistry: sampling and analytical methods

Hydrochemical analyses were performed on two groups of samples. The first group, for ionic determinations, were withdrawn by an automatic liquid sampler (ALS Mk 4B) installed at BRR (Figure 2). Throughout summer 1989, the device took $640 \times 600\,ml$ samples of river water at a 1-8 h frequency. Samples were filtered through Whatman 52 filter papers (nominal initial pore size 4·5 μm). A subset of filtrate samples was re-filtered through 0·45 μm membranes and subsamples of 100 ml were refrigerated at $\approx 5°C$ in clean plastic bottles for 1–21 days for hydrochemical analysis at Orkustofnun (160 km distant). Acquisition of ions from suspended sediment during pre-filtration storage may contribute noise to ionic time series (e.g. Brown *et al.*, 1994). However, despite large and variable suspended sediment concentrations here (typically 800–4000 mg l^{-1} in summer; Lawler and Brown, 1992; Lawler *et al.*, 1992), such problems were minimized by low water and air temperatures (typical maxima of 2 and 13°C, respectively) and ensuring that samples remained in the ALS sampling carousel for a maximum of two days before filtration. Logistics constrained to 47 the number of ALS samples selected for ionic analyses. This subset was chosen to include samples taken around the times of the diurnal discharge maxima and minima. Concentrations of four base cations and three anions were quantified: Ca^{2+}, Mg^{2+}, Na^+, K^+, Cl^-, NO_3^- and SO_4^{2-}. All samples were refiltered through 0·22 μm membranes before analysis. Cation concentrations were determined with an atomic absorption spectrometer (Perkin-Elmer AAS370). Typical precisions and detection limits were: Ca^{2+} ($\pm2\%$, 5 μequiv. l^{-1}); Mg^{2+} ($\pm2\%$, 0·1 μequiv. l^{-1}), Na^+ ($\pm4\%$, 0·4 μequiv. l^{-1}); K^+ ($\pm4\%$, 0·3 μequiv. l^{-1}). Anions were determined on 5-ml samples by ion chromatography using a Dionex 2010i ion chromatograph. Typical precisions and detection limits were: Cl^- ($\pm2\%$, 3 μequiv. l^{-1}); NO_3^- ($\pm3\%$, 0·2 μequiv. l^{-1}), SO_4^{2-} ($\pm4\%$, 2 μequiv. l^{-1}). Both machines were calibrated against appropriate standards solutions immediately before and after each suite of analyses.

The second, smaller group of samples was taken specifically for the purpose of resolving the concentration of the more volatile compound H_2S and determining on selected samples pH and total dissolved carbonate, C_T. These species, together with SO_4^{2-} in this type of environment, are normally taken as the key geothermal indicators (e.g. Stefánsson and Björnsson, 1982; Sigvaldason; 1963, 1981). The important aim here was to minimize oxidation of H_2S by sampling meltwaters very close to the portal, isolating the samples from air and analysing shortly after collection. Samples were taken manually every 1–3 days in the

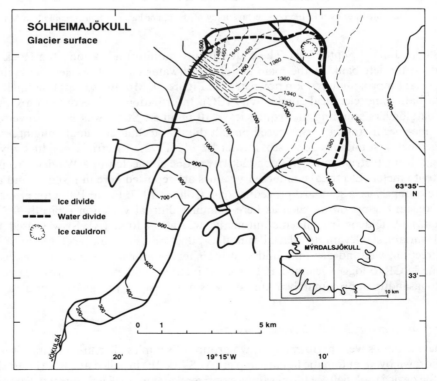

Figure 4. Surface map of Sólheimajökull, adapted from the US Defence Mapping Agency map of 1988. Ice catchment basin and the water drainage basin of Jökulsá á Sólheimasandi are also shown (see text). Note ice cauldron at the basin-head indicative of geothermal activity (upper right). Contours in metres

mid-afternoon (near the time of the diurnal maximum discharge) from the eastern (left) bank of the Jökulsá as near as possible to the Sólheimajökull portal (approximately 150 m downstream). Samples were collected using the standard method for geothermal fluids (e.g. Olafsson, 1988: 7–8; see also Ellis and Mahon, 1977: 166). River water was sucked via rubber tubing into a clean glass gas tube until the tube was full and all air bubbles had been excluded. The tube was then sealed by taps at both ends and taken back to the local Skógar base (Figure 2) for analysis within four hours of collection. Between 6 June and 17 August 1989, 32 such samples were taken from the portal for H_2S analyses, with a further four taken at BRR or BGS (Figure 2). This includes 12 samples used for additional determinations of C_T and pH.

Meltwater pH and temperature were determined in the laboratory on filtered samples using a portable pH meter calibrated by pH 4 and pH 7 buffers. The H_2S concentration was measured on 50-ml samples by titration to a red endpoint with 0.001 M mercury (II) acetate solution, $\alpha Hg(CH_3COO)_2$, using dithizone as indicator (Olafsson, 1988: 16–17; Armannsson et al., 1989). Precision was $\pm 5\%$ and the detection limit was $0.6 \, \mu mol \, l^{-1}$. Reported values are the means of two titrations on each sample. The importance of determining H_2S concentrations shortly after sample collection (Ellis and Mahon, 1977) is illustrated by the decrease in measured concentration obtained in analyses carried out on three gas-tube water samples taken from the same site within a few seconds of each other, but analysed 4, 8 and 10 h later: the respective H_2S values were 3.1, 2.8 and $1.8 \, \mu mol \, l^{-1}$.

C_T concentrations were determined to detect the presence of dissolved CO_2 in the bulk meltwaters (Arnórsson, 1979; Raiswell, 1984: 52–53; Olafsson, 1988), where, using Stumm and Morgan's (1970: 120) terminology

$$C_T = [H_2CO_3^*] + [HCO_3^-] + [CO_3^{2-}]$$

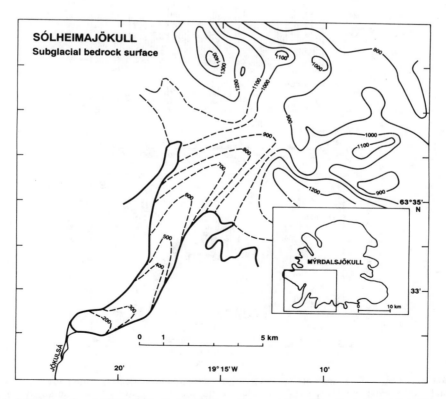

Figure 5. Subglacial topography of Sólheimajökull, showing breach in complex caldera lip (upper right). Contours in metres

$H_2CO_3^*$ is taken to represent both CO_2(aq.) and H_2CO_3, although the former species is much more abundant (Raiswell, 1984; Raiswell *et al.*, 1980: 24). In practice, the CO_3^{2-} form was absent from these waters, because sample pH values lay below 8·3. C_T analyses were carried out on 50-ml samples at approximately 20°C by acid titration with hydrochloric acid (0·1 M HCl) using automatic pipettes and the standard method of Olafsson (1988: 15–16) and Golterman and Clymo (1969: 145). Sample pH values were first increased to 8·2 by adding 0·1 M NaOH, before titrating to pH 3·8 (to convert most of the HCO_3^- to H_2CO_3). Precision was ±5%, and detection limit was $2\,\mu\mathrm{mol\,l^{-1}}$. In geothermal investigations (e.g. Ellis and Mahon, 1977; 63; Arnórsson, 1979), C_T values are normally expressed as equivalent CO_2 concentrations in ppm and derived thus (after Olafsson, 1988: 16)

$$\mathrm{ppm}\,C_T = (88 \times \mathrm{ml\,HCl}) - 7\cdot92 - (1\cdot182 \times \mathrm{ppm\,H_2S})$$

For this study, C_T values (the means of two titrations) are expressed as molar concentrations to allow easier comparison with other glacial hydrochemistry work (e.g. Raiswell, 1984).

RESULTS

Ice surface and subglacial topography

The 1990 GPS and echosounding survey work facilitates the drawing of the border of the ice catchment that drains ice towards the Jökulsá (Figure 4). The boundary of the ice drainage basin was drawn perpendicular to smoothed elevation contours. The location of the central flow divide of the ice-cap corresponds to the highest ice surface profile (Figure 4). The survey also reveals evidence of current geothermal

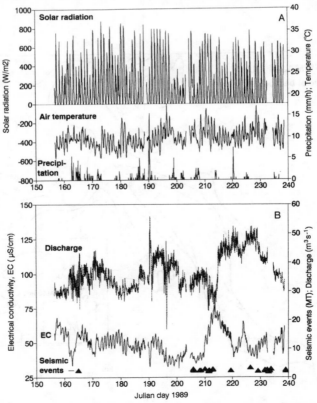

Figure 6. Time series of (A) basin meteorological variables recorded by the automatic weather station on Skógasandur and (B) seismic events beneath south-west Myrdalsjökull (M_T magnitude values plotted are based on average event durations: see Table I) and meltwater discharge and EC for the Jökulsá á Sólheimasandi glacial river in the 1989 melt season. Julian day 150 = 30 May

processes in the form of a small depression (ice cauldron), $\approx 1\,km^2$ in area, in the surface of the ice in the Sólheimajökull head area — one of a number located on the Myrdalsjökull ice-cap. Ice is flowing into the cauldron from a surrounding area of $1\cdot5\,km^2$ (Figure 4).

The new subglacial topographic map is shown in Figure 5. Below 1300 m glacier elevation, Sólheimajökull was inaccessible because of huge crevasses, so the bedrock topography in this area was sketched after estimating an average glacier thickness (h) from the glacier slope (α), assuming perfect plasticity and a basal yield stress of $\tau = 1$ bar, i.e. $h = (\tau/\rho g)/\sin\alpha$, where g is the gravitational acceleration and ρ is ice density. These new data clarify the position, size and geometry of a huge caldera that lies beneath Myrdalsjökull, the southern edge of which is shown in Figure 5. The caldera is 600–750 m deep and the highest rims rise to 1400 m a.s.l. Three outlet glaciers, including Sólheimajökull, have eroded 300–500 m breaches in the caldera rim (Figure 5). The southern part of the caldera floor, near the basin head of Sólheimajökull, is more rugged and elevated, suggesting a greater scale of subglacial vulcanism here. A small dome-shaped mountain is located beneath the ice cauldron, rising above 1000 m elevation and covered by ice 300 m thick (Figures 4 and 5). Maximum ice thickness in the basin head area is about 400 m.

The drainage divide of the Jökulsá basin on the glacier is drawn as a continuation of the divide outside the glacier (Figure 4). Beneath a glacier the water divide is located where there is no gradient in the potential driving water along the bed. This potential is the sum of gravitational component and the basal water pressure. The *actual* water divide may fluctuate inter- and intra-annually due to fluctuations in the subglacial

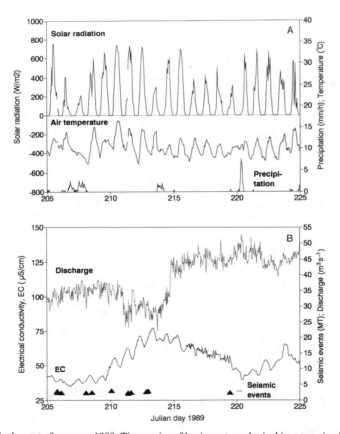

Figure 7. Major hydrochemical event of summer 1989. Time series of basin meteorological inputs, seismic events and meltwater quality and quantity for the Jökulsá á Sólheimasandi glacial river for Julian days 205–225 (24 July–13 August). (A) Solar radiation receipts, air temperature and precipitation from the automatic weather station on Skógasandur; and (B) meltwater discharge and electrical conductivity (EC) together with seismic events beneath south-west Myrdalsjökull (M_T magnitude values plotted are based on average event durations: see Table I)

water pressure. Here, however, we have taken the water pressure as equal to the ice overburden pressure (see Björnsson, 1988). This assumption is realistic close to the water divides where the flow velocity of water is low. The predicted outline of the water basin clearly encloses an ice cauldron (Figure 4).

Melt variables and EC signals

The Jökulsá basin displays many typical features of cool, oceanic, high-latitude environments in summer. These include air temperatures in the range 5–18°C, long 'day' lengths, solar radiation peaks up to 900 W m^{-2} and copious amounts of liquid precipitation (Figure 6A). The flow and EC signals in Figure 6B show two features which are classically associated with proglacial rivers: (a) the progressive lowering of EC through the melt season [to around day 200 (19 July) here] as discharge increases; and (b) diurnal EC cycling (e.g. see Collins, 1977; Fenn, 1987). The intention here, however, is to focus on the major late season peak in EC during a virtually rain-free period between Julian day 209 and 220 (28 July–8 August), which represents a clear departure from these two classical features (Figures 6 and 7). The peak EC value achieved (78 μS cm^{-1}) on Julian day 213 represents the summer maximum (Figure 6B). Also, the EC recession limb is characterized by numerous 'disturbance features' (Figures 6B and 7B).

Hydrochemical time series

Figure 8A shows a general seasonal decrease in Na$^+$ and Ca^{2+} concentrations over the summer period, in line with discharge-dilution effects. Generally, cation concentrations (in μequiv. l^{-1}) are such that:

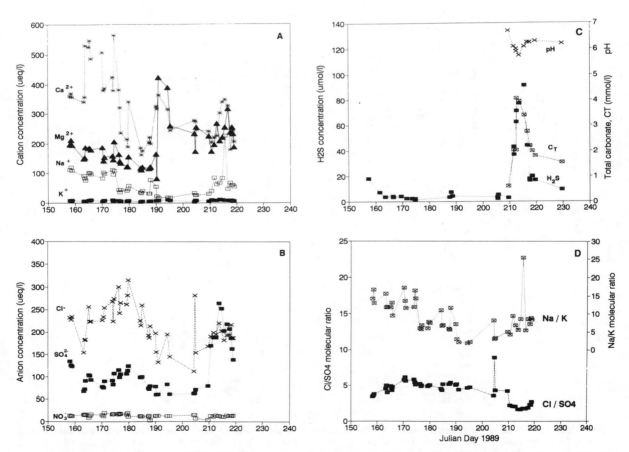

Figure 8. Hydrochemical time series for the 1989 melt season: (A) cations; (B) anions; (C) key geothermal indicators; and (D) molar ratios for Na$^+$/K$^+$ and Cl$^-$/SO$_4^{2-}$. Julian day 150 = 30 May

$Ca^{2+} > Mg^{2+} > Na^+ > K^+$ (calcium is usually the dominant cation in glacial meltwaters; Raiswell, 1984). From Julian day 190, however, Mg^{2+} concentrations tend to equal or exceed Ca^{2+} concentrations (Figure 8A). The late season event is characterized by significant increases in Ca^{2+} and Mg^{2+} and a smaller increase in Na^+ (Figure 8A). NO_3^- is relatively invariant, but Cl^- and SO_4^{2-} concentrations show two step increases on Julian days 210 and 213 (Figure 8B). The strong SO_4^{2-} peak on day 213 represents the melt season maximum (Figure 8B). C_T also rises and falls dramatically at this time, whereas the pH initially decreases from 6·7 to 5·75 before largely recovering (Figure 8C). These are relatively low pH values by glacial meltwater standards (e.g. see Raiswell, 1984, his figure 1; Raiswell and Thomas, 1984; Thomas and Raiswell, 1984). H_2S concentrations rise from background summer levels of around $5-8\,\mu mol\,l^{-1}$ to a maximum of $92\,\mu mol\,l^{-1}$, before decreasing very abruptly (Figure 8C). Other departures are evident. Note that SO_4^{2-} and Cl^- concentrations positively covary, although Cl^- is the more responsive (Figure 8B). However, during the late season event, meltwater SO_4^{2-} concentrations increase strongly with respect to a relatively conservative Cl^- value (Figure 9A). Pre-event SO_4^{2-} concentrations are also inversely related to meltwater discharge (Figure 9B). During the hydrochemical perturbation, however, we observed significant departures from, and strongly clockwise hysteretic effects within, this relation. Note the *increase* in SO_4^{2-} values when discharge increases to day 210 and the sustained high concentrations as discharge subsequently falls then rises rapidly from day 213 (Figure 9B). A similar clockwise hysteresis loop can be seen for the H_2S–discharge relationship (Figure 9C).

DISCUSSION

Interpretation of water quality perturbations

We interpret this late season hydrochemical perturbation as a 'geothermal event', in which solute-rich geothermal fluids were injected into the basal drainage system of Sólheimajökull and thence to the main proglacial channel. The H_2S peak (Figure 8C) is especially illuminating as this is usually taken as the key geothermal indicator (e.g. Sigvaldason, 1963), with a magma body assumed to provide the source of sulphur (e.g. Arnórsson, 1974; Armannsson *et al.*, 1989). H_2S is a volcanic gas and increased concentrations of dissolved H_2S and CO_2 in geothermal fluids are commonly taken to represent enhanced magmatic activity (Armannsson *et al.*, 1989). The pronounced H_2S spike (Figure 8C) is thus likely to represent degassing of a magma reservoir (e.g. Sigvaldason, 1981). It is unlikely simply to reflect a release of englacially or subglacially stored water rich in sulphurous compounds, because H_2S is very volatile, oxidizing easily to sulphate. Indeed, the parallel SO_4^{2-} increase (Figure 8B) probably reflects oxidation of some of the H_2S supplied by geothermal processes and release of magmatic SO_2 into the hydrothermal system (Stefánsson and Björnsson, 1982: 127). This is reinforced by the sharp decrease at this time in the Cl^-/SO_4^{2-} molar ratio (Figure 8D), low values of which are often indicative of near-surface oxidation of newly supplied H_2S to SO_4^{2-} in geothermally influenced environments (Ellis and Mahon, 1977: 66).

Similarly, the strong peak in total carbonate (Figure 8C) is interpreted as a substantial injection of geothermal CO_2 into the subglacial drainage system. The situation here in which C_T responds and peaks slightly ahead of H_2S concentrations (Figure 8C) mirrors events in the Krafla geothermal field in northern Iceland (Figure 1), when such chemical changes in the geothermal fluid were known to be associated with magmatic activity and a rifting episode (Stefánsson and Björnsson, 1982: 127). The substantial decrease in meltwater pH at this time (Figure 8C) is a likely response to the dissolution of geothermally derived acidic species H_2S and CO_2 within circulating groundwaters, and consequent release of H^+ ions (see later).

The pattern for cations is more complex. The subsidiary peaks in the concentration of Ca^{2+}, Mg^{2+} and Na^+ at this time (Figure 8A) may partly relate to enhanced subglacial solute acquisition driven by increases in protons supplied during the injection of geothermal fluids into the subglacial drainage network. Increased levels of dissolved CO_2 and decreasing pH values usually imply vigorous weathering rates if supplies of reactive particulate debris abound (Raiswell, 1984; Sharp, 1991); this is very likely at the glacier sole here, given the very high meltwater suspended sediment concentrations (Tómasson, 1990; Lawler, 1991; 1994a). Interestingly, the initial increase in Cl^- on day 205 (24 July), followed by pulses of Na^+

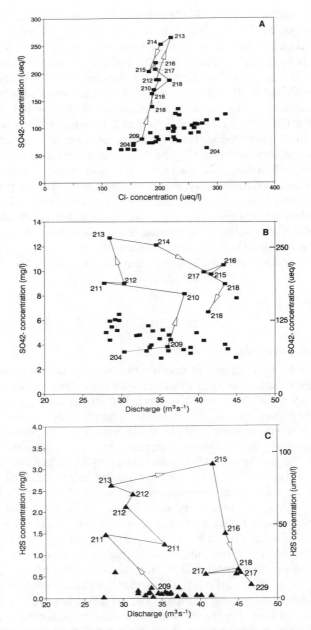

Figure 9. Excursions in the behaviour of the concentration of key geothermal indicators during the geothermal event, relative to the 1989 melt season as a whole (numerals indicate Julian day of sample collection during the geothermal event). (A) Radical SO_4^{2-} departures with respect to chloride; (B) excursions and hysteresis in the SO_4^{2-}–discharge relation; and (C) excursion and hysteresis in the H_2S–discharge relation

and C_T on days 212 and 216 when the flow increases (Figure 8), mirrors at a smaller scale features of the geothermally driven jökulhlaup in the Skeidará river (see Steinthórsson and Oskarsson, 1983: 77).

Links between the subglacial geothermal system and meltwater quality

We can qualify the processes involved using geothermal and chemical theory. Glacial meltwater is free to percolate down through faults, dykes and fissures in porous rocks towards the geothermally active zone

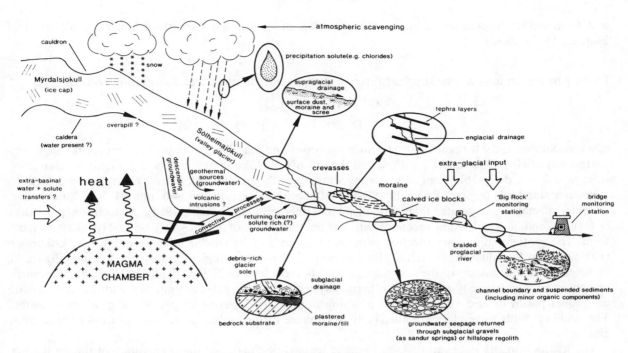

Figure 10. Schematic representation of the solute acquisition and delivery system for the Jökulsá á Sólheimasandi basin

around Katla (e.g. Figure 10). Percolation is especially probable underneath ice cauldrons. Water is then heated at depth and heat energy and mass are transferred upwards towards the subglacial area by hydro-thermal convection (see Björnsson *et al.*, 1982). It is possible that late season hydrochemical perturbations recur each summer and simply reflect the 'arrival' of a pulse of circulating melt-generated groundwater which has first descended and 'swept' the geothermal area before being returned — chemically enriched — to the subglacial drainage network. However, this is unlikely to be the only process, as dispersion phenomena are likely to have resulted in much less pronounced, but more sustained, EC spikes than those in evidence here (Figure 6B).

The nature and location of the heat source is important in driving geothermal circulations, of whatever type. Two-dimensional undershooting geophysical surveys by Gudmundsson *et al.* (1994: 277) have shown that the Katla system consists of a magma chamber containing 'excessively molten rock'. The chamber appears to be ellipsoidal in shape, approximately 5 km in diameter north–south (4 km east–west), 1 km thick, 10 km^3 in volume and shallow, with its top about 1·5 km below the ice-cap sole. Thin, dense, crystalline areas are thought to border the magma chamber (to the south especially) and are interpreted as intrusives, whereas rhyolitic sills are hypothesized above the chamber proper (Gudmundsson *et al.*, 1994: 288). Indeed, Fridleifsson (1979) argues that central volcanoes of the Katla type have large numbers of such intrusions and rifting structures and these may increase groundwater hydraulic conductivities.

Geothermal waters in volcanic areas acquire solutes through the hydrothermal alteration processes of hydration, carbonation and sulphide formation (Ellis and Mahon, 1977: 91; Fournier, 1981), and especially wallrock leaching (Sigvaldason, 1981). Volcanic gases, especially H_2S and CO_2, are also usually dissolved in the circulating water. The warm solute-rich fluids mix with fresher, 'recently' infiltrated, water nearer the surface (Figure 10), which renews proton supplies and lowers pH. For example, acidification occurs from dissolution of H_2S in groundwater, i.e.

$$H_2O + H_2S \rightarrow H_3O^+ + HS^-$$

$$H_2O + HS^- \rightarrow H_3O^+ + S^{2-}$$

and from oxidation reactions observable at the top of geothermal circulation systems (Gíslason and Eugster, 1987a: 2838)

$$H_2S + 2O_2 \rightarrow 2H^+ + SO_4^{2-}$$

Further hydrogen ions can be supplied through carbonate dissolution and dissociation reactions

$$CO_2 + H_2O \rightarrow H_2CO_3$$

$$H_2CO_3 \leftrightarrow H^+ + HCO_3^-$$

(see also Stumm and Morgan, 1970: 71). Such processes and reactions probably account for the high concentrations of H_2S, SO_4^{2-} and C_T within, and the low pH of, Jökulsá meltwaters (e.g. Figure 8), characteristics which are detectable even after strong dilution with surface meltwaters. C_T values are high by meltwater standards (e.g. see Raiswell, 1984) and the peak value achieved ($4\cdot07$ mmol l^{-1}) is, for example, an order of magnitude greater than at Fjallsjökull (Raiswell and Thomas, 1984). At such low pH values ($<6\cdot4$), carbonate equilibrium reactions will ensure that most of the C_T will be in H_2CO_3 [CO_2(aq.)] form. The geothermal and acidification processes may be broadly similar to those inferred by Oskarsson (1978; cited in Sigvaldason, 1981: 183) for the Námafjall/Krafla complex in northern Iceland (Figure 1), in which 'intense degassing of the magma, as it intrudes to a level at 2–3 km depth in the crust, temporarily exceeds the buffer capacity of the hydrothermal system. Reaction rates are not fast enough to cope with occasional pulses of acid volcanic gases resulting in a drastic lowering of pH of the geothermal water'. The acidified fluids themselves are also likely to enhance further solute acquisition processes beneath the glacier.

The hydrochemistry data can be used to make tentative estimates of the *temperature* of the geothermal reservoir, using standard chemical geothermometric techniques (e.g. Fournier, 1981). These are based on the assumption — now validated by theory and experimental rock–water interaction studies — that for temperatures less than around 300°C most dissolved elements in geothermal fluids are in solution equilibrium 'with a mineral assemblage stable at the respective environmental conditions' (Sigvaldason 1981: 183). One commonly used example is the Na–K geothermometer, which 'is based on the temperature dependence of the partitioning of sodium and potassium between solution and alkali feldspars' (Arnórsson, 1979: 199). The logarithm of the Na^+/K^+ ratio of the fluid (in units of mg kg^{-1}; Fournier, 1981: 114) has been found via comparisons with direct measurements to vary inversely with temperature. For waters deriving from high-temperature areas, the Na^+/K^+ method 'generally gives excellent results ... and is less affected by dilution and steam separation than other commonly used geothermometers, provided there is little Na^+ and K^+ in the diluting water compared to the reservoir water' (Fournier, 1981: 119) — and this is likely to be the case in glacierized basins. The preliminary temperature estimates in the following have not been corrected for precipitation-derived sodium or for any complex mixing with cold waters (see also Steinthórsson and Oskarsson, 1983). Meteoric sodium is a *potentially* significant source of error, because the ice-cap is less than 20 km from the North Atlantic coastline. However, initial inspection of subdued Na^+ responses in the bulk meltwaters after rainfall events in summer (Figures 6A and 8A) and low total dissolved solids values for snow and ice samples (Lawler, unpublished data), suggest that results may not be unduly influenced by this. Moreover, molar Na/K ratios (Figure 8D) are in the typical range for geothermal systems (e.g. Ellis and Mahon, 1977: 65). In this instance, we assume dilution of geothermal fluids with pure (melt) water, as Steinthórsson and Oskarsson (1983) did, and limited modification of the Na^+/K^+ ratio as the waters pass through the glacial drainage network.

We used three geothermometric models to predict geothermal reservoir temperature, $T_{Na/K}$ (°C), as proposed, respectively, by Fournier (1981), Arnórsson (1980) (cited in Steinthórsson and Oskarsson, 1983) and Truesdell (cited in Fournier, 1981: 114). These are

$$T_{Na/K} = \{1217/[\log(Na^+/K^+) + 1\cdot483]\} - 273\cdot15 \tag{1}$$

$$T_{Na/K} = \{933/[\log(Na^+/K^+) + 0\cdot993]\} - 273\cdot15 \tag{2}$$

$$T_{Na/K} = \{855\cdot6/[\log(Na^+/K^+) + 0\cdot8573]\} - 273\cdot15 \tag{3}$$

The models were applied to the 47 individual Jökulsá meltwater analyses for 1989 (Figure 8A) and the temperature values averaged. The Fournier (1981) model (1) gives the most consistent results, with a mean reservoir temperature of 304°C (s.d. 77°C). Arnórsson's (1980) formula (2) yields a mean temperature of 308°C (s.d. 106°C), whereas the Truesdell equation (3) returns a mean $T_{Na/K}$ estimate of 311°C (s.d. 119°C). To adjust for slight skewness in the distributions, a median value for the Fournier (1981) estimates was also calculated ($T_{Na/K} = 289$°C). This indicates that the contributing geothermal reservoir beneath the Sólheimajökull basin head is a high-temperature system [see Bodvarsson's (1961) classification, cited in Stefánsson and Björnsson (1982: 125)]. This is consistent with its location in the active volcanic zone (Pálmason, 1974).

Some of the variability in $T_{Na/K}$ values reflects sensitivity of the calculations to the low absolute concentrations of Na^+ and K^+ involved (Figure 8A) and it does appear that some $T_{Na/K}$ values in mid-summer are overestimated (cf. Ellis and Mahon, 1977: 103). There is a suggestion, however, of systematic temporal change in the Na^+/K^+ ratio, with the reservoir temperature apparently increasing steadily throughout the melt season to Julian day 190–205 (Figure 8D). This may simply represent shifts in the main source of groundwater within the hydrothermal circulation, i.e. later in the melt season, the 'wave' of descending groundwater has penetrated to deeper (and therefore hotter) parts of the geothermal reservoir before it is returned to the glacial drainage system. It could simply mean a decreasing supply of Na^+ from other sources. The pattern could also imply that the geothermal system itself heats up (and therefore reinvigorates convectively) towards mid-July (Figure 8D). Certainly, the strong peak in C_T (Figure 8C) is consistent with higher temperatures and enhanced convective instability in geothermal systems (Stefánsson and Björnsson, 1982).

It is feasible, therefore, that hydrothermal circulation change has contributed to enhanced geothermal fluid delivery to the subglacial drainage network first detected hydrochemically on day 209. Moreover, the subsequent increase in flow-rate on day 214 — in the absence of high air temperatures, strong solar radiation or *heavy* rainfall inputs (Figures 6A and 7A) — may be a delayed response to this increase in subglacial heat production. The rapidity of the increase in discharge could also suggest emptying of a subglacial reservoir. Such a reservoir might be associated with the basin-head cauldron (Figure 4) or the subglacial lake plotted in this area by Björnsson (1992). However, during the event, Jökulsá meltwater was much less chemically enriched than, for example, the jökulhlaup waters from the Grímsvötn cauldron beneath Vatnajökull (see Steinthórsson and Oskarsson, 1983; Björnsson and Kristmannsdóttir, 1984). This may mean that geothermal fluids are leaked fairly continuously into, and evacuated out of, the Jökulsá drainage system — as shown also by the presence of H_2S in the bulk meltwaters throughout the melt season (Figure 8C) and indeed throughout the year (Sigvaldason, 1963) — and do not accumulate to any *great* extent in the breached subglacial caldera complex (Figure 5). Also, the fact that Dolan (unpublished data) did not find any temperature changes in the Jökulsá river during the event (cf. Fenn and Ashwell, 1985) could be indicative of the low volume of geothermal fluid additions relative to melt inputs.

Relationship of water quality perturbations with seismic events

There is, moreover, a degree of correspondence between some of the pronounced water quality fluctuations and some of the seismic events recorded in the Sólheimajökull basin-head area. Figures 6B and 7B show that six significant seismic events occurred over Julian days 205–206, including a Richter scale 3·3 event at 2000 h GMT on day 205. A further two seismic events occurred on Julian day 208 and, a little over 24 h later, the EC in the Jökulsá started its sharp increase (Figures 6B and 7B). More local earthquakes were recorded on days 210–211, but the sequence abruptly ends with two moderate events late on day 212 (31 July), just before the EC reaches its peak (Figure 7B). No further seismic activity was then recorded until days 219 and 226, and both episodes appear to have been followed by significantly increased and disturbed EC levels (Figure 6B).

It is known that seismicity is highly seasonal in south-west Myrdalsjökull (Brandsdóttir and Einarsson, 1992; Gudmundsson et al., 1994), with peak activity in the second half of the year, beginning around July. This is possibly associated with seasonal crustal pore water pressure changes, related to summer glacial meltwater inputs which reduce the frictional strength of faults (e.g. Brandsdóttir and Einarsson, 1992; Einarsson,

1991: 275). Some argue, however, that a direct ice unloading effect associated with melt-driven loss of mass in summer is responsible (Tryggvason, 1973; Einarsson and Björnsson, 1979; Ragnar Stefánsson, pers. comm.). It is significant that the large hydrochemical perturbation characterized by substantial changes in the key geothermal indicators occurred immediately after the start of the main period of annual seismicity (Figures 6–8).

Seismic events in Iceland can be associated with magmatic intrusive (inflationary) activity (Brandsdóttir and Einarsson, 1992), periodic refilling of magma reservoirs (Sigvaldason, 1981), dyke injection (Stefánsson and Björnsson, 1982), deflationary events related to magma withdrawal (e.g. as at Krafla; Stefánsson and Björnsson, 1982: 130) and as suspected in south-east Myrdalsjökull (Einarsson, 1991: 275), degassing of magma, crustal movement along fissures and fault boundaries, unloading expansionary effects, or transient geothermal heat production episodes. Many of these phenomena could cause physical disruption, increased convective activity or pressure surges within the subglacial groundwater and hydrothermal systems. This is likely to increase the dissolution of evolved gases within circulating groundwaters [e.g. as in the 1992 Lander earthquake in California (Linde, pers. comm.); see also Darling and Armannson (1989)], enhance solute acquisition and encourage the upward migration of enriched waters to the glacial drainage system. Interesting feedback mechanisms may therefore be in operation here. For example, if seismic events lead to the injection of warm fluids into the subglacial area, these may enhance rates of basal ice melt and groundwater percolation, causing increased pore water pressures in the subglacial crust, and increase the likelihood of further seismic events. In summary, therefore, the sequence of events appears to be (without any presumption of direct causality): geothermal reservoir 'warming' → seismic burst → meltwater hydrochemical perturbation → increase in meltwater flow.

System response dynamics: estimation of intraglacial throughflow velocities

If seismic events result in pulsed solute injection into the subglacial drainage network, it is possible to estimate intraglacial throughflow velocities by timing the resultant hydrochemical perturbations downstream at BRR monitoring station (Figure 2). In this sense, each seismic event is analogous to a tracing experiment, except that the *precise* slug injection details (e.g. location, time, volume and concentration) are uncertain. We take here the injection location and time as the position and moment of the seismic event and ignore the possibility of elapsed time involved in any intervening groundwater solute transport. Thus *intraglacial* velocities may be underestimated, but errors may be partly offset by inclusion in the calculations of the 'fast' open channel section (1 km) between portal and BRR monitoring station (Figure 2). Maximum and mean intraglacial throughflow velocities, $U_{i\max}$ and $U_{i\mathrm{mean}}$, respectively, are estimated as follows

$$U_{i\max} = D/(t_{\mathrm{ECinit}} - t_{\mathrm{s}}) \tag{4}$$

$$U_{i\mathrm{mean}} = D/(t_{\mathrm{ECpeak}} - t_{\mathrm{s}}) \tag{5}$$

where D is the curvilinear distance between the monitoring station and the site of the seismic event, measured along the dog-legged valley axis [cf. the straight line distance of Willis *et al.* (1990) and Fountain (1993)]; t_{s} is the time of the seismic event; and t_{ECinit} and t_{ECpeak} are, respectively, the time of initial EC response and time of EC peak at BRR. [Slight asymmetry in the EC spike, as a result of a steeper rising limb (Figure 7B), produced only small differences between t_{ECpeak} and the time centroid of the EC chemograph which could have been used to produce $U_{i\mathrm{mean}}$ estimates.] The use of time to peak concentration to estimate mean velocities is the normal procedure for glaciological dye tracing investigations (e.g. Willis *et al.*, 1990; Fountain, 1993: 143). Earthquake origin times (t_{s}) are published to the nearest 0·01 s. Earthquake positions, calculated using a single crustal model for the whole of Iceland, are considered to be accurate to within 3–5 km of the published horizontal location, but depths are unknown at present (Brandsdóttir, pers. comm). For those seismic events plotting just outside the basin, additional velocity estimates are produced by taking the value of D in Equations (4) and (5) to be the valley distance between BRR and the point on the water divide nearest the earthquake site.

Table I. Estimates of intraglacial through-flow velocities from meltwater EC responses to basin-head seismic events. For each earthquake, M_L is the Richter magnitude and M_T is an index of magnitude based on event durations, averaged from a number of seismometric stations, which is occasionally more reliable than M_L values. BRR is Big Rock Reach water quality monitoring station. $U_{i\text{max}}$ and $U_{i\text{mean}}$ are maximum and mean intraglacial through-flow velocities, respectively. GMT is identical to Icelandic local time

Seismic event date	Time (GMT)	Julian day	Latitude (degs)	Longitude (degs)	Richter magnitude M_L	Duration M_T	Distance (m): BRR to event	BRR to divide	EC response (Julian day) Start	Peak	Lag (days) to Start of EC rise	EC peak	Intraglacial throughflow velocities (m s^{-1}) BRR to event location $U_{i\text{max}}$	$U_{i\text{mean}}$	BRR to basin divide $U_{i\text{max}}$	$U_{i\text{mean}}$
24 July 1989	20·00	205·83	63·62	19·07	3·3	2·9	22 500	16 925	209·622	213·372	3·789	7·539	0·069	0·035	0·052	0·026
27 July 1989	01·43	208·07	63·60	19·14	?	2·4	17 625	—	209·622	213·372	1·550	5·300	0·132	0·038	—	—

A cluster of five seismic events took place just outside the notional basin divide on Julian day 205 (24 July) and the EC responded and peaked, respectively, 3·79 and 7·54 days later (Figure 7B; Table I). These time lags translate to maximum and mean intraglacial velocities of 0·069 and 0·035 m s^{-1}, respectively (Table I). If, instead, the proximal seismic event of the two occurring on day 208 is taken to 'produce' the EC spike (Figure 7B), then $U_{i\text{max}}$ values almost double to 0·132 m s^{-1}, although mean velocities are largely unaffected (Table I). These $U_{i\text{mean}}$ estimates are within the range of velocities inferred for distributed subglacial hydrological networks or linked cavity systems (e.g. Iken and Bindschadler, 1986; Willis *et al.*, 1990; Fountain, 1993: 145). A distributed system is not inconsistent with Lawler's (1991) observation of high catchment solute yields here (700 t km^{-2} a^{-1}), which tend to reflect long meltwater residence times (e.g. Tranter *et al.*, 1993). This inference is also consistent with the disturbance features noted on the recessional limb of the EC spike (Figure 7B), which may be longer time-scale analogues of dispersion phenomena often observed in artificial tracing experiments. For example, the multiple tracer peaks observed in stream 2 of South Cascade Glacier were interpreted by Fountain (1993: 155) to indicate a 'distributed flow system with multiple preferred flow paths'.

Solute sources and pathways in the Jökulsá á Sólheimasandi catchment

Figure 10 summarizes the range of possible solute acquisition sites and delivery pathways envisaged for the Jökulsá basin. Quantifying the relative dominance of each is beyond the scope of the present study and, in particular, further work on meltwater chemical evolution, solute sourcing and spatially distributed load calculation is needed before we can confirm that geothermal components play a dominant part in *annual* solute *budgets* here. High solute loads here (Lawler, 1991) are believed to reflect a range of catchment attributes (Figure 10), including: (i) a debris-rich advancing glacier; (ii) a highly crevassed ice surface which encourages copious englacial and subglacial drainage, thereby maximizing contact between meltwater and reactive englacial debris, including tephra; (iii) a distributed subglacial hydrological system conducive to long meltwater residence times; (iv) significant in-channel solute acquisition in the subglacial and proglacial zones related to high suspended sediment concentrations (e.g. see Brown *et al.*, 1994); (v) vigorous subglacial solute acquisition rates influenced by substantial proton supplies delivered by groundwater enriched with geothermal fluids and volcanic gases (especially CO_2 and H_2S) which are likely to maintain open system weathering conditions (see Raiswell, 1984); and (vi) significant all-year round river flow and solute transport (Lawler, 1991), probably assisted by geothermal heat flux in winter and substantial groundwater flows.

CONCLUSIONS

The following conclusions can be drawn from this study.

1. A new echosounding and GPS survey has revealed considerable fresh detail of the scale, geometry and position of the large subglacial caldera in the basin-head area of Sólheimajökull and firmly places at least one ice surface cauldron within the Jökulsá catchment. This represents strong evidence of localized, geothermally driven, subglacial melting.

2. Current geothermal activity is also demonstrated through data on geothermal fluid additions in the meltwater quality signal. In particular, H_2S is detectable throughout the melt season in the bulk meltwaters. One specific period was also marked by dramatic increases in meltwater EC, and SO_4^{2-} and C_T concentrations, and decreases in pH. Hydrochemical profiles are consistent with the dissolution in meltwater of the volcanic gases H_2S and CO_2, and these are probably the main source of hydrogen ions during such geothermal events. A vigorous circulation system is envisaged in which percolating glacial meltwaters descend to depth, where they are heated in contact with hot rocks and acquire gases and solutes. Enriched waters are then returned by convective processes to the subglacial drainage system, mixing with cold, dilute, groundwater en route. Further solutes may be acquired as fresh supplies of reactive debris and protons become available in the subglacial and proglacial channels. Chemical geothermometric techniques suggest a mean geothermal reservoir temperature of ≈289–304°C, tentatively confirming its suspected high-temperature status.

3. More specifically, there is some evidence of the links between seismicity and subsequent meltwater hydrochemical response. After a long period of seismic quiescence, clusters of subglacial earthquakes were followed a few days later by strong hydrochemical excursions and a sudden meltwater flow increase. A possible warming of the geothermal reservoir may have preceded the seismic burst. The prior seismic and/or geothermal activity is thought to be associated with physical or thermal disruption of the subglacial hydrothermal circulation which enhances injection of solute-rich fluids into the basal glacial drainage network. Moreover, time lags between seismic events and EC responses suggested mean and maximum intraglacial throughflow velocities of $0.035-0.132 \, \mathrm{m \, s^{-1}}$, respectively, values consistent with a distributed glacial drainage network.

The following implications arise from these findings.

1. The potential exists to use meltwater EC variation to forecast volcanically and geothermally driven floods, because distinctive EC spikes tend to lag seismic events and lead discharge increases.
2. Subglacial areas have often been considered as 'closed systems' (cf. Raiswell, 1984; Tranter *et al.*, 1993) because they are isolated from fresh *atmospheric* CO_2 supplies to replace those consumed in weathering reactions (Souchez and Lemmens, 1987: 297). In geothermally influenced subglacial systems, however, copious supplies of *volcanically derived* CO_2 can be available from below. Indeed, the application of solute acquisition models in such environments may have to recognize the probability that the dominant source of H^+ ions necessary for weathering reactions to proceed will derive from the dissolution of CO_2 and H_2s released during subterranean magmatic activity, which help to reinitiate 'open system' tendencies conducive to vigorous solute acquisition rates. This reinforces the need highlighted by Tranter *et al.* (1993) for a greater focus on dissolved gases in glacial hydrochemical investigations.
3. The study has generated a number of further questions which need addressing. Firstly, simultaneous monitoring of subglacial heat flux, (glacio)hydrological processes and hydrochemical perturbations over longer time-scales would help to explain the precise impact of volcanic and seismic processes on the solute response dynamics of geothermal systems such as the Jökulsá á Sólheimasandi. Further work should aim to 'restore' the composition of the geothermal fluids, estimate the dilution factors involved and analyse seasonal and event-scale changes in runoff sources using chemical and isotopic hydrograph separation methods. Secondly, estimates of intraglacial throughflow velocities and inferences about the character of drainage networks in glacio-geothermal basins could be refined through (i) independent knowledge of boundary conditions, especially valley glacier ice thicknesses and substrate character, (ii) improved three-dimensional locational information for seismic events based on *regional* crustal models (e.g. see Gudmundsson *et al.*, 1994) and (iii) a comparative series of dye tracing tests and/or investigations of meltwater isotope signatures. Thirdly, *geothermally driven* meltwater discharge departures may be important and could be quantified with hydrograph separation methods or with respect to flows predicted by melt- and rain-driven glaciohydrological models which are calibrated during hydrochemically defined 'non-geothermal' periods and constrained with reference to neighbouring non-geothermal basins.

ACKNOWLEDGEMENTS

We greatly appreciate the research permits issued by the National Research Council of Iceland and the Icelandic Council of Science, and the great help provided by Orkustofnun (Icelandic National Energy Authority). Enormous thanks are due, in particular, to: Arni Snorrason, Snorri Zóphóníasson, Hrefna Kristmannsdóttir, Halldor Armannsson, Oddur Sigurdsson, Svanur Pálsson, Haukur Tómasson and Erla Sigthórsdóttir of Orkustofnun for providing much patient advice, data and technical support; Bryndis Brandsdóttir and Ragnar Stefánsson for supplying and discussing seismic event data processed by the Icelandic Meteorological Office; and Thordur Tómasson for local fieldwork accommodation. The UK Natural Environment Research Council (NERC), through studentship GT4/88/AAPS/1 to M.D., is gratefully acknowledged for supporting this work, as is the NERC Equipment Pool and Institute of Hydrology for the loan of the Automatic Weather Station. Heather Lawler is thanked for enthusiastic assistance with all parts of the work. Adrian Merrylees, Roddy Gosden and Mike Sheppard kindly helped with fieldwork. M.D. is very grateful for the help and encouragement provided by his parents and Marie Dolan. Louise Bull is heartily thanked for patiently carrying out trace digitizing work. We also record our sincere thanks to the late Jean Dowling for the cartography. We are extremely grateful for the very useful comments received from Ian Fairchild and the two anonymous referees on an earlier draft.

REFERENCES

Armannsson, H., Benjamínsson, J., and Jeffrey, A. W. A. 1989. 'Gas changes in the Krafla geothermal system, Iceland', *Chem. Geol.*, **76**, 175–196.

Arnórsson, S. 1974. 'The composition of thermal fluids in Iceland and geological features related to thermal activity' in Kristjansson (Ed.), *Geodynamics of Iceland and the North Atlantic Area*. D. Reidel, Dordrecht. pp. 307–323.

Arnórsson, S. 1979. 'Hydrochemistry in geothermal investigations in Iceland: techniques and applications', *Nordic Hydrol.*, **10**, 191–224.

Björnsson, H. 1975. 'Subglacial water reservoirs, jökulhlaups and volcanic eruptions', *Jökull*, **25**, 1–12.

Björnsson, H. 1977. 'The cause of jökulhlaups in the Skaftá river, Vatnajökull', *Jökull*, **27**, 71–77.

Björnsson, H. 1979. 'Glaciers in Iceland', *Jökull*, **29**, 74–80.

Björnsson, H. 1988. 'Hydrology of ice caps in volcanic regions', *Societas Scientiarum Islandica*, Rit 45, 139 pp.

Björnsson, H. 1992. 'Jökulhlaups in Iceland: prediction, characteristics and simulation', *Ann. Glaciol.*, **16**, 95–106.

Björnsson, H., Björnsson, S., and Sigurgeirsson, Th. 1982. 'Penetration of water into hot rock boundaries of magma at Grímsvötn, *Nature*, **295**, 580–581.

Björnsson, H. and Kristmannsdóttir, H. 1984. 'The Grímsvötn geothermal area, Vatnajökull, Iceland', *Jökull*, **34**, 25–50.

Björnsson, S. and Stefánsson, V. 1987. 'Heat and mass transport in geothermal reservoirs' in Bear, J. and Yavuz Corapcioglu, M. (Eds), *Advances in Transport Phenomena in Porous Media*. Martinus Nijhoff, Dordrecht. pp. 145–183.

Brandsdóttir, B. and Einarsson, P. 1992. 'Volcanic tremor and low-frequency earthquakes in Iceland' in Johnson, R. W., Mahood, G., and Scarpa, R. (Eds), *IAVCEI Proc. Volcanol.* Vol. 3. *Volcanic Seismology*. pp. 212–222.

Brown, G. H., Sharp, M. J., Tranter, M., Gurnell, A. M., and Nienow, P. W. 1994. 'Impact of post-mixing chemical reactions on the major ion chemistry of bulk meltwaters draining the Haut Glacier D'Arolla, Valais, Switzerland', *Hydrol. Process.*, **8**, 465–480.

Carswell, D. A. 1983. 'The volcanic rocks of the Sólheimajökull area, south Iceland', *Jökull*, **33**, 61–71.

Collins, D. N. 1977. 'Hydrology of an alpine glacier as indicated by the chemical composition of meltwater', *Z. Gletscherk. Glazialgeol.*, **13**, 219–238.

Darling, W. G. and Armannson, H. 1989. 'Stable isotopic aspects of fluid flow in Krafla, Námafjall and Theistareykir geothermal systems of northeast Iceland', *Chem. Geol.*, **76**, 197–213.

Einarsson, P. 1991. 'Earthquakes and present-day tectonism in Iceland', *Tectonophysics*, **189**, 261–279.

Einarsson, P. and Björnsson, S. 1979. 'Earthquakes in Iceland', *Jökull*, **29**, 37–43.

Ellis, A. J. and Mahon, W. A. J. 1977. *Chemistry and Geothermal Systems*. Academic Press, New York. 392 pp.

Fairchild, I. J., Bradby, L., Sharp, M., and Tison, J-L. 1994. 'Hydrochemistry of carbonate terrains in alpine glacial settings', *Earth Surf. Process. Landforms*, **19**, 33–54.

Fenn, C. R. 1987. 'Electrical conductivity' in Gurnell, A. M. and Clark, M. J. (Eds), *Glacio-Fluvial Sediment Transfer: an Alpine Perspective*. Wiley, Chichester, pp. 377–414.

Fenn, C. and Ashwell, I. 1985. 'Some observations on the characteristics of the drainage system of Kverkjökull, Central Iceland', *Jökull*, **35**, 79–82.

Fournier, R. O. 1981. 'Application of water geochemistry to geothermal exploration and reservoir engineering' in Rybach, L. and Muffler, L. J. P. (Eds), *Geothermal Systems: Principles and Case Histories*. Wiley, Chichester. pp. 109–143.

Fridleifsson, I. B. 1979. 'Geothermal activity in Iceland', *Jökull*, **29**, 47–56.

Gíslason, S. R. and Eugster, H. P. 1987a. 'Meteoric water-basalt interactions. I: A laboratory study', *Geochim. Cosmochim. Acta*, **51**, 2827–2840.

Gíslason, S. R. and Eugster, H. P. 1987b. 'Meteoric water-basalt interactions. II: A field study in N.E. Iceland', *Geochim. Cosmochim. Acta*, **51**, 2841–2855.

Golterman, H. L. and Clymo, R. S. 1969. *Methods for Chemical Analysis of Fresh Waters*. *Int. Biol. Progr. IBP Handbook No. 8*. Blackwell, Oxford. 166 pp.

Gudmundsson, O., Brandsdóttir, B., Menke, W., and Sigvaldason, G. E. 1994. 'The crustal magma chamber of the Katla volcano in south Iceland revealed by 2-D seismic undershooting', *Geophys. J. Int.*, **119**, 277–296.

Gurnell, A. and Fenn, C. R. 1985. 'Spatial and temporal variations in electrical conductivity in a pro-glacial stream system', *J. Glaciol.*, **31**, 108–114.

Iken, A. and Bindschadler, A. 1986. 'Combined measurements of subglacial water pressure and surface velocity of Findelengletscher, Switzerland: conclusions about drainage system and sliding mechanism', *J. Glaciol.*, **32**, 101–119.

Jónsson, J. 1982. 'Notes on the Katla volcanoglacial debris flows', *Jökull*, **32**, 61–68.

Jonsson, G., Kristjansson, L., and Sverrisson, M. 1991. 'Magnetic surveys of Iceland', *Tectonophysics*, **189**, 229–247.

Lawler, D. M. 1991. 'Sediment and solute yield from the Jökulsá á Sólheimasandi glacierized river basin, southern Iceland' in Maizels, J. K. and Caseldine, C. (Eds), *Environmental Change in Iceland: Past and Present*. Kluwer, Dordrecht. pp. 303–332.

Lawler, D. M. 1994a. 'Recent changes in rates of suspended sediment transport in the Jökulsá á Sólheimasandi glacial river, southern

Iceland' in Olive, L. *et al.* (Eds), *Variability in Stream Erosion and Sediment Transport. Proc. Canberra Symp. 12–16 December 1994. IAHS Publ.*, **224**, 335–342.

Lawler, D. M. 1994b. 'The link between glacier velocity and the drainage of ice-dammed lakes: comment on a paper by Knight and Tweed', *Hydrol. Process.*, **8**, 447–456.

Lawler, D. M. and Brown, R. M. 1992. 'A simple and inexpensive turbidity meter for the estimation of suspended sediment concentrations', *Hydrol. Process.*, **6**, 159–168.

Lawler, D. M., Dolan, M., Tómasson, H., and Zóphóníasson, S. 1992. 'Temporal variability in suspended sediment flux from a subarctic glacial river, southern Iceland' in Bogen, J., Walling, D. E., and Day, T. J. (Eds), *Erosion and Sediment Transport Monitoring Programmes in River Basis. Proc. Oslo Symp. IAHS Publ.*, **210**, 233–243.

Maizels, J. 1989. 'Sedimentology and palaeohydrology of Holocene flood deposits in front of a jökulhlaup glacier, South Iceland' in Beven, K. and Carling, P. (Eds), *Floods, Hydrological, Sedimentological and Geomorphological Implications*. Wiley, Chichester. pp. 239–252.

Maizels, J. 1991. 'The origin and evolution of Holocene sandur deposits in areas of jökulhlaup drainage, Iceland' in Maizels, J. and Caseldine, C. (Eds), *Environmental Change in Iceland: Past and Present*. Kluwer, Dordrecht. pp. 267–279.

Olafsson, M. 1988. 'Sampling methods for geothermal fluids and gases', *Orkustofnun Publ.*, **OS-88041/JHD-06**, Orkustofnun (Icelandic National Energy Authority), Reykjavík, 22 pp.

Pálmason, G. 1974. 'Heat flow and hydrothermal activity in Iceland' in Kristjansson (Ed.), *Geodynamics of Iceland and the North Atlantic Area*. D. Reidel, Dordrecht. pp. 297–306.

Raiswell, R. 1984. 'Chemical models of solute acquisition in glacial meltwaters', *J. Glaciol.*, **30**, 49–57.

Raiswell, R. and Thomas, A. G. 1984. 'Solute acquisition in glacial meltwaters. I. Fjallsjökull (South East Iceland): bulk meltwaters — with closed system characteristics', *J. Glaciol.*, **30**, 35–43.

Raiswell, R., Brimblecombe, P., Dent, D. L., and Liss, P. S. 1980. *Environmental Chemistry*. Edward Arnold, London. 184 pp.

Rist, S. 1967. 'Jökulhlaups from the ice cover of Myrdalsjökull on June 25, 1955 and January 20, 1956', *Jökull*, **17**, 243–248.

Sharp, M. J. 1991. 'Hydrological inferences from meltwater quality data: the unfulfilled potential' in Gurnell, A. M. (Ed.), *British Hydrological Society 3rd National Hydrology Symposium*. pp. 5.1–5.8.

Sigvaldason, G. E. 1963. 'Influence of geothermal activity on the chemistry of three glacier rivers in southern Iceland', *Jökull*, **13**, 10–17.

Sigvaldason, G. E. 1965. 'The Grímsvötn thermal area. Chemical analysis of jökulhlaup water', *Jökull*, **15**, 125–128.

Sigvaldason, G. E. 1981. 'Fluids in volcanic and geothermal systems', in Rickard, D. T. and Wickman, F. E. (Eds), *Chemistry and Geochemistry of Solutions at High Temperatures and Pressures. Phys. Chem. Earth*. Vol. 13–14. pp. 179–195.

Souchez, R. A. and Lemmens, M. M. 1987. 'Solutes' in Gurnell, A. M. and Clark, M. J. (Eds), *Glacio-fluvial Sediment Transfer: an Alpine Perspective*. Wiley, Chichester. pp. 285–303.

Stefánsson, V. and Björnsson, S. 1982. 'Physical aspects of hydrothermal systems' in Pálmason, G. (Ed.), *Continental and Oceanic Rifts. Geodyn. Ser. Vol. 8*. American Geophysical Union, Washington, and Geological Society of America, Boulder. pp. 123–145.

Steinthorsson, S. and Oskarsson, N. 1983. 'Chemical monitoring of jökulhlaup water in Skeidará and the geothermal system in Grímsvötn, Iceland', *Jökull*, **33**, 73–86.

Stumm, W. and Morgan, J. J. 1970. *Aquatic Chemistry: an Introduction Emphasizing Chemical Equilibria in Natural Waters*. Wiley, Chichester. 583 pp.

Sverrisson, M., Jóhannesson, Æ., and Björnsson, H., 1980. 'A radio-echo equipment for depth sounding of temperate glaciers', *J. Glaciol.*, **25**, 477–486.

Sæmundsson, K. 1979. 'Outline of the geology of Iceland', *Jökull*, **29**, 7–28.

Thomas, A. G. and Raiswell, R. 1984. 'Solute acquisition in glacial meltwaters. II. Argentière (French Alps): bulk melt waters with open-system characteristics', *J. Glaciol.*, **30**, 44–48.

Thorarinsson, S. 1967. 'Hekla and Katla. The share of acid and intermediate lava and tephra in the volcanic products through the geological history of Iceland' in Björnsson, S. (Ed.), *Iceland and Mid-Ocean Ridges*. Report of Symposium held by the Geoscience Society of Iceland, 27 February–8 March 1967, Reykjavik, Iceland. Vísindafélag Islandinga (Societas Scientiarum Islandica), **38**, 190–199.

Thorarinsson, S. 1975. 'Katla og annáll Kötlugosa', *Arbók Ferdafélags Islands*, Reykjavík. pp. 125–149.

Tómasson, H. 1990. 'Glaciofluvial sediment transport and erosion' in Gjessing, Y., Hagen, J. O., Hassel, K. A., Sand, K., and Wold, B. (Eds), *Arctic Hydrology. Present and Future Tasks. Hydrology of Svalbard — Hydrological Problems in Cold Climate. Norwegian National Committee for Hydrology, Report.* **23**, 27–36.

Tranter, M. and Raiswell, R. 1991. 'The composition of the englacial and subglacial components in bulk meltwaters draining the Gornergletscher', *J. Glaciol.*, **37**, 59–66.

Tranter, M., Brown, G., Raiswell, R., Sharp, M., and Gurnell, A. 1993. 'A conceptual model of solute acquisition by Alpine meltwaters', *J. Glaciol.*, **39**, 573–581.

Tryggvason, E. 1973. 'Seismicity, earthquake swarms, and plate boundaries in the Iceland region', *Bull. Seismol. Soc. Am.*, **63**, 1327–1348.

Willis, I. C., Sharp, M. J., and Richards, K. S. 1990, 'Configuration of the drainage system of Midtdalsbreen, Norway, as indicated by dye-tracing experiments', *J. Glaciol.*, **36**, 89–101.

7

VELOCITY–DISCHARGE RELATIONSHIPS DERIVED FROM DYE TRACER EXPERIMENTS IN GLACIAL MELTWATERS: IMPLICATIONS FOR SUBGLACIAL FLOW CONDITIONS

PETER W. NIENOW

Department of Geography, University of Edinburgh, Drummond Street, Edinburgh EH8 9XP, UK

MARTIN SHARP

Department of Earth and Atmospheric Sciences, University of Alberta, Edmonton, Alberta, T6G 2E3, Canada

AND

IAN C. WILLIS

Department of Geography, University of Cambridge, Downing Place, Cambridge CB2 3EN, UK

ABSTRACT

Repeated dye tracer tests were undertaken from individual moulins at Haut Glacier d'Arolla, Switzerland, over a number of diurnal discharge cycles during the summers of 1989–1991. It was hoped to use the concepts of at-a-station hydraulic geometry to infer flow conditions in subglacial channels from the form of the velocity–discharge relationships derived from these tests. The results, however, displayed both clockwise and anticlockwise velocity–discharge hysteresis, in addition to the simple power function relationship assumed in the hydraulic geometry approach.

Clockwise hysteresis seems to indicate that a moulin drains into a small tributary channel rather than directly into an arterial channel, and that discharges in the two channels vary out of phase with each other. Anticlockwise hysteresis is accompanied by strong diurnal variations in the value of dispersivity derived from the dye breakthrough curve, and is best explained by hydraulic damming of moulins or sub/englacial passageways.

Despite the complex velocity–discharge relationships observed, some indication of subglacial flow conditions may be obtained if tributary channels comprise only a small fraction of the drainage path and power function velocity–discharge relationships are derived from dye injections conducted during periods when the supraglacial discharge entering the moulin and the bulk discharge vary in phase. Analyses based on this premise suggest that both open and closed channel flow occur beneath Haut Glacier d'Arolla, and that flow conditions are highly variable at and between sites.

INTRODUCTION

During the summers of 1989–1991 we conducted 533 dye tracer experiments at Haut Glacier d'Arolla, Switzerland (Figure 1). Results from these experiments demonstrate that, during a summer melt season, a system of major drainage channels, characterized by strongly peaked dye breakthrough curves and flow velocities in the range $0.3–0.5\,\mathrm{m\,s^{-1}}$, develops within the lower $3.3\,\mathrm{km}$ of the 4.0-km long glacier (Nienow, 1993). This channel system replaces a distributed drainage system, which is associated with diffuse dye breakthrough curves and flow velocities $<0.15\,\mathrm{m\,s^{-1}}$, as the principal means of evacuating water from the lower glacier. The distributed system remains the primary means of drainage throughout the year in the upper $0.7\,\mathrm{km}$ of the glacier. In this paper, we utilize velocity–discharge relationships derived from a series of closely spaced dye tracer injections in an attempt to determine whether flow in these channels occurs under atmospheric (open) or pressurized (closed) conditions. The conditions of flow have important

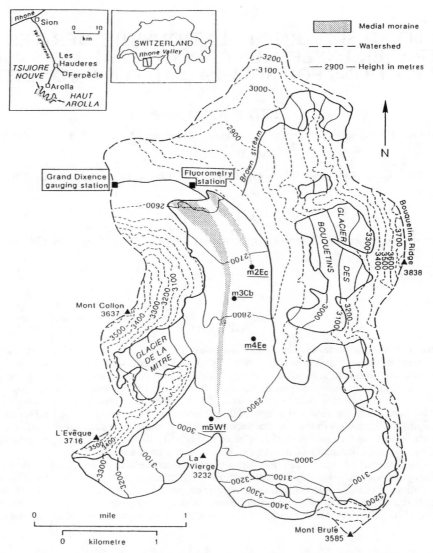

Figure 1. Map of Haut Glacier d'Arolla, showing locations of the fluorometry station, Grande Dixence SA gauging station and moulins cited in the text

implications for both ice dynamics and meltwater hydrochemistry. Basal sliding and ice dynamics are affected by subglacial water pressures (Iken, 1981), whilst processes and rates of solute acquisition are affected by the degree to which waters have access to sources of atmospheric CO_2 and O_2 (Raiswell, 1984; Tranter *et al.*, 1993).

BACKGROUND THEORY

Various workers have used the relationship between discharge and mean flow velocity derived from dye tracer tests to infer subglacial flow conditions (e.g. Seaberg *et al.*, 1988; Willis *et al.*, 1990; Fountain, 1993; Hock and Hooke, 1993). The form of the relationship is dependent on the 'hydraulic geometry' of the flow system. Changes in discharge, Q, result in changes in the mean flow velocity (u), width (w) and depth (d) (Leopold and Maddock, 1953). These adjustments can be described by the power functions:

$$w = aQ^b \qquad d = cQ^f \qquad u = kQ^m$$

Since continuity requires that $wdu = Q$, it follows that $b + f + m = 1$ and $ack = 1$. The exponent set in any given system reflects the shape of the stream cross-section, which is dependent upon the granulometry, strength and hydraulic properties of the bed and bank materials (Richards, 1982). Since the three exponents sum to unity, however, it follows that if an en/subglacial channel is full and its cross-sectional area, wd, is constant, any change in Q is accommodated by changes in u, requiring $u = kQ^{1\cdot0}$. Thus, under conditions of 'closed channel' flow, $m = 1\cdot0$. Under 'open channel' flow conditions, where water does not fill the channel, however, increases in Q are partly accommodated by increases in the channel cross-sectional area and $m < 1$. This suggests that the value of m should be a good indicator of flow conditions in subglacial channels.

However, most natural river cross-sections display more complex relationships between discharge and mean width, depth and velocity than are suggested by the simple power functions given above (Richards, 1977). Non-uniform changes in channel roughness during discharge variations, and complex channel cross-sections result in discontinuities in the velocity–discharge relationship (Knighton, 1979). In an englacial or subglacial channel, the possibility of a transition between 'open' and 'closed' channel flow is an additional complication. Departures from simple power function relationships are illustrated by velocity–discharge hysteresis. This is commonly observed during the passage of a flood wave, during which higher velocities occur on the rising limb of the hydrograph than at corresponding discharges on the falling limb (Richards, 1982). Possible causes for such hysteresis include the presence of different bedforms at similar discharges on rising and falling stages (Simons and Richardson, 1962) and greater hydraulic gradient on the rising limb of the hydrograph than after the peak has passed (Richards, 1982). Values of m derived by fitting power functions to velocity–discharge relationships that are hysteretic in form are likely to be erroneous, and will not provide a good indication of subglacial flow conditions.

An additional problem associated with the use of velocity–discharge relationships to infer subglacial flow conditions is the assumption that hydraulic geometry relationships are constant throughout the time period covered by the data set used to derive the relationships. For subglacial channels, the geometries of which adjust relatively rapidly to discharge variations by a combination of wall melting and ice deformation (Röthlisberger, 1972; Shreve, 1972), this assumption is only likely to be valid if injections are carried out at intervals of a few hours. This is not the case in several previous studies (Seaberg *et al.*, 1988; Willis *et al.*, 1990), in which velocity–discharge relationships have been derived using the results of tracer experiments conducted from several different moulins over the course of two complete melt seasons. The potential limitations of this approach are illustrated in Figure 2, which shows the data set dependence of the derived power law exponents and inferred subglacial flow regime. Thus, as Collins (1982) noted, if reliable inferences are to be made concerning the hydraulics of flow under glaciers using dye tracers, 'complex studies using many moulins ... with frequent tests in closely spaced 24 h cycles' are required. Ideally, these studies should be conducted from single injection sites.

METHODS

Dye injection and detection

During the 1989–1991 melt seasons, a number of regular dye injection series were undertaken from different locations on Haut Glacier d'Arolla (Figure 1). Rhodamine-B was used as a tracer in all three seasons and fluorescein was also used in 1990. Dye injection usually involved manually flushing a known quantity of dye in solution into a freely draining moulin. In some cases, however, automated multiple injections were made on a single day using procedures described by Sharp *et al.* (1993). Dye emergence was detected by fluorometry at the location shown on Figure 1. Dye detection involved the analysis of manually collected gulp samples in 1989, and continuous flow fluorometry in 1990 and 1991 (Sharp *et al.*, 1993).

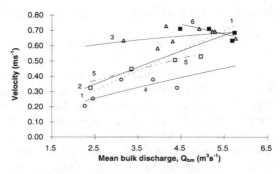

Line no.	Symbol	Injection dates	Velocity exponent, m	Regression, r^2	Inferred flow regime
1	All	All tests	1·040	0·640	Closed
2	□ and ■	1989	0·830	0·800	Open
3	△	1990	0·160	0·170	Open
4	○	1991	0·730	0·600	Open
5	□	20/07/89	0·680	0·960	Open
6	■	20/08/89	−0·320	0·460	Closed

Figure 2. Variations of throughflow velocity with mean bulk discharge, Q_{bm}, for various samples of injections from moulin m4Ee

Discharge measurement

Measurements of bulk discharge (Q_b) were obtained from the Grande Dixence SA hydroelectric intake structure located 950 m downstream from the glacier snout (Figure 1). The accuracy of these measurements is $\pm 4\%$ (Brown and Tranter, 1990). The local meltwater discharge (Q_1) through the glacier cross-section at each injection site was estimated using a knowledge of the glacier hypsometry and observed patterns of variation in melt rates with surface elevation (Nienow, 1993). Mean discharge between the injection site and the glacier snout was taken to be $Q_m = [(Q_{b1} + Q_{b2}/2) + (Q_{11} + Q_{12}/2)]/2$, where the subscripts 1 and 2 refer to discharges at the beginning and end of the tracer test, respectively. Since Q_1 is a constant fraction of Q_b, it follows that temporal variations in Q_m follow those in Q_b. The rationale for calculating Q_m is that by using this parameter, rather than Q_b, to calculate the cross-sectional area of the drainage system (see below), one is less likely to overestimate the cross-sectional area.

The discharge of supraglacial streams (Q_s) draining into moulins was measured by monitoring stage changes within the stream using a Druck PDCR 830 pressure transducer held at the bed of the channel with a boss and clamp. Stage values were read every 30 seconds and averaged and recorded at 15-minute intervals on a CSL 21X data logger. The stream cross-section at the transducer location was measured each morning and assumed to remain constant throughout the day (it is difficult to prove the validity or otherwise of this assumption, given the problems involved in establishing a fixed survey datum on the ablating glacier surface). Stream velocity was determined at several stage levels each day by calculating, from four measurements, the average time for a float to pass along a 10 m length of channel. Velocity–stage and stage–cross-sectional area relationships were used to estimate discharge from stage measurements. The estimated accuracy of these discharge values ($\pm 15\%$) takes into account the difficulty of measuring mean flow velocities in very small channels, and allows for the possibility that stream cross-section did vary slightly during a day. This level of accuracy was considered acceptable, given that the principal requirements of our analysis were knowledge of the order of magnitude and general temporal pattern of supraglacial discharges (see below).

Analysis of dye return curves

Five main parameters were estimated from the dye breakthrough curves.

(i)   The time between dye injection and peak dye concentration at the detection site (t).
(ii)   A minimum estimate of the mean flow velocity during the test ($u = x/t$, where x is the straight-line distance between injection and detection sites). We recognize that englacial/subglacial channels are probably sinuous and that the true value of u may be substantially greater than this estimate. It is unlikely, however, that the ratio between the true value of u and our minimum estimate will change markedly over the short periods of time during which individual injection series were conducted. Our estimate should therefore be regarded as an index measure of the true velocity.
(iii) The dispersion coefficient (D, $m^2 s^{-1}$), calculated according to the method used by Seaberg et al. (1988, Eq. 4, p. 222), which describes the rate of dispersion of the dye cloud during its passage through the glacier.
(iv)   The dispersivity ($d = D/u$), which represents the rate of spreading of a dye cloud relative to the rate of advection of the dye during transit through a flow system, and which has been used to infer the complexity of the drainage path followed by the dye (e.g. Seaberg et al., 1988; Hock and Hooke, 1993; Hooke and Pohjola, 1994).
(v)   The apparent mean cross-sectional area ($A_m = Q_m/u$) of the en/subglacial drainage system between injection site and the glacier snout. Given the assumptions that underlie our estimates of Q_m and u, A_m provides an index, rather than a true measure, of cross-sectional area.

RESULTS

Diurnal variations in velocity–discharge relationships

Figures 3a–d show velocity–discharge plots derived from nine series of closely spaced tracer tests made at three separate moulins. The discharge plotted, $Q_m/2$, is an estimate of the mean discharge between the test site and the snout during each test, based on the assumption that there were two equal-sized channels in the catchment into which each moulin drained (Sharp et al., 1993). There are significant variations in the form of the velocity–discharge relationship, both at a single site and between sites. Both linear and hysteretic (clockwise and anticlockwise) relationships occur. We analyzed a number of these sets of results in an attempt to establish the implications of each type of relationship for en/subglacial flow conditions.

Clockwise velocity–discharge hysteresis. Six injection series showed clockwise hysteresis (Figures 3a and b), such as is commonly observed during the passage of flood waves through open channels. Since diurnal discharge hydrographs in proglacial outwash streams are analogous to diurnal flood waves, we attempted to determine whether the two principal causes of hysteresis in such flood waves, changes in roughness and/or hydraulic gradient, could account for the observed behaviour. The most extreme hysteresis resulted from the injection series conducted at moulin m2Ec on 13 August 1991 (Figure 3a), so this series was analyzed in greatest detail. If realistic variations in roughness and hydraulic gradient can account for this hysteresis, then the less extreme hysteresis loops can probably be similarly explained.

The roughness coefficient, n, required to generate the velocities observed in each test can be obtained by rearranging the Gauckler–Manning–Strickler equation, such that

$$n = (R^{2/3} S^{1/2})/u_{act} \tag{1}$$

Here, R is the hydraulic radius of the channel and u_{act} is the observed throughflow velocity. A maximum estimate of n was obtained by assuming a semi-circular channel with a sinuosity of 2, and a hydraulic gradient determined by the ice surface slope between moulin and detection site ($S = 0.146$; equivalent to assuming subglacial water flows at the pressure of overlying ice). A minimum estimate was obtained by assuming a rectangular channel with a width/depth ratio of 1:10 (comparable to the proglacial channel) and sinuosity of 3, and a hydraulic gradient given by the bedrock slope between moulin and detection site [as determined by radio echo sounding (Sharp et al., 1993), $S = 0.05$, equivalent to assuming open channel flow] (Table I).

Over the diurnal cycle, the maximum estimates of n ranged from 0.10 to $0.74 \, m^{-1/3} \, s$ (Table I). This latter

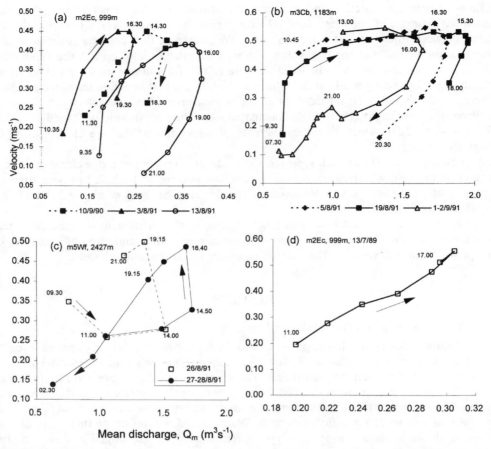

Figure 3. Variation of velocity with mean discharge assuming flow through two channels, $(Q_m/2)$, for a series of tracer tests made over the course of diurnal discharge cycles from three separate moulins. Date and time of injection and travel distance from injection site to fluorometry station are given on the figure

value is much higher than $0.29 \, \mathrm{m}^{-1/3} \, \mathrm{s}$, which is the maximum n value that can be estimated for roughness in a channel using the computational method of roughness determination of Cowan (1956), and is therefore considered implausible. The minimum estimates of n varied between 0.04 and $0.31 \, \mathrm{m}^{-1/3} \, \mathrm{s}$ (Table I). Although these values seem more realistic, and might be taken to suggest that the observed velocity–discharge hysteresis results from diurnal changes in hydraulic gradient and roughness of a rectangular sub-glacial channel, several lines of evidence suggest that this is not the case. Firstly, although the estimated values of n are physically plausible, the *range* of values seems rather large given the short time period over which the data were collected. Secondly, our calculations imply that both cross-sectional area and roughness continued to increase after 1800, while discharge decreased (Figure 4, Table I). It is difficult to see how this could occur in a rectangular open channel (the channel would have to be open for cross-sectional area to change). Bedforms created at peak flow are unlikely to be substantially modified as discharge and velocity decline. Hence, an increase in channel cross-sectional area would be likely to drown out roughness elements (Richards, 1977), and one would expect n to decrease, rather than increase, if cross-sectional area increased while velocity fell. It thus seems unlikely that changes in hydraulic gradient and channel roughness alone can explain the extreme velocity–discharge hysteresis observed.

Another possible explanation for the hysteresis is that the common practice of using bulk discharge (or some fraction thereof) for investigating u–Q relationships is inappropriate. This practice assumes that discharge variations are synchronous throughout the drainage system and that their rhythm can be

Table I. Moulin m2Ec injection series, 13 August 1991

Injection time	Mean Velocity $u_m (u_{act})$ (ms⁻¹)	Snout discharge at time of injection $t=1$ Q_{b1} (m³ s⁻¹)	Snout discharge at time of dye peak $t=2$ Q_{b2} (m³ s⁻¹)	Mean discharge* $Q_m/2$ (m³ s⁻¹)	Mean cross-sectional area $(Q_m/2)/u$ (m²)	Roughness #1† n (m⁻¹/³ s)	Roughness #2‡ n (m⁻¹/³ s)	Temperature (°C)	Supraglacial discharge Q_s (m³ s⁻¹)	Cross-sectional area in tributary channel Q_s/u_m (m²)
9·35	0·13	2·637	3·199	0·172	1·344	0·36	0·15	1·69	0·003	0·026
11·00	0·25	2·670	3·457	0·181	0·716	0·15	0·06	11·72	0·033	0·132
12·00	0·32	3·375	4·084	0·220	0·687	0·12	0·05	17·31	0·049	0·152
13·00	0·36	4·190	4·809	0·265	0·733	0·10	0·04	18·37	0·049	0·136
14·00	0·41	4·997	5·648	0·314	0·773	0·10	0·04	19·62	0·055	0·135
15·00	0·42	5·750	6·117	0·350	0·841	0·10	0·04	20·56	0·063	0·151
16·00	0·42	6·204	6·373	0·371	0·891	0·11	0·04	20·23	0·081	0·194
17·00	0·40	6·560	6·639	0·389	0·982	0·14	0·06	19·75	0·060	0·150
18·00	0·33	6·716	6·538	0·391	1·198	0·22	0·09	11·59	0·043	0·131
19·00	0·22	6·506	5·849	0·364	1·642	0·41	0·17	8·25	0·018	0·082
20·00	0·14	6·027	4·764	0·318	2·333	0·74	0·31	6·15	0·010	0·076
21·00	0·08	5·315	3·753	0·268	3·213			3·98	0·005	0·056
22·00		4·780						2·65	0·000	0·000
23·00		4·307						1·85		
24·00		3·891						1·28		

* Assuming flow through two channels
† Assuming flow in a full, sinuous, semi-circular R-channel with a sinuosity, w, of 2, and hydraulic gradient, S, of 0·146
‡ Assuming flow in a free surface, rectangular channel with a width/depth ratio of 10, sinuosity, w, of 3, and hydraulic gradient, S, of 0·05

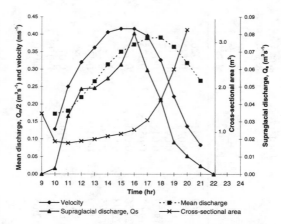

Figure 4. Variations of mean discharge ($Q_m/2$), velocity, cross-sectional area and supraglacial discharge (Q_s) with time for injections at moulin m2Ec on 13 August 1991. The shape of the supraglacial discharge hydrograph reflects the diurnal pattern of air temperature variation (Table I)

represented by measurements of bulk discharge. If, however, dye is injected into a moulin with a tributary link to a major channel, the average flow velocity determined from the dye breakthrough curve will depend upon the discharge histories and velocity–discharge relationships in both the tributary and the main channel. If the discharges are out of phase, the two-component flow system may produce a hysteretic relationship between average velocity and main channel discharge.

The discharge into m2Ec on 13 August 1991 was indeed out of phase with the bulk discharge from the glacier (Figure 4), and the average velocity determined from dye injections was more closely related to the discharge into the moulin, Q_S, than to the mean discharge, Q_m (Figure 5). Regression of velocity against discharge produced:

$$u = 0.69Q_m^{0.80} \qquad (r^2 = 0.18)$$

and

$$u = 1.53Q_S^{0.49} \qquad (r^2 = 0.92, p < 0.01)$$

The strength of the u–Q_S relationship and the value of the exponent (0.49) suggest that the observed velocity variations might be explicable in terms of flow through a single open channel which transported only the discharge into this one moulin. This view is supported by a plot of the flow velocity predicted from measurements of Q_S against Q_m (Figure 5b), which shows velocity–discharge hysteresis very similar to that observed. The discharge into m2Ec was, however, much lower than that of the stream in which dye emerged, so the stream draining m2Ec clearly coalesced with other channels prior to emergence at the snout. The flow path, therefore, probably contains at least two components, with discharges in the components fluctuating out of phase.

In a two-component drainage system, the average flow velocity, m, determined from a dye tracing experiment is:

$$u = x/[(x_1/u_1) + (x_2/u_2)] \qquad (2)$$

where x, x_1 and x_2 are the total flow distance and the flow distances in the tributary and main channels, respectively. The flow velocities in the tributary, u_1, and main channel, u_2, are given by the relationships $u_1 = aQ_S^X$ and $u_2 = bQ_m^y$, the form of which can be varied to reflect different sets of channel geometries and flow conditions. This model can be used to investigate the causes of velocity–discharge hysteresis, and to simulate the hysteresis observed.

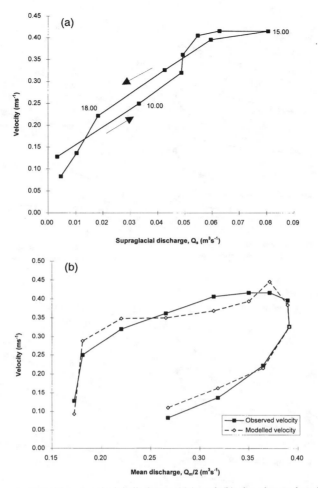

Figure 5. (a) Variation of flow velocity with supraglacial discharge (Q_s) and (b) the observed and predicted flow velocity [from Equation (8)] as a function of mean discharge, $Q_m/2$, for injections at moulin m2Ec on 13 August 1991

Taking input values for Q_S and Q_m from the 13 August 1991 test series from m2Ec, diurnal variations in u were simulated using a variety of assumptions about flow conditions and channel lengths. Figure 6a shows the results of four model runs where $x = 1000$ m, x_1 and x_2 are varied, and u_1 and u_2 both conform to the relationship $u = 0.71Q^{0.43}$, which was obtained from measurements of flow velocity and discharge in the proglacial stream at the fluorometry station in 1989. Significant velocity–bulk discharge hysteresis resulted, even when x_1 was assumed to be as low as 50 m. Varying the velocity–discharge relationships to simulate different flow conditions in the two channels illustrates the range of hysteresis plots that can be obtained (Figure 6b and c). When the flow conditions and absolute distance travelled in the tributary were held constant, the degree of hysteresis varied with the total travel distance (compare Figure 6a and d).

This simple model therefore shows that clockwise velocity–bulk discharge hysteresis occurs when the following conditions are met: (i) dye is injected into a tributary channel which subsequently drains into an arterial channel; (ii) the diurnal discharge hydrograph in the stream draining into the injection site peaks before Q_b. To detect this hysteresis, injections must be made at regular intervals throughout a diurnal discharge cycle.

Assuming different velocity–discharge relationships and distances of flow in the two components of the drainage system, we attempted to simulate the velocity–discharge (Q_m) hysteresis obtained from the 13 August 1991 test series. The discharges, Q_s and Q_m, assumed to occur through each component of

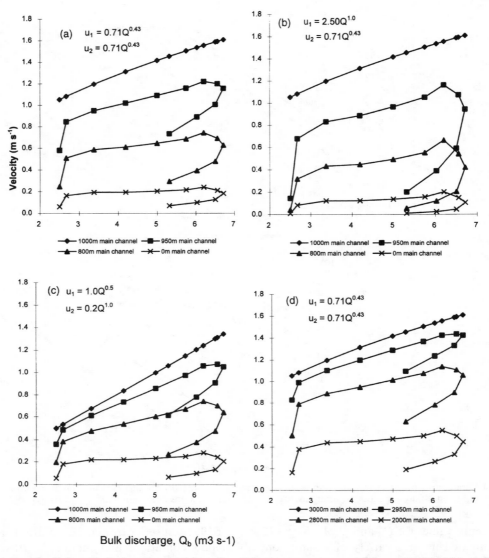

Figure 6. Variation of velocity with discharge resulting from flow through a two-component flow system in which both the velocity–discharge relationship and flow distance in each component can be varied. Total flow distance is 1000 m in (a), (b) and (c) and 3000 m in (d). The flow distance in the main channel component, x_2, is given, as are the velocity–discharge relationships assumed for each component

the drainage system during each test are given in Table I. Since neither of the requisite velocity–discharge relationships is known, we initially utilized the directly measured relationship for the proglacial stream, $u = 0.71Q_m^{0.43}$, as an estimate for the main channel. Assuming main channel lengths of 200 and 600 m, the observed velocities could best be predicted with velocity exponents for the tributary channel of 0.59 and 0.89, respectively (Figure 7a). In both cases, correlation of observed and predicted velocities produced $r^2 > 0.95$.

When the main channel flow distance was assumed to be > 700 m, velocity exponents of > 0.43 for the main channel were required to obtain good predictions of the overall velocity. With $x_2 = 800$ m, velocity exponents were varied to simulate the following combinations of flow conditions in the main and tributary channels: open:open (both exponents < 1); open(< 1):closed(1); closed(1):open(< 1); and closed(1):

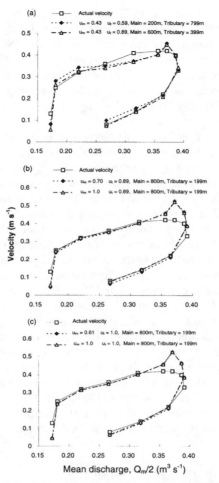

Figure 7. Simulations of the velocity–discharge hysteresis obtained from the 13 August 1991 injection series at moulin m2Ec assuming flow over the 999 m between moulin and snout through a two-component drainage system. The flow distances and velocity exponents in the main (u_m) and tributary (u_t) components are given

closed (1) (Figure 7b and c). The observed velocity–mean discharge hysteresis could be predicted (with $r^2 > 0.95$) using any of the above combinations of flow conditions. Thus, in a two-component system, the precise form of the velocity–discharge hysteresis appears to be rather insensitive to variations in channel length and flow conditions, at least in so far as we have been able to characterize them.

However, an estimate of flow conditions in the main channel may still be possible. The hysteresis observed in injections conducted at moulin m2Ec most likely results from phase differences between the supraglacial and subglacial discharge hydrographs. However, between 1100 and 1600, the two hydrographs rise in phase (Figure 4, Table I). The velocity–discharge (Q_m) relationship for this part of the hysteresis loop may provide information on the flow conditions in the main channel if this dominates the flow path and water spends only a small proportion of the total travel time in the tributary (see below). Regression of velocity on discharge (Q_m) for the period 1100–1600 yields the best fit power function:

$$u = 0.89 Q_m^{0.68} \qquad (r^2 = 0.94)$$

which suggests open channel flow in the main channel below m2Ec on 13 August 1991. Velocity–discharge relationships derived using the same method for the five additional test series that exhibited clockwise velocity–discharge hysteresis (Figure 3a and b) have exponents of 0.20–1.05. Only the results from the injection series from m2Ec on 10 August 1990 ($m = 1.05$) suggest closed channel flow.

Theoretical considerations (Röthlisberger, 1972; Shreve, 1972; Hooke, 1984) support the suggestion that tributary channels are relatively short. When flow is pressurized, larger channels capture discharge from smaller channels because of the pressure difference between them (Shreve, 1972) and 'tributary channels join into increasingly large trunk passages' (Paterson, 1981, p. 138). When flow is at atmospheric pressure, water drains towards topographic lows on the glacier bed and tributaries drain into larger arterial channels (Sharp *et al.*, 1993). Furthermore, mean flow velocities were high (0.25–$0.42\,\mathrm{m\,s^{-1}}$) during the tests conducted at m2Ec between 1100 and 1600 on 13 August 1991, suggesting that water was not delayed for a long period in the tributary part of the system.

Anticlockwise velocity–discharge hysteresis. Injections into m5Wf (Figure 1, Table II) on 26 and 27 and 28 August 1991 exhibited anticlockwise velocity–discharge (Q_m) hysteresis (Figure 3c). As in the cases described above, supraglacial discharge (Q_s) peaked prior to Q_m, but in this case velocity was very poorly correlated with Q_s (Figure 8). Hydraulic reconstructions using Equation (1) indicate that realistic changes in roughness and hydraulic gradient cannot account for the observed hysteresis. Instead, it seems likely that the cause of hysteresis was a water input to the moulin that initially exceeded its drainage capacity and caused water to back up and be stored within it (as has been observed visually and with pressure transducers in moulins on numerous occasions at Haut Glacier d'Arolla, Richards *et al.*, 1996). Thus, injections conducted between 1055 and 1450 (when supraglacial discharges were rising) resulted in low (and even decreasing) velocities as water backed up in the moulin. The resultant increase in hydraulic gradient, coupled with the drop in supraglacial discharge after 1500, allowed the moulin to drain more freely and the flow velocity to increase despite the fall in input discharge. This form of 'hydraulic damming' (Smart, 1990) has also been inferred from the results of tracer tests by Fountain (1993) at South Cascade Glacier.

Additional support for this interpretation is provided by the dispersivities, *d*, derived from the breakthrough curves (Table II). Breakthrough curves from the injections carried out at 1915 and 2100 on 26 August 1991 were strongly peaked, with dispersivities of 3.92 and $2.71\,\mathrm{m}$, respectively. By contrast, curves from injections made at 1055 and 1400 on the same day were less peaked and had dispersivities of 30.04 and $22.91\,\mathrm{m}$, respectively. Since it seems unlikely that the basic structure of the drainage system between the moulin and the glacier snout could change substantially over the course of a single day, the high dispersivities characteristic of the earlier tests are best explained by a diurnal change in the relative

Table II. Moulin m5Wf injection series data, 26–28 August 1991

Date	Injection time	Snout discharge at time of injection $t=1$ Q_{b1} $(\mathrm{m^3\,s^{-1}})$	Snout discharge at time of dye peak $t=2$ Q_{b2} $(\mathrm{m^3\,s^{-1}})$	Mean discharge* $Q_m/2$ $(\mathrm{m^3\,s^{-1}})$	Mean cross-sectional area $(Q_m/2)/u$ $(\mathrm{m^2})$	Mean velocity $u_m(u_{act})$ $(\mathrm{ms^{-1}})$	Supraglacial discharge Q_s $(\mathrm{m^3\,s^{-1}})$	Dispersion coefficient D $(\mathrm{m^2\,s^{-1}})$	Dispersivity d (m)
26/8	9·30	2·001	2·958	0·751	2·154	0·35	0·014		
26/8	10·55	2·556	4·360	1·048	4·041	0·26	0·103	7·790	30·04
26/8	14·00	4·637	5·325	1·509	5·410	0·28	0·080	6·390	22·91
26/8	19·15	4·664	4·201	1·343	2·689	0·50	0·003	1·960	3·92
26/8	21·00	4·083	3·727	1·183	2·545	0·46	0·002	1·260	2·71
27/8	11·00	2·303	4·544	1·037	3·949	0·26	0·067	6·110	23·26
27/8	13·00	4·073	5·681	1·478	5·261	0·28	0·102	6·550	23·32
27/8	14·50	5·471	5·866	1·718	5·223	0·33	0·112	4·260	12·95
27/8	16·40	5·856	5·165	1·670	3·426	0·49	0·083		
27/8	18·25	5·283	4·593	1·496	3·329	0·45	0·026		
27/8	19·15	4·887	4·176	1·373	3·394	0·40	0·014	1·670	4·13
27/8	22·00	3·565	2·617	0·937	4·446	0·21	0·003		
28/8	2·30	2·348	1·785	0·626	4·489	0·14	0·001		

* Assuming flow through two channels

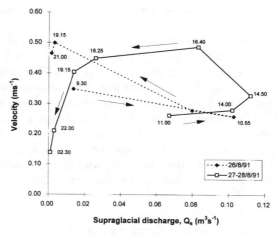

Figure 8. Variation in velocity as a function of supraglacial discharge (Q_s) for injections at moulin m5Wf on 26–28 August 1991

importance of dispersion and advection of dye. The importance of dispersion is maximized whilst water and dye are stored in the moulin, while that of advection is maximized when the moulin drains freely and allows rapid flow of water to a major drainage channel. If, as we propose, anticlockwise velocity–discharge hysteresis is a consequence of hydraulic damming and water backing up in moulins, it provides evidence for pressurized water flow within the glacier. It is, however, impossible to be certain of the extent of the area over which pressurized flow occurred.

Linear velocity–discharge relationships. Eight test series (those listed in Table III that did not show hysteresis) generated statistically significant ($p<0.1$) linear velocity–discharge (Q_m) relationships. An example is shown in Figure 3d. Two explanations can be suggested for the absence of hysteresis in the results from these test series. (i) The time period over which the tests in each series was conducted was such that supraglacial and bulk discharges varied in phase throughout the series. Since all tests in these series were conducted between 1030 and 1715, this would seem a viable explanation. (ii) The moulins at which

Table III. Velocity exponents derived from velocity–discharge relationships for 15 dye tracer series at the Haut Glacier d'Arolla. The temporal span of the tests used to derive the exponents and the flow conditions inferred from these exponents are indicated

Injection site	Distance (m)	Date	Temporal span of tests	Velocity exponent	Flow conditions	Hysteresis exhibited	Number of tests	r^2	Significance
m2Ec	999	13/7/89	1100–1700	2·19	Closed		7	0·98	0·01
m2Ec	999	3/8/91	1245–1530	0·62	Open	Y	3	0·99	0·05
m2Ec	999	10/8/90	1130–1430	1·05	Closed	Y	4	0·98	0·02
m2Ec	999	13/8/91	1100–1600	0·68	Open	Y	6	0·94	0·01
m3Cb	1183	10/8/90	1145–1645	0·18	Open		5	0·97	0·01
m3Cb	1183	5/8/91	1045–1630	0·2	Open	Y	10	0·86	0·01
m3Cb	1183	19/8/91	1100–1530	0·34	Open	Y	10	0·87	0·01
m3Ef	1444	16/8/90	1100–1400	0·35	Open	Y	4	0·99	0·01
m4Ee	1751	20/7/89	1030–1630	0·68	Open		4	0·96	0·05
m5Cb	2310	21/7/89	1045–1700	0·2	Open		4	0·99	0·01
m5Wf	2427	27/8/91	1100–0230	1·08	Closed	Y	8	0·8	0·02
m5Cg	2479	27/7/90	1115–1715	0·46	Open		4	0·99	0·01
m6Ca	2541	24/8/91	1030–1500	1·35	Closed		4	0·94	0·05
m7Cb	2773	24/7/89	1045–1630	0·72	Open		4	0·99	0·01
m8Ca	3602	30/8/91	1335–2045	1·45	Closed		4	0·91	0·1

these injection series were made may have drained directly into a major subglacial channel rather than via a tributary.

Five of these injection series produced velocity exponents of <1·0, indicative of open channel flow, while the remaining three generated exponents in the range 1·35–2·19 (Table III). Exponents >1·0 imply that the cross-sectional area of the flow decreased as discharge increased. Such a situation seems implausible over such a short time span because rising discharges would generate increasing amounts of heat to counteract channel closure by ice deformation. If we assume that the true value of the exponent for these test series was 1·0 (i.e. the channel cross-section remained constant over the duration of the test series), the variability in the empirically derived estimates can be explained if our estimates of Q_m are in error by no more than ±30% (mean required error = +12%). Given the assumptions made in estimating Q_m, such a margin of error seems possible. Whilst such an error would not affect our conclusions about the existence and likely causes of velocity–discharge hysteresis, it would have consequences for our estimates of *m*.

CONCLUSIONS

It is clearly extremely difficult to infer flow conditions in subglacial drainage channels using velocity–discharge relationships derived from tracer experiments, even when these relationships are constructed from injection series conducted at individual moulins over the course of single diurnal discharge cycles. Velocity–discharge (Q_m) plots are commonly hysteretic in form, and do not conform to simple power function relationships which would allow flow conditions to be classified as 'open' or 'closed' using the concepts of at-a-station hydraulic geometry. However, when velocity data obtained from such dye tracer tests are combined with a knowledge of both supraglacial and bulk discharges, the following conclusions can be drawn.

(i)   The regular occurrence of hysteresis demonstrates that the practice of estimating velocity exponents by fitting power functions to the results from injections made at different moulins on different days and even in different melt seasons (Seaberg *et al.*, 1988; Willis *et al.*, 1990) is inappropriate. Conclusions drawn from the magnitude of exponents derived in this way are probably unreliable.

(ii)   Clockwise hysteresis probably indicates that a moulin drains into a tributary channel prior to an arterial channel and that discharge in the tributary channel varies out of phase with that in the arterial channel. The tributary channel probably occupies only a small proportion (<20%) of the total flow path length, but flow through it may have a disproportionate influence on the overall travel time from moulin to snout. Since discharges through tributary channels are much lower than through arterial channels, flow velocities are also much lower. Hence the proportion of the travel time spent in the tributary tends to be much greater than its proportional length.

(iii)   Anticlockwise hysteresis appears to occur when water backs up in a moulin or tributary channel as a result of hydraulic damming at some point in the drainage system. It therefore indicates at least local pressurized water flow.

(iv)   Despite the hysteretic form of the relationship between velocity and bulk discharge (or derivative therefrom), it may still be possible to use velocity–discharge relationships to determine flow conditions in the major sub/englacial channels if they are derived from injections made when Q_s and Q_m are rising in phase.

(v)   If this is true, flow conditions in channels at Haut Glacier d'Arolla cannot be classified as solely 'open' or 'closed'. Flow conditions vary both between sites and at-a-site. Recognition of the diversity of subglacial flow conditions is important for both glacier hydrochemistry and ice dynamics. Variable flow conditions will affect accessibility to atmospheric CO_2 and O_2 sources (which affects the availability of protons and oxidants, and thus the processes and rates of solute acquisition). Thus, it is inappropriate to assume a constant subglacial flow condition when trying to explain the hydro-chemistry of glacial meltwaters. Similarly, varying flow conditions indicate varying subglacial water pressures which may locally affect basal sliding rates and, thus, ice dynamics. The spatial and temporal complexity of the channelized flow conditions suggest basal sliding events may be extremely localized on a diurnal time-scale.

A word of caution is required concerning the conclusions drawn from analysis of the velocity–discharge relationships at Haut Glacier d'Arolla. Using the exponents obtained from these relationships to infer subglacial flow conditions is questionable even when extreme care is exercised in selecting appropriate data sets to obtain the exponent. The flow conditions encountered by a parcel of water may change during passage through the system. Thus, water flowing under closed system conditions in a tributary channel may encounter open flow conditions in a main channel. It is not yet possible to infer the extent to which flow through the system occurs under open or closed conditions. Similarly, velocity exponents of less than one, indicative of flow through open channels, may occur even when flow is through closed channels for part of the travel distance. The velocity exponent can therefore only be used as a best estimate of the average flow conditions encountered.

ACKNOWLEDGEMENTS

This work was supported by the Natural Environment Research Council through grant GR3/7004a, and by grants from Earthwatch. P. Nienow acknowledges receipt of a NERC Studentship (GT4/89/AAPS/53) and Fellowship (GT5/93/AAPS/1). We thank W. H. Theakstone (University of Manchester) for the loan of his fluorometer, Grande Dixence SA for logistic support and Yvonne Bams for her help and support in Arolla. Field assistance was provided by Keith Richards, Chris Hill, Bryn Hubbard, Neil Arnold, Wendy Lawson, Julia Branson, Jim Strike, Jean-Louis Tison, Stuart Lane, Mark Skidmore, Nick Spedding and 36 Earthwatch volunteers.

REFERENCES

Brown, G. H. and Tranter, M. 1990. 'Hydrograph and chemograph separation of bulk meltwaters draining the Upper Arolla Glacier, Switzerland', in Lang, H. and Musy, A. (Eds), *Hydrology in Mountain Regions. I. Hydrological Measurements; the Water Cycle (Proceedings of the Second Lausanne Symposium, August 1990)*, Int. Assoc. Hydrol. Sci. Publ., **193**, 429–437.

Collins, D. N. 1982. 'Flow-routing of meltwater in an alpine glacier as indicated by dye tracer tests', *Beiträge zur Geologie der Schweiz-Hydrologie*, **28**, 523–534.

Cowan, W. L. 1956. 'Estimating hydraulic roughness coefficients', *Agric. Engng*, **37**, 473–475.

Fountain, A. G. 1993. 'Geometry and flow conditions of subglacial water at South Cascade Glacier, Washington State, U.S.A.; an analysis of tracer injections', *J. Glaciol.*, **39**, 143–156.

Hock, R. and Hooke, R. Le B. 1993. 'Evolution of the internal drainage system in the lower part of the ablation area of Storglaciären, Sweden', *Geol. Soc. Am. Bull.*, **105**, 537–546.

Hooke, R. Le B. 1984. 'On the role of mechanical energy in maintaining subglacial water conduits at atmospheric pressure', *J. Glaciol.*, **30**, 180–187.

Hooke, R. Le B. and Pohjola, V. A. 1994. 'Hydrology of a segment of a glacier situated in an overdeepening, Storglaciären, Sweden', *J. Glaciol.*, **40**, 140–148.

Iken, A. 1981. 'The effect of the subglacial water pressure on the sliding velocity of a glacier in an idealised numerical model', *J. Glaciol.*, **27**, 407–421.

Knighton, A. D. 1979. 'Comments on log-quadratic relations in hydraulic geometry', *Earth Surf. Proc.*, **4**, 205–209.

Leopold, L. B. and Maddock, T. 1953. 'The hydraulic geometry of stream channels and some physiographic implications', *US Geol. Surv. Prof. Pap.*, **252**, 1–57.

Nienow, P. W. 1993. 'Dye tracer investigations of glacier hydrological systems', *PhD Thesis*, University of Cambridge, Cambridge. 337 pp.

Paterson, W. S. B. 1981. *The Physics of Glaciers*, 2nd edn. Pergamon Press, Oxford. 380 pp.

Raiswell, R. 1984, 'Chemical models of solute acquisition in glacial meltwaters', *J. Glaciol.*, **30**, 49–57.

Richards, K. S. 1977. 'Channel and flow geometry: a geomorphological perspective', *Prog. Phys. Geog.*, **1**, 65–102.

Richards, K. S. 1982. *Rivers, Form and Process in Alluvial Channels*. Methuen, London. 358 pp.

Richards, K., Sharp, M., Arnold, N., Gurnell, A., Clark, M., Tranter, M., Nienow, P., Brown, G., Willis, I., and Lawson, W. 1996. 'An integrated approach to modelling hydrology and water quality in glacierised catchments', *Hydrol. Process.*, **10**, 479–508.

Röthlisberger, H. 1972. 'Water in intra- and subglacial channels', *J. Glaciol.*, **11**, 177–203.

Seaberg, S. Z., Seaberg, J. Z., Hooke, R. Le B., and Wiberg, D. W. 1988. 'Character of the englacial and subglacial drainage system in the lower part of the ablation area of Storglaciären, Sweden, as revealed by dye-trace studies', *J. Glaciol.*, **34**, 217–227.

Sharp, M., Richards, K., Willis, I., Arnold, N., Nienow, P., Lawson, W., and Tison, J.-L. 1993. 'Geometry, bed topography and drainage system structure of the Haut Glacier d'Arolla, Switzerland', *Earth Surf. Proc. Landforms.*, **18**, 557–572.

Shreve, R. L. 1972. 'Movement of water in glaciers', *J. Glaciol.*, **11**, 205–214.

Simons, D. B. and Richardson, E. V. 1962. 'The effect of bed roughness on depth–discharge relations in alluvial channels', *Water Supply Paper, US Geol. Surv.*, **26**, 1498E. 26 pp.

Smart, C. C. 1990. 'Comments on: "Character of the englacial and subglacial drainage system in the lower part of the ablation area of Storglaciären, Sweden, as revealed by dye-trace studies" ', *J. Glaciol.*, **36**, 126–128.

Tranter, M., Brown, G. H., Raiswell, R., Sharp, M. J., and Gurnell, A. M. 1993. 'A conceptual model of solute acquisition by alpine glacial meltwaters', *J. Glaciol.*, **39**, 573–581.

Willis, I. C., Sharp, M. J., and Richards, K. S. 1990. 'Configuration of the drainage system of Midtdalsbreen, Norway, as indicated by dye-tracing experiments', *J. Glaciol.*, **36**, 89–101.

8

LINKS BETWEEN PROGLACIAL STREAM SUSPENDED SEDIMENT DYNAMICS, GLACIER HYDROLOGY AND GLACIER MOTION AT MIDTDALSBREEN, NORWAY

IAN C. WILLIS AND KEITH S. RICHARDS

Department of Geography, University of Cambridge, Downing Place, Cambridge CB2 3EN, UK

AND

MARTIN J. SHARP

Department of Earth and Atmospheric Sciences, University of Alberta, Edmonton, Alberta T6G 2E3, Canada

ABSTRACT

Two-hourly suspended sediment concentration variations observed during the summer of 1987 in the proglacial stream draining Midtdalsbreen, Norway are modelled using multiple regression and time series techniques. Suspended sediment fluctuations are influenced by stream discharge variations, diurnal hysteresis effects, medium-term sediment supply and transport variations and the recent suspended sediment concentration history of the stream. They do not appear to be influenced by seasonal exhaustion or rainfall variations. Possible reasons for this are discussed. Large positive residuals from the fitted models are major pulses of suspended sediment unrelated to discharge variations; these sediment flushes correlate with periods of enhanced glacier motion. They cannot be explained by enhanced sediment production by subglacial erosion, but are probably due to the tapping of subglacially stored sediment during sudden changes in the hydraulics and/or configuration of the subglacial hydrological system. Seasonal changes in the lag between glacier motion peaks and suspended sediment flushes suggest that the subglacial hydrological system evolves over the summer from a distributed to a more channelized configuration.

INTRODUCTION

Proglacial stream suspended sediment dynamics have practical significance where glacially derived waters are used for crop irrigation (Butz, 1989) or hydroelectric power production (Bezinge *et al.*, 1989). Scientifically, proglacial stream suspended sediment concentrations are of interest because they reflect the combined effects of sediment production by subglacial erosion (quarrying, shearing, crushing and abrasion) and sediment transport by subglacial water, both of which are important areas of glaciological research (Drewry, 1986; Fenn, 1987). However, the links between sediment production and transport are complicated because they operate interdependently, differentially and intermittently in ways that are not fully understood (Fenn, 1987).

Sediment production and transport are both influenced by the nature of the subglacial hydrological system, which is likely to consist of a distributed component (e.g. thin water film, linked cavities, sediment pores) and a channelized component (e.g. channels incised up into ice and/or down into bedrock or sediment) (Hooke, 1989). The relative proportion of these components has important implications for sediment production via their effect on subglacial water pressure and basal sliding (Fowler, 1987; Harbor, 1992). Basal sliding variations can control rates of quarrying (Robin, 1976; Röthlisberger and Iken, 1981; Shoemaker, 1986; Iverson, 1991), shearing and crushing (Boulton *et al.*, 1974) and abrasion (Boulton, 1974; Hallet, 1979; Metcalf, 1979; Iverson, 1990). The nature of the subglacial drainage system places important controls on sediment transport because a distributed system will cover a larger area of glacier bed and will be more dynamic than a channelized system and might therefore be expected to have access to greater supplies of sediment. Conversely, water velocities are likely to be less in a distributed system than in a channelized system and so sediment transport might be limited by the ability of the water to mobilize the sediment to which it has access. Thus studies of proglacial stream suspended sediment dynamics have the potential to provide useful information about sediment production and transport and therefore subglacial hydrology, water pressures and sliding.

PREVIOUS WORK

Many previous studies of proglacial stream suspended sediment dynamics have made inferences about temporal changes in subglacial hydrology and/or basal motion (e.g. Collins, 1979; 1989; Gurnell and Fenn, 1984; Humphrey *et al.*, 1986). However, most of these inferences have been rather speculative because variations in sediment concentrations, glacier drainage system characteristics and basal motion have not been investigated simultaneously. One aspect of proglacial stream suspended sediment dynamics to receive particular attention is the occurrence of large short-term pulses of sediment. These have been variously associated with short-term increases in stream discharge resulting from rainstorms (Bezinge, 1987), the bursting of intraglacial water pockets (Gurnell, 1982; Beecroft, 1983) or the emptying of ice-dammed lakes (Collins, 1986), but may also be unrelated to short-term increases in discharge (Østrem, 1975; Collins, 1979; Bogen, 1980; Gurnell, 1982; Humphrey *et al.*, 1986). In the last instance, such pulses may be related to the reorganization of the subglacial hydrological system (Collins, 1979; 1989) and could accompany short periods of enhanced basal motion (Gurnell, 1982; Gurnell and Warburton, 1990). However, there are no previous studies of non-surge-type glaciers which have demonstrated direct links between changes in the hydraulics and structure of the subglacial hydrological system, variations in subglacial motion and sediment flushes in proglacial streams.

Only one study has shown directly that sediment flushes are associated with major changes in subglacial hydrology and enhanced basal motion, but this was on the surge-type Variegated Glacier in Alaska, USA during the two years preceding the 1982–1983 surge (Humphrey *et al.*, 1986; Kamb and Engelhardt, 1987). Of the seven observed 'mini-surges', six were accompanied by dramatic peaks in proglacial stream suspended sediment concentration lasting for 2–3 h and superimposed on a more gentle rise and fall of background concentrations lasting up to one day. A more recent study on the surge-type Trapridge Glacier in Yukon Territory, Canada has shown that subglacial sediment flushes, measured by turbidity sensors lowered down boreholes to the bed, occur in response to large changes in basal water pressure (Stone *et al.*, 1993). Unfortunately, it is not known whether such flushes also produced sediment pulses in the proglacial stream. There is therefore a need to investigate the links between proglacial stream suspended sediment dynamics, glacier hydrology and glacier motion, particularly on non-surge-type glaciers, and this was the general purpose of the research reported in this paper.

AIMS AND APPROACH OF THIS WORK

The specific aims and approach of the paper are as follows: (1) to model the variations of suspended sediment concentration in the proglacial stream of a non-surge-type glacier in terms of proglacial and subglacial hydrological processes; (2) to relate the model structure to subglacial hydrological processes interpreted from dye tracing results; (3) to analyse the residuals from the model to identify short-term flushes of glacially derived suspended sediment; (4) to identify the correlation between sediment pulses and variations in subglacial motion determined by short-term surveys of glacier surface velocities; (5) to examine possible causal mechanisms linking sediment pulses with variations in basal motion; and (6) to use the nature of the relationships between glacier motion and sediment flushing to infer characteristics of the subglacial hydrological system.

FIELD SITE

The study was conducted at Midtdalsbreen, a northern outlet glacier from Hardangerjökulen in southern Norway (Figure 1). Midtdalsbreen is largely underlain by granitic gneisses, phyllite and mica schists (Sigmond *et al.*, 1984). During the Neoglacial, it attained a maximum extent at around 1750, since when it has retreated by about 1·5 km leaving a gently undulating forefield covered with a thin veneer of till and fluvial gravels, but with many bedrock outcrops (Andersen and Sollid, 1971). The total catchment area is 9·6 km^2, of which 7·4 km^2 (77%) is glacierized. Midtdalsbreen is drained by three

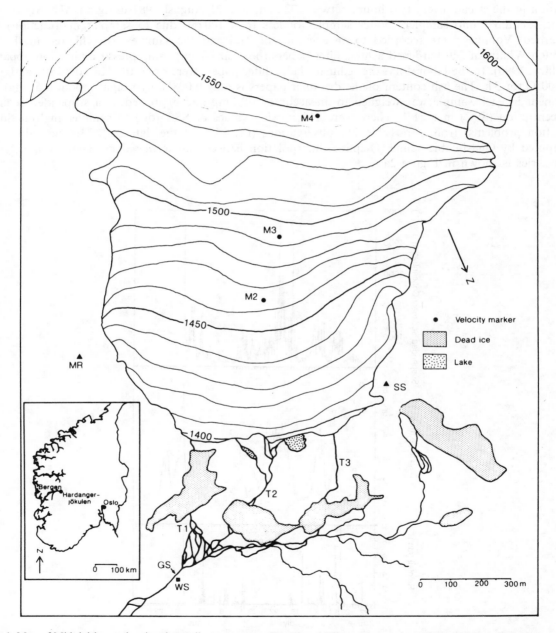

Figure 1. Map of Midtdalsbreen showing three tributary streams (T1, T2 and T3), gauging station (GS), weather station (WS), survey station (SS), survey reference marker (MR) and glacier velocity markers (M2, M3 and M4). Contours are in metres a.s.l.

main meltwater streams (T1, T2 and T3) which merge into one stable channel about 500 m from the glacier snout (Figure 1).

METHODS

Suspended sediment concentration

A gauging station (GS) was established on the proglacial stream about 500 m from the glacier snout just after the three tributary streams had converged (Figure 1). Water samples were taken from a fixed position in the stream every two hours between 22 June and 21 August 1987 using an ALS Mk4b automatic vacuum pump sampler. Water samples of less than 100 ml (due to sampler malfunction) were discarded. Volumes were recorded to an accuracy of 2 ml and the samples were filtered in the field through Whatman 540 hardened ashless filter papers (pore size 8 μm) using pressure filtration apparatus (Collins, 1990). In the laboratory, the sediment-laden filter papers were dried for 12 h at 105°C and ashed at 800°C for 3 h. The ash content of the dry filter papers is only 0·008% by weight (about 0·65 mg). The sediment in each sample was weighed to an accuracy of 1 mg and expressed as a suspended sediment concentration (SSC) in mg l^{-1}. There were 110 missing values of SSC (due to sampler malfunction or filtration problems) from a possible 717 observations (i.e. 15% of the data set). Missing values were computed by Quasi-cubic spline (Q-spline) interpolation between adjacent values. The two-hourly SSC time series is shown in Figure 2a.

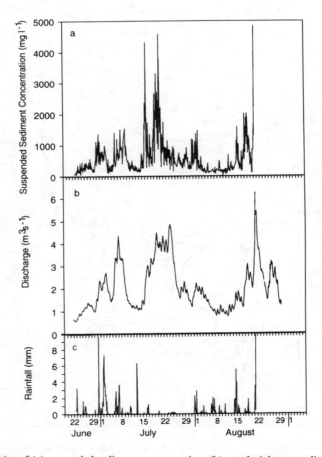

Figure 2. Two-hourly time-series of (a) suspended sediment concentration, (b) proglacial stream discharge and (c) rainfall in 1987

Discharge and rainfall

Water stage was monitored using a Druck PDCR 830 pressure transducer and a Campbell Scientific (CS) 21X datalogger. Between 22 June and 28 August 1987, hourly averages of water stages measured every 10 s were recorded by the logger and converted to water discharges using stage–discharge curves constructed from 42 discharge measurements made over a wide range of flow conditions throughout the summer (see Willis *et al.*, 1990). Hourly totals of rainfall were measured from 22 June to 28 August 1987 using a CS tipping-bucket raingauge and 21X datalogger at a weather station also situated 500 m from the glacier snout (Figure 1). To aid comparison with the SSC data, the discharge and rainfall data were re-sampled to produce two-hourly time series (Figure 2b and 2c).

Glacier velocity

The surface motion of the lower 1 km of Midtdalsbreen was monitored between 7 July and 29 August 1987. Motion was determined by repeatedly surveying to three markers drilled into the glacier surface along its centreline from a survey station (SS) located on bedrock on the western side of the valley (Figure 1). The markers consisted of Kern targets and reflectors screwed onto 3 m long aluminium stakes. They were surveyed using a Wild 2 Theodolite and a Kern electro-optical distance meter (EDM) mounted on a fixed tripod. The markers were initially placed so that about 0·5 m protruded above the glacier surface. Before a marker ablated out of its hold, it was surveyed, immediately re-drilled and then re-surveyed. In this way, continuity of measurements was maintained. Surveys consisted of measuring horizontal angle, vertical angle and distance to each marker. Horizontal angles were measured relative to a reference marker (MR) fixed on bedrock on the eastern side of the valley (Figure 1). The errors associated with horizontal and vertical angle measurements were typically $\pm2\cdot5''$ ($0\cdot0007°$) and $\pm5\cdot0''$ ($0\cdot0014°$), respectively and the error in the distance measurements was typically $\pm2\cdot5$ mm. This gives an accuracy in positioning marker 4 (which is furthest from the SS) of ±2 cm. The accuracy in positioning markers 2 and 3 is about ±1 cm. When weather conditions allowed, surveys were made five times a day (approximately every 3 h from 0900 to 2100).

Horizontal velocities of the markers were determined as follows. Firstly, 'line of sight velocities' towards the SS were calculated from successive EDM distance measurements. These were converted to horizontal 'down-glacier velocities' along the azimuths of motion by trigonometry using the horizontal and vertical angle measurements. Owing to errors in the horizontal angle measurements, this method gave spurious results when each consecutive pair of angle measurements was used. To overcome this problem, the azimuths of marker motion were assumed to be constant from the time immediately after a marker had been drilled to the time just before it was reset. Thus the azimuths of marker motion were calculated using the angle measurements made immediately after resetting and before subsequently resetting a marker. For each setting of a marker, a 'correction factor' was obtained by dividing the distance travelled along the azimuth of motion by the distance travelled according to the EDM measurements. The 'line of site velocities' were converted to 'down-glacier velocities' by multiplying them by the relevant 'correction factor' (cf. Iken, 1977). The error associated with the velocity calculations was typically less than ±0.5 cm d^{-1} (cf. Hooke *et al.*, 1989). The full details of the error analysis will be presented elsewhere.

To aid comparison with the SSC, discharge and rainfall data, the glacier velocity data were converted into two-hourly time series as follows. Q-splines were fitted to the three velocity data sets. This procedure connects successive irregularly spaced glacier velocity values (joints) with a fifth-order polynomial. It then interpolates 10 glacier velocities at regular intervals between any two contiguous joints (Hazony, 1979). The resulting data sets were then re-sampled to produce two-hourly time series of glacier speed for each of the three markers (cf. Walters and Dunlap, 1987). Errors in the time base were generally small ($< \pm6$ min). The resulting series are shown in Figure 3a, 3b and 3c.

MODELLING VARIATIONS OF SUSPENDED SEDIMENT CONCENTRATION

Modelling SSCs in proglacial streams provides a means of forecasting SSCs and loads from discharge data alone (Østrem *et al.*, 1967; Østrem, 1975; Ferguson, 1984; Gurnell and Fenn, 1984), and of examining

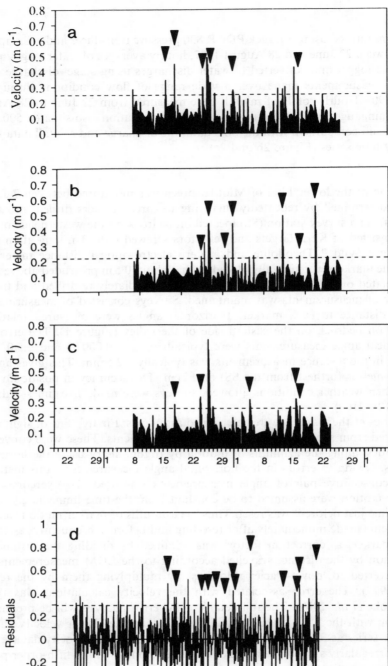

proglacial hydrological processes (Fahnestock, 1963; Church, 1972; Church and Ryder, 1972; Hammer and Smith, 1983; Richards, 1984) or subglacial hydrological and erosional processes (Mathews, 1964; Metcalf, 1979; Collins, 1979; Gurnell and Fenn, 1984). In this section, the temporal pattern of SSC in Midtdalsbreen's proglacial stream is modelled in terms of variables representing proglacial and subglacial hydrological processes. The model residuals are then analysed in terms of processes associated with basal motion. Two methods used in previous attempts to model proglacial stream SSC variations, regression and time series techniques, are used in this study.

Linear regression techniques

Linear regression models were fitted to concurrent discharge and SSC values to produce suspended sediment rating curves (cf. Fenn, 1989). The data were first transformed to linearize the trend and reduce the heteroscedasticity (Ferguson, 1977). Using a Box–Cox procedure, a $\log_{10}$ transformation was identified as the most suitable (Box and Cox, 1964; Fenn *et al.*, 1985). Figure 4 illustrates the relationship between the transformed data. Several rating relationships were fitted to these transformed data as outlined in the following and summarized in Table I. The suitability of each model was assessed on the basis of the significance of the model parameters, the magnitude of the residual sum of squares (RSS) and coefficient of determination (r^2) and the absence of pattern in the autocorrelation function (acf) and partial autocorrelation function (pacf) of the residuals (Chatfield, 1984).

An ordinary rating curve (ORC) (Figure 4) had an r^2 of 53% and intercept and slope parameters that were significantly different from zero ($\alpha = 0.01$), but it had a high RSS (39·8) and there was high serial autocorrelation in the residuals (Figure 5a) (Table II). One way of trying to account for this autocorrelation involves fitting a rating curve after removing the trends from the two variables by first differencing (ORC1D) (Gurnell and Fenn, 1984). This relates rates of change of SSC to rates of change of discharge. When applied to Midtdalsbreen data, this model reduced the RSS to 32·6 and removed the serial auto-correlation in the residuals (Figure 5b), but reduced the r^2 to just 2% (Table II), showing that rates of change were not significantly correlated. Another way of trying to account for autocorrelation is to fit an ordinary rating curve to the data after weighting the first differencing of the variables by the residual autocorrelation at lag = 1 from the ORC model (ORCW1D) (Cochrane and Orcutt, 1989; Fenn *et al.*, 1985). When this was applied to the Midtdalsbreen data, the RSS was reduced even further to 26·0 and there was no serial autocorrelation left in the residuals (Figure 5c), but the r^2 of 23% was still less than half the ORC model value (Table II), showing that weighted first-differenced discharge is a poor predictor of weighted first-differenced SSC.

Some workers have derived separate rating curves for different sub-periods by splitting the observations into the rising and falling limbs of daily discharge hydrographs (Østrem, 1975; Church and Gilbert, 1975; Collins, 1979; Fenn *et al.*, 1985), early and late ablation season (Hammer and Smith, 1983; Fenn *et al.*, 1985) or rainfall and rain-free periods (Richards, 1984; Fenn *et al.*, 1985). Because such arbitrary distinctions are subjective, we did not pursue this option further. The largest cross-correlation between discharge and SSC was at a lag of zero showing that lagging the series before linear regression analysis would not reduce the high residual autocorrelation observed in the ORC.

Multiple regression techniques

The ORC method assumes that the sediment concentrations are largely controlled by the capacity of the stream rather than by sediment supply and storage (as affected, for example, by subglacial processes). The inclusion of variables other than discharge might therefore remove some of the scatter around the ORC and some of the residual autocorrelation. This scatter may be linked with several factors which can be represented by surrogate variables for use in regression analysis.

Figure 3. Two-hourly time-series of (a) surface velocity at marker 2, (b) surface velocity at marker 3, (c) surface velocity at marker 4 and (d) SSC residuals after fitting the multiple regression and integrated autoregressive-moving average (MR–ARIMA) model. For each series, the top 5% of the observations between 2200 on 7 July and 1000 on 21 August lie above the broken line. Arrows show glacier motion peaks which correspond with sediment flushes

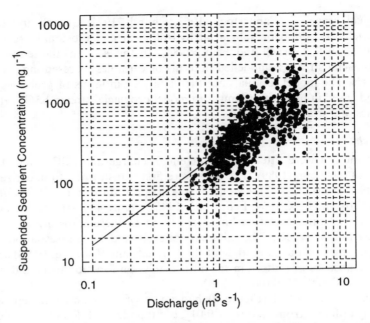

Figure 4. Scatter-plot of logarithmically transformed discharge and suspended sediment concentration. The line is the ordinary rating curve calculated using simple linear regression techniques

1. Long-term seasonal exhaustion of sediment. This may occur if extra-glacial and subglacial weathering occurs throughout the year, but sediment transport is mainly confined to the summer (Østrem, 1975; Hammer and Smith, 1983), or if a distributed subglacial hydrological system develops into a channelized system through the summer (Röthlisberger and Lang, 1987). The latter reduces subglacial water pressures, sliding and therefore erosion (Willis, 1995; Iverson, 1990; 1991) and also the area of turbulent water in contact with the products of glacial erosion (Collins, 1989). To account for possible seasonal exhaustion, a variable measuring 'days since the beginning of summer (22 June)' was defined.
2. Medium-term sediment supply variations. These may depend on the extent to which recent discharges

Table I. Models for explanation of suspended sediment concentration

Model	Equation
ORC (ordinary rating curve)	$\log_{10} SSC_t = 2\cdot331 + 1\cdot182 \log_{10} Q_t + e_t$ $(0\cdot014)(0\cdot042)$
RDV (regression on 1st differenced variables)	$\log_{10} SSC_t - \log_{10} SSC_{t-1} = 0\cdot001 + 1\cdot506(\log_{10} Q_t - \log_{10} Q_{t-1}) + e_t$ $(0\cdot008)(0\cdot427)$
GLS (generalized least squares)	$\log_{10} SSC_t - 0\cdot59 \log_{10} SSC_{t-1} = 0\cdot964 + 1\cdot193(\log_{10} Q_t - 0\cdot59 \log_{10} Q_{t-1}) + e_t$ $(0\cdot011)(0\cdot081)$
MR (multiple regression)	$\log_{10} SSC_t = 2\cdot330 + 1\cdot118 \log_{10} Q_t + 1\cdot190\Delta Q + 0\cdot006\, \mathrm{d}Q \Rightarrow_t + e_t$ $(0\cdot013)(0\cdot043)(0\cdot499)(0\cdot001)$
MRARIMA (multiple regression then time series)	$\log_{10} SSC_t = 2\cdot330 + 1\cdot118 \log_{10} Q_t + 1\cdot190\Delta Q + 0\cdot006\, \mathrm{d}Q \Rightarrow_t$ $(0\cdot013)(0\cdot043)(0\cdot499)(0\cdot001)$ $+0\cdot055(r_{t-12} - r_{t-13}) + 0\cdot111(r_{t-24} - r_{t-25}) + e_t + 0\cdot680e_{t-1}$ $(0\cdot039)(0\cdot039)(0\cdot029)$

SSC, Suspended sediment concentration; Q, discharge; ΔQ, rate of change of discharge, $\mathrm{d}Q \Rightarrow$, days since discharge is equalled or exceeded; r, residual from MR model; e, error term; subscript t is a time identifier. Numbers in parentheses are the standard errors of the coefficients

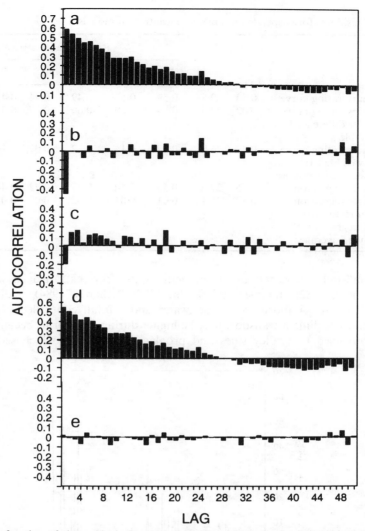

Figure 5. Autocorrelation function of the residuals from (a) the ordinary rating curve (ORC) model, (b) the ORC after first differencing, (c) the ORC after weighted first differencing, (d) the multiple regression (MR) model and (e) the multiple regression and integrated autoregressive-moving average (MR–ARIMA) model

have exhausted proglacial and subglacial sediment (Gurnell, 1987). Rising discharges mobilize sediment deposited by previous lower stage flows. The amount of sediment accumulated in storage locations depends largely on the time elapsed since the last discharge event of a similar magnitude. To account for this possibility, a variable measuring 'days since discharge was equalled or exceeded' was defined. This variable was measured from the beginning of the measurement period (22 June) and values ranged from 0·083 (on falling stages) to 60·417 during the rainstorm on 21 August which produced the highest discharge of the season (Figure 6a).

3. Short-term diurnal sediment supply variations. These reflect the flushing and exhaustion of sediment on the rising and falling limbs of the daily discharge hydrograph (Østrem et al., 1967; Liestol, 1967; Church, 1972; Collins, 1979; Bogen, 1980; Richards, 1984). Clockwise hysteresis was common at Midtdalsbreen on days with a well-defined diurnal discharge cycle (e.g. Figure 7). To account for this, a variable measuring 'rate of change of discharge' was obtained by subtracting the preceding discharge from each recorded flow (Liestol, 1967; Richards, 1984). This variable is positive during rising discharges and negative during falling discharges (Figure 6b).

Table II. Summary of residuals for suspended sediment concentration models

Model		r^2	RSS	Residual autocorrelation at lag							
				1	2	3	4	5	6	7	8
ORC	Ordinary rating curve	0·53	39·8	0·59	0·53	0·49	0·44	0·45	0·42	0·37	0·34
ORC1D	Ordinary rating curve on 1st differenced variables	0·02	32·6	−0·44	0·00	−0·00	−0·07	0·06	0·00	−0·01	0·03
ORCW1D	Ordinary rating curve on weighted 1st differenced variables	0·23	26·0	−0·17	0·14	0·12	0·07	0·16	0·11	0·09	0·11
MR	Multiple regression	0·55	37·6	0·55	0·50	0·47	0·42	0·44	0·40	0·36	0·33
MR–ARIMA	Multiple regression then integrated autoregressive-moving average	0·82	23·1	0·01	−0·01	−0·02	−0·07	0·05	0·01	0·01	−0·01

4.   Short-term rainfall-induced variations in sediment supply from extra-glacial sources (Church, 1972; Mills, 1979; Gurnell, 1982; Hammer and Smith, 1983; Richards, 1984). Sediment transport varies according to the relative importance of meltwater and rainfall controlled discharges. For a given discharge, suspended sediment transport may be higher during rainstorms because a higher proportion of runoff has traversed the valley sides and proglacial zone from which sediment can be readily

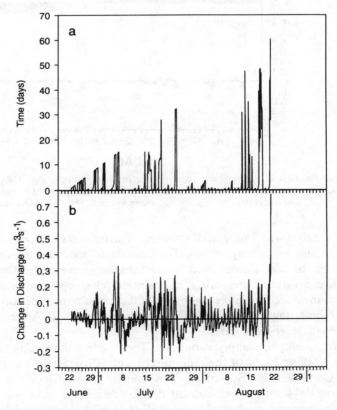

Figure 6. Time series of (a) the variable 'days since discharge was equalled or exceeded' ($dQ \Rightarrow$) used in the multiple regression model to represent medium-term sediment supply versus transport effects and (b) the variable 'rate of change of discharge' (ΔQ) used to represent daily hysteresis effects

removed. To account for this possibility, variables were defined representing the amounts of rainfall in the previous 1, 2, 3, 4 and 5 h.

These eight variables, together with discharge itself, were entered into a multiple regression (MR) model. Only three of the variables were significant in explaining the variation in SSC ($\alpha = 0.05$): discharge (Q), 'rate of change of discharge' (ΔQ) and 'days since discharge was equalled or exceeded' ($dQ =>$) (Table I). The two additional variables only accounted for an extra 2% of the SSC variance (Table II). Moreover, the residuals remained autocorrelated (Figure 5d; Table II).

Time series techniques

Autocorrelation clearly exists in the SSC data such that SSC is influenced not only by various hydrological characteristics, but also by past values of SSC itself. This is presumably because once sediment is mobilized at a given discharge, it is likely to remain transported in suspension even if discharge drops because the entrainment velocity of sediment is greater than the settling velocity (Richards, 1982: 80). This can be represented by autoregressive (AR) and/or moving average (MA) structures in the data.

For this reason, a Box–Jenkins seasonally integrated AR and MA model (ARIMA model) was fitted to the residuals from the MR model in an attempt to account for the remaining structure (Box and Jenkins, 1970). ARIMA models are expressed in the form ARIMA $(pdq)(PDQ)_s$ where p and q represent the orders of the AR and MA process, respectively, and d is the level of differencing; P and Q refer to the order of a 'seasonal' periodic AR and MA process, respectively, and D is the level of 'seasonal' differencing; s is the frequency of the 'seasonal' component. The acf and pacf of the residuals from the MR model (see Figure 5d) were used to identify suitable ARIMA models. Several models were found, but the ARIMA $(011)(200)_{12}$ model was the simplest which produced the lowest RSS (Table II) and removed all the residual serial autocorrelation (Figure 5).

IMPLICATIONS OF THE MODEL

The MR model has three variables which are significant in explaining SSC variations: discharge (Q), 'rate of change of discharge' (ΔQ) and 'days since discharge was equalled or exceeded' ($dQ =>$) (Table I). The positive sign of the partial regression coefficient for ΔQ indicates that higher SSCs occur on the rising limbs of diurnal discharge cycles than on the falling limbs. This may be because sediment (from subglacial erosion and from settling out of suspension in water) accumulates subglacially during low discharges (usually at night) and is flushed from subglacial locations during rising discharges (usually in the afternoon). The positive coefficient for $dQ =>$ suggests that for a given flow magnitude, SSCs increase with the length of time that has elapsed since a flow of comparable magnitude because a longer time period increases the storage of glacially eroded sediment in subglacial and proglacial areas.

Rainfall did not contribute significantly to the explanation of SSC variation. There were only 18 completely rain-free days during the 60 days between 22 June and 21 August and the rainfall variables were therefore not totally independent of stream discharge. Richards (1984) suggested that a rainfall variable might help to explain some of the variation in SSC in the proglacial stream draining Storbreen, Norway. In that study, however, the gauging station was located 1·2 km downstream from the glacier snout. At Midtdalsbreen, the proglacial zone (i.e. the area for supplying sediment during rainstorms) is much smaller. This might also account for the lack of significance of rainfall variables.

The variable measuring 'days since beginning of summer' also failed to contribute significantly to the explanation of SSC variation. There was therefore no gradual exhaustion of sediment supply over the summer. This could reflect reduced sediment production during the winter months when basal motion is reduced by about 50% (Willis, 1995). Alternatively, it might be explained if a dynamic distributed hydrological system remained over large parts of the glacier bed throughout the summer and did not collapse into a channelized system. Dye tracing experiments suggest that although a distributed system appeared to collapse into a channelized system beneath the eastern half of the glacier, a distributed system did remain beneath the western half (Willis et al., 1990).

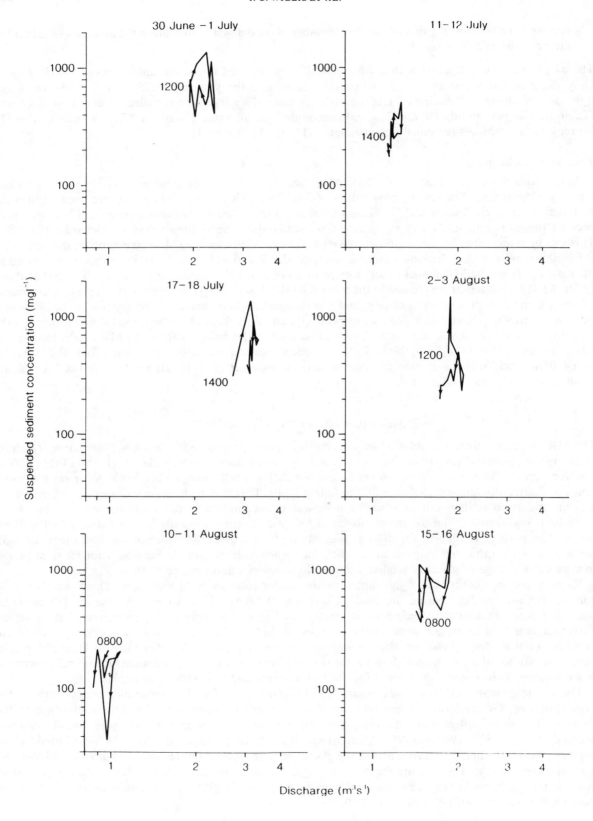

The ARIMA model fitted to the residuals from the MR model accounts for the remaining structure in the suspended sediment data. The model has a first-order moving average and a second-order seasonal auto-regressive component. Thus at each time interval, rates of change in SSC are affected by a random SSC change 2 h previously and changes in SSC 24 and 48 h previously. That SSCs are influenced by SSCs during the previous two days might reflect the typical length of time required for subglacial water to exhaust a new supply of sediment or for a new supply of sediment to become isolated again from the main subglacial drainage pathways (e.g. by channel migration). Interestingly, the same form of ARIMA model best described the proglacial stream SSC pattern at Glacier de Tsidjiore Nouve, Switzerland in the summer of 1981 (Gurnell and Fenn, 1984).

ANALYSIS OF THE MODEL RESIDUALS

Figure 3d shows the pattern of residuals from the combined MR and ARIMA model (MR–ARIMA model). Large positive residuals represent short-lived sediment flushes, whereas large negative residuals represent periods of short-term sediment exhaustion. As expected, some short-term sediment pulses are followed immediately by sediment exhaustion events. However, many large positive and negative residuals appear completely unrelated to one another. The rest of this paper is concerned with analyzing and explaining the large positive residuals (short-term sediment flushes). Short-term sediment flushes in proglacial streams have previously been linked to proglacial sources (i.e. bank slumping) (Fahnestock, 1963; Gurnell, 1982; Hammer and Smith, 1983; Gurnell, 1987; Gurnell and Warburton, 1990), or to sub-glacial sources (i.e. enhanced glacier erosion and/or tapping of sediment during changes in the hydraulics/configuration of the subglacial hydrological network) (Humphrey et al., 1986). Studies at Glacier de Tsidjiore Nouve and Bas Glacier d'Arolla, both in Switzerland, suggest that bank collapse causes only minor sediment pulses, which are short-lived and attenuate rapidly downstream (Gurnell and Warburton, 1990). One of the Midtdalsbreen pulses was observed in the field. It lasted for about 20 min and persisted downstream. For these reasons, we believe that most large positive residuals left in the pattern of SSC after model fitting are caused by periods of enhanced glacier erosion and/or changes in the hydraulics/configuration of the subglacial hydrological network. Both of these would be expected to be associated with increases in glacier motion.

CORRELATION BETWEEN SEDIMENT PULSES AND VARIATIONS IN GLACIER MOTION

In this section, the SSC residual series (Figure 3d) is compared with the glacier velocity series at M2, M3 and M4 (Figure 3a, 3b and 3c) to relate large positive residuals (short-term sediment flushes) to high glacier velocities. Frequency histograms for the SSC residual series and the glacier velocity series are shown for the period of overlap (i.e. between 2200 on 7 July and 1000 on 21 August) in Figure 8. To quantify 'large' positive residuals and 'high' glacier velocities, 27 observations representing the upper 5% of these frequency distributions were isolated. Large SSC values are those with greater than 0·296 residual SSC units and high glacier velocities are where surface speeds are over $0·243\,\mathrm{m\,d}^{-1}$ at M2, over $0·215\,\mathrm{m\,d}^{-1}$ at M3 and over $0·257\,\mathrm{m\,d}^{-1}$ at M4.

Most of the large positive SSC residuals occur as isolated observations, but eight occur in four groups of two observations separated by up to 4 h on 15 July, 23 July, 18 August and 20 August (Table III). There are therefore really 23 separate episodes of sediment flushing between 7 July and 21 August (Figure 3d). Most of the high glacier velocities occur consecutively in groups and represent large diurnal cycles or 'short-term motion events' (Willis, 1995). There are therefore really 11 separate episodes of fast glacier motion at M2, 12 at M3 and nine at M4 between 7 July and 21 August (Figure 3a, 3b and 3c). The association between the peaks in sediment flushing episodes and the peaks associated with fast motion episodes can be seen by comparing Figures 3a, 3b, 3c and 3d and a summary of the association is given in Table III. Many of

Figure 7. Diurnal variations in discharge and suspended sediment concentration showing clockwise hysteresis loops

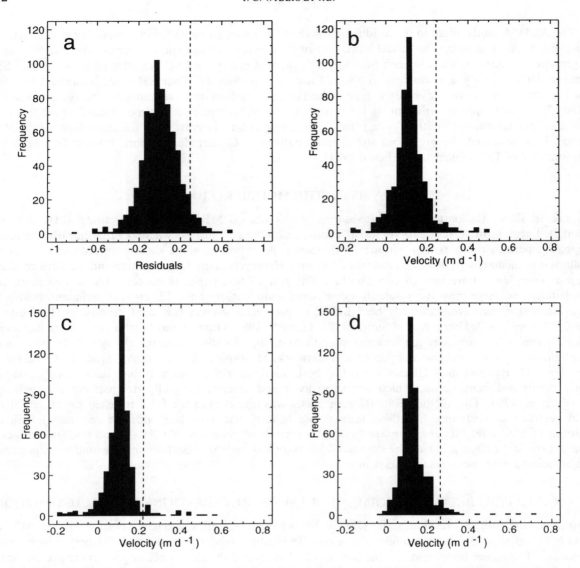

Figure 8. Frequency distributions for the time series between 2200 on 7 July and 1000 on 21 August of (a) MR–ARIMA model residuals, (b) glacier velocity at M2, (c) glacier velocity at M3 and (d) glacier velocity at M4. 5% of the observations in each series lie to the right of the broken lines

the sediment and motion peaks are associated with one another, although some sediment peaks occur independently of motion peaks and similarly some motion peaks occur independently of sediment peaks. The details of the associations are now discussed.

Of the 23 episodes of sediment flushing, five occur 'simultaneously' with peaks of fast motion episodes at one or more of the markers, nine occur up to 2 h after peaks in fast motion at one or more markers, 11 occur up to 4 h after fast motion peaks and 13 occur up to 6 h after motion peaks. Thus 57% of the sediment flushing episodes are correlated, albeit with varying time lags, with episodes of enhanced glacier motion. This considerably exceeds the percentage that would be expected if suspended sediment and glacier motion peaks occurred purely randomly. The theoretical probability distribution of association between randomly occurring sediment and motion peaks is not known. We therefore adopted a Monte Carlo simulation approach to generate an empirical probability distribution with which to compare the

Table III. Correlation between sediment flushes and enhanced glacier motion

Residual peak		Relative time of velocity peak (h)		
Day	Time	Marker 2	Marker 3	Marker 4
9 July	1800	—	—	—
15 July	2000/2200	−6	—	—
17 July	1000	—	—	—
	2000	−2	—	−2
19 July	2000	—	—	—
20 July	1800	—	—	—
23 July	2000/2400	—	−2	−2
25 July	1600	0	0	—
27 July	1200	0	—	—
29 July	1600	—	0	—
31 July	1600	—	−2	—
1 August	0200	0	—	—
2 August	0200	—	—	—
	1400	—	—	—
4 August	2200	—	—	−4
8 August	0800	—	—	—
12 August	1800	—	−2	—
16 August	0400	—	—	−4
	1600	0	0	0
17 August	1200	—	—	—
18 August	2200/2400	—	—	—
20 August	1800/2400	—	−6	—
21 August	0800	—	—	—

observed association (cf. Besag and Diggle, 1977). There are 535 two-hourly time intervals between 2200 on 7 July and 1000 on 21 August. There are 23 sediment peaks and 32 motion peaks. Our Monte Carlo algorithm generates 23 random numbers between 1 and 535 to represent a hypothetical distribution of sediment peaks and another 32 random numbers between 1 and 535 to represent a hypothetical distribution of motion peaks. It then calculates the number of sediment peaks which are associated with motion peaks, where an association is defined as a sediment peak occurring 0, 1, 2 or 3 time intervals after a motion peak. This procedure was repeated 1000 times to produce a probability distribution of the association between randomly occurring sediment and motion peaks. The mean number of associations was 5·4 and the standard deviation was 2·3. Only two of the 1000 simulations produced 13 or more associations. Thus there is a probability of 0·002 that the number of associations between suspended sediment pulses and glacier motion peaks observed at Midtdalsbreen would have happened by chance. We conclude, therefore, that the observed association is highly significant and that the sediment flushes observed in Midtdalsbreen's proglacial stream are caused by mechanisms associated with enhanced glacier motion.

Of the 23 sediment flushing episodes, three are associated with enhanced motion at M2 only, four with accelerated movement at M3 only and two with high speeds at M4 only (Figure 3; Table III). Thus proglacial stream sediment pulses *may* be most likely if enhanced motion is confined to the middle ablation area, less likely if an event is limited to the lower ablation area and least likely if an event occurs only higher up on the glacier, although these differences are probably not significant. Four of the sediment flushes are associated with glacier motion peaks at more than one marker: one with enhanced motion at both M2 and M3, one with fast motion at M3 and M4, one with fast motion at M2 and M4 and one with motion at all three markers (Table III).

Of the 11 high glacier velocity peaks at M2, six (55%) are linked with sediment flushes; seven of the 12 velocity peaks at M3 (58%) are related to sediment flushes; and five of the nine velocity peaks at M4 (56%) correspond with sediment flushes. Thus, on average, 56% of the 32 highest glacier velocity peaks recorded between 7 July and 21 August were associated with short-lived pulses of SSC in the proglacial stream between 0 and 6 h later. It is possible that more velocity peaks were associated with sediment flushes which were not recorded due to the two-hourly sampling interval and missing SSC data.

POSSIBLE CAUSAL MECHANISMS LINKING SEDIMENT FLUSHES WITH GLACIER MOTION

Short-term pulses of high SSC could reflect increased supply of sediment to the subglacial hydrological system in the region affected by enhanced glacier motion and its subsequent transport to the proglacial stream. Possible mechanisms for the generation of sediment pulses are considered in the following.

1. Increased subglacial water pressures may cause rapid basal sliding and enhanced subglacial erosion (quarrying, shearing, crushing and abrasion) (e.g. Boulton *et al.*, 1974; Hallet, 1979; Röthlisberger and Iken, 1981; Shoemaker, 1986). This effect will be more likely in a high-pressure distributed hydrological system and less likely in a low-pressure channelized hydrological system.
2. Increased subglacial water pressures in a subglacial hydrological system will steepen hydraulic gradients within the system. This may increase subglacial water velocities and may lead to the increased mobilization and transport of sediment (e.g. Clarke, 1987). This may happen in all types of subglacial hydrological system.
3. Increased pore water pressures in parts of a subglacial sediment layer may decrease sediment strength and increase the rate at which sediment deforms into nearby channels incised into the sediment (Boulton and Hindmarsh, 1987; Clarke, 1987; Alley, 1992; Walder and Fowler, 1994).
4. Increased subglacial water pressures in parts of a linked-cavity system may cause an increase in the number of active orifices linking subglacial cavities (Sharp *et al.*, 1989). This may lead to the tapping of sediment in new hydraulically activated cavities.
5. Increased subglacial water pressures and glacier sliding may cause the subglacial hydrological system to extend onto new areas of sediment down-glacier. This mechanism will be most effective if the drainage routes lie transverse to the direction of ice flow and may therefore be greater for a distributed system than a channelized system.
6. High subglacial water pressures may cause the collapse of parts of a distributed hydrological system into a channel system (Kamb, 1987; Boulton and Hindmarsh, 1987). This will reorientate the piezometric surface and steepen the hydraulic gradients within the remaining distributed system towards the channel. This may lead to greater water velocities and sediment transport as water flows from the distributed system into the channel or to greater sediment deformation into the channel.

Combinations of any of these mechanisms may generate sediment flushes. The collapse of a distributed system into a channelized system is likely to affect large areas of a glacier bed over fairly long periods of time (greater than a few hours) (e.g. Collins, 1989). For this reason, mechanism (6) is not likely to be relevant for sediment flushes lasting just a few hours. Thus most of the flushes are likely to be produced by either increased subglacial erosion [mechanism (1)], increased mobilization and transport of sediment within an existing hydrological network [mechanisms (2) and (3)] or increased mobilization and transport of sediment from new areas of the glacier bed [mechanisms (4) and (5)].

Theoretical calculations using Hallet's (1979) model of subglacial abrasion suggest that only a small fraction of the extra sediment measured during the flushes is derived directly from glacier erosion (Willis, unpublished data). Thus most of the sediment is removed from the subglacial store due to changes in the hydraulics and/or configuration of the subglacial hydrological system.

STRUCTURE OF THE SUBGLACIAL HYDROLOGICAL SYSTEM

Mobilized sediment will be transported to the proglacial stream provided the instantaneous vertical velocity components of turbulent flow remain greater than the settling velocity of the grains (about 10^{-6} m s^{-1} for clay and 10^{-2} m s^{-1} for very fine sand). In this respect, the transport of sediment through a subglacial drainage system is different from the transport of dye. Dye in slow-moving or stagnant subglacial water is removed by diffusion once the dye concentration in the main flow pathway drops below the dye concentration in the storage zones. Once sediment is stored in pools or cavities it may settle out of suspension and is only removed when turbulent flow in the storage zones is re-established and the sediment is reactivated. Thus, sediment pulses differ from dye-return curves. They are short-lived with a rapid rise to, and fall from, peak concentration; they lack a gently falling concentration limb resulting from storage retardation.

Time lags between glacier motion peaks and sediment flushes represent the average time taken for sediment mobilized by subglacial water to be transported to the glacier snout and thence to the GS. In the 1987 summer, the average proglacial travel times were about 13 min if water travelled down T1 and 31 min if water travelled along T3 (see Figure 1) (Willis *et al.*, 1990). We now use the lags between the glacier velocity peaks and the sediment flushes to compute the average sediment transport velocity in the subglacial hydrological system after each period of enhanced motion. Two velocities were calculated for each sediment flush associated with fast glacier motion, assuming, respectively, straight line distances from the affected marker to the emergence points of T1 (525, 750 and 1185 m from M2, M3 and M4, respectively) and T3 (510, 700 and 1090 m from M2, M3 and M4, respectively). Clearly, entrainment of sediment beneath the glacier cannot produce an instantaneous sediment pulse in the proglacial stream. The apparent synchroneity of some motion peaks and sediment pulses is an artefact of the two-hourly sampling interval; in these instances, the sediment travelled down-glacier in less than 2 h, implying a velocity from M2 of at least $0.08-0.10 \, \mathrm{m \, s^{-1}}$ depending on the route taken (T1 or T3). Sediment velocities suggested by the time lags between motion and sediment peaks are given in Table IV. Thus if sediment enters the subglacial hydrological system at the time and location of the motion event peak, it travelled down-glacier from M2 at velocities of between 0.03 and $0.30 \, \mathrm{m \, s^{-1}}$, from M3 at velocities of between 0.04 and $0.40 \, \mathrm{m \, s^{-1}}$ and from M4 at velocities of between 0.09 and $0.76 \, \mathrm{m \, s^{-1}}$. All these velocities are similar orders of magnitude (i.e. 10^{-2} and $10^{-1} \, \mathrm{m \, s^{-1}}$) to those observed in dye tracing experiments during 1987 (Willis *et al.*, 1990).

Throughout the 1987 summer, water velocities through a sinuous R-channel were in the range 0.65 to $1.42 \, \mathrm{m \, s^{-1}}$ (Willis *et al.*, 1990). Only one of the estimated sediment velocities fell within this range: the pulse which could have travelled between M4 and the GS on 16 August at 1600 (Table IV). However, this pulse was also associated with glacier motion peaks at M2 and M3. Thus if the sediment had been mobilized beneath these markers rather than M4, lower velocities of between 0.19 and $0.40 \, \mathrm{m \, s^{-1}}$ are suggested. Given the range of velocities observed during the dye tracing experiments, it seems more likely that the sediment associated with the 1600 pulse on 16 August was mobilized beneath the lower and/or middle ablation area rather than the upper ablation zone. All other sediment velocities were substantially less than $0.65 \, \mathrm{m \, s^{-1}}$, suggesting that the sediment flushes travelled at least part of their length through a

Table IV. Range of subglacial water (and sediment) velocities $(\mathrm{m \, s^{-1}})$ calculated from the time between glacier velocity peak and associated proglacial stream sediment flush

Day	Time	Marker 2	Marker 3	Marker 4	Range
15 July	2000/2200	0.026 ± 0.001	—	—	0.025–0.027
17 July	2000	0.089 ± 0.007	—	0.195 ± 0.010	0.082–0.205
23 July	2000/2400	—	0.056 ± 0.001	0.195 ± 0.010	0.055–0.205
25 July	1600	0.240 ± 0.054	0.334 ± 0.068	—	0.186–0.402
27 July	1200	0.240 ± 0.054	—	—	0.186–0.294
29 July	1600	—	0.334 ± 0.068	—	0.266–0.402
31 July	1600	—	0.124 ± 0.007	—	0.117–0.131
1 August	0200	0.240 ± 0.054	—	—	0.186–0.294
4 August	2200	—	—	0.087 ± 0.000	0.087–0.087
12 August	1800	—	0.124 ± 0.007	—	0.117–0.131
16 August	0400	—	—	0.087 ± 0.000	0.087–0.087
16 August	1600	0.240 ± 0.054	0.334 ± 0.068	0.755 ± 0.179	0.186–0.934
20 August	1800/2000	—	0.036 ± 0.001	—	0.035–0.037

hydrological network more distributed than a meandering R-channel. During the 1987 summer, water velocities in a linked-cavity system have been estimated to lie between 0·03 and 0·07 m s^{-1} (Willis *et al.*, 1990). The observed sediment velocities can therefore be interpreted in terms of flow through such a system when velocities are less than 0·07 m s^{-1}, or through a combined linked-cavity and sinuous R-channel system when velocities are greater than this.

The first sediment pulse associated with a peak in glacier velocity on 15 July travelled very slowly beneath the glacier at a velocity of about 0·03 m s^{-1} (Table IV), consistent with transport through a linked-cavity system. By 17 July, the lag between glacier motion peaks and proglacial stream sediment pulses had shortened such that the observed velocities can only be explained with recourse to a combined linked-cavity/channel system (Table IV). The evidence also implies that sometime between 16 and 20 August, the lag increased again and the velocity decreased.

An interpretation of these changes is that before 17 July, Midtdalsbreen was entirely drained by a distributed hydrological system consisting, perhaps, of linked-cavities. This is supported by dye tracing experiments which showed that before 17 July, dye moved slowly and became highly dispersed as it travelled through the glacier (Willis *et al.*, 1990). It is also supported by glacier velocity measurements which were generally higher in early July than they were later in the season (Figure 3).

On 15 July, a short-term glacier motion event occurred on Midtdalsbreen which resembled the motion events observed on Findelengletscher, Switzerland (Iken and Bindschadler, 1986) and the mini-surges of Variegated Glacier, Alaska (Kamb and Engelhardt, 1987). The event involved the down-glacier propagation of a wave of high velocities, during which water was released from subglacial storage (Willis, 1995). During this event, unstable orifice growth probably occurred as a result of high discharges and water pressures and the existing linked-cavity system evolved into an arborescent channel system over the next few days. After 15 July, therefore, Midtdalsbreen was underlain by both a distributed linked-cavity system (particularly beneath the western half) and a channelized system (mainly beneath the eastern half) (Willis *et al.*, 1990; Willis, 1995). The increased lag between the motion peak and sediment pulse on 20 August might indicate that the source of the sediment pulse lay further from the main drainage pathways than in the early summer. It might also indicate the collapse of large conduits and the restoration of linked-cavity drainage over certain parts of the bed in response to a reduction in discharge throughout the basal hydrological network.

CONCLUSIONS

Two-hourly variations in proglacial stream SSC can be explained in terms of two-hourly variations in stream discharge, diurnal hysteresis effects, medium-term variations in sediment supply and transport and changes in the recent SSC history of the stream. Large pulses of SSC unrelated to discharge were observed throughout the year in the proglacial stream. Most of these sediment flushes were correlated with enhanced glacier motion. They cannot be explained by enhanced sediment production by subglacial erosion (Willis, unpublished data), but are probably due to the tapping of subglacially stored sediment during sudden changes in the subglacial hydrological system associated with high subglacial water pressures. Changes in the hydrological system probably involve the mobilization and transport of sediment in existing linked-cavities in response to increased hydraulic gradients and the hydraulic integration of new cavities into the system by an increase in the number of active orifices linking the cavities and by the down-glacier growth of existing cavities.

These findings justify an integrated approach to glacier hydrology. Dye tracing studies, glacier velocity measurements and suspended sediment data can all be interpreted to provide a picture of the seasonally evolving nature of the hydrological system. They also have important implications for the prediction of suspended sediment yields from glacierized catchments. Most existing methods use linear regression or transfer function techniques to predict sediment yields on the basis of discharge fluctuations alone (Fenn, 1989). An improvement to existing models might result if changing patterns of sediment supply to proglacial streams in response to subglacial drainage changes were incorporated. Thus variables such as subglacial water pressure or glacier velocity might prove useful additional predictors of suspended sediment concentrations in proglacial streams.

ACKNOWLEDGEMENTS

We gratefully acknowledge the financial support of the Natural Environment Research Council (Studentship GT4/AAPS/86/10 to I.C.W.), Earthwatch, Emmanuel College, Cambridge, the Worts Travelling Scholars Fund and the Scandinavian Studies Fund. The extensive data set could not have been collected without the valiant help of many Cambridge undergraduates and Earthwatch volunteers. Special thanks are due to J. Campbell Gemmel for help in the field, Erika Leslie and the Universities of Oslo and Bergen for use of the Finse Field Station, the staff of the Finsehytte for allowing us occasional escape from our camp, and Norwegian State Railways for granting us road access to Finse. Damian Lawler commented on an earlier draft of the paper which resulted in significant improvements to it.

REFERENCES

Alley, R. B. 1992. 'How can low-pressure channels and deforming tills coexist subglacially?' *J. Glaciol.*, **38**, 200–207.

Andersen, J. L. and Sollid, J. L. 1971. 'Glacial chronology and glacial geomorphology in the marginal zone of the glaciers Midtdalsbreen and Nigardsbreen, South Norway', *Norsk Geogr. Tidssk.*, **25**, 1–38.

Beecroft, I. 1983. 'Sediment transport during an outburst from Glacier de Tsidjiore Nouve, Switzerland 16–19 June 1981', *J. Glaciol.*, **29**, 185–190.

Besag, J. E. and Diggle, P. J. 1977. 'Simple Monte Carlo tests for spatial pattern', *Appl. Statis.*, **26**, 327–333.

Bezinge, A. 1987. 'Glacial meltwater streams, hydrology and sediment transport: the case of the Grande Dixence hydroelectricity scheme' in Gurnell, A. M. and Clarke, M. J. (Eds), *Glacio-fluvial Sediment Transfer: an Alpine Perspective*. Wiley, Chichester. pp. 473–498.

Bezinge, A., Clarke, M. J., Gurnell, A. M., and Warburton, J. 1989. 'The management of sediment transported by glacial melt-water streams and its significance for the estimation of sediment yield', *Ann. Glaciol.*, **13**, 1–5.

Bogen, J. 1980. 'The hysteresis effect of sediment transport systems', *Norsk Geogr. Tidssk.*, **34**, 45–54.

Boulton, G. S. 1974. 'Processes and patterns of glacial erosion' in Coates, D. R. (Ed.), *Glacial Geomorphology*. State University, New York. pp. 41–87.

Boulton, G. S. and Hindmarsh, R. C. A. 1987. 'Sediment deformation beneath glaciers: rheology and geological consequences', *J. Geophys. Res.*, **92**(B9), 9059–9082.

Boulton, G. S., Dent, D. L., and Morris, E. M. 1974. 'Subglacial shearing and crushing and the role of water pressures in tills from south-east Iceland', *Geogr. Ann.*, **56A**, 135–145.

Box, G. E. P. and Cox, D. R. 1964. 'An analysis of transformations', *J. Roy. Statis. Soc. Ser. B*, **26**, 211–252.

Box, G. E. P. and Jenkins, G. M. 1970. *Time Series Analysis, Forecasting and Control*. Holden-Day, San Francisco.

Butz, D. 1989. 'The agricultural use of meltwater in Hopar settlement, Pakistan', *Ann. Glaciol.*, **13**, 35–39.

Chatfield, C. 1984. *The Analysis of Time Series: an Introduction*. 3rd edn. Chapman and Hall, London.

Church, M. 1972. 'Baffin Island sandurs: a study of arctic fluvial processes', *Geol. Surv. Can. Bull.*, **216**, 1–208.

Church, M. and Gilbert, R. 1975. 'Proglacial, fluvial and lacustrine environments' in Jopling, A. V. and Macdonald, B. C. (Eds), *Glaciofluvial and Glaciolacustrine Sedimentation*. Soc. Econ. Palaeontol. Mineral. Spec. Publ., **23**, 22–100.

Church, M. and Ryder, J. M. 1972. 'Paraglacial sedimentation: a consideration of fluvial processes conditioned by glaciation', *Bull. Geol. Soc. Am.*, **83**, 3059–3072.

Clarke, G. K. C. 1987. 'Subglacial till: a physical framework for its properties and processes', *J. Geophys. Res.*, **92**(B9), 9023–9036.

Cochrane, D. and Orcutt, G. H. 1949. 'Applications of least squares regression to relationships containing autocorrelated error terms', *J. Am. Statis. Assoc.*, **44**, 32–61.

Collins, D. N. 1979. 'Sediment concentration in meltwaters as an indicator of erosion processes beneath an Alpine glacier', *J. Glaciol.*, **23**, 247–259.

Collins, D. N. 1986. 'Characteristics of meltwater draining from the portal of an Alpine glacier during the emptying of a marginal ice-dammed lake', *Mater. Glyatsiolog. Issl. Khron. Obs.*, **58**, 224–232.

Collins, D. N. 1989. 'Seasonal development of subglacial drainage and suspended sediment delivery to melt waters beneath an Alpine glacier', *Ann. Glaciol.*, **13**, 45–50.

Collins, D. N. 1990. 'Glacial processes' in Goudie, A. (Ed.), *Geomorphological Techniques*. 2nd edn. George Allen and Unwin, London.

Drewry, D. 1986. *Glacial Geologic Processes*. Edward Arnold, London.

Fahnestock, M. 1963. 'Morphology and hydrology of a glacial stream, White River, Mount Rainer, Washington', *U.S. Geol. Surv. Prof. Pap.*, **422-A**.

Fenn, C. R. 1987. 'Sediment transfer processes in Alpine glacier basins' in Gurnell, A. M. and Clarke, M. J. (Eds), *Glacio-fluvial Sediment Transfer: an Alpine Perspective*. Wiley, Chichester. pp. 59–85.

Fenn, C. R. 1989. 'Quantifying the errors involved in transferring suspended sediment rating equations across ablation seasons', *Ann. Glaciol.*, **13**, 64–68.

Fenn, C. R., Gurnell, A. M., and Beecroft, I. 1985. 'An evaluation of the use of suspended sediment rating curves for the prediction of suspended sediment concentration in a proglacial stream', *Geogr. Ann.*, **67A**, 71–82.

Ferguson, R. I. 1977. 'Linear regression in geography', *Concepts and Techniques in Modern Geography*, **15**, Geo-Abstracts, University of East Anglia.

Ferguson, R. I. 1984. 'Sediment load of the Hunza River' in Miller, K. J. (Ed.), *International Karakoram Project 2*. Cambridge University Press, Cambridge. pp. 581–598.

Fowler, A. C. 1987. 'Sliding with cavity formation', *J. Glaciol.*, **33**, 255–267.

Gurnell, A. M. 1982. 'The dynamics of suspended sediment concentration in an Alpine proglacial stream network', *IAHS Publ.*, **138**, 319–330.

Gurnell, A. M. 1987. 'Suspended sediment' in Gurnell, A. M. and Clarke, M. J. (Eds), *Glacio-fluvial Sediment Transfer: an Alpine Perspective*. Wiley, Chichester. pp. 305–354.

Gurnell, A. M. and Fenn, C. R. 1984. 'Box–Jenkins transfer function models applied to suspended sediment concentration–discharge relationships in a proglacial stream', *Arctic Alpine Res.*, **16**, 93–106.

Gurnell, A. M. and Warburton, J. 1990. 'The significance of suspended sediment pulses for estimating suspended sediment load and identifying suspended sediment sources in Alpine glacier basins', *IAHS Publ.*, **193**, 463–470.

Hallet, B. 1979. 'A theoretical model of glacial abrasion', *J. Glaciol.*, **23**, 39–50.

Hammer, K. M. and Smith, N. D. 1983. 'Sediment production and transport in a proglacial stream: Hilda Glacier, Alberta, Canada', *Boreas*, **12**, 91–106.

Harbor, J. M. 1992. 'Application of a general sliding law to simulating flow in a glacier cross-section', *J. Glaciol.*, **38**, 182–190.

Hazony, Y. 1979. 'Algorithms for parallel processing: curve and surface definitions with Q-splines', *Comput. Graphics*, **4**, 165–176.

Hooke, R. Le B. 1989. 'Englacial and subglacial hydrology: a qualitative review', *Arctic Alpine Res.*, **21**, 221–233.

Hooke, R. Le B., Calla, P., Holmlund, P., Nilsson, M., and Stroeven, A. 1989. 'A three year record of seasonal variations in surface velocity, Storglaciären, Sweden', *J. Glaciol.*, **35**, 235–247.

Humphrey, N., Raymond, C., and Harrison, W. 1986. 'Discharges of turbid water during the mini-surges of Variegated Glacier, Alaska, U.S.A.', *J. Glaciol.*, **29**, 28–47.

Iken, A. 1977. 'Variations of surface velocities of some Alpine glaciers measured at intervals of a few hours. Comparison with Arctic glaciers', *Z. Gletscherk. Glazialgeol.*, **13**, 23–35.

Iken, A. and Bindschadler, R. A. 1986. 'Combined measurements of subglacial water pressure and surface velocity of the Findelengletscher, Switzerland: conclusions about drainage system and sliding mechanism', *J. Glaciol.*, **32**, 101–119.

Iverson, N. R. 1990. 'Laboratory simulations of glacial abrasion: comparisons with theory', *J. Glaciol.*, **36**, 304–314.

Iverson, N. R. 1991. 'Potential effects of subglacial water-pressure fluctuations on quarrying', *J. Glaciol.*, **37**, 27–36.

Kamb, B. 1987. 'Glacier surge mechanism based on linked cavity configuration of the basal water conduit system', *J. Geophys. Res.*, **92(B9)**, 9083–9100.

Kamb, B., and Engelhardt, H. F. 1987. 'Waves of accelerated motion in a glacier approaching surge: the mini-surges of Variegated Glacier, Alaska, U.S.A.', *J. Glaciol.*, **33**, 27–46.

Liestol, O. 1967. 'Storbreen glacier in Jotunheimen, Norway', *Norsk. Polarinst. Skr.*, **141**, 1–63.

Mathews, W. H. 1964. 'Sediment transport from Athabasca Glacier, Alberta', *IAHS Publ.*, **65**, 155–165.

Metcalf, R. C. 1979. 'Energy dissipation during subglacial abrasion at Nisqually Glacier, Washington, USA', *J. Glaciol.*, **23**, 233–246.

Mills, H. H. 1979. 'Some implications of sediment studies for glacial erosion on Mt. Rainer', *Northwest Sci.*, **53**, 190–199.

Østrem, G. 1975. 'Sediment transport in glacial meltwater streams' in Jopling, A. V. and McDonald, B. C. (Eds), *Glaciofluvial and Glaciolacustrine Sedimentation. Soc. Econ. Palaeontol. Mineral. Spec. Publ.*, **23**. pp. 101–122.

Østrem, G., Bridge, C. W., and Rannie, W. F. 1967. 'Glacio-hydrology, discharge and sediment transport in the Decade Glacier area, Baffin Island, N.W.T.', *Geogr. Ann.*, **49A**, 268–282.

Richards, K. S. 1982. *Rivers, Form and Process in Alluvial Channels*. Methuen. London.

Richards, K. S. 1984. 'Some observations on suspended sediment dynamics in Storbregrova, Jotunheimen', *Earth Surf. Process. Landforms*, **9**, 101–112.

Robin, G. de Q. 1976. 'Is the basal ice of a temperate glacier at the pressure melting point?' *J. Glaciol.*, **16**, 183–196.

Röthlisberger, H. and Iken, A. 1981. 'Plucking as an effect of water pressure variations at the glacier bed', *Ann. Glaciol.*, **2**, 57–62.

Röthlisberger, H. and Lang, H. 1987. 'Glacial hydrology' in Gurnell, A. M. and Clarke, M. J. (Eds), *Glacio-fluvial Sediment Transfer: an Alpine Perspective*. Wiley, Chichester. pp. 207–284.

Sharp, M. J., Gemmell, J. C., and Tison, J-L. 1989. 'Structure and stability of the former subglacial drainage system of the glacier de Tsanfleuron, Switzerland', *Earth Surf. Process. Landforms*, **14**, 119–134.

Shoemaker, E. M. 1986. 'Subglacial hydrology for an ice sheet resting on a deformable aquifer', *J. Glaciol.*, **32**, 20–30.

Sigmond, E. M. O., Gustavson, M., and Roberts, D. 1984. *1 : 1 000 000 Bedrock Map of Norway*. Norges Geologiske Unders Ýkelse.

Stone, D. B., Clarke, G. K. C., and Blake, E. W. 1993. 'Subglacial measurements of turbidity and electrical conductivity', *J. Glaciol.*, **39**, 415–420.

Walder, J. S. and Fowler, A. 1994. 'Channelised subglacial drainage over a deformable bed', *J. Glaciol.*, **40**, 3–15.

Walters, R. A. and Dunlap, W. W. 1987. 'Analysis of time series of glacier speed: Columbia Glacier, Alaska', *J. Geophys. Res.*, **92(B9)**, 8969–8975.

Willis, I. C. 1995. 'Intra-annual variations in glacier motion: a review', *Progr. Phys. Geogr.*, **19**, 61–106.

Willis, I. C., Sharp, M. J., and Richards, K. S. 1990. 'Configuration of the drainage system of Midtdalsbreen, Norway as indicated by dye-tracing experiments', *J. Glaciol.*, **36**, 89–101.

IMPACT OF POST-MIXING CHEMICAL REACTIONS ON THE MAJOR ION CHEMISTRY OF BULK MELTWATERS DRAINING THE HAUT GLACIER D'AROLLA, VALAIS, SWITZERLAND

G. H. BROWN AND M. J. SHARP

Department of Geography, University of Cambridge, Downing Place, Cambridge CB2 3EN, UK

M. TRANTER

Department of Geography, University of Bristol, Bristol BS8 1SS, UK

A. M. GURNELL

Department of Geography and GeoData Institute, University of Southampton, Highfield, Southampton SO9 5NH, UK

AND

P. W. NIENOW

Department of Geography, University of Cambridge, Downing Place, Cambridge CB2 3EN, UK

ABSTRACT

Until now, alpine glacial meltwaters have been assumed to consist of two components, dilute quickflow and concentrated delayed flow, the mixing of which has been regarded as chemically conservative for the major dissolved ions and electrical conductivity. Dye tracing results suggest that this two-component model adequately represents the subglacial hydrology of the Haut Glacier d'Arolla, Switzerland. However, laboratory dissolution experiments in which various concentrations of glacial rock flour are placed in dilute solutions show that this rock flour is highly reactive and suggest that bulk meltwaters may acquire significant amounts of solute through rapid chemical reactions with suspended sediment which occur after mixing of the two components. This view is supported by detailed analysis of variations in the hydrochemistry of meltwaters draining from the Haut Glacier d'Arolla over three diurnal cycles during the 1989 melt season. Variations in the composition of bulk meltwaters are controlled by two main factors: dilution of the delayed flow component by quickflow, and the extent of post-mixing reactions. The latter depends on the suspended sediment concentration in bulk meltwaters and on the duration of contact between these waters and suspended sediment. Seasonal changes in the magnitude of these factors result in changes in the character and causes of diurnal variations in meltwater chemistry. In June, these variations reflect discharge-related variations in residence time within a distributed subglacial drainage system; in July, when a channelized drainage system exists beneath the lower glacier, they primarily reflect the dilution of delayed flow by quickflow; in August, when suspended sediment concentrations are particularly high, they reflect varying degrees of solute acquisition by post-mixing reactions with suspended sediment that take place in arterial channels at the glacier bed.

INTRODUCTION

It is well known that meltwaters acquire solutes during transit through a glacier (Collins, 1978). The ionic strength (I) of alpine glacial meltwaters has a broad spectrum of values, ranging from dilute supraglacial waters ($I \approx 10\,\mu$equiv. 1^{-1}) to the concentrated bulk meltwaters characteristic of recession flows ($I \approx 1000\,\mu$equiv. 1^{-1}). Solutes are acquired by reactions such as sulphide oxidation and carbonation.

However, the rates of these reactions are known only in qualitative terms (Reynolds and Johnson, 1972; Raiswell, 1984; Tranter *et al.*, in press). There has been little detailed study of the sources of solute transported by meltwaters (Fairchild *et al.*, in press) or of the way in which these are determined by the nature of the hydrological environment at the glacial bed (Sharp, 1991).

Geochemical research in glaciated regions has had two main aims (Eyles *et al.*, 1982). The first has been to characterize water quality and to quantify geochemical weathering rates (Rainwater and Guy, 1961; Keller and Reesman, 1963; Reynolds and Johnson, 1972; Slatt, 1972; Church, 1974; Sasserville *et al.*, 1977; Eyles *et al.*, 1982). These studies document high rates of chemical weathering in glaciated catchments, a result of the rapid dissolution of freshly ground glacial rock flour. The second aim has been to identify chemical signatures which may differentiate runoff components and hydrological pathways through glacierized basins (e.g. Collins, 1978; 1979; 1981; Oerter *et al.*, 1980; Collins and Young, 1981; Gurnell and Fenn, 1984; Brown and Tranter, 1990; Tranter and Raiswell, 1991). Most of the flow routing studies make the pragmatic assumptions that individual runoff components have constant compositions and that they mix in a manner which is conservative with respect to parameters such as electrical conductivity. When applied to turbid glacial meltwaters, however, these assumptions seem inconsistent with the results of water quality studies in that they do not recognize the geochemical reactivity of rock flour (Slatt, 1972; Tranter *et al.*, in press).

Dye tracing experiments carried out in 1989 and 1990 at the Haut Glacier d'Arolla, Switzerland, indicate that the subglacial drainage system contains two principal components: a distributed system in which flow velocities are of the order of $0.05 \, \text{m s}^{-1}$, and a channelized system in which flow velocities are in the range $0.5–1.0 \, \text{m s}^{-1}$. During the melt season the channelized system expands headwards at the expense of the distributed system, so that by the end of the season it occupies the lowermost 3–3.5 km of the 4 km long glacier (Nienow *et al.*, in prep.). Headwards growth of the subglacial drainage channels closely follows the upglacier retreat of the transient summer snowline on the glacier surface. This drainage configuration suggests that a two-component model should adequately describe the subglacial transport of meltwaters at the Haut Glacier d'Arolla for much of the summer ablation season. We therefore suggest that bulk (or total) discharge consists of two components. The first, quickflow, is derived largely from icemelt, which passes rapidly through the glacier in ice-walled channels (englacial flow *sensu stricto*), and has only limited contact with sediments. Hence quickflow maintains the composition of icemelt and is extremely dilute. The second, delayed flow, is derived largely from snowmelt, and drains relatively slowly through the distributed system at the ice–bedrock interface, where it can pick up significant amounts of solutes (Tranter *et al.*, in press). Before draining from the glacier portal, the two components mix in the major arterial channels at the glacier bed (Sharp *et al.*, in prep.). It is only in these large subglacial channels that the bulk meltwaters encounter significant sediment concentrations (early season runoff from the distributed system transports minimum amounts of suspended sediment). We suggest that bulk meltwaters may acquire a significant proportion of their solute load by so-called 'post-mixing reactions', which involve the weathering of suspended sediments, within this environment.

The aim of this paper is to present the results of simple laboratory dissolution experiments which allow us to define the types of reactions which occur when dilute waters come into contact with glacial rock flour, and to quantify the dependence of solute acquisition rates on such factors as the suspended sediment concentration and water–rock contact time. We limit this study to the behaviour of the major ions and pH. We then use the results of these experiments to evaluate the potential contribution of post-mixing reactions between meltwaters and suspended sediments to seasonal and diurnal variations in bulk meltwater chemistry, as observed over three diurnal melt cycles at the Haut Glacier d'Arolla during the 1989 melt season.

LABORATORY DISSOLUTION EXPERIMENTS

A suite of dissolution experiments were conducted to simulate chemical weathering by dilute meltwaters using boundary conditions approximating those in the field at the Haut Glacier d'Arolla. The time-scale of the experiments, three hours, is representative of the transit time of bulk meltwaters through basal arterial channels during the later part of the ablation season, as indicated by dye tracing experiments (Nienow *et al.*, in prep.). The temperature ($0–3\cdot1°C$) is realistic for field conditions (bulk meltwater temperatures ranged between $0\cdot5$ and $2\cdot5°C$ during the diurnal cycles sampled). Three sediment concentrations were used ($0\cdot3$, $1\cdot0$ and $4\cdot0\,g\,l^{-1}$) to simulate the maximum concentrations observed during the three diurnal cycles sampled in the field (field suspended sediment concentrations ranged from $0\cdot1$ to $4\cdot0\,g\,l^{-1}$). Deionized water was chosen as an analogue for icemelt.

The choice, collection and treatment of natural particulate materials for use in dissolution experiments is always problematic as the surface properties of the particulates may be modified during transport and treatment (Tranter, 1982). We adopted the following approach to minimize potential experimental artefacts. Silt- and clay-sized sediment, which had been deposited on the proglacial outwash plain of the Haut Glacier d'Arolla, was collected and transported without pretreatment from the field in plastic containers. It was dried for 36 hours at $50°C$ before weighing.

A pre-cleaned plastic beaker containing $950 \pm 2\,ml$ of deionized water was agitated continuously by a polyethylene covered magnetic bar which was rotated by a magnetic stirrer beneath the beaker. The apparatus was contained within a fridge, which maintained the temperature of the solution between $0\cdot0$ and $3\cdot1°C$. The water temperature was monitored with a Grant thermistor which has an accuracy of $\pm0\cdot2°C$. The pH was measured continuously during the first 60 minutes of each experiment using a Jenway PWA2 water analyser and Jenway combination glass electrode, calibrated using standard Gallenkamp pH buffer powders ($7\cdot11$ and $9\cdot49$ at $0°C$). Subsequently, the pH was recorded simultaneously with the extraction of samples (see later), using one-point recalibration of the electrode. Recalibration of the electrode showed only minor drift ($\pm0\cdot1$ pH) during the continuous monitoring. Continuous data were recorded on a ΔT Devices datalogger. To begin an experiment, the thermistor and pH electrode were allowed to equilibrate for five minutes before a known weight of sediment was added to the solution. The stirred system was then allowed to evolve for three hours with free access to the atmosphere. Aliquots ($10\,ml$) of the reaction mixture were collected at intervals indicated by the data points in Figure 2. Samples were immediately syringe-filtered through Whatman $0\cdot45\,\mu m$ cellulose nitrate membranes and refrigerated at $4°C$. The reduction in the volume of stirred solution over the duration of the experiment was about 17%.

The water samples were used for the following analyses. Total alkalinity was determined using a Radiometer MTS 800 Multi-Titration System, titrating to an end-point of pH $4\cdot5$ with $0\cdot36$ mmol HCl. Sample volumes were $4\,ml$, and titre volumes ranged from $0\cdot1$ to $2\cdot0\,ml$. Precision was $\pm5\%$. Ca^{2+}, Mg^{2+}, Na^+ and K^+ were determined by atomic absorption spectrophotometer on a Perkin-Elmer 2280 instrument using an air–acetylene flame. Samples were acidified with HNO_3 and $210\,\mu equiv.\,l^{-1}$ of the releasing agent $La(NO_3)_3$ was added to all samples and standards before analysis. Precision was $\pm5\%$. SO_4^{2-} was determined by ion chromatography on a Dionex 2000i. Precision was $\pm3\%$.

AREA OF FIELD STUDY

Field sampling was undertaken at the Haut Glacier d'Arolla, the most southerly glacier in the Arolla valley, Val d'Hérens, Switzerland (Figure 1). Table I summarizes the geomorphological characteristics of the

Table I. Catchment characteristics of the Haut Glacier d'Arolla (after Warburton, 1989)

Catchment area	$12\cdot00\,km^2$
Glaciated area	$6\cdot30\,km^2$
Percentage of catchment glacierized	$54\cdot00\%$
Orientation of accumulation area	NNE
Orientation of ablation area	NNW
Maximum length of glacier	$4\cdot20\,km$
Maximum catchment altitude	$3838\,m$
Maximum glacier altitude	$3498\,m$
Altitude of snout	$2560\,m$

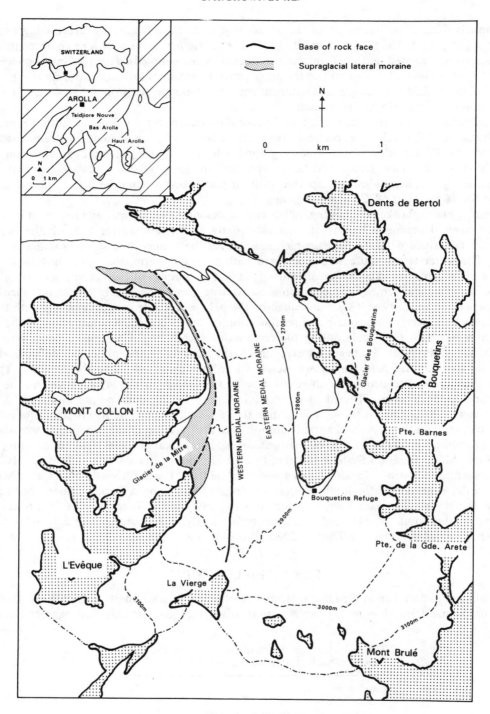

Figure 1. Location map for the Haut Glacier d'Arolla, Switzerland

catchment. The bedrock geology is varied and consists of metamorphic and igneous rocks of the Arolla series of the Dent Blanche nappe, the highest tectonic unit of the Valais Alps (Mazurek, 1986). The accumulation zone and upper tongue of the glacier are underlain predominantly by schistose granite. Chloritic schist and gneiss underlie the south-eastern corner, whereas the western side of the catchment below Glacier de la Mitre is dominated by amphibolites and chloritic schist. Amphibolites underlie the lower glacier to the north-east. To the west, Mont Collon consists of a diorite/gabbro intrusion. Geo-chemically reactive minerals, such as pyrite and calcite, have been identified by microscopy and are present in trace amounts in many of the rocks throughout the catchment (Brown, 1991).

SAMPLING METHODS AND TECHNIQUES

Three diurnal discharge cycles were sampled intensively during the 1989 ablation season (22–23 June, 17–18 July, 16–17 August), when samples were collected hourly for between 13 and 24 hours, starting at 1000 hours.

Full details of discharge and suspended sediment monitoring can be found in Gurnell *et al.* (in press). Full details of water sample collection and treatment may be found in Brown (1991). In summary, alkalinity (largely HCO_3^-) was determined in the field on filtered solutions by colorimetric titration to an end-point of 4·5 using 1 mmol HCl and BDH 4·5 mixed indicator solution. pH was measured in the field laboratory. The major cations (Ca^{2+}, Mg^{2+}, Na^+ and K^+) and the other major anion, SO_4^{2-}, were determined on return to the laboratory by atomic absorption spectrophotometry and ion-exchange chromatography, respectively (Brown, 1991).

The charge balance errors, defined as

$$(\Sigma M^+ - \Sigma A^-)/(\Sigma M^+ + \Sigma A^-) \times 100 \qquad (1)$$

where M^+ is the sum of positively charged cations and A^- the sum of negatively charged anions, were June, $-8·6$ to $-4·7\%$ ($x = -6·8\%$); July, -13 to -7% ($x = -10\%$); and August, -11 to $-5·4\%$ ($x = -9\%$). These negative charge balance errors result primarily from the systematic overestimation of HCO_3^- in the field.

The partial pressure of CO_2 in solution, $p(CO_2)$, and the saturation index of the solution with respect to calcite, $SI_{calcite}$, were calculated using the following formulae (Ford and Williams, 1989)

$$\log p(CO_2) = \log(HCO_3^-) - pH + pKCO_2 + pK_1 \qquad (2)$$

$$SI_{calcite} = \log(Ca^{2+}) + \log(HCO_3^-) + pH - pK_2 + pK_c \qquad (3)$$

where concentrations are in $mol\,l^{-1}$. The constants used ($T = 0°C$) are $pKCO_2 = 1·12$, $pK_1 = 6·58$, $pK_2 = 10·63$ and $pK_c = 8·38$ (Ford and Williams, 1989). Atmospheric $p(CO_2)$ is assumed to be $10^{-3.5}$ atm in the laboratory and $10^{-3.61}$ atm in the field, as the altitude of the glacier snout is 2560 m.

RESULTS AND DISCUSSION

Laboratory dissolution experiments

On a diurnal time-scale, it seems likely that there are two principal controls on the variability in composition of bulk glacial meltwaters. These are (i) the dilution of concentrated delayed flow by dilute quick-flow and (ii) the extent of post-mixing reactions between bulk meltwaters and suspended sediment, which will depend on the concentration of suspended sediment in the bulk meltwaters and on the duration of contact between these waters and suspended sediments. On a seasonal time-scale, additional factors include variations in the composition of the delayed flow resulting from changing residence times within the distributed component of the drainage system, and changes in drainage system structure which alter

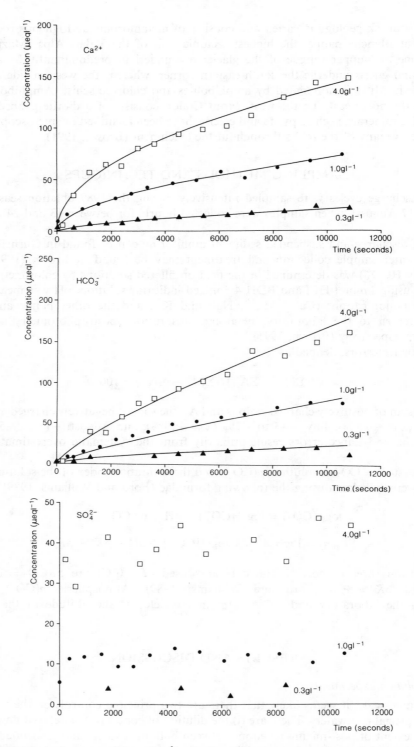

Figure 2. Temporal variations in the concentrations of Ca^{2+}, HCO_3^-, SO_4^{2-} and pH during laboratory dissolution experiments using different rock to water ratios. Hatched areas indicate 30 second monitoring interval

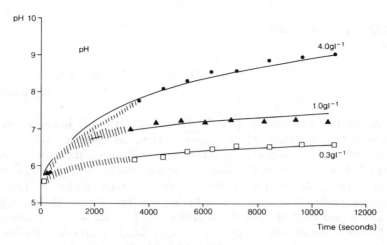

Figure 2 (continued)

the relative proportions of quickflow and delayed flow. Our laboratory experiments were designed to quantify the potential contribution of post-mixing reactions to this variability in bulk meltwater composition.

Measurements of pH and concentrations of Ca^{2+}, HCO_3^- and SO_4^{2-} from the dissolution experiments are presented in Figure 2. The amounts of Ca^{2+} and HCO_3^- released, and of protons consumed (denoted by increasing pH), increased continuously during each experiment and were positively associated with the concentration of sediment. However, the rates of increase in concentration and pH tended to decrease over time during each experiment. Similar patterns were found for the other base cations (Mg^{2+}, K^+ and Na^+). Final concentrations of Ca^{2+} and HCO_3^- (the equivalent ratio of which approximated $1:1$) ranged from 24 to 164 μequiv. l^{-1} after three hours for experiments using suspended sediment concentrations of 0·3 and 4·0 g l^{-1}, respectively. Rate laws fitted to the experimental data for Ca^{2+}, HCO_3^-, and pH are presented in Table II. Power curves were estimated by applying linear regression analysis to the $\log_e$ transformed dissolved ion concentration (dependent) and time (independent) variables and then back-transforming the estimated relationship (Williams, 1986). These laws resemble those presented elsewhere in the dissolution literature (e.g. Lerman, 1979). Similar rate laws apply to concentrations of Mg^{2+}, Na^+ and K^+, but these cations comprise 20% and usually 10% of the sum of base cations. Ca^{2+} and HCO_3^- are the dominant ions, and their approximately $1:1$ ratio indicates that the carbonation reactions, examples of which are shown in Equations 3 and 4, characterize the dissolution experiments

$$CaCO_3(s) + CO_2(aq) + H_2O(aq) \rightleftharpoons Ca^{2+}(aq) + 2HCO_3^-(aq) \qquad (4)$$

calcite

Table II. Dissolution rate parameters of proglacial sediment derived from the Haut Glacier d'Arolla in deionized water, 0·0–3·1°C, atmospheric CO_2 pressure

	0·3 g l^{-1}	1·0 g l^{-1}	4·0 g l^{-1}
Ca^{2+}	$\hat{Y} = 0.07X^{0.62}$	$\hat{Y} = 0.54X^{0.53}$	$\hat{Y} = 1.55X^{0.49}$
HCO_3^-	$\hat{Y} = 0.11X^{0.58}$	$\hat{Y} = 0.05X^{0.80}$	$\hat{Y} = 0.14X^{0.77}$
pH	$\hat{Y} = 4.32X^{0.05}$	$\hat{Y} = 4.53X^{0.05}$	$\hat{Y} = 2.40X^{0.14}$

$$CaAl_2Si_2O_8(s) + 2CO_2(aq) + 2H_2O(aq) \rightleftharpoons$$

anorthite

$$Ca^{2+}(aq) + 2HCO_3^-(aq) + H_2Al_2Si_2O_8(s)$$

partially weathered feldspar   (5)

As in other simple dissolution experiments (e.g. Wollast, 1967), the pH increased immediately after the introduction of sediment to the system. The maximum pH values recorded ranged from 6·6 to 9·01 for experiments using 0·3 and 4·0 g l^{-1}, respectively. At 0·3 g l^{-1}, $p(CO_2)$ values deviated little from the atmospheric value; at 1·0 g l^{-1} they were lowered slightly below the atmospheric value, but at 4·0 g l^{-1} they were lowered significantly, reaching 10^{-5} atm after three hours of contact (Figure 3). Lowering of the $p(CO_2)$ indicates that the consumption of protons in weathering reactions exceeded the supply by the diffusion of CO_2 across the gas–liquid interface. As the solution had free access to the atmosphere, this implies that the supply of protons to the dissolution system was kinetically controlled. This disequilibrium between the supply and consumption of protons was most marked at higher sediment concentrations, and tended to

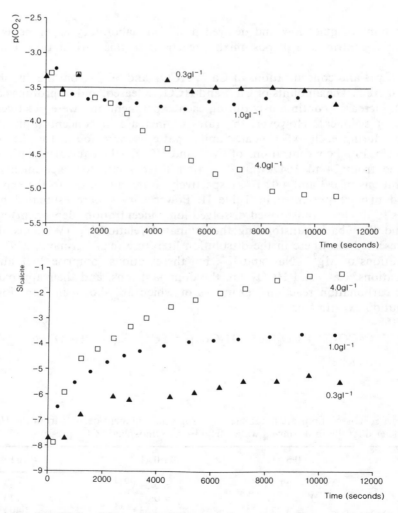

Figure 3. Temporal variations in $p(CO_2)$ and $SI_{calcite}$ during laboratory dissolution experiments using different rock to water ratios

increase with time over the duration of the experiments. At all times the solutions were undersaturated with respect to calcite, though values of the saturation index tended to increase over time during each experiment and were higher in the experiments with higher sediment concentrations. Maximum values of $SI_{calcite}$ approached -0.7 at the end of the experiment using $4.0\,g\,l^{-1}$ of rock flour (Figure 3).

There was some release of SO_4^{2-} during each experiment (Figure 2). Much of this release occurred immediately on addition of the suspended sediment to the solution. Thereafter there was minimal release. For example, in the $4\,g\,l^{-1}$ experiment, only $10\,\mu equiv.\,l^{-1}$ SO_4^{2-} was liberated after the initial release of $35\,\mu equiv.\,l^{-1}$. This compares with the release of $150\,\mu equiv.\,l^{-1}$ of HCO_3^- over the same period. We believe that the rapid initial release of SO_4^{2-} is an experimental artefact. During transport and storage, slow sulphide oxidation may occur at or near the mineral–grain surface. The SO_4^{2-} so formed is available for immediate leaching on contact with solution. Thereafter, the release of SO_4^{2-} is limited. This suggests that the release of sulphate by the oxidation of sulphides in suspended sediment is a slow process which would only be likely to contribute significantly to the chemistry of waters which spent long periods at the glacier bed. This supports the contention of Tranter *et al.* (in press) that sulphate is added to meltwaters primarily by pyrite oxidation which takes place within the distributed system

$$4FeS_2(s) + 15O_2(aq) + 14H_2O(aq) \rightleftharpoons 16H^+(aq) + 8SO_4^{2-}(aq) + 4Fe(OH)_3(s) \qquad (6)$$

Thus our experiments suggest that chemical weathering by dilute waters which spend relatively short periods in contact with suspended sediment is likely to be dominated by the carbonation reactions. Ca^{2+} and HCO_3^- are the principal ions released by such reactions, their concentrations increasing as suspended sediment concentrations and contact times increase. Disequilibrium between rates of proton consumption and rates of proton supply results in lowering of $p(CO_2)$ to subatmospheric values, an effect which is most marked at higher sediment concentrations. Solutions remain undersaturated with respect to calcite at residence times and sediment concentrations typical of basal arterial channels. The release of sulphate by sulphide oxidation is minimal under these conditions.

Field data

Figure 4 shows the discharge and suspended sediment concentration records for much of the 1989 ablation season (June to August) and indicates the position of each diurnal cycle sampled. Discharge ($< 2\,m^3\,s^{-1}$) and diurnal discharge amplitude ($0.6\,m^3\,s^{-1}$) were low during early June, but they increased as the season progressed, giving rise to higher, more strongly peaked hydrographs of high diurnal amplitude ($> 3.0\,m^3\,s^{-1}$) during July and August. This trend was accompanied by increased asymmetry of the diurnal hydrograph as the time to peak between the onset of daily ablation and maximum bulk discharge was reduced. The significance of these hydrograph parameters as indicators of drainage system evolution is explored in detail by Sharp *et al.* (in prep.).

The diurnal variations in the magnitude and pattern of discharge, suspended sediment concentration, Ca^{2+}, Mg^{2+}, Na^+, K^+, HCO_3^-, SO_4^{2-}, pH and electrical conductivity from three diurnal cycles spanning the 1989 meltseason are shown in Figure 5. The summary characteristics of each diurnal cycle are given in Tables III and IV. The $p(CO_2)$, $SI_{calcite}$ and SO_4^{2-}/HCO_3^- variations are illustrated in Figure 6. As with discharge, the magnitude and form of the individual diurnal chemographs evolved as the meltseason progressed.

June. During the June cycle, discharges and sediment concentrations were relatively low ($2\,m^3\,s^{-1}$ and $300\,mg\,l^{-1}$, respectively) and there was limited diurnal variation in discharge (range $1.4–2.0\,m^3\,s^{-1}$). Consequently, there was little variation in the chemical properties of the meltwater. Solute concentrations were relatively high ($200–240\,equiv.\,l^{-1}\,Ca^{2+}$) and the pH relatively low ($7.3–7.8$). The ratio SO_4^{2-}/HCO_3^-, which provides a measure of the relative importance of dissolution of CO_2 and sulphide oxidation as proton sources, was relatively high ($0.29–0.33$), implying significant proton supply from sulphide oxidation. The waters were undersaturated with respect to calcite and showed $p(CO_2)$ values close to atmospheric. The latter result is consistent with laboratory weathering experiments with low suspended sediment concentrations, but the solute concentrations, pH and values of $SI_{calcite}$ attained would require a

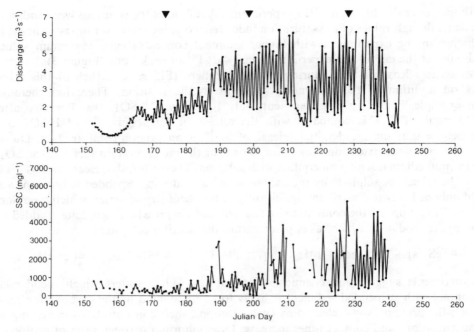

Figure 4. Variability in meltwater discharge and suspended sediment concentration in runoff from the Haut Glacier d'Arolla over the 1989 ablation season. Arrows indicate the diurnal cycles sampled for detailed chemical analyses

Table III. Range in concentration of the major ions, pH and electrical conductivity (EC) determined for each of the diurnal cycles [units are μequiv. l^{-1}, except Q (m^3 s^{-1}), EC (μS cm^{-1}), SSC (g l^{-1}) and pH]

	June 1989		July 1989		August 1989	
	Minimum	Maximum	Minimum	Maximum	Minimum	Maximum
Q	1·4	2·0	2·0	4·5	3·4	6·9
EC	17	20	15	24	14	18
SSC	0·01	0·29	0·16	0·73	1·40	4·00
Ca^{2+}	200	240	170	290	180	240
Mg^{2+}	26	43	16	31	15	31
Na^+	9·2	17	7·2	17	7·7	14
K^+	6·2	14	5·6	12	7·5	13
HCO_3^-	210	260	190	320	240	280
SO_4^{2-}	64	83	47	100	26	57
pH	7·3	7·8	8·2	9·0	8·5	9·2

considerably longer period of weathering than was allowed in the experiments. Assuming a suspended sediment concentration of $0·3\,g\,l^{-1}$, application of the rate law for Ca^{2+} presented in Table II suggests residence times in the range 103–139 hours. These figures should, however, be taken as no more than illustrative in view of the degree of extrapolation required, and questions about whether the suspended sediment concentration provides a good measure of the rock–water ratio within the environment in which weathering takes place (i.e. water passing through the distributed system may come into contact with, or

Figure 5. Variations in meltwater discharge, suspended sediment concentration and the concentration of major ions over diurnal cycles sampled in June (22–23), July (17–18) and August (16–17) 1989

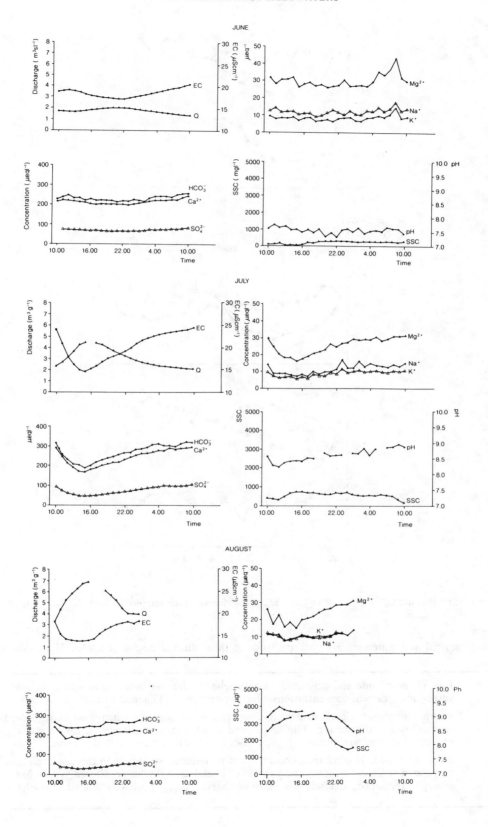

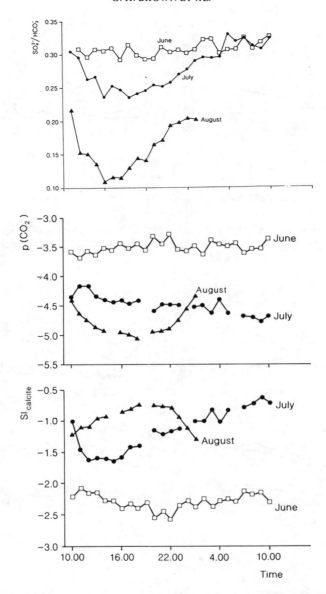

Figure 6. Variations in SO_4^{2-}/HCO_3^-, $p(CO_2)$ and $SI_{calcite}$ for diurnal cycles sampled in June, July and August 1989

Table IV. Hydrological and water quality characteristics of three diurnal discharge cycles, Haut Glacier d'Arolla, Switzerland

22–23 June	Low Q_b amplitude and magnitude. Low/negligible diurnal variation in solute species. Low suspended sediment concentrations. Open system $p(CO_2)$ characteristics. $SI_{calcite}$ low
17–18 July	Diurnal amplitude clearly evident in discharge, accompanied by increased magnitude. Dissolved species inverse to discharge. Higher suspended sediment concentrations. Lower $p(CO_2)$ values, but of limited diurnal amplitude. $SI_{calcite}$ higher and clearly inverse with discharge
16–17 August	Large amplitude in discharge, accompanied by reduced amplitude in the dissolved species. High suspended sediment concentrations positively related to discharge. Closed system low $p(CO_2)$ strongly developed, inversely related to discharge. $SI_{calcite}$ positively related to discharge

even flow through, immobile sediments, but be unable to entrain them due to its low flow velocities). The limited variability of parameters such as pH, $p(CO_2)$ and $SI_{calcite}$ over the diurnal cycle suggests that dilution of delayed flow by quickflow contributed only slightly to the diurnal variability in solute concentration. This variability may therefore be due to discharge-related variations in the residence time of waters within the glacier; the asymptotic form of the rate laws would make solute concentrations insensitive to such variations when the mean residence time is relatively long. These conclusions are consistent with the results of dye tracing investigations into the seasonal evolution of the subglacial drainage system (carried out in 1990), which suggest that in early June the whole glacier is underlain by a distributed drainage system (Nienow *et al.*, in prep.).

July. By the time of the July cycle, discharges and suspended sediment concentrations were higher and more variable than in June ($2 \cdot 0$–$4 \cdot 5 \, m^3 \, s^{-1}$ and 160–$730 \, mg \, l^{-1}$, respectively). Not surprisingly, therefore, solute concentrations were also more variable, with higher peaks and lower minima than in June (Table III). The pH and $SI_{calcite}$ were higher, and $p(CO_2)$ lower than in June (Figure 6). There were clear diurnal cycles in $SI_{calcite}$ and the ratio SO_4^{2-}/HCO_3^-, with waters at high discharge being more undersaturated with respect to calcite and containing relatively more bicarbonate than those at low discharge (Figure 6). Waters leaving the glacier at low discharges were thus similar in character to those sampled at all discharges in June. In contrast, waters sampled at high discharges showed evidence of dilution by quickflow (lower solute concentrations and $SI_{calcite}$) and greater solute acquisition by carbonation reactions [lower SO_4^{2-}/HCO_3^- ratio and $p(CO_2)$]. These characteristics imply a significant reduction in mean water residence time between the low and high discharge in July, and between June and July. Assuming that all the solute was derived from the weathering of suspended sediment, extrapolation on the basis of the rate law for Ca^{2+} for a sediment concentration of $1 \, g \, l^{-1}$ suggests a mean residence time at peak discharge of around 14 hours. This figure is not really meaningful, however, as the bulk waters consist of a mixture of long residence time delayed flow and short residence time quickflow.

These results suggest that the exposure of ice at the glacier surface in the lower glacier introduced a component of quickflow which was responsible for the bulk of the diurnal runoff peak, but which made only a limited contribution to runoff at minimum flow. This quickflow was able to mobilize increased amounts of suspended sediment, with the result that the carbonation reactions had a larger impact on the chemistry of bulk meltwaters, particularly at peak daily discharge. Disequilibrium between rates of proton supply and proton consumption in weathering reactions resulted in lowering of $p(CO_2)$ to sub-atmospheric values. Again, these results are entirely consistent with those of dye tracer investigations of drainage system evolution, which suggest that by this stage of the melt season a conduit system exists beneath the lowest 2 km of the glacier (Nienow *et al.*, in prep.). At this time, residence times within the conduit system are of the order of one hour or less at peak discharge, and three hours or less at minimum discharge. Given this independent evidence for water–rock contact times, and a knowledge of suspended sediment concentrations, some indication of the potential for solute acquisition by post-mixing reaction can be obtained from the rate laws presented in Table II. Assuming a sediment concentration of $0 \cdot 3 \, g \, l^{-1}$ at minimum discharge, three hours of rock–water contact would allow the acquisition of about $22 \, \mu equiv. \, l^{-1} \, Ca^{2+}$, about 8% of the total. With $1 \, g \, l^{-1}$ at maximum discharge, the potential acquisition would be $42 \, \mu equiv. \, l^{-1} \, Ca^{2+}$, or 25% of the total. These figures are maximum estimates because post-mixing reactions take place within diluted delayed flow rather than in quickflow alone. Nevertheless, it seems that by mid-July at peak daily discharge, solute acquisition by post-mixing reaction with suspended sediment can significantly offset the dilution of concentrated delayed flow by quickflow. This will make it unsafe to assume that the component waters mix conservatively when attempting hydrograph separation using chemically based mixing models.

August. By 16 August, discharge and suspended sediment concentration had risen still further, and they showed even greater diurnal variability than in July (Table III). Surprisingly, however, solute concentrations showed less diurnal variation than in July (a range of only $60 \, \mu equiv. \, l^{-1} \, Ca^{2+}$ over a discharge range of $3 \cdot 5 \, m^3 \, s^{-1}$, compared with $120 \, \mu equiv. \, l^{-1} \, Ca^{2+}$ over a range of $2 \cdot 5 \, m^3 \, s^{-1}$ in July). $p(CO_2)$ and the ratio SO_4^{2-}/HCO_3^- were lower than in July, suggesting that the carbonation reactions made a larger contribution to solute acquisition. Waters were closer to saturation with respect to calcite. The relationships between the last three parameters and discharge were, however, different from those apparent over the July cycle. $p(CO_2)$ and SO_4^{2-}/HCO_3^- showed well defined minima at peak discharge, but $SI_{calcite}$ peaked at the same time as discharge. This suggests that the contribution of the carbonation reactions to solute acquisition was greatest at maximum discharge when suspended sediment concentration was also at its peak. Whereas in July dilution of delayed flow by quickflow caused $SI_{calcite}$ to decrease as discharge increased, during August this effect was substantially offset by the impact of post-mixing reactions with suspended sediment. This is remarkable, given that dye tracing data indicate that by 16 August the conduit system extended over 3 km upstream from the glacier snout, so that quickflow presumably contributed a much larger runoff volume to the diurnal hydrograph peak than in July (Nienow et al., in prep.).

Some indication of the contribution of post-mixing reactions to the solute load of the bulk meltwaters can be gained from the rate laws given in Table II. Dye tracing data suggest that at this stage of the season, residence times within the conduit system are two hours or less at peak discharge, and six hours or less at minimum discharge. Given suspended sediment concentrations of $1 \, g \, l^{-1}$ at minimum discharge and $4 \, g \, l^{-1}$ at peak discharge, the rate laws predict the release of $107 \, \mu equiv. \, l^{-1} \, Ca^{2+}$ at minimum discharge (45% of the total), and $120 \, \mu equiv. \, l^{-1} \, Ca^{2+}$ at peak discharge (67% of the total). Although they may be maximum estimates, these figures suggest that post-mixing reactions may play a major part in reducing the amplitude of diurnal variations in solute concentration at this stage of the ablation season, and that they may make a major contribution to the total solute flux from the glacier at all stages of the discharge cycle. If they are broadly correct, they imply that the assumptions of constant component composition and conservative mixing made in most chemically based studies of glacier hydrology are totally untenable by this stage of the season.

CONCLUSIONS

Laboratory dissolution experiments indicate that glacial flour is more geochemically reactive than hitherto appreciated in meltwater routing studies, and shed light on the processes by which dilute meltwaters acquire solutes by weathering of suspended sediment. Field studies show that post-mixing reactions with suspended sediment clearly occur when quickflow waters dilute delayed flow waters in arterial conduits at the glacier bed. These reactions become increasingly important as the melt season progresses, because the arterial conduit system becomes more extensive, and suspended sediment concentrations in bulk meltwaters increase. In post-mixing reactions, solutes are acquired by carbonation. The magnitude of solute acquisition depends on the concentration of suspended sediment, the residence time of bulk waters in basal arterial conduits, and the availability of atmospheric CO_2, which may be kinetically limited by the rate of diffusion across the air–liquid interface when sediment concentrations are high.

At the Haut Glacier d'Arolla there is a seasonal progression in the factors controlling variations in meltwater chemistry over diurnal discharge cycles. In June, the basal conduit system is poorly developed and runoff is composed primarily of long residence time delayed flow. Diurnal variations in water chemistry are manifest primarily in the concentrations of anions and base cations, and appear to reflect discharge-related variations in residence time at the glacier bed. By mid-July, the conduit system is well developed, and delayed flow is diluted significantly at peak daily discharge by quickflow. Suspended sediment concentrations at peak discharge are considerably higher than in June, however, and post-mixing reactions may furnish up to 25% of the Ca^{2+} load at this stage of the cycle. The contribution at minimum discharge is minimal, however, because suspended sediment concentrations are then low. As a result of these diurnal variations in the degree of dilution and post-mixing reaction, there are diurnal cycles in the SO_4^{2-}/HCO_3^- ratio and $SI_{calcite}$ of bulk meltwaters as well as in solute concentration. By mid-August the conduit system is even more extensive, and peak daily discharges and sediment concentrations are larger than in July. The impact of post-mixing reactions on the chemistry of bulk meltwaters is therefore pronounced. Up to 67% of the solute load leaving the glacier may come from this source. The amplitude of the diurnal cycle in solute concentration is significantly reduced compared with what would be expected from dilution alone, because sediment concentrations (and hence post-mixing reactions) are greatest at peak discharge. Strong diurnal cycles in the $p(CO_2)$, $SI_{calcite}$ and SO_4^{2-}/HCO_3^- ratio of bulk meltwaters attest to the importance of post-mixing reactions at maximum discharge.

These results have profound implications for the use of measurements of meltwater chemistry for studies of flow routing through glaciers. They clearly undermine the commonly made assumptions that the chemical composition of meltwater discharge components is constant over a melt season and that mixing of these components is geochemically conservative. They therefore bring into question the use of chemically based mixing models for hydrological inference in glacierized catchments (Collins, 1979; Gurnell and Fenn, 1984). Specifically, the following problems are likely to arise: (i) a failure to recognize that meltwaters acquire solutes by post-mixing reaction with suspended sediment is likely to result in the overestimation of the contribution of delayed flow to total runoff; and (ii) this problem will be most marked late in the ablation season and during the daily discharge peak, when sediment concentrations are highest.

It should also be noted that it may be problematic to use a two-component runoff model to represent the hydrology of a glacier in the early melt season before basal conduits are well developed. At this stage of the season there is a risk that solute concentration variations arising from discharge-related variations in residence time within a single hydrological 'reservoir' may be wrongly interpreted as a product of variable dilution of delayed flow by quickflow. If variable dilution is indeed a major influence on the bulk water chemistry, there should be diurnal cycles in parameters such as the SO_4^{2-}/HCO_3^- ratio, but these do not become apparent at Haut Glacier d'Arolla until the July cycle.

Finally, we wish to sound a note of caution. Our laboratory dissolution experiments are not precise analogues of the type of reactions which occur in basal arterial channels. There are a number of important differences to be considered. Firstly, the suspended sediment used in the experiments has already passed through the subglacial drainage system and therefore the most reactive parts of the mineral surfaces have already been dissolved. Thus solute acquisition during post-mixing reactions in the field may be even more rapid than we have suggested. Secondly, it is likely that different mineral assemblages will be found in different size fractions of glacial flour, and that the reactivity of individual mineral species may be dependent on size (Fairchild *et al.*, in press). Therefore, any changes in the particle size distribution that occur over an ablation season will alter the potential for solute acquisition by post-mixing reactions. Thirdly, post-mixing reactions take place in diluted delayed flow, which will initially be more concentrated than the deionized water used in these experiments. Finally, the suspended seidment concentrations being monitored in the bulk meltwaters may not be representative of the rock to water ratios which are effective in basal arterial channels. Given these reservations, the dissolution experiments give results which facilitate the interpretation of the chemical characteristics shown by bulk meltwaters during three diurnal melt cycles at the Haut Glacier d'Arolla, and hence are useful analogues for the chemical weathering reactions which fix the chemical composition of meltwaters.

ACKNOWLEDGEMENTS

This work was supported by a NERC Fellowship (GT5/F/91/AAPS/3) to GHB, an NERC Studentship (GT4/89/AAPS/53) to PWN and by NERC grants (GR3/7004 and GR3/8114). Chris Hill provided invaluable assistance in the collection of field samples, and Grande Dixence SA kindly provided discharge data. We thank Jenny Wyatt for drawing the diagrams.

REFERENCES

Brown, G. H. 1991. 'Solute provenance and transport pathways in Alpine glacial environments', *Unpublished PhD Thesis*, University of Southampton. 283 pp.

Brown, G. H., and Tranter, M. 1990. 'Hydrograph and chemograph separation of bulk meltwaters draining the Upper Arolla Glacier, Valais, Switzerland' in *Hydrology in Mountainous Regions. I. Hydrological Measurements; the Water Cycle* (*Proceedings of the Second Lausanne Symposium, August 1990*), Int. Assoc. Hydrol. Sci. Publ. No. 6. **193**. 429–437.

Church, M. 1974. 'On the quality of some waters on Baffin Island, Northwest Territories', *Can. J. Earth Sci.*, **11**, 1676–1688.

Collins, D. N. 1978. 'Hydrology of an alpine glacier as indicated by the chemical composition of meltwater', *Z. Gletscherk. Glazialgeol.*, **13**, 219–238.

Collins, D. N. 1979. 'Quantitative determination of the subglacial hydrology of two alpine glaciers', *J. Glaciol.*, **23**, 347–361.

Collins, D. N. 1981. 'Seasonal variation of solute concentration in meltwaters draining from an alpine glacier', *Ann. Glaciol.*, **2**, 11–16.

Collins, D. N., and Young, G. J. 1981. 'Meltwater hydrology and hydrochemistry in snow- and ice-covered mountain catchments', *Nordic Hydrol.*, **12**, 319–334.

Eyles, N., Sasserville, D. R., Slatt, R. M., and Rogerson, R.J. 1982. 'Geochemical denudation rates and solute transport mechanisms in a maritime temperate glacier basin', *Can. J. Earth Sci.*, **19**, 1570–1581.

Fairchild, I. J., Bradby, L., and Spiro, B. 'Reactive carbonate in glacial sediments: a preliminary synthesis of its creation, dissolution and reincarnation' in M. Deynoux, J. G. Miller, E. Dornack, N. Eyles, I. J. Fairchild and G. M. Young (Eds), *Earth's Glacial Record*. Cambridge University Press, Cambridge, in press.

Ford, D.C. and Williams, P. 1989. *Karst Geomorphology and Hydrology*. Unwin Hyman, London, 601 pp.

Gurnell, A. M., and Fenn, C. R. 1984. 'Flow separation, sediment source areas and suspended sediment transport in a pro-glacial stream' in A. P. Schick (Ed.), *Channel Processes: Water, Sediment, Catchment Controls, Catena*, Suppl. **5**, 109–119.

Gurnell, A. M., Brown, G. H., and Tranter, M. 'A sampling strategy to describe the temporal hydrochemical characteristics of an alpine proglacial stream', *Hydrol. Process.*, in press.

Keller, W. D., and Reesman, A. L., 1963. 'Glacial milks and their laboratory-simulated counterparts', *Geol. Soc. Am. Bull.*, **74**, 61–76.

Lerman, A. 1979. *Geochemical Processes Water and Sediment Environments*. Wiley, Chichester. 481 pp.

Mazurek, M. 1986. 'Structural evolution and metamorphism of the Dent Blanche nappe and the Combin zone west of Zermatt (Switzerland)', *Ecol. Geol. Helv.*, **79**, 41–56.

Nienow, P., Sharp, M. J., Willis, I. C., and Richards, K. S. 'Dye tracer investigations of the spatial structure of the subglacial drainage system of the Haut Glacier d'Arolla, Valais, Switzerland', in preparation.

Oerter, H., Beherens, H., Hibsch, G., Ravert, W., and Stichler, W. 1980. 'Combined environmental isotope and electrical conductivity investigations of the runoff of Vernagtferner (Oetzal Alps)' in *International Symposium on the Computation and Prediction of Runoff from Glaciers and Glacierised Areas, Thbilisi, USSR, 3–11 September 1978,* Akademiia Nauk SSSR, Institut Geografii, Materialy Gliatsiologicheskikh Issledovanii Khronika, Obsuzhdeniia, **39**, pp. 86–91 and 157–161.

Rainwater, F. H. and Guy, H. P. 1961. 'Some observations on the hydrochemistry and sedimentation of the Chamberlain Glacier area, Alaska', *US Geol. Surv. Prof. Pap. No. 414-c*, 14 pp.

Raiswell, R. 1984. 'Chemical models of solute acquisition in glacial meltwaters', *J. Glaciol.*, **30**, 49–57.

Reynolds, R. C., and Johnson, N. M. 1972. 'Chemical weathering in the temperate glacial environment of the Northern Cascade Mountains', *Geochim. Cosmochim. Acta*, **36**, 537–554.

Sasseville, D. R., Eyles, N., Slatt, R. M., and Rogerson, R. 1977. 'Chemical transport mechanisms in an actively glaciated basin, British Columbia' [abstract], *Geol. Soc. Am. Abstr. Prog.*, **9**, 1156.

Sharp, M. J. 1991. 'Hydrological inferences from meltwater quality data: the unfulfilled potential', *Proceedings of the British Hydrological Society National Symposium, Southampton, UK, 16–18 September 1991*. Institute of Hydrology, Wallingford 5.1–5.8.

Sharp, M. J., Richards, K. S., Willis, I. C., and Nienow, P. 'The shape of diurnal runoff hydrographs from glaciers — implications for the temporal evolution of glacier drainage systems', *J. Glaciol.*, submitted.

Sharp, M. J., Tranter, M., Brown, G. H., Nienow, P., and Willis, I. C. 'Hydrological behaviour of the Haut Glacier d'Arolla, Valais, Switzerland, as inferred from the hydrochemistry of glacial meltwaters', *J. Glaciol.*, in preparation.

Slatt, R. M. 1972. 'Geochemistry of meltwater streams from nine Alaskan glaciers', *Geol. Soc. Am. Bull.*, **83**, 1125–1132.

Tranter, M. 1982. 'Controls on the chemical composition of Alpine glacial meltwaters', *Unpublished PhD Thesis*, University of East Anglia.

Tranter, M. and Raiswell, R. 1991. 'The composition of the englacial and subglacial components in bulk meltwaters draining the Gornergletscher', *J. Glaciol.*, 37, 59–66.

Tranter, M., Brown, G. H., Raiswell, R., Sharp, M. J., and Gurnell, A. M. 'A conceptual model for solute acquisition by alpine glacial meltwaters', *J. Glaciol.*, in press.

Warburton, J. 1989. 'Alpine proglacial fluvial sediment transfer', *Unpublished PhD Thesis*, University of Southampton. 444 pp.

Williams, R. B. G. 1986. *Intermediate Statistics for Geographers and Earth Scientists*. Macmillan Education, London. 712 pp.

Wollast, R. 1967. 'Kinetics of the alteration of K-feldspar in buffered solutions at low temperature', *Geochim. Cosmochim. Acta*, **31**, 635–648.

10

EXPERIMENTAL INVESTIGATIONS OF THE WEATHERING OF SUSPENDED SEDIMENT BY ALPINE GLACIAL MELTWATER

GILES H. BROWN

Centre for Glaciology, Institute of Earth Studies, The University of Wales, Aberystwyth, Dyfed, SY23 3DB, UK

M. TRANTER

Department of Geography, University of Bristol, University Road, Bristol BS8 1SS, UK

AND

M. J. SHARP

Department of Earth and Atmospheric Sciences, University of Alberta, Edmonton, Alberta, Canada T6G 2E3

ABSTRACT

The magnitude and processes of solute acquisition by dilute meltwater in contact with suspended sediment in the channelized component of the hydroglacial system have been investigated through a suite of controlled laboratory experiments. Constrained by field data from Haut Glacier d'Arolla, Valais, Switzerland the effects of the water to rock ratio, particle size, crushing, repeated wetting and the availability of protons on the rate of solute acquisition are demonstrated. These 'free-drift' experiments suggest that the rock flour is extremely geochemically reactive and that dilute quickflow waters are certain to acquire solute from suspended sediment. These data have important implications for hydrological interpretations based on the solute content of glacial meltwater, mixing model calculations, geochemical denudation rates and solute provenance studies.

INTRODUCTION

Alpine hydroglacial systems appear to be adequately described by two major discharge components (Collins, 1978; 1979b; Oerter *et al.*, 1980; Brown and Tranter, 1990; Tranter and Raiswell, 1991; Nienow *et al.*, Unpublished data). Dilute *quickflow* waters are derived largely from icemelt and pass rapidly through the hydrological system in ice-walled channels and major arterial conduits (Sharp, 1991; Brown et al., 1994a). Conversely, concentrated *delayed flow* waters, derived largely from snowmelt, are transported slowly at the ice–bedrock interface through a distributed system (such as a system of linked cavities; Walder, 1986), where rock–water contact is prolonged and intimate (Tranter *et al.*, 1993). These two components mix in major subglacial arterial channels to form *bulk* meltwaters.

Studies of the water quality characteristics of meltwater draining from alpine glaciers offer the potential to provide important information about the character and evolution of subglacial drainage systems (Sharp, 1991). Hydrological interpretation of the solute content of glacial meltwater is dependent on a knowledge of the processes and rates at which meltwater acquires solute in the subglacial environment. To date, this has been limited because there are no independent studies of (i) the chemical *processes* involved in solute acquisition by glacial meltwater, (ii) the *rate* at which they occur and (iii) the parameters that *control* the reaction rates. There is a need for laboratory dissolution experiments directly relevant to hydroglacial studies to improve hydrological interpretations such as: (a) the prediction of meltwater residence times in subglacial reservoirs; (b) definition of the rate and magnitude of post-mixing reactions (the possibility of solute uptake from suspended sediments is a serious limitation to the validity of simple mixing models that assume conservative mixing); (c) improving hydrograph separation techniques, solute provenance studies and calculations of chemical denudation rates; (d) assessment of the controls on chemical weathering processes, so that the nature of the different subglacial environments (open channels, linked cavities, discrete water pockets) can be defined in terms of open or closed system characteristics (Raiswell, 1984), redox potential, effective gas pressure and turbulence; and (e) assessment of the impact of the 'acid-flush', associated with preferential elution during spring snowmelt of the winter snowpack, on the rate and magnitude of subglacial weathering processes.

Previous geochemical dissolution studies are not easily applicable to the glacial weathering system, as experimental conditions have not been representative of the glacial regime. Studies have been carried out under high $p(CO_2)$ conditions (Busenberg and Clemency, 1975), in buffered (Wollast, 1967; Petrovic *et al.*, 1976; Nicholson *et al.*, 1988; Zhang *et al.*, 1993) and acidified solutions (Deju and Bhappa, 1965; Holdren and Berner, 1979; Chou *et al.*, 1989; McKibben and Barnes, 1986; Acker and Bricker, 1992; Casey *et al.*, 1988) and temperature has varied between 25 and 200°C (e.g. Busenberg and Clemency, 1975; Petrovic *et al.*, 1976; Lagache, 1965; 1976; Nagy and Lasaga, 1992). A further difficulty in relating previous work to meltwater systems is that rock to water ratios are generally too high (frequently $50–100\,g\,l^{-1}$ compared with typical loads in meltwater streams of $0\cdot5–4\,g\,l^{-1}$; Collins, 1979a; Gurnell, 1987; Brown *et al.*, 1994a) and the size fraction may be unrealistic compared with meltwater suspended sediment loads (usually considerably larger than the typical size range of meltwater streams where >95% by weight may be $<32\,\mu m$; Thomas, Unpublished data). Even those studies which attempted to compare meltwater quality with laboratory simulations (Keller and Reesman, 1963; Keller *et al.*, 1963a; 1963b) failed to reproduce field conditions accurately (e.g. experiments were conducted at 25°C, water to rock ratios were $100\,g\,l^{-1}$, abrasion pH was measured and anions were not determined). Another contrast between meltwater and laboratory systems is that meltwaters are of low ionic strength and poorly buffered. Therefore laboratory systems with low ionic strength and a pH which is allowed to vary during dissolution should provide a better analogue of field conditions than the buffered, high ionic strength systems.

Usually field chemical weathering conditions are difficult to reproduce in laboratory studies because of difficulties in defining the rock to water ratio and reactive grain size, the presence of organic acids, microbiological effects and the influence of microparticles (Tranter, Unpublished data). By contrast, interactions between glacial flour and meltwater are relatively easy to simulate, because both natural and laboratory systems contain freshly comminuted, fine-grained material containing microparticles, low dissolved organic carbon concentrations and minor biological influences. Further, reasonable estimates of the water to rock ratio, proton availability and rock to water contact time have been published and can be used to constrain the choice of experimental variables.

Recent models of chemical weathering in alpine glacial systems suggest that solute acquisition by meltwater results from reactions involving species in the solid, liquid and gas phases (Tranter *et al.*, 1993). Chemical weathering in the channelized component of the hydroglacial system gives rise to carbonation reactions, promoting solute acquisition as a result of the influx of $CO_2(g)$ [Equations (1) and (2)]

$$CaCO_3(s) + CO_2(aq) + H_2O(aq) \rightleftharpoons Ca^{2+}(aq) + 2HCO_3^-(aq) \tag{1}$$
calcite

$$CaAlSi_2O_8(s) + 2CO_2(aq) + 2H_2O(aq) \rightleftharpoons Ca^{2+}(aq) + 2HCO_3^-(aq) + H_2Al_2Si_2O_8(s) \tag{2}$$
anorthite   partially weathered feldspar

Conversely, meltwater chemistry in the distributed component is dominated by coupled reactions between sulphide and carbonate minerals. Protons derived from the oxidation of sulphide minerals are utilized in the dissolution of carbonates, which are common in the bedrocks of Alpine glaciers (Tranter *et al.*, 1993)

[Equation (3)]

$$4FeS_2(s) + 14H_2O(aq) + 15O_2(aq) + 16CaCO_3(s) \rightleftharpoons$$
pyrite   calcite

$$4Fe(OH)_3(s) + 8SO_4^{2-}(aq) + 16Ca^{2+}(aq) + 16HCO_3^-(aq) \tag{3}$$
ferric oxyhydroxides

The aim of this paper is to examine the processes and magnitude of the weathering reactions most likely to occur when dilute meltwater comes into contact with suspended sediment in the channelized component of the subglacial drainage system. The weathering experiments have five main objectives which investigate the effect of changing a single parameter on the rate and magnitude of solute acquisition. The objectives are to evaluate (i) the effect of particle size (reflecting variations in the size distribution of suspended sediment on seasonal/diurnal timescales), (ii) the effect of the water to rock ratio (reflecting variations in suspended sediment load), (iii) the effect of crushing the rock flour (reflecting the physical processes of glacier erosion on mineral surface properties), (iv) the effect of repeated wetting (reflecting deposition and re-mobilization of sediment) and (v) the effect of changing the composition of the initial solution (reflecting seasonal variations in proton availability). Experimental variables (e.g. the duration of the water to rock interaction, water to rock ratio, mineralogy) have been constrained by field data from Haut Glacier d'Arolla, Valais, Switzerland to facilitate a direct evaluation of potential weathering rates and reactions operating in a known hydroglacial system.

METHODS AND TECHNIQUES

Table I provides a summary of the methods and techniques adopted during the laboratory dissolution experiments. A more detailed description of the experimental and analytical details may be found in Brown *et al.* (1994a) and Brown (Unpublished data). The solution–sediment mixture was in free contact with the laboratory atmosphere for the duration of the experiments (so-called 'free-drift'). All samples were immediately filtered through $0.45\,\mu m$ cellulose nitrate membranes on collection.

The methods and techniques may be summarized as follows. The duration of the experiments, 2–3 hours, is representative of bulk meltwater residence times in arterial channels during the latter part of the ablation season. The temperature (0–3°C) is realistic of field conditions. Water to rock ratios (0.3–$4.0\,g\,l^{-1}$) were derived from the maximum concentrations sampled during three diurnal cycles in the field. Deionized water

Table I. Summary of the methods and techniques adopted during 'open-drift' laboratory dissolution experiments involving sediment from Haut Glacier d'Arolla proglacial plain

Experiment	Sediment	
	Concentration $(g\,l^{-1})$	Size fraction (μm)
Particle size	4	<63, 63–125, 125–250, 500–1000
Water to rock ratio	0·3–4·0	Mixed
Crushing	4	63–125/<63
Repeated wetting	4	63–125
Solution composition	4	63–125

All experiments were run for 3 h except the crushing/repeated wetting suite, which were run for 2 h. Deionized water was used as a surrogate for icemelt in all experiments except the solution composition suite. All sediments were from Haut Glacier d'Arolla proglacial plain. Parameters measured: Ca^{2+}, Mg^{2+} Na^+, K^+, HCO_3^-, Cl^-, NO_3^-, SO_4^{2-}, pH, T°. Methods of analysis: Perkin-Elmer 2280 AAS, colorimetric titration, Dionex 2000i IC, Jenway PWA2 water analyser, Grant thermistor, ΔT Devices datalogger.

was used as a surrogate for icemelt. Sediment deposited on the proglacial plain of Haut Glacier d'Arolla was collected and transported without pretreatment from the field in plastic containers. This was dried at 50°C before weighing (Brown *et al.*, 1994a). Size fractions of <63, 63–125, 125–250 and 500–100 μm for the particle size experiments were derived by dry-sieving this material through Endecott laboratory test sieves using a Fritsch sieve shaker for 10 minutes. Experiments were conducted in a plastic beaker containing 950 ml ($\pm$2 ml) of deionized water, stirred continuously by a polyethylene covered magnetic bar rotated by a magnetic stirrer beneath the beaker. The pH was measured continuously during the first

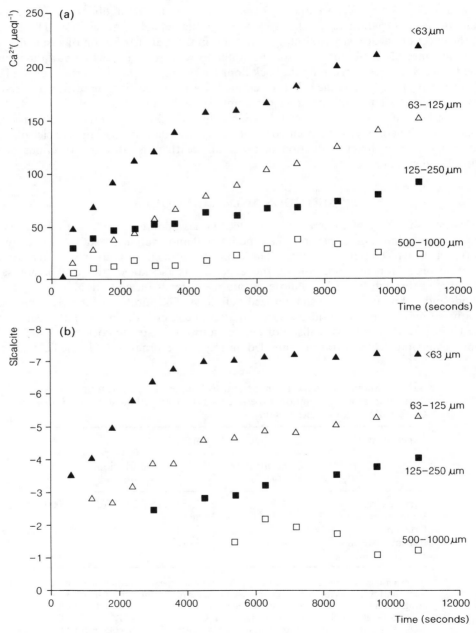

Figure 1. Variation over time of (a) Ca^{2+} and (b) $SI_{calcite}$ in dissolution experiments using <63, 63–125, 125–250 and 500–1000 μm size fractions of sediment from the proglacial plain of Haut Glacier d'Arolla. Sediment concentration in all experiments was 4 g l^{-1}

60 minutes of each experiment using a Jenway PWA2 water analyser and Jenway combination glass electrode linked to a ΔT Devices datalogger. Subsequently, pH was recorded simultaneously with the collection of samples. These samples (10 ml) were collected at set intervals (see Figure 1) during the course of the experiment, immediately syringe-filtered through Whatman 0·45 μm cellulose nitrate membranes and refrigerated at 4°C.

The following determinations were performed on the water samples. Total alkalinity was measured using a Radiometer MTS 800 multi-titration system, titrating to an endpoint of pH 4·5 using 0·36/1·0 mmol HCl. Sample volumes were 4 ml and titre volumes were between 0·1 and 2·0 ml. Cations were determined by atomic absorption spectrophotometry on a Perkin-Elmer 2280 instrument using an air–acetylene flame. Samples were acidified with HNO_3 and releasing agent $[La(NO_3)_3]$ and ionization suppressant (CsCl) added to all samples and standards. The precision was ±5%. SO_4^{2-} was determined on a Dionix 2000i ion chromatograph. The precision was ±3%.

The effect of the initial solution composition on the rate of solute release was investigated using acids present in the seasonal snowcover at Haut Glacier d'Arolla. Solutions which were a mixture of 0·11 (solution 1), 0·04 (solution 2), 0·02 (solution 3) and 0 mM (solution 4) AnalaR sulphuric acid and 0·3, 0·008, 0 and 0 mM AnalaR nitric acid, respectively, in deionized water. Repeated wetting involved adding glacial flour of grain size 63–125 μm to deionized water to give a water to rock ratio of 4·0 g l^{-1}. After 120 minutes the flour was separated from the solution and re-introduced to a similar volume of deionized water (to give the same water to rock ratio). The effects of crushing were demonstrated by comminuting a volume of the original glacial flour (63–125 μm) with a pestle and mortar until it all passed through the 63 μm sieve. This preserves the mineralogical composition of the parent size fraction while changing the grain size and surface properties. This size fraction was added to deionized water to give a water to rock ratio of 4·0 g l^{-1} also.

EFFECTS OF PARTICLE SIZE

Surface area increases, relative to volume, as the particle or grain size decreases and with increasing intricacy of particle shape (Parks, 1990). Increases in the total external surface area exposed to solution may result from physical and chemical processes. The former may lead to disaggregation and fragmentation of rocks, whereas the latter may loosen grain to grain contacts and expose fresh surfaces through dissolution (Lerman, 1979). In glacial environments mechanical comminution processes are especially effective, resulting from both abrasion (Boulton, 1974) and crushing (Boulton, 1978; Whalley and Krinsley, 1974) of subglacial bedrock surfaces (see later). This results in finely ground particles with fresh mineral surfaces and microparticles.

Other things being equal, the overall rate of transfer of solutes into the solution is proportional to the interface (surface) area of the solid (Petrovic, 1981b). For example, for the first-order dissolution reaction (after Lerman, 1979)

$$\frac{dC}{dt} = k(C_s - C) \tag{4}$$

where dC/dt = rate of change (g cm^{-3} s^{-1}); k = reaction rate parameter (s^{-1}); C_s = equilibrium or steady-state concentration (g cm^{-3}); and C = concentration in solution at time t (g cm^{-3}).

The rate constant k is defined as

$$k = k_s \frac{S}{V} \tag{5}$$

k is a product of two terms: (i) the specific surface area of the solid

$$\frac{S}{V} \tag{6}$$

where S = reactive solid surface area (cm^2) and V = solution volume (cm^3), and (ii) k_s, the transport or reaction rate parameter that refers to a unit of area of the solid (which is independent of S), defined by

$$k_s \equiv \frac{D}{h} \qquad (7)$$

where D = diffusion coefficient of dissolved species D ($cm^2 s^{-1}$) and h = thickness of the laminar layer (cm).

For a given volume of solution and a given mass of solid, the specific surface area (S/V) can be increased by communition, creating more particles of smaller radius, thus increasing the reaction rate parameter (k). In addition to the increased surface area, the influence of very fine rock material has additional positive effects on the rate of dissolution, resulting from the Kelvin or Gibbs–Thomson effect (where the surface free energy of sub-micron particles becomes a significant fraction of the total free energy of the grain to which they are adhered) and the presence of sharp edges and blades producing very small radius of curvature of their surfaces (see Petrovic, 1981b).

The preceding discussion suggests that variations in the grain size distribution of suspended sediment transported in bulk meltwater may have an important influence on the magnitude of post-mixing solute enrichment. Recent studies of suspended sediment variability in proglacial streams (e.g. Richards, 1984; Gurnell, 1987; Fenn and Gomez, 1989; Karlsen, 1991) suggest that particle size distributions in glacial meltwater streams are poorly defined. Fenn and Gomez (1989) illustrate subtle shifts around the mean grain size of medium silt ($14\cdot6 \pm 2\cdot3\,\mu$m), rather than the inclusion and exclusion of sand-sized particles (c.f. Rainwater and Guy, 1961). However, fluctuations in median grain size occur broadly in phase (though not in magnitude) with changes in discharge and suspended sediment concentration (SSC). Additionally, variations in the grain size series may be related to distinct hydrological periods. For example, coarser particle size distributions have been recognized on the ascending limb of the diurnal hydrograph, during snowmelt and during precipitation events after the removal of the seasonal snowcover (Fenn and Gomez, 1989). Factors such as the composition of morainic or stream bed material may also be important in controlling variations in the grain size distribution in glacial environments (Karlsen, 1991).

Spatial variability in grain size distribution in the subglacial hydrological network may also influence rates of solute acquisition from suspended sediment. Data from proglacial streams draining the Storbreen catchment (Richards, 1984) suggest coarse material is derived from the margins of the dominant meltstream at high discharges, whereas smaller streams transport a finer suspended load as a result of lower discharge, lower hydraulic gradients and inefficient cross-sections. This spatial variability in grain size may enhance hysteresis in diurnal particle size–discharge relations, increasing the proportion of fines on the descending limb of the hydrograph.

Figure 1 shows the effect of particle size on the release of Ca^{2+} from Haut Glacier d'Arolla flour, using a water to rock ratio of $4\,g\,l^{-1}$. The rate of change and magnitude of Ca^{2+} concentration is clearly an inverse function of grain size, with concentrations after 3 h of water to rock interaction ranging between $38\,\mu$equiv.l^{-1} (500–1000 μm fraction) and $220\,\mu$equiv.l^{-1} (<63 μm fraction). This suggests that the surface area (and hence the number of reactive surface sites), which increase with decreasing grain size, directly affects the rate at which meltwater acquires solute. As dissolution proceeds, the rate of solute acquisition (and resultant proton consumption) slows. This reflects (i) the dissolution and disappearance of micro-particles (see later), (ii) the disappearance of reactive calcite and aluminosilicate sites and (iii) the progression towards saturation of the solution (Figure 1b).

After the removal of microparticles (c.f. Holdren and Berner, 1979), studies of mineral surfaces by scanning electron microscopy have illustrated the occurrence of dissolution at preferential sites, producing characteristic lens-shaped etch pits and cross-hatchings (Tazaki, 1976; Berner, 1981). The preferential dissolution of these sites of atomic uncoordination (e.g. steps, kinks, discontinuities; see Hochella, 1990) results in a less reactive mineral surface where weathering proceeds more uniformly and at a slower rate. The rapid release of Ca^{2+} during the early stages of weathering may also reflect the preferential dissolution of more reactive carbonate minerals (e.g. calcite). Calcite is present in trace amounts throughout the bedrock of the Haut Glacier d'Arolla catchment (Brown, Unpublished data). This liberation of Ca^{2+} into solution results in the solution moving towards equilibrium with respect to calcite [Equation (4); Figure 5]. The initial preferential release of Ca^{2+} can be seen in Figure 2.

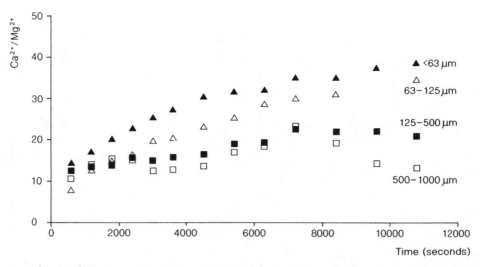

Figure 2. Ratio of Ca^{2+} to Mg^{2+} over time during dissolution experiments using <63, $63–125$, $125–250$ and $500–1000$ μm size fractions of sediment from the proglacial plain of the Haut Glacier d'Arolla. Sediment concentration in all experiments was $4\,g\,l^{-1}$

EFFECT OF ROCK TO WATER RATIO

Temporal variations in the SSC of alpine proglacial streams are well documented and illustrate three major characteristics (Gurnell, 1987)

1. SSC exhibits a broadly positive association with bulk discharge at both a seasonal and diurnal time-scale.
2. Hysteresis is common between SSC and discharge, with the peak in SSC frequently occurring before maximum discharge and declining more rapidly than the discharge from peak levels at a diurnal time-scale. This suggests sediment exhaustion effects are operative.
3. The well-defined suspended sediment–discharge relations are interrupted by apparently random, sudden changes in the concentration of the suspended load, often without a noticeable change in discharge. These pulses in SSC are primarily generated subglacially, especially during meltwater outburst events and during seasonal re-organization of the subglacial hydrological network (Fenn and Gomez, 1989; Sharp, 1991; Willis et al. this issue).

Spatial variations in SSC transported by the various elements of the alpine hydroglacial system are marked, ranging by several orders of magnitude from a few $mg\,l^{-1}$ in supraglacial channels to thousands of $g\,l^{-1}$ in proglacial streams late in the ablation season (Gurnell, 1987). This reflects the *source* of suspended sediment in glaciated alpine catchments, as large amounts of sediment are produced by the physical processes of glacial erosion in the subglacial environment (Collins, 1979a; Hammer and Smith, 1983; Fenn and Gomez, 1989). This variability in sediment concentration has important implications for the location and magnitude of solute acquisition from the suspended load. It is only when quickflow and delayed flow waters mix in major subglacial arterial conduits that bulk meltwaters acquire sufficient power to mobilize sediment as suspended load. It has been suggested that bulk meltwater may acquire a significant proportion of their solute load by so called 'post-mixing reactions' (Tranter and Raiswell, 1991; Brown et al., 1994a), which involve the weathering of suspended sediment in this environment, and it is this premise which will now be explored using laboratory dissolution experiments.

At a diurnal time-scale, it seems likely that there are two principal controls on the variability in composition of bulk glacial meltwater (Brown et al., 1994a): (i) dilution of the concentrated delayed flow component by dilute quickflow; and (ii) post-mixing solute acquisition by bulk meltwater from suspended sediment.

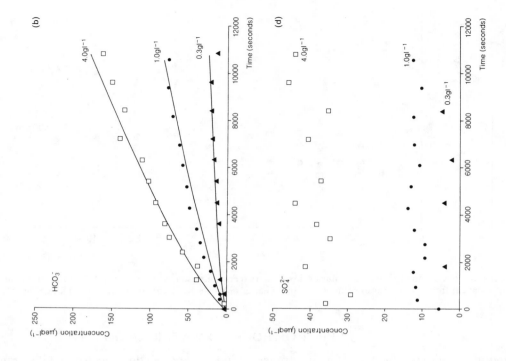

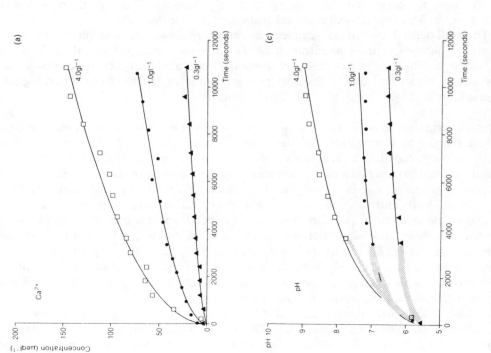

Figure 3. Temporal variations in the concentrations of (a) Ca^{2+}, (b) HCO_3^-, (c) pH and (d) SO_4^{2-} during laboratory dissolution experiments using different water to rock ratios. Shaded areas indicate 30 second monitoring interval

Table II. Dissolution rate parameters of Haut Glacier d'Arolla proglacial sediment in deionized water, $0.0-3.1°C$, atmospheric CO_2 pressure

	$0.3\,g\,l^{-1}$	$1.0\,g\,l^{-1}$	$4.0\,g\,l^{-1}$
Ca^{2+}	$\hat{Y} = 0.02X^{0.74}$ [0.98]	$\hat{Y} = 0.21X^{0.62}$ [0.99]	$\hat{Y} = 2.80X^{0.42}$ [0.98]
HCO_3^-	$\hat{Y} = 0.06X^{0.62}$ [0.88]	$\hat{Y} = 0.20X^{0.65}$ [0.99]	$\hat{Y} = 0.06X^{0.86}$ [0.93]
pH	$\hat{Y} = 4.56X^{0.04}$ [0.93]	$\hat{Y} = 4.83X^{0.05}$ [0.99]	$\hat{Y} = 3.03X^{0.12}$ [0.97]

Power curves were estimated by applying linear regression analysis to the $\log_e$ transformed dissolved ion concentration (dependent) and time (independent) variables and then back-transforming the estimated relationship (Williams, 1986). Coefficient of determination (R^2) values are given in brackets

The latter is a function of the *concentration* of suspended sediment in the bulk meltwater and the *duration* of contact between these waters and suspended sediment. The aim of the rock to water ratio experiments was to examine the magnitude of these post-mixing reactions.

The release of Ca^{2+}, HCO_3^-, and the consumption of protons (denoted by increasing pH) are depicted in Figure 3a–c and the rate laws fitted in the experimental data are detailed in Table II. The dissolved ion concentrations are positively associated with the concentration of sediment and the duration of rock to water contact. However, the rates of increase in concentration and pH during individual experiments tend to decrease over time. Similar patterns were found for the other base cations (Mg^{2+}, Na^+, K^+) (Figure 4), though these cations comprise <20% and usually <10% of the sum of cations. The approximate 1:1 ratio of Ca^{2+} : HCO_3^- suggests that so-called '*carbonation*' reactions (Reynolds and Johnson, 1972; Tranter *et al.*, 1993) characterize these dissolution experiments and give rise to HCO_3^- as the dominant anion and (usually) Ca^{2+} as the dominant cation [Equations (1) and (2)]. Values of $p(CO_2)$ and $SI_{calcite}$ during the dissolution experiments are shown in Figure 5 and are also clearly a function of the rock to water ratio and duration of contact. Although the dissolved ion concentrations indicate the *magnitude* of post-mixing solute acquisition, $p(CO_2)$ values may be used to interpret the *processes* controlling chemical weathering.

At $0.3\,g\,l^{-1}$, $p(CO_2)$ values deviate little from the atmospheric equilibrium value of $10^{-3.5}$ atm. However, a slight lowering to sub-atmospheric values occurs at a sediment concentration of $1.0\,g\,l^{-1}$, and at $4.0\,g\,l^{-1}$ they are lowered significantly, reaching 10^{-5} atm after 3 h of rock to water interaction. Such a *low* $p(CO_2)$ system (Raiswell, 1984; Tranter *et al.*, 1993) suggests that carbonation weathering reactions consume

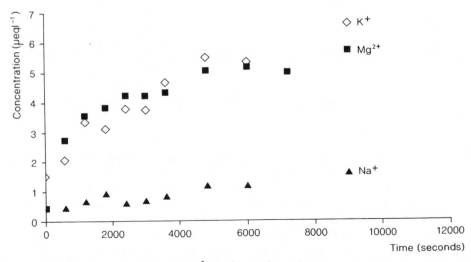

Figure 4. Temporal variations in the concentrations of Mg^{2+}, Na^+ and K^+ during a laboratory dissolution experiment using a water to rock ratio of $4\,g\,l^{-1}$

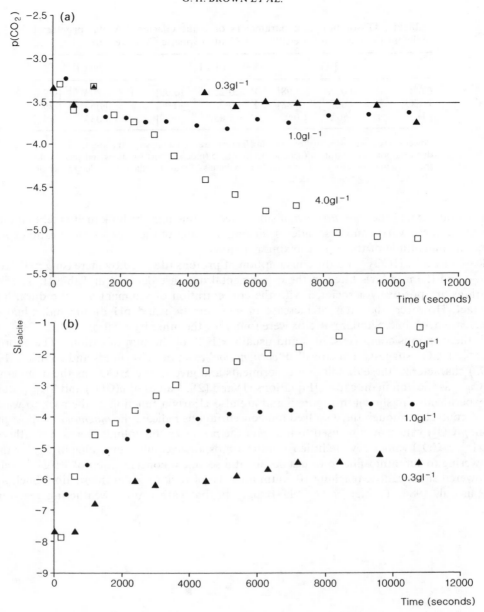

Figure 5. Temporal variations in (a) $p(CO_2)$ and (b) $SI_{calcite}$ during laboratory dissolution experiments using different water to rock ratios

protons, supplied from the dissolution and dissociation of atmospheric carbon dioxide, at a rate faster than that at which CO_2 can diffuse into solution across the gas–liquid interface. As the solution had free access to the atmosphere, this implies that the supply of protons to the dissolution system is kinetically controlled.

It has been suggested that chemical weathering by dilute waters which spend relatively short periods in contact with suspended sediment is likely to be dominated by carbonation reactions, as proton supply by sulphide oxidation is limited to the distributed system [Tranter and Raiswell, 1991; Tranter *et al.*, 1993; Brown *et al.*, 1994a; Equation (3)]. This is further substantiated when the results in Figure 3a and 3b are compared with the release of SO_4^{2-} during the same experiments (Figure 3d). Although there is some SO_4^{2-} release during the early phase of rock to water contact using $4.0\,g\,l^{-1}$ of rock flour ($\approx 35\,\mu equiv.\,l^{-1}$), only $\approx 10\,\mu\mathrm{equiv.}\,\mathrm{l}^{-1}$ was liberated subsequently. Water at $0°\mathrm{C}$ holds $13\,\mathrm{ml}\,O_2$ (i.e. $0.59\,\mathrm{mM}\,\mathrm{l}^{-1}$). The stoichiometry of Equation (3) indicates that 15 mol of oxygen gives rise to 8 mol (and therefore 16 equivalents) of SO_4^{2-}. Therefore, $0.59\,\mathrm{mM}\,\mathrm{l}^{-1}$ of $O_2(\mathrm{aq})$ gives rise to $633\,\mu\mathrm{equiv.}\,\mathrm{l}^{-1}$ of SO_4^{2-}. This suggests that the availability of dissolved oxygen is not a limiting factor in the oxidation of sulphide minerals during these experiments, which were conducted in free contact with the laboratory atmosphere. In comparison, $150\,\mu\mathrm{equiv.}\,\mathrm{l}^{-1}$ of HCO_3^- released during the same period (Figure 3b). This may reflect slow sulphide oxidation during transport and storage of the sediment, making the SO_4^{2-} so formed readily available for immediate leaching on contact with solution (Brown *et al.*, 1994a). The repeated wetting experiments described below further substantiate this hypothesis, as zero SO_4^{2-} is released during the second wetting compared with a significant release of Ca^{2+} and increase in HCO_3^-.

These data suggest that dissolution is faster in meltwater with higher SSC. Figure 6 illustrates the *overall* reaction rate parameter k ($\mu\mathrm{equiv.}\,\mathrm{l}^{-1}\,\mathrm{s}^{-1}$) (Equation 5) versus the water to rock ratio ($\mathrm{g}\,\mathrm{l}^{-1}$) for the release of Ca^{2+} from Haut Glacier d'Arolla sediment. The least-squares regression line indicates that the reaction rate is clearly a function of the solid to solution ratio and hence the reactive solid surface area [assuming transport through the laminar layer (h) in the continuously stirred solution is not a rate limiting factor and the diffusion coefficient (D) is constant].

EFFECTS OF CRUSHING

Very large tractive forces operate at the base of a glacier, resulting in crushing and grinding of subglacial bedrock and sediments (Drewry, 1986). The combination of basal sliding and the incorporation of tractive particles in basal ice results in two types of erosion: *abrasion*, producing fine-grained sediment (rock flour) generally $<0.5\,\mathrm{mm}$ across and *crushing*, producing a variety of particle sizes, generally $>0.5\,\mathrm{mm}$ (Boulton, 1978).

The processes and controls on mechanical weathering in the subglacial environment are beyond the scope of this paper. Detailed reviews may be found in Drewry (1986), Sugden and John (1976) and Gurnell and Clark (1987). In relation to the current discussion, the result of these subglacial physical weathering processes is to produce an abundance of fine particles with fresh, reactive mineral surfaces which are

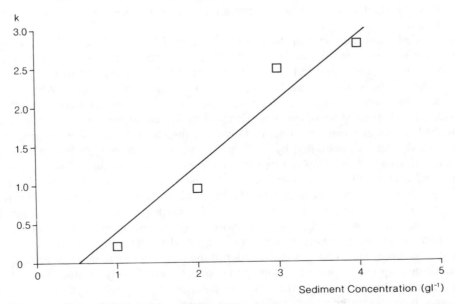

Figure 6. Overall reaction rate constant (k) derived from the rate laws fitted to the experimental data in Table II as a function of sediment concentration for laboratory experiments involving 0.3, 1, 2, 3, and $4\,\mathrm{g}\,\mathrm{l}^{-1}$ of sediment from Haut Glacier d'Arolla

susceptible to dissolution (Fairchild *et al.*, 1994). In addition to the effects of mechanical comminution on the grain size, and hence the specific surface area discussed previously, crushing and abrasion have important effects on the morphology of the individual mineral grains which may enhance the solubility of the solid and the rate constant for dissolution (Petrovic, 1981a; 1981b). These include: (i) *brittle fracture*, which may affect the kinetics of dissolution by increasing the density of surface reaction sites through cleavage fracture, dislocation and abrasion; (ii) localized *plastic deformation*, where grains are subject to substantial compressive and shear stresses producing significant subsurface structural damage to the mineral lattice (amorphous zones); and (iii) the production of *sub-micron, super-soluble* particles which adhere tightly to larger grain surfaces (Holdren and Berner, 1979). These grains are highly unstable in solution, and dissolve at an accelerated rate as the surface free energy of the particle becomes a significant fraction of the total free energy of the grain to which it adheres (Petrovic, 1981b).

The preferential dissolution of the microparticles manifests itself in high initial rates of dissolution, which slow as the microparticles are consumed and dissolution is transferred to the parent grains themselves. This appears to be the primary process controlling the characteristic curvilinear dissolution kinetics widely reported (e.g. Wollast, 1967; Holdren and Berner, 1979; Holdren and Speyer, 1985; Lerman, 1979). Microparticles have been observed on quartz grains from a variety of glacial locations (Krinsley and Doornkamp, 1973; Tranter, Unpublished data) and, as glacial comminution of quartz grains produces microparticles, it is probable that comminution of other rock minerals also produces microparticles (Tranter, Unpublished data). Petrovic (1981b) notes that crushing under dry- and wet-based glaciers should be analogous to dry crushing and crushing under water. Microparticles ($0.01–0.1$ μm) of quartz adhere strongly to micrometre-sized fragments produced during dry crushing. Conversely, grinding under water followed by immediate elutriation left relatively few microparticles adhered to larger grains (Bergman *et al.*, 1963). Wet or dry crushing conditions may also affect the degree of brittle fracture relative to plastic deformation, as water acts as a coolant and dipole, reducing the formation of strained bonds (Petrovic, 1981a; 1981b). Hence, dry crushing produces more reactive particles than wet crushing.

The preceding discussion suggests firstly that mechanically crushed rock material should be comparable with the products of crushing and abrasion of subglacial bedrock and sediments (Petrovic, 1981b; Tranter, Unpublished data). Both types of grains are strained, have fresh surfaces and microparticles are present in both systems. Secondly, crushing rock flour to simulate glacially induced comminution may have an important influence on the rate of dissolution and hence solute acquisition from suspended sediment. Such mechanical crushing simulates the effect of either bedrock comminution or of renewed glacially induced grinding of subglacial sediments.

Measurements of pH and the concentration of Ca^{2+} released from crushed and uncrushed sediment in deionized water are presented in Figure 7a and 7b. As in the experiments described previously, the rate of increase in concentration and pH decrease over time. However, the amount of Ca^{2+} released is related to the pretreatment of the sediment involved (Figure 7a). Final concentrations were between 105 and 240 μequiv. l^{-1} after two hours of water to rock interaction using untreated and crushed flour, respectively. Characteristically, the consumption of protons (denoted by increased pH) occurs immediately after sediment is introduced to the experimental system. However, the effect of crushing on proton consumption is even more marked than that exhibited by Ca^{2+}. pH increases by more than four pH units in 1 minute (Figure 7a; points during the first 3600 s represent readings taken every 30 s). The maximum pH values recorded were 7·67 and 10·09 for the untreated and crushed experiments, respectively. It is clear that crushing leads to enhanced rates of chemical weathering. This is consistent with theory in that mechanical comminution gives rise to an increased number of reactive surface sites on the rock flour, both through increasing the surface area and by changing the morphology/microtopography of the mineral surfaces. Additionally, the *form* of the rate laws which may be applied to the untreated and crushed data would also be different to accommodate changes in the order of the reaction rate to describe adequately the release of Ca^{2+} and the consumption of protons. Thus the introduction of freshly ground sediment to bulk meltwater draining alpine glaciers (for instance, during periods of subglacial hydrological reorganisation; Sharp, 1991) may have a greater impact on post-mixing solute acquisition than the re-mobilization of previously weathered sediments deposited in the channelized system during lower discharge conditions.

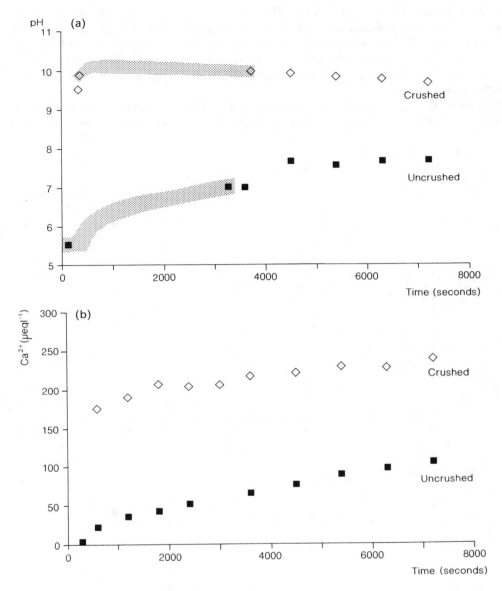

Figure 7. Change over time in the (a) pH and (b) Ca^{2+} concentration in waters used in dissolution experiments using $4\,g\,l^{-1}$ of un-treated 63–$125\,\mu$m sediment from the proglacial plain of Haut Glacier d'Arolla and sediment of the same initial size fraction crushed to $<63\,\mu$m. Stippled areas denote continuous monitoring of pH

EFFECT OF REPEATED WETTING

The characteristic rate laws which define solute release during dissolution experiments may change over time (Lerman, 1979). Power law dissolution kinetics are associated with high initial solute release, followed by slower quasi-linear release as the duration of rock to water contact increases. This reflects the destruction of microparticles adhering to the mineral grains, preferential dissolution of amorphous zones, the removal of fresh reactive surface sites such as dislocations and blades, the dissolution of rapidly weathered minerals such as calcite and the evolution towards to equilibrium of the solution.

This suggests that the 'wetting history' of sediment in the hydroglacial system may also influence rates of solute acquisition. Seasonal and diurnal variations in discharge magnitude result in cycles of sediment mobilization and deposition within channels as stream power and wetted perimeters respond to discharge variations (Gurnell, 1987). This is especially true of proglacial environments, as not all sediment produced during a given time span is necessarily removed concurrently, and not all the sediment removed from the basin has been produced contemporaneously [Church and Ryder's (1972) '*paraglacial*' environment; Fenn, Unpublished data].

Figure 8a shows the effect of repeated wetting on Ca^{2+} release from Haut Glacier d'Arolla flour. It is clear that there is significant release of calcium during the second wetting, although the amount released

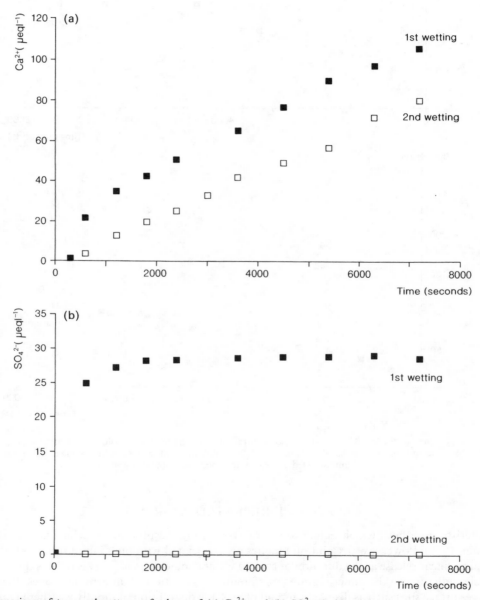

Figure 8. Comparison of temporal patterns of release of (a) Ca^{2+} and (b) SO_4^{2-} during dissolution experiments which involved repeat wetting of the same sample of 62–125 μm sediment from the proglacial plain of Haut Glacier d'Arolla. Debris concentration was $4\,g\,l^{-1}$

is less than during the first wetting (80 relative to 105 μequiv. l^{-1} after 2 h). In contrast, it is clear from Figure 8b that SO_4^{2-} is released only during the initial phase of wetting. Thereafter, the release of SO_4^{2-} is zero, within analytical error. This confirms the suggestion by Brown *et al.* (1994a) that release during the first wetting reflects the release of readily available SO_4^{2-} from slow sulphide oxidation during transport and storage of the sediment. This suggests that under field conditions the release of SO_4^{2-} from suspended sediment will be extremely limited, indicating that the oxidation of sulphide minerals in Haut Glacier d'Arolla sediments is largely run to completion in the distributed component of the hydroglacial system. This suggests SO_4^{2-} may provide an exclusive marker of the distributed system for the separation of individual components of the bulk hydrograph (see Tranter and Raiswell, 1991) after elution of the over-lying winter snowcover and helps explain the dominance of so called 'carbonation' reactions in controlling chemical weathering of sediments in the channelized component of the hydroglacial system (Tranter *et al.*, 1993).

EFFECT OF THE INITIAL COMPOSITION OF THE SOLUTION

The effect of initial water composition on the rate of dissolution operates mainly through the hydrogen ion concentration of the solution, that promote acid hydrolysis reactions (Raiswell *et al.*, 1980; Raiswell, 1984; Curtis, 1976). These aqueous protons are derived from two main sources in the glacier system; the dissolution and dissociation of atmospheric carbon dioxide [Equation (8)], and the oxidation of sulphide minerals such as pyrite [Equation (9)].

$$CO_2(aq) + H_2O(aq) \rightleftharpoons H^+(aq) + HCO_3^-(aq) \tag{8}$$

$$4FeS_2(s) + 15O_2(aq) + 14H_2O(aq) \rightleftharpoons 16H^+(aq) + 8SO_4^{2-}(aq) + 4Fe(OH)_3(s) \tag{9}$$
$$\text{pyrite} \qquad\qquad\qquad\qquad\qquad\qquad\qquad\qquad \text{ferric oxyhydroxides}$$

Protons may also be derived from the dissolution of acid deposition [e.g. $H_2SO_4(s)$, $HNO_3(s)$] (Tranter *et al.*, 1993), especially early in the snowmelt season when snowmelt is most acidic (Tranter, Unpublished data). Another important dissolved substance in the glacial meltwater is dissolved oxygen, as the decay of oxidizable minerals such as sulphides [Equation (9)] requires the presence of dissolved O_2 (Tranter *et al.*, 1993; Brown *et al.*, 1994b).

The role of solution chemistry in laboratory dissolution studies has been widely reported (e.g. Wollast, 1967; Deju and Bhappa, 1965; Petrovic *et al.*, 1976; Holdren and Berner, 1979). Such studies show that the initial reaction rate is a simple function of pH and hence distinctly lower rate laws apply to higher pH solutions. Thus variability in the source and abundance of protons may have important implications for the rate and magnitude of solute acquisition from suspended sediment in glacial environments.

Table III illustrates the variability in pH associated with samples of pre-melt snow from the Arolla valley during the 1992–1993 winter and suggests that a significant supply of protons may be associated with waters input to the hydroglacial system during periods when snowmelt dominates meltwater production. Figure 9a and 9b clearly shows that the release of Ca^{2+} is enhanced by more acidic solutions. Concentrations

Table III. Pre-melt snow pH values, Arolla valley, winter 1992–1993. Bulk meltwater values, 1 June–31 August 1989 are presented for comparison

Location	Altitude (m)	pH	
		Minimum	Maximum
Haut Glacier d'Arolla	3000	4·47	6·05
	2600	4·77	6·25
La Monta village	1900	4·35	5·91
Bulk meltwaters	2500	6·81	9·23

range between 270 and 142 μequiv. l^{-1} after 3 h water to rock contact for initial starting solutions of pH 3·8 (sol 1) and 5·4 (deionized water), respectively. This suggests that chemical weathering in these open solutions is not only controlled by the number of reactive surface sites, but also by the availability of protons. This is consistent with the results of the repeated wetting experiments described earlier, in which the second wetting of a weight of sediment at the same water to rock ratio also liberated significant quantities of Ca^{2+}, suggesting that the proton supply was limiting solute release during the first wetting.

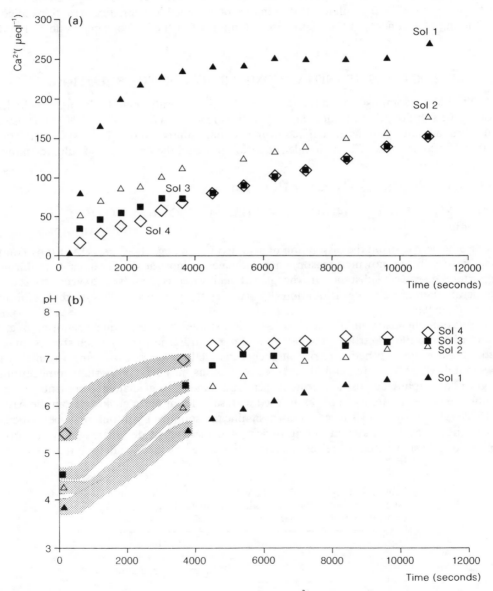

Figure 9. Effect of variable initial solution chemistry on the release of (a) Ca^{2+} and (b) proton consumption (denoted by pH) during dissolution experiments involving constant water to rock ratio (4 g l^{-1}) and grain size distribution (63–125 μm) of Haut Glacier d'Arolla sediment. Stippled areas denote continuous monitoring of pH. Initial solution chemistries were a mixture of 0·11 (solution 1), 0·04 (solution 2), 0·02 (solution 3) and 0 mM (solution 4) analar sulphuric acid and 0·03, 0·008, 0 and 0 mM nitric acid, producing starting solution pH values of 3·9 (solution 1), 4·3 (solution 2), 4·6 (solution 3) and 5·4 (solution 4)

CONCLUSIONS

The experiments detailed here illustrate the range of parameters which may influence the rate and magnitude of post-mixing reactions in open channels beneath Haut Glacier d'Arolla. These are the water to rock ratio, the duration of water to rock interaction, the number of reactive surface sites (which is a function of particle size, the degree of comminution, the wetting history and mineralogy) and the availability of aqueous protons. A clear conclusion from these experiments is that glacial flour is more geochemically reactive than previously assumed (Souchez and Lemmens, 1987) and that dilute waters flowing in open channels are certain to acquire solute.

This conclusion has important implications for the application of water quality data to hydroglacial investigations. For example, the assumption that no post-mixing evolution of water chemistry occurs when the two components of hydroglacial system mix to form bulk meltwater has been implicit in the adoption of a constant composition for the quickflow component in many mixing model calculations (e.g. Collins, 1978; 1979b; Gurnell and Fenn, 1984). The experimental data presented here clearly illustrate that this is unrealistic for ions involved in chemical weathering reactions (i.e. base cations, pH and HCO_3^-), particularly when the composition of bulk meltwater is dilute.

The experiments provide little evidence for sulphide oxidation during chemical weathering reactions between deionized water and suspended sediment from Haut Glacier d'Arolla. The approximate 1 : 1 ratio of the dominant ions (Ca^{2+} and HCO_3^-), allied with the closed system low $p(CO_2)$ characteristics shown by many of the free drift experiments, suggests that chemical weathering reactions providing solute to dilute meltwater in the channelized component of the hydroglacial system are dominated by carbonation reactions. Hence solute acquisition in these environments may be controlled by kinetic rather than equilibrium processes, as the consumption of protons involved in the dissolution of the geochemically reactive glacial flour exceeds the diffusion of CO_2 across the gas–liquid interface.

The experiments suggest that post-mixing chemical weathering of suspended sediment may supply a significant proportion of the total solute concentration in the bulk meltwater under certain hydrological conditions. Given a knowledge of rock to water contact times from dye tracer investigations (Nienow et al., Unpublished data; Brown et al., 1994a) and SSC in the bulk meltwater at Haut Glacier d'Arolla (Brown, Unpublished data; Gurnell et al., 1991), some indication of the potential for solute acquisition by post-mixing reactions can be obtained from the rate laws derived from the experimental data. Dye tracing data suggest meltwater residence times of $\leqslant 2\,h$ at maximum diurnal discharge by mid-August, with a corresponding suspended sediment concentration of $4\,g\,l^{-1}$. Application of this field data to the rate laws gives a maximum release of 65% of the total Ca^{2+} concentration at maximum discharge. Indeed, post-mixing reactions may make a significant contribution to the total solute flux from a glacier at all stages of the discharge cycle late in the ablation season (Brown et al., 1994a). This is contrary to the common assumption that the bulk of the solute load is acquired as meltwater is routed slowly through a distributed network.

Brown et al. (1994a) caution against the uncritical application of the laboratory dissolution data to simulations of chemical weathering processes in basal arterial channels. These is because

1. The sediment used in the experiments has already passed through the subglacial drainage system, dissolving the most reactive parts of the mineral surface. Thus the rate and magnitude of solute acquisition may be even more significant than we have suggested.
2. Different mineral assemblages may be found in the different size fractions of the suspended sediment. Thus differences in the rate of solute release during the particle size experiments may not be a simple function of surface area.
3. Post-mixing reactions occur in diluted delayed flow, which is more concentrated than the deionized water used in these experiments. Thus the rate of solute acquisition may be enhanced in the laboratory [Equation (4)].
4. The difficulty in defining suspended sediment concentrations in basal arterial conduits from bulk meltwater records.

However, it has been shown that laboratory experiments constrained by field data reproduce features of results which are valuable in the interpretation of chemical processes operative in bulk meltwater during three diurnal cycles at the Haut Glacier d'Arolla during 1989 (Brown *et al.*, 1994a). Natural variability in other parameters (e.g. grain size, processes of sediment delivery to the channelized component, quality of input waters) are currently less well defined and require further field study before the results of the laboratory experiments may be fully exploited.

ACKNOWLEDGEMENTS

This work was supported by a NERC Fellowship (GT5/F/AAPS/3) to G.H.B. and by NERC Grants (GR3/7004 and GR3/8114). We thank the Head of Department, Department of Geography, University of Cambridge for support of this project.

REFERENCES

Acker, J. G. and Bricker, O. P. 1992. 'The influence of pH on biotite dissolution and alteration kinetics at low temperature', *Geochim. Cosmochim. Acta*, **56**, 3073–3092.
Bergman, I., Cartwright, J., and Casswell, C. 1963. 'The disturbed layer on ground quartz powders of respirable size', *Br. J. Appl. Phys.*, **14**, 399–404.
Berner, R. A. 1981. 'Kinetics of weathering and diagenesis' in Lasaga, A. C. and Kirkpatrick, R. J. (Eds), *Kinetics of Geochemical Processes. Mineral. Soc. Am. Rev. Mineral.*, **8**, 111–134.
Boulton, G. S. 1974. 'Processes and patterns of glacial erosion' in Oates, R. D. (Ed.), *Glacial Geomorphology*. State University, New York. pp. 41–87.
Boulton, G. S. 1978. 'Boulder shapes and grain-size distributions of debris as indicators of transport paths through a glacier and till genesis', *Sedimentology*, **25**, 773–799.
Brown, G. H. and Tranter, M. 1990. 'Hydrograph and chemograph separation of bulk meltwaters draining the Upper Arolla Glacier, Valais, Switzerland' in *Hydrology in Mountainous Regions. I. Hydrological Measurements; the Water Cycle (Proceedings of the Two Lausanne Symposium, August, 1990). IAHS Publ.*, **193**, 429–437.
Brown, G. H., Sharp, M. J., Tranter, M., Gurnell, A. M., and Nienow, P. W. 1994a. 'Impact of post-mixing chemical reactions on the major ion chemistry of bulk meltwaters drainage the Haut Glacier d'Arolla, Valais, Switzerland', *Hydrol. Process.*, **8**, 465–480.
Brown, G. H., Tranter, M., Sharp, M. J., Davies, T. D., and Tsiouris, S. 1994b. 'Dissolved oxygen variations in alpine glacial meltwaters', *Earth Surf. Process. Landforms*, **19**, 247–253.
Busenberg, E. and Clemency, C. V. 1975. 'The dissolution kinetics of feldspars at 25°C and 1 atms CO_2 partial pressure', *Geochim. Cosmochim. Acta*, **40**, 41–49.
Casey, W. H., Westrich, H. R., and Holdren, G. R. Jr. 1988. 'Dissolution rates of plagioclases at pH = 2 and 3', *Am. Mineral.*, **76**, 211–217.
Chou, L., Garrels, R. M., and Wollast, R. 1989. 'Comparative study of the kinetics and mechanisms of dissolution of carbonate minerals', *Chem. Geol.*, **78**, 269–282.
Church, M. and Ryder, J. M. 1972. 'Paraglacial sedimentation: a consideration of fluvial processes conditioned by glaciation', *Geol. Soc. Am. Bull.*, **83**, 3059–3071.
Collins, D. N. 1978. 'Hydrology of an alpine glacier as indicated by the chemical composition of meltwater', *Z. Gletscherk. Glazialgeol.*, **13**, 219–238.
Collins, D. N. 1979a. 'Sediment concentration in melt waters as an indicator of erosion processes beneath an alpine glacier', *J. Glaciol.*, **23**, 247–257.
Collins, D. N. 1979b. 'Quantitative determination of the subglacial hydrology of two alpine glaciers', *J. Glaciol.*, **23**, 347–361.
Curtis, C. D. 1976. 'Chemistry of rock weathering: fundamental reactions and controls' in Derbyshire, E. (Ed.), *Geomorphology and Climate*. Wiley, New York. pp. 25–27.
Deju, R. A. and Bhappa, R. B. 1965. 'Surface properties of silicate minerals', *N. M. Bur. Mines, Min. Resourc., Circ.*, **82**, 6pp.
Drewry, D. 1986. *Glacial Geologic Processes*. Edward Arnold, London. 276pp.
Fairchild, I. J., Bradby, L., and Spiro, B. 1994. 'Reactive carbonate in glacial sediments: a preliminary synthesis of its creation, dissolution and reincarnation' in Deynoux, M., Miller, J. G., Dornack, E., Eyles, N., Fairchild, I. J., and Young, G. M. (Eds), *Earth's Glacial Record*. Cambridge University Press, Cambridge, 176–192.
Fenn, C. R. and Gomez, B. 1989. 'Particle size analysis of the suspended sediment in a proglacial stream: Glacier de Tsidjiore Nouve, Switzerland', *Hydrol. Process.*, **3**, 123–135.
Gurnell, A. M. 1987. 'Suspended sediment' in Gurnell, A. M. and Clark, M. J. (Eds), *Glacio-Fluvial Sediment Transfer: an Alpine Perspective*. Wiley, Chichester. pp. 305–354.
Gurnell, A. M. and Clark, M. J. (Eds) 1987. *Glacio-Fluvial Sediment Transfer: an Alpine Perspective*. Wiley, Chichester. 524pp.
Gurnell, A. M. and Fenn, C. R. 1984. 'Flow separation, sediment source area and suspended sediment transport in a pro-glacial stream' in Schick, A. P. (Ed.), *Channel Processes: Water, Sediment, Catchment Controls. Catena Suppl.*, **5**, 109–119.
Gurnell, A. M., Clark, M. J., Tranter, M., Brown, G. H., and Hill, C. T. 1991. 'Alpine glacier hydrology inferred from a proglacial river monitoring programme' in *Proc. British Hydrological Society National Symposium, Southampton, 16–18 September 1991.* pp. 5.9-5.16.
Hammer, K. M. and Smith, N. D. 1983. 'Sediment production and transport in a proglacial stream: Hilda glacier, Alberta, Canada', *Boreas*, **12**, 91–106.

Hochella, M. F., Jr 1990. 'Atomic structure, microtopography, composition, and reactivity of mineral surfaces' in Hochella, M. F. and White, A. F. (Eds), *Mineral–Water Interface Geochemistry. Mineral. Soc. Am. Rev. Mineral.*, **23**, 87–132.

Holdren, G. R. and Berner, R. A. 1979. 'Mechanism of feldspar weathering. I. Experimental studes', *Geochim. Cosmochim. Acta*, **43**, 1161–1171.

Holdren, G. R. Jr and Speyer, P. M. 1985. 'Reaction rate-surface area relationships during the early stages of weathering. I. Initial observations, *Geochim. Cosmochim. Acta*, **49**, 675–681.

Karlsen, E. 1991. 'Variations in grain-size distribution of suspended sediment in a glacial meltwater stream, Austre Okstindbreen, Norway', *J. Glaciol.*, **37**, 113–119.

Keller, W. D. and Reesman, A. L. 1963. 'Glacial milks and their laboratory-simulated counterparts', *Geol. Soc. Am. Bull.*, **74**, 61–76.

Keller, W. D., Baglord, W. D. and Reesnam, A. L. 1963. 'Dissolved products of artificially pulverised silicate minerals and rocks: part 1', *J. Sedim. Petrol.*, **33**, 191–204.

Krinsley, D. H. and Doornkamp, J. C. 1973. *Atlas of Quartz Sand Surface Textures.* Cambridge University Press, Cambridge. 91pp.

Lagache, M. 1965. 'Contribution a l'étude de l'alteration des feldspaths dans l'eau, entre 100 et 200°C, sous diverses pressions de CO_2, et application a la synthèse des numeraux argileux', *Bull. Soc. Fr. Mineral. Cristallogr.*, **88**, 223–253.

Lagache, M. 1976. 'New data on the kinetics of the dissolution of alkali feldspars at 200°C in CO_2 charged water', *Geochim. Cosmochim. Acta*, **40**, 157–161.

Lerman, A. 1979. *Geochemical Processes Water and Sediment Environments.* Wiley, Chichester. 481pp.

McKibben, M. A. and Barnes, H. L. 1986. 'Oxidation of pyrite in low temperature acidic solutions: rate laws and surface textures', *Geochim. Cosmochim. Acta*, **50**, 1509–1520.

Nagy, K. L. and Lasaga, A. G. 1992. 'Dissolution and precipitation kinetics of gibbsite at 80°C and pH 3: the dependence on solution saturation state', *Geochim. Cosmochim. Acta*, **56**, 3093–3111.

Nicholson, R. V., Gillham, R. W., and Reardon, E. J. 1988. 'Pyrite oxidation in carbonate-buffered solution: 1. Experimental kinetics', *Geochim. Cosmochim. Acta*, **50**, 1077–1085.

Oerter, H., Beherens, H., Hibsch, G., Ravert, W., and Stichler, W. 1980. 'Combined environmental isotope and electrical conductivity investigations of the runoff of Vergnagtferner (Oetzal Alps)' in *Int. Symp. Computation and Prediction of Runoff from Glaciers and Glacierised Areas, Thbilisi, USSR, 3–11 September 1978. Akad. Nauk SSSR, Inst. Geogr. Mater. Gliatsiologichesk. Issl. Khr. Obsuzhd.*, **39**, 86–91 and 157–161.

Parks, G. A. 1990. 'Surface energy and adsorption at mineral/water interfaces: an introduction' in Hochella, M. F. and A. F. White (Eds), *Mineral–Water inface Geochemistry. Mineral. Soc. Am. Rev. Mineral.*, **23**, 133–176.

Petrovic, R. 1981a. 'Kinetics of dissolution of mechanically comminuted rock-forming oxides and silicates–I. Deformation and dissolution of quartz under laboratory conditions', *Geochim. Cosmochim. Acta*, **45**, 1665–1674.

Petrovic, R. 1981b. 'Kinetics of dissolution of mechanically comminuted rock-forming oxides and silicates–II. Deformation and dissolution of oxides and silicates in the laboratory and at the Earth's surface', *Geochim. Cosmochim. Acta*, **45**, 1675–1686.

Petrovic, R., Berner, R. A., and Goldhaber, M. B. 1976. 'Rate control in dissolution of alkali feldspars. I. Study of residual grains by X-ray photoelectron spectroscopy', *Geochim. Cosmochim. Acta*, **40**, 537–548.

Plummer, L. N., Wigley, T. M. L., and Parkhurst, D. L. 1978. 'The kinetics of calcite dissolution in CO_2–water systems at 5° to 60°C and 0·0 to 1·0 atm CO_2, *Am. J. Sci.*, **278**, 179–216.

Rainwater, F. H. and Guy, H. P. 1961. 'Some observations on the hydrochemistry and sedimentation of the Chamberlain Glacier area, Alaska', *U.S. Geol. Surv. Prof. Pap.*, **414-c**, 14pp.

Raiswell, R. 1984. 'Chemical models of solute acquisition in glacial meltwaters', *J. Glaciol.*, **30**, 49–57.

Raiswell, R. W., Brimblecombe, P., Dent, D. L., and Liss, P. S. 1980. *Environmental Chemistry.* Edward Arnold, London. 184pp.

Reynolds, R. C. and Johnson, N. M. 1972. 'Chemical weathering in the temperate glacial environment of the Northern Cascade Mountains', *Geochim. Cosmochim. Acta*, **36**, 537–554.

Richards, K. S. 1984. 'Some observations on suspended sediment dynamics in Stobregrova, Jotunheimen', *Earth Surf. Process. Landforms*, **9**, 101–112.

Sharp, M. J. 1991. 'Hydrological inferences from meltwater quality data: the unfulfilled potential' in *Proc. British Hydrological Society National Symposium, Southampton, 16–18 September 1991.* pp. 5.1–5.8.

Souchez, R. A. and Lemmens, M. M. 1987. 'Solutes' in Gurnell, A. M. and Clark, M. J. (Eds), *Glacio-Fluvial Sediment Transfer: an Alpine Perspective.* Wiley, Chichester. pp. 285–303.

Sugden, D. E. and John, B. S. 1976. *Glaciers and Landscape.* Edward Arnold, London. 376pp.

Tazaki, K. 1976. 'Scanning electron microscopic study of formation of gibbsite from plagioclase', *Papers, Inst. Therm. Sprung. Res. Okayama Univ.*, **45**, 11–24.

Tranter, M. and Raiswell, R. 1991. 'The composition of the englacial and subglacial components in bulk meltwaters draining the Gornergletscher', *J. Glaciol.*, **37**, 59–66.

Tranter, M., Brown, G. H., Raiswell, R., Sharp, M. J., and Gurnell, A. M. 1993. 'A conceptual model of solute acquisition by Alpine glacial meltwaters', *J. Glaciol.*, **39**, 573–581.

Walder, J. S. 1986. 'Hydraulics of subglacial cavities', *J. Glaciol.*, **32**, 439–445.

Whalley, W. B. and Krinsley, D. H. 1974. 'A scanning electron microscope study of surface textures of quartz grains from glacial environments', *Sedimentology*, **21**, 87–105.

Wollast, R. 1967. 'Kinetics of the alteration of K-feldspar in buffered solutions at low temperature', *Geochim. Cosmochim. Acta*, **31**, 635–648.

Zhang, H., Brown, P. R., and Nater, E. A. 1993. 'Change in surface area and dissolution rates during hornblende dissolution at pH 4·0', *Geochim. Cosmochim. Acta*, **57**, 1681–1689.

11

STATISTICAL EVALUATION OF GLACIER BOREHOLES AS INDICATORS OF BASAL DRAINAGE SYSTEMS

C. C. SMART

Department of Geography, University of Western Ontario, London, Ontario N6A 5C2, Canada

ABSTRACT

Between 1988 and 1992 closely spaced arrays of boreholes were drilled at Small River Glacier, British Columbia. The borehole arrays have been used to investigate the interannual and spatial consistency of patterns of basal hydraulics beneath the glacier. A simple robust classification was devised identifying *unconnected*, *high standing*, *low standing* and *dry* base water levels in boreholes. Spatial and interannual comparisons were made using a simple nearest neighbour statistic, corrected for differences in frequency of different borehole types and evaluated using Monte Carlo confidence intervals to compensate for array form. Arrays in the lower ablation zone showed spatial and interannual coherence, with three distinct areas characterized by low water pressure, till-associated non-connection and high pressure. There was no indication of a dominant conduit. Slightly higher up-glacier borehole patterns were less coherent, and varied from year to year, probably a result of subglacial karst capturing basal waters at a number of low pressure points at the bed. Therefore both the upper and lower arrays at Small River Glacier appear to encompass unusual drainage conditions. The nearest neighbour analysis provides valuable constraints on more specific interpretation.

INTRODUCTION

Boreholes provide a unique window on conditions at the bed of glaciers and have provided important information for investigations of the geometry, hydraulics and chemistry of basal drainage systems. Basal water pressures inferred from borehole water levels have shown considerable departure from theoretical expectations (e.g. Hooke *et al.*, 1990). Boreholes allow the injection of tracer dyes directly into random points in the basal drainage system, rather than relying on moulins (Fountain, 1993). Until boreholes allowed *in situ* sampling (Lamb *et al.*, Unpublished data), basal water quality had to be inferred from measurements on supraglacial and proglacial streams (e.g. Collins, 1979).

Despite their significance in glaciology, there has been surprisingly little critical appraisal of boreholes. For example, boreholes open to the surface may not represent a meaningful basal water pressure because a flux of water is necessary to alter the water level. Boreholes penetrating more active conduits will be more responsive to pressure fluctuations than those drilled into more isolated areas of the bed. The latter will also be sensitive to the influx of supraglacial water. Iken and Bindschadler (1986) recognized this problem and suggested installing surface feeders to maintain a steady flux into the borehole. They rejected data from boreholes showing high sensitivity to surface fluxes. Stone and Clarke (this issue) have pointed out that pressure transducers sealed in boreholes by refreezing in cold ice provide better basal pressure data than open boreholes. However, hot water drilling under any conditions results in the application of an over-pressurized water column to the bed and this in itself may induce artificial linkages and cause bed disturbance.

Unfrozen boreholes are generally classified as 'connected' and 'unconnected', with the former exhibiting water levels below the glacier surface. Borehole water levels are generally taken to indicate basal hydraulic conditions. For example, it is considered that conduits would be indicated by a longitudinal thread of low pressure with a subordinate zone or 'axis' of higher pressure grading into more remote unconnected areas

(Alley, this issue; Hantz and Lliboutry, 1983; Iken and Bindschadler, 1986; Iken *et al.*, 1993; Fountain, 1994).

Difficulties in drilling, instrumenting and maintaining boreholes mean that they are often placed many tens or hundreds of metres apart. A sparse borehole array has a low probability of intersecting a discrete conduit a few metres wide, although more extensive 'braided' systems (e.g. Hock and Hooke, 1993) might be more accessible. An irregularly shaped conduit or a distributed drainage system (e.g. Walder, 1986) may contain abrupt heterogeneities which will be difficult to detect from borehole records. Bedform and composition might provide useful constraints on the problem, but borehole videophotography has not yielded much information (Kamb *et al.*, 1978). To compensate for inadequate borehole coverage and density, data from successive years have been aggregated to provide a fuller record (e.g. Hodge, 1979; Hock and Hooke, 1993; Fountain, 1994). It is not known if basal conditions are sufficiently consistent from year to year to allow such aggregation. Bedform and composition are best inferred from extraglacial areas, but these are often far removed from borehole sites.

Such difficulties were a concern in early drilling programmes on Blue Glacier (Engelhardt *et al.*, 1978) and South Cascade Glacier (Hodge, 1979), which led to the conclusion that 'subsole drift' caused hydraulic isolation of much of the bed and high pressures (>50% overburden) away from relatively rare, low pressure conduits. In contrast, Fountain (1994) has inferred subsole drift to be rather permeable (hydraulic conductivity 10^{-4}–10^{-7} m s^{-1}) at South Cascade Glacier, allowing a discrete conduit to 'draw down' water pressures over a very broad area of bed. Unfortunately, the sparse and irregular distribution of boreholes prevented systematic analysis of the spatial heterogeneity evident from close juxtaposition of connected and unconnected holes. In addition, Hodge inferred a significant change in basal conditions over a number of years, although he did not redrill in a fixed location.

Two simple questions arise. Firstly, is conventional borehole spacing well matched to the characteristic scale of basal drainage systems? Secondly, are similar basal drainage systems re-established every year? Answers are necessarily indirect, as actual patterns of basal drainage have yet to be established. Geomorphological evidence from exposed glacier beds suggests that conduit footprints are relatively narrow (Walder and Hallet, 1979) and incised Nye channels imply regular re-establishment over a number of years (e.g. Sharp *et al.*, 1989). Hydraulic arguments are less helpful, currently supporting a lively debate on the relative significance of discrete conduits versus braided or anabranching systems (Hock and Hooke, 1993; Nienow *et al.*, Unpublished data).

Unusually dense borehole arrays established over both hard and soft beds at Small River Glacier, British Columbia provide a basis for evaluating the questions of scale and replication. This paper describes a simple, robust borehole classification scheme and novel techniques used to search for spatial and temporal patterns in the borehole types.

SITE AND METHODS

Small River Glacier lies in the Western Rocky Mountains of British Columbia (Figure 1). The 2·5 km^2 glacier overlies massive, north-dipping, karstified carbonates of the Cambrian Mural Formation into which most of the upper glacier is drained. Lower down, a small, hydrologically isolated ablation zone of 800 m length, 0·25 km^2 in area and up to 100 m in thickness drains directly to two small proglacial streams (Huntley, Unpublished data; Carr, Unpublished data). The valley hosting the Central Lobe is aligned along the strike of a unit of weak, till-producing siltstone. North-dipping tabular carbonates form the glacier bed on the south and an upper carbonate unit forms a subglacial ridge and cliff on the north side (Figure 1). The discrete geology results in distinctive bed lithologies and unusual correspondence between subglacial and forefield topography (Huntley, Unpublished data). Proglacial carbonate surfaces are characterized by 1–10 m scale roches moutonnées. Basal velocities of between 5 and 8 cm d^{-1} (Carr, Unpublished data) imply obstacles are traversed by basal ice in between 10 and 200 days.

A high speed hot water drill, combined with thin ice, allowed rapid drilling and allowed extensive borehole arrays to be established and monitored. Boreholes were spaced at ≈15–20 m on a crude grid aligned across the glacier. Wider spacing was used for more peripheral, exploratory drilling. Exact positioning was

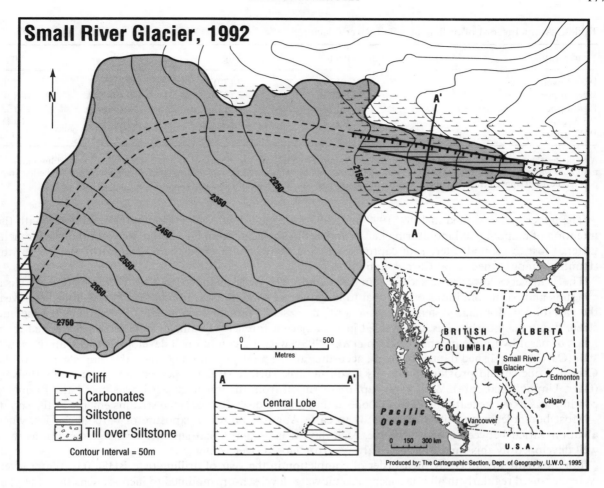

Figure 1. Small River Glacier with underlying geology and cross-section through Central Lobe

arbitrary, depending on local topography such as supraglacial channels and crevasses. Sediment build-up from dirty ice was not a serious impediment to drilling in the thin, clean ice at Small River, so that virtually all boreholes are believed to have reached the bed. On carbonate areas, the bed could be tapped with the drillstem. Over the siltstone the bed was much softer and till could be recovered adhering to the drill tip. The dominant association of hydraulic connection with the drill reaching the bed leads to interpretations founded on *basal* drainage systems only in this paper, although englacial drainage may be locally important in areas among and down-glacier from crevasses. Table I summarizes the annual drilling record and Figure 2 shows the borehole arrays from four years of drilling. More boreholes were drilled each year in response to improved drilling and monitoring capacity. At the maximum, it is believed that these arrays constituted the densest and largest arrays of simultaneously monitored boreholes reported. Borehole verticality was not determined. However, bed elevations determined from boreholes were consistent to within ±15 cm between years. Inclinometry measurements in boreholes drilled at Haut Glacier d'Arolla, Switzerland with a less sensitive drill and less experienced drillers suggest a depth error of $\sigma = 0.52$ m in boreholes less than 100 m deep (Willis *et al.*, Unpublished data). Deeper boreholes at Arolla show a consistent down-glacier displacement of the borehole bottom by 2–3 m, with easting and northing 1σ errors of 2.5 and 3.1 m, respectively. These errors are probably excessive for shallow boreholes drilled with sensitive equipment and experienced operators, but nevertheless are still much less than the spacing between boreholes at Small River.

Table I. Annual record of drilling at Small River Glacier

Year array	N	Unconnected	High	Low	Dry
1988 lower	10	3 (30)	2 (20)	3 (30)	2 (20)
1989 upper	28	7 (25)	7 (25)	0 (0)	14 (50)
1990 upper	39	6 (15)	5 (13)	15 (38)	13 (33)
1992 lower	24	9 (38)	4 (17)	10 (42)	1 (4)
1992 upper	21	5 (24)	5 (24)	5 (25)	11 (52

N indicates the number of holes completed to the bed, unconnected indicates number (%) of water-filled holes, low and high indicate minimum water levels below and above 50% of ice thickness and dry indicates number (%) of holes which drained to the bed at some point. The 1989 upper array data include 13 boreholes outside the area outlined in Figure 2

The upper arrays drilled in 1989, 1990 and 1992 were placed across a zone of subglacial karst, with the intention of mapping and monitoring karst capture of subglacial drainage. About 70% of total glacier meltwater is lost to groundwater, with capture occurring above and around the position of the upper swaths (Huntley, unpublished data; Carr, Unpublished data).

The transverse lower arrays drilled in 1988 and 1992 were intended to reveal the position of basal conduits and basal conditions orthogonal to longitudinal conduit axes. In addition, it was hoped to reveal the variation in drainage systems across a contiguous set of tabular carbonates, till-mantled siltstone and steep, hummocky carbonates, as revealed in the adjacent forefield. Drainage system models derived from maps of basal hydraulic potential (Sharp *et al.*, 1993) suggest a conduit will develop part way up the south slab (Carr, Unpublished data). However, according to the model of Hooke (1984), the thin, steep ice of the Central Lobe should provide conditions favouring free surface streams focused along the valley thalweg (Carr, Unpublished data). With karst capture of runoff from the upper glacier it was uncertain if a major conduit was likely beneath the Central Lobe, although two discrete proglacial streams emerge at the snout.

Most boreholes were drilled in mid-July to early August, following exposure of bare glacier ice, when ablation varied between 0 and 8 cm w.e. each day. It is assumed that summer drainage conditions were well developed by this time. Boreholes always froze closed over winter.

Boreholes are identified by their order of completion in the year of drilling (e.g. BH88-03). Water levels were measured regularly in all holes using groundwater level sensors modified to increase sensitivity. Higher frequency data were obtained in selected holes using up to 30 Drück PDCR and Omega 100–300 lb in^{-2} pressure transducers positioned ≈ 1 m (or more depending on cable length) above the bed to avoid destruction by surface lowering and incorporation into basal ice. Transducers were read by Campbell Scientific CR21X dataloggers every 10 s, with averages recorded every 5 min. The logger data were converted to borehole water level elevation (hydraulic head) using on-site calibration in water-filled boreholes and mapped borehole position. Figures 3 and 4 show example sets of borehole water level data, with the borehole elevation range indicated to the right. Manual readings are shown as point symbols.

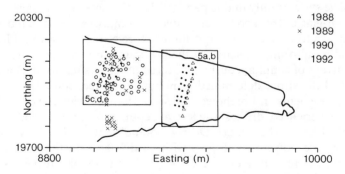

Figure 2. Map of the Central Lobe with all boreholes and outline boxes for 1989–1992 drilling data. Map coordinates are based on a local grid oriented approximately to magnetic north

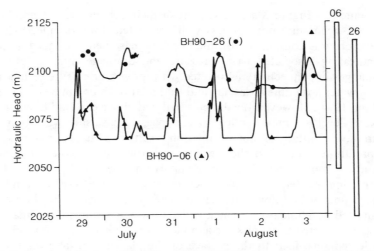

Figure 3. Example of borehole water level data from BH90-06 and BH90-26 with manual (symbols) and high frequency data. Borehole depth range is shown to the right

Supraglacial water may enter a borehole and cause diurnal water-level fluctuations unrelated to basal hydrology, especially where hydraulic conductivity is low. Efforts to prevent surface water from entering boreholes were only moderately successful, but, in any instance, slight seepage occurred in the permeable upper zones of most boreholes. Freezing of seepage water sometimes led to borehole constriction, typically at a depth of about 5–10 m. Discordant high point measurements and rises associated with reaming to remove such obstructions were purged from water-level records. Water levels cannot be measured below the transducer tip, except by manual probe. As a result, the fully drained condition may not always be recognized, despite its importance in indicating free surface basal streams. Five minute averaging may lead to loss of detail in exceptionally dynamic holes.

CLASSIFICATION

There are difficulties in inferring basal drainage conditions from borehole water-level records. The water

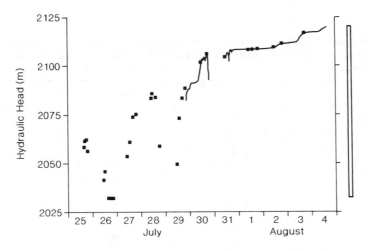

Figure 4. Manual and high frequency water level data from BH90-11. The window of classification (1–4 August) encompasses an intermediate stage as the borehole converts from low base–high amplitude to unconnected. Manual data are indicated by squares and borehole depth range is shown to the right

level time series presented in Figure 3 are consistent in their behaviour, but boreholes can shift in their performance over time, becoming more or less well connected, for example, BH90-11 (Figure 4). The time-scale of such evolution is consistent with the borehole base being dragged across bumps at the bed, with water levels indicating associated basal pressure and drainage changes. However, such evolution may also indicate changes in connectivity associated with growth or closure of linking passageways.

Spatial similarity in water level time series, lag and lead relationships can be revealed by cross-correlation analysis (e.g. Fountain, 1994). However, the difficulties noted above can lead to errors. For example, where trend in borehole water level is a major source of variance, a phase shift will occur in the timing of peak correlation. In addition, the pattern of diurnal ablation results in a characteristic recharge signature which makes spatial interpretation of open borehole water levels difficult. Cross-correlation is not applicable to inter-year comparisons. A more robust basis for comparison is required.

Rather than interpreting the detail of water-level records, a simple classification scheme was devised to describe systematic features of borehole water levels. The primary hydraulic characteristic of boreholes is 'connection'. A connected borehole is one which is dry or which exhibits a (normally variable) water level below the ice surface. Connected boreholes are assumed to be linked into the local basal drainage system. An unconnected hole remains filled to the surface of the glacier and, as a first approximation, can be assumed to be isolated from the basal drainage system, despite implicit over-pressurization at the bed. The pattern of connected and unconnected holes will reveal if large areas of bed are hydraulically isolated or if basal drainage is pervasive.

Consistent, connected boreholes can be classified on an ordinal scale using the minimum and maximum observed water levels. Minimum or base water level is defined by the common minimum level obtained on a daily basis. Base character can be dry (drained), low (below 50% of ice overburden) or high (>50% of ice overburden). In connected holes, maximum water levels are associated with superimposition of a diurnal impulse on the base water level. The amplitude of the diurnal response can also be arbitrarily classified as zero (no diurnal signal), low (rising to less than 50% of the free space above the base level) and high (rising to occupy more than 50% of the free space above base water level). Table II summarizes this classification and presents the symbols used to represent the different classes of borehole. For example, BH90-06 and BH90-26 (Figure 3) are classified as low base–high amplitude and high base–high amplitude, respectively. Formal classification is indicated in the text by capitalization of the descriptive adjective.

A significant advantage of the simple classification is that it remains effective when applied to manual water level data (consider Figure 3). However, some systematic difficulties remain. Pressure transducers suspended above the bed cannot identify dry conditions; the distinction from low holes relies on manual

Table II. Symbols used to represent borehole water level behaviour. Base is the typical minimum water level, Amplitude indicates the relative amplitude of diurnal fluctuations above that base

Base	Amplitude		
	0	Low	High
Dry	▯	▯	▮
Low	▯	▯	▮
High	▮	▮	▮
Unconnected	▮		

observations, which (at Small River) are less common during early water level minima. As an example, BH90-06 shows flat (zero pressure) overnight readings as water levels drop below the transducer. The borehole almost certainly became dry overnight, but there is only a single sounding on 1 August to indicate this. Some Low classifications may therefore be artefacts. The manual water level obtained for BH90-06 on 3 August (Figure 3) is significantly different from the corresponding transducer record. Such occasional anomalies, resulting from errors in sounding, perched water or a flaw in data compilation only become apparent when several days' data are considered. More significant is a concern that, in some instances, the magnitude of the diurnal pulse is determined by the balance of surface inflow and drainage capacity of the borehole. In addition, Alley (this issue), has shown that diurnal amplitude and minimum water level may well be correlated. The present analysis therefore considers only the base (minimum) of borehole water levels. This fortuitously increases the sample size in each ordinal class.

A preliminary analysis suggests that non-stationary (evolving) behaviour is not systematic across large areas of the glacier, but is localized. There is no accommodation in the classification (Table II) for such boreholes (e.g. Figure 4). A narrow time window of 2–3 days was therefore used to characterize the system.

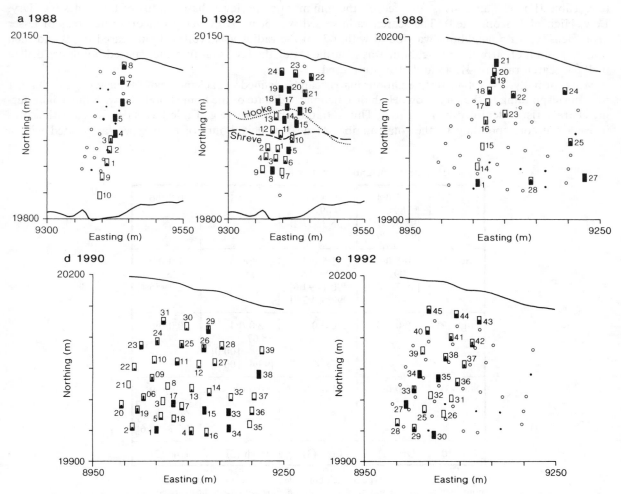

Figure 5. Spatial patterns of borehole types for the central lobe of Small River Glacier based on the classification in Table II. Borehole patterns for other years are shown as small circles. Open symbols are connected, closed symbols are unconnected. (a) Lower array 1988. (b) Lower array 1992 (1988 background). (c) Upper array 1989 (1990 background). (d) Upper array 1990. (e) Upper array 1992 (1990 background)

As drilling continued for most of the season, this time window lies in the last few days of complete borehole monitoring when the maximum number of boreholes was available. Figure 5 shows the classified arrays for the upper and lower drill sites between 1988 and 1992. A few outlying boreholes have not been included in the analysis, especially a group in thin, crevassed ice to the south side of the 1989 set (Figure 2). To assist inter-annual comparison, the borehole array from other years has been provided as background using small open circles to indicate connection and closed circles to represent unconnected holes.

NEAREST NEIGHBOUR ANALYSIS

Visual assessment of spatial pattern is a powerful, but subjective analytical tool. More objectivity is gained from 'nearest neighbour analysis' (e.g. Ebdon, 1985), which relates to the problem of determining spatial relations between phenomena, with more formal analysis considering 'spatial autocorrelation' (e.g. Cliff and Ord, 1973). However, a novel technique was required for handling nominal-ordinal data of the type derived from borehole classification. Here, a search radius was defined around each borehole and the characteristics of 'neighbours' lying within this circle were tabulated with respect to the status of the central borehole. Tallies of borehole status were developed to produce a nearest neighbour frequency matrix (e.g. Table III). The analysis is founded on the minimum water level (base) status of boreholes (i.e. Dry, Low, High or Unconnected). The search radius used was 30 m. However, to incorporate areas of lesser array density, if no neighbour was found within 30 m the radius was arbitrarily increased to 40 and then to a maximum of 50 m if no neighbour was identified. For the few, slightly wider spaced boreholes drilled in 1988, search radii of 41, 45 and 50 m were used.

More formally, a 4×4 nearest neighbour matrix can be defined with elements of frequency $f_{i,j}$, with $i = 1$ to 4, corresponding to the Dry, Low, High or Unconnected state of the home borehole, with the neighbouring characteristic identified by $j = 1$ to 4. Thus a frequency matrix (e.g. Table III) is developed with scores in each cell corresponding to the total number of nearest neighbours of each type j associated with a

Table III. Nearest neighbour matrix for 1988

Dry (2)	1·5/0·4 57·89 [98·5–99·3] [98·7–99·3]	0·5/0·6 −9·09 [35·3–37·4]	0·0/0·4 *	0·0/0·6 *
Low (3)	0·5/0·6 −9·09 [35·7–38·5]	1·5/0·9 25·00 [90·1–92·6] [90·6–92·7]	0·0/0·6 *	1·0/0·9 5·26 [52·7–55·4]
High (2)	0·0/0·4 *	0·0/0·6 *	2·0/0·4 66·67 [99·4–100·0] [99·4–100·0]	0·0/0·6 *
Unc (3)	0·0/0·6 *	1·0/0·9 5·26 [55·2–52·6]	0·0/0·6 *	2·0/0·9 37·93 [96·7–97·6] [96·7–97·6]
$N = 10$	Dry (2)	Low (3)	High (2)	Unc (3)

The rows relate to the type of each borehole and the columns show the neighbouring types. At the top of each cell of the table are shown the aggregate observed/expected frequency of each type of neighbour. Beneath this is the nearest neighbour statistic and (in brackets) the confidence level of that statistic in per cent. Two confidence intervals are shown when two independent simulations have been run.

Table IV. Nearest neighbour statistics for borehole types

(a) Lower array 1988 ($n = 10$)

Borehole type	Neighbour			
	Dry (2)	Low (3)	High (2)	Unc (3)
Dry (2)	57·89++	−9·09	*	*
Low (3)	−9·09	25·00++	*	5·26
High (2)	*	*	66·67++	*
Unc (2)	*	5·26	*	37·93++

(b) Lower array 1992 ($n = 24$)

Borehole type	Neighbour			
	Dry (0·64)	Low (10·14)	High (3·33)	Unc (9·89)
Dry (1)	*	22·43+	*	−10·54
Low (10)	18·92+	18·99++	*	−9·53
High (4)	*	*	71·43++	−42·39−−
Unc (9)	1·82	−7·57	*	19·33++

(c) Upper array 1989 ($n = 11$)

Borehole type	Neighbour			
	Dry (4·5)	Low (0)	High (4·67)	Unc (1·83)
Dry (5)	13·18+		−22·81−	9·09
Low (0)				
High (4)	−32·52−		15·79++	11·11
Unc (2)	10·00		8·20	*

(d) Upper array 1990 ($n = 39$)

Borehole type	Neighbour			
	Dry (12·45)	Low (17·35)	High (3·83)	Unc (5·37)
Dry (13)	−1·84	−1·02	−4·55	−9·55
Low (15)	0·47	−2·61	−22·56+	−15·83
High (5)	−15·55	22·28+	−19·17	*
Unc (6)	12·25	−18·13	*	34·62++

(e) Upper array 1992 ($n = 21$)

Borehole type	Neighbour			
	Dry (7)	Low (5)	High (5)	Unc (4)
Dry (7)	31·09++	−32·71−	−33·33−	−13·04
Low (5)	−32·26−	37·28++	−17·65	−51·16−
High (5)	−41·87−−	−2·75	32·43++	−7·44
Unc (4)	−13·10	−38·54	2·44	41·57++

−−, −, + and ++ indicate where the observed statistic falls below 10 and 20% and above 80 and 90% of the stimulated arrays, respectively. Cells with no observations are indicated by *. Blank rows or columns indicate the absence of a particular borehole type.

particular borehole type i. Where more than one neighbour occurs, fractional scores are allocated. Thus a Dry hole ($i = 1$) with Dry ($j = 1$) and High ($j = 3$) neighbours would increment frequency scores $f_{1,1}$ by one-half and $f_{1,3}$ by one-half. The matrix is the aggregate of scores obtained by taking each borehole in an array in turn.

The resulting table is heavily biased by the frequency of borehole types and the pattern of drilling. Bias due to frequency can be removed by comparing the observed frequency with a frequency $f'_{i,j}$ expected if borehole types were randomly distributed. Thus

$$f'_{i,j} = \sum f_i \sum f_j / N \tag{1}$$

where $\sum f_i$ and $\sum f_j$ are the column and row frequencies, respectively, and N is the grand total $\sum \sum f_{i,j}$. Anomalous scores can be indicated by a simple error statistic

$$E_{ij} = 100(f_{ij} - f'_{ij})/(f_{ij} + f'_{ij}) \tag{2}$$

which has a range between plus and minus 100. Values greater and less than zero indicate a greater or lesser observed frequency than might be expected, respectively. An asterisk is used in the tables to identify when there are no observations in a particular class. Similarly, values of 100 are not possible as they imply finite observation with zero expected frequency, which is not possible based on the definition of f'. As an example based on the lower borehole array in 1988, Table III shows the observed and expected frequencies and resulting E_{ij}.

Nearest neighbour analyses for the lower array in 1988 and 1992 show exceptionally high values for E_{ij} in the positive diagonal (Table IVa and IVb), implying considerable coherence in the spatial distribution of borehole types. Less clear results for the upper array (Table IVc–IVe) suggest that the absolute value of E_{ij} is not a singular indicator of spatial correlation, but is also governed by the form of the array and the position of boreholes in the array. Significance levels for E_{ij} from arbitrary arrays cannot be readily calculated analytically. Instead, a Monte Carlo simulation technique was developed in which borehole status was randomly assigned based on the proportion of borehole types in any given array. Simulated values for $E_{ij(sim)}$ were obtained for 10 000 realizations, ranked and the position of the observed E_{ij} in the ranked set taken as an indicator of significance level. Figure 6 shows cumulative probability distributions for the simple 1988 borehole array. The distribution of cumulative probability $p(E_{ij})$ for each borehole type depends on the numbers

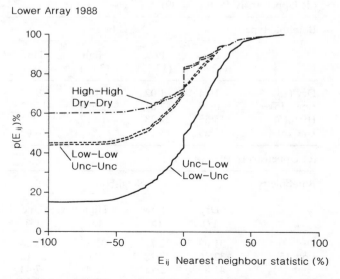

Figure 6. Probability distributions of the nearest neighbour statistic (E_{ij}), based on Monte Carlo simulations of selected array elements for the 1988 borehole array. Critical levels of E_{ij} can be derived from the figure. Contrasts in curves depend on the frequency of borehole types and the particular array configuration. Absolute values of E_{ij} cannot be interpreted without the simulation

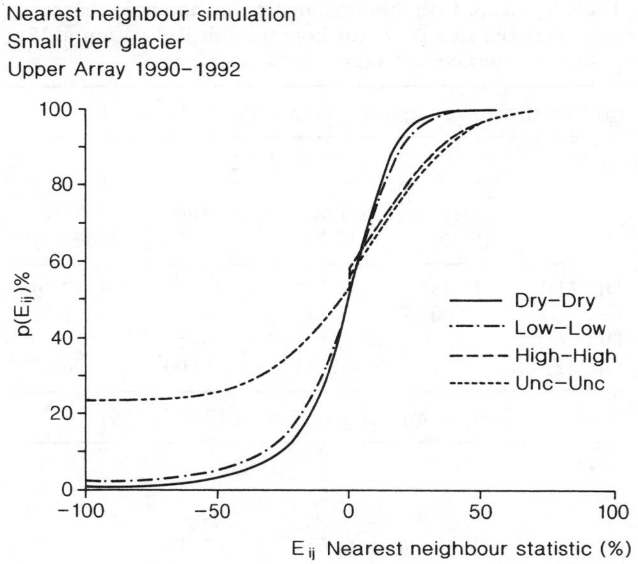

Nearest neighbour simulation
Small river glacier
Upper Array 1990–1992

Figure 7. Probability distributions of E_{ij}, based on Monte-Carlo simulations of four positive diagonal array elements for comparison of 1990 and 1992 borehole array patterns. Larger numbers of boreholes generate a superior probability distribution

observed and expected with slight differences arising from the pattern of random numbers. The form of the probability function depends on the particular borehole configuration and improves in form as borehole frequency increases, as is shown in Figure 7 for simulations comparing 1990–1992 borehole arrays. Critical levels of E_{ij} are drawn from the simulation tables.

Table III shows observed/expected frequencies and E_{ij} for the lower array in 1988. In square brackets beneath these values are given the range of probabilities, $p(E_{ij})$ for one or two independent simulations for E_{ij} derived from Figure 6. Replication is excellent. In practice, rather than using the probabilities, the null hypothesis $E_{ij} = 0$ was tested. A significant association between boreholes types was identified when $p(E_{ij}) > 90\%$ (indicated in subsequent tables by $++$). Significant *absence* of association between borehole types was given by $p(E_{ij}) < 10\%$ (indicated by $--$). Respective lower levels of significance are shown for $p(E_{ij}) > 80\%$ $(+)$ and $<20\%$ $(-)$.

Changes in borehole pattern from year to year are equally amenable to the nearest neighbour approach. Taking boreholes in a particular (reference) year, nearest neighbours are now drawn from boreholes drilled in the same area in some comparison year. However, anisotropy arises from the selection of the base year. Tables Va and Vb allow comparison of 1988 versus 1992 in the lower array. Taking 1988 (Table Va) as the reference year gives slightly different results from using 1992 (Table Vb). Recalling that the latter table should be rotated about the positive axis (top left–bottom right) for comparison, the differences are at the 20 and 80% level and are associated with low holes (0.00 becomes 13.92^+, -11.76 becomes -38.78^- and -12.73 becomes -36.17^-). The ambiguity in separating dry and low holes may partially account for this. Working at the 10 and 90% confidence level will assuage this difficulty for now.

RESULTS

A highly significant positive diagonal (top left to bottom right) in Table IVa and IVb indicates significant spatial coherence in both 1988 and 1992 in the lower borehole array. Moreover, this pattern proves consistent for the two years (Table Va and Vb). The south side of the lower lobe exhibits coherent Low borehole water levels, the north side High water levels, with an unconnected zone in between. Figure 5b shows that two Unconnected boreholes lie within the Low and Dry zone, but this feature is not revealed in the analysis.

Patterns in the upper borehole array are much less clear-cut. Some elements of the positive diagonal

Table V. Nearest neighbour statistics for borehole characteristics between two years for boreholes drilled within a 25 m
radius in a subsequent year

(a) Lower array comparing 1988 with 1992 ($n = 9$)

1988	1992			
	Dry (0·75)	Low (2·58)	High (2·50)	Unc (3·17)
Dry (1)	71·43[++]	*	*	17·39[+]
Low (3)	0·00	38·00[++]	*	−11·76
High (2)	*	*	56·52[++]	*
Unc (3)	*	−12·73	−25·00	26·92[++]

(b) Lower array comparing 1992 with 1988 ($n = 9$)

1992	1988			
	Dry (1·50)	Low (5·67)	High (2·50)	Unc (5·33)
Dry (1)	66·67[++]	13·92[+]	*	*
Low (5)	*	37·61[++]	*	−36·17[−]
High (3)	*	*	66·67[++]	−36·17
Unc (6)	25·00[+]	−38·78[−]	*	30·43[++]

(c) Upper array comparing 1989 with 1990 ($n = 13$)

1989	1990			
	Dry (4·25)	Low (3·83)	High (2·83)	Unc (2·08)
Dry (5)	10·05	0·86	−4·29	−23·15
Low (0)				
High (6)	−10·67	−14·05[−]	16·73	13·04
Unc (2)	0·97	25·81[++]	*	1·96

(d) Upper array comparing 1989 with 1992 ($n = 13$)

1989	1992			
	Dry (4·25)	Low (3·83)	High (2·83)	Unc (2·08)
Dry (5)	27·54[++]	7·72	*	25·71[+]
Low (0)				
High (6)	*	5·77	31·62[++]	*
Unc (2)	4·76	−49·62	24·53[+]	11·86

(e) Upper array comparing 1990 with 1992 ($n = 25$)

1990	1992			
	Dry (11·25)	Low (6·00)	High (4·00)	Unc (3·75)
Dry (10)	16·28[+]	−65·52[−−]	−3·23	7·69
Low (8)	−18·03[−]	21·95[++]	−12·28	11·11
High (5)	−20·00	25·00[+]	30·43[+]	*
Unc (2)	5·26	2·04	*	25·00[+]

elements are significant, but only in the 1992 study were these consistently significant at the 90% level. In 1989 sample size was 11, with no Low holes. Holes 24–28 were too widely spaced for inclusion. The map of the 1989 array (Figure 5c) suggests grouping of Dry and High holes, with unconnected holes towards the margins, but only the High–High association is picked up in the analysis (Table IVc). In 1990, Unconnected holes were the only holes statistically associated (Table IVd), despite apparent consistency in the map (Figure 5d).

It is possible that the nearest neighbour technique might be misrepresenting the array, or that the form of the borehole array could influence the pattern. Various experiments were carried out on the technique applied to the 1989 and 1990 analysis. For example, the minimum nearest neighbour radius criteria were varied between 15 and 50 m. Unconnected BH90-17 lies out of alignment in the general array and constitutes an obvious anomaly, so was removed from an analysis. The 1989 array was stripped to a single transverse line (BH89-01 to BH89-21). None of these changes appeared to dramatically alter the resulting nearest neighbour statistics, although significance tests were not made for these runs.

The temporal comparisons of the lower array in 1988 and 1992 show exceptionally strong replication, at the 90% confidence level (Tables Va and Vb). The upper array is less coherent, including some anticorrelation, e.g. 1989–1990: Low–Unc. There are better indications of replication for 1989–1992 and 1990–1992. Such ambiguity is not surprising, given that the patterns in 1989 and 1990 were themselves incoherent.

Overall, the nearest neighbour technique was reasonably robust, but might contradict visual impressions where complex associations are found. Local patterns may be lost in the ensemble of a full array and therefore careful visual inspection remains important. Borehole behaviour in the lower array appeared to be coherent and well replicated. The upper array was much less coherent and was not well replicated.

DISCUSSION

What can be inferred about the basal drainage system from the classification and analysis? The south side of the lower array displays a zone, about 300 m wide, of low hydraulic head, with a small diurnal signal (Figure 5a and 5b). The southernmost, shallowest boreholes are dry, suggesting that their bottoms lie above a crude potentiometric surface, common to all these boreholes.

Digital terrain models similar to those developed by Sharp et al. (1993) were used to predict the position of major basal conduits under the central lobe. Two models were run, one based on free-surface conditions (described by Hooke, 1984) and the other based on hydrostatic conditions (Shreve, 1972). The drainage axes identified by these models both run towards the northern limit of the low pressure zone (Figure 5b), although the free surface system overlaps part of the unconnected area. A conduit in this area might be indicated by local water level depression (e.g. Fountain, 1994), or marked diurnal signals (i.e. dry–high borehole types). There is no evidence from borehole water levels that a dominant conduit exists beneath this part of the glacier, unless it passes through without influencing the surrounding glacier bed and has been missed by boreholes.

Rather than pointing to a discrete conduit, the present evidence suggests that there is an extensive drainage system beneath the southern half of the lower central lobe. This may be a broad anastomosing or braided conduit system or a distributed system. The coherent, low water pressure and lack of a strong diurnal signal suggest effective interconnection. Within this area, boreholes BH92-8 and BH92-5 are unconnected (Figure 5b). It is inferred that they represent points of enhanced contact pressure on the bed, a consequence of extensive low pressures in the drainage system. This drainage system is unlike any previously described from borehole data. In planform, it resembles a low pressure version of the broad, braided systems described by Hock and Hooke (1993) to account for high water pressures beneath Storglaciären.

The central unconnected area of the lower central lobe (Figure 5b) is closely associated with basal till. It is possible that sediment melted out by drilling through dirty basal ice (up to 1·5 m has been observed in exposures at the snout) may have prevented penetration to the bed in this area. However, extraction of distinct lumps of till on the drillstem suggests that a soft bed has indeed been reached. If this is so, there is no evidence for a drainage system capable of draining a borehole with limited surface surface influx. 'Walder–Fowler canals' (Walder and Fowler, 1994) are thus excluded, although a soft-bed drainage system with limited capacity (e.g. Alley, 1989; Kamb, 1991) is possible. It is also implied that the basal till has a low hydraulic conductivity compared with that inferred by Fountain (1993; 1994).

The area of high water pressures on the north side of the lower array (Figure 5b) may represent a distributed drainage system of linked cavities. There are few moulins in this area and modelling showed little propensity for the formation of integrated drainage systems.

The upper arrays show less obvious systematics. Unconnected holes show no clear association with the till-producing basal siltstone. A few unconnected holes are scattered within broad connected zones of dry–low or low–low holes in all years (Figure 5c–5e). Assuming interconnection between the connected holes, the basal drainage system is implicitly braided. There are no data on up-glacier bed topography to allow conduit modelling, but again there is no compelling borehole evidence for dominant conduits in this area. However, dye traces from this part of the glacier show widespread capture by karst with travel times similar to those from adjoining karst areas. This implies uninhibited drainage from the glacier bed into underlying karst.

Basal conditions revealed by boreholes in the upper array are more heterogeneous in character than those in the lower array. Any underlying pattern is not revealed by the current borehole spacing. It is inferred that karst openings (and associated Nye channels) at the glacier bed act as sustained low pressure points exploited by subglacial streams. Intermixed with the low pressure areas are zones of high contact pressure seen as unconnected boreholes. Adjoining low pressure areas are boreholes with high water levels. These are possibly separated by high pressure seals along channel margins (Weertman, 1972).

CONCLUSIONS

The borehole classification used here shows the limitations and advantages of all simple descriptive statistics in being robust and allowing straightforward inspection and interpretation of complex data, but with considerable loss of information. The analysis does allow reasonably objective spatial characterization of borehole character. However, there is little merit in extending the classification to incorporate additional characteristics (such as water level trend or diurnal variability) as sample size is already a limiting factor.

The nearest neighbour analysis appears to be reasonably robust and provides important restraint on interpretations of borehole data. It does not entirely support subjective impressions, but instead introduces questions about borehole spacing, pattern and reliability which have so far been generally absent from published work. Spatial correlation not only depends on borehole spacing, but also on the characteristic scale of basal drainage systems. Discrete, longitudinal conduits may not be readily identified with analyses of this kind. In part this is because the borehole spacing is greater than the expected width of a conduit, but also because consensus on the pattern of drainage expected in and around a conduit has not yet been reached. Areas of glacier bed exhibiting less spatial coherence are more temporally variable; thus the coherent lower array shows greater inter-annual replication than the less coherent upper array.

The Small River basal drainage systems appear to be different from those previously described. Unconnected or high pressure holes a few metres from dry holes indicate basal hydraulic gradients considerably greater than unity. Karst underdraining is unusual and is associated with extensive areas of very low basal water pressure. This emphasizes the need for caution in drawing general conclusions from the drainage patterns described from exposed carbonate glacier beds. The coherent pattern of the lower array suggests sealing of the bed by till (where present) and spatially extensive low and high pressure drainage systems. This association is unlike that predicted from modelling either free surface or pressurized conduits and requires further analysis. It may be that the absence of a dominant conduit allows unperturbed distributed drainage systems to be sustained. There are virtually no field data on distributed drainage, so this area may merit further attention. Further work will focus on potentiometric surfaces, water tracing and hydrochemical analysis of boreholes at Small River Glacier.

ACKNOWLEDGEMENTS

Research work at Small River has been supported by the Natural Sciences and Engineering Research

Council of Canada and by the University of Western Ontario Academic Development Fund. The citizens of Valemount, British Columbia have provided welcome and support for the work. Numerous individuals have assisted in the field projects. David Huntley, Brian Fowle and Keith Carr were instrumental in field campaigns in 1988–1989, 1990 and 1992, respectively. Stuart Lane, University of Cambridge provided the topographic map of the upper glacier. An anonymous reviewer suggested the Monte Carlo approach to nearest neighbour analysis.

REFERENCES

Alley, R. 1989. 'Water-pressure coupling of sliding and bed deformation I: water system', *J. Glaciol.*, **35**, 108–118.

Cliff, A. D. and Ord, J. K. 1973. *Spatial Autocorrelation*. Pion, London. 235 pp.

Collins, D. N. 1979. 'Quantitative determination of the subglacial hydrology of two alpine glaciers', *J. Glaciol.*, **23**, 347–362.

Ebdon, D. 1985. *Statistics in Geography*. Blackwell, Oxford. 232 pp.

Engelhardt, H. 1978. 'Water in glaciers: observations and theory of the behaviour of water levels in boreholes', *Z. Gletscherk. Glazialgeol.*, **14**, 35–60.

Fountain, A. 1993. 'Geometry and flow conditions of subglacial water at South Cascade Glacier, Washington State, U.S.A.; an analysis of tracer injections', *J. Glaciol.*, **39**, 143–156.

Fountain, A. G. 1994. 'Borehole water-level variations and implications for the subglacial hydraulics of South Cascade Glacier, Washington State, U.S.A.' *J. Glaciol.*, **40**, 293–304.

Hantz, D. and Lliboutry, L. 1983. 'Waterways, ice permeability at depth and water pressures at Glacier d'Argentière, French Alps', *J. Glaciol.*, **29**, 227–239.

Hodge, S. M. 1979. 'Direct measurement of basal water pressures: progress and problems', *J. Glaciol.*, **23**, 309–319.

Hock, R., and Hooke, R. Le B. 1993. 'Evolution of the internal drainage system in the lower part of the ablation area of Storglaciären, Sweden', *Geol. Soc. Am. Bull.*, **105**, 537–546.

Hooke, R. Le B. 1984. 'On the role of mechanical energy in maintaining subglacial water conduits at atmospheric pressure', *J. Glaciol.*, **30**, 180–187.

Hooke, R. Le B., Laumann, T., and Kohler, J. 1990. 'Subglacial water pressures and the shape of subglacial conduits', *J. Glaciol.*, **36**, 67–71.

Iken, A. and Bindschadler, R. A. 1986. 'Combined measurements of subglacial water pressure and surface velocity of Findelengletscher, Switzerland: conclusions about drainage system and sliding mechanism', *J. Glaciol.*, **32**, 101–119.

Iken, A., Echelmeyer, K., Harrison, W., and Funk, M. 1993. 'Mechanisms of fast flow at Jakobshavns Isbrae, West Greenland: part I: measurements of temperature and water level in deep boreholes', *J. Glaciol.*, **39**, 15–25.

Kamb, B. 1991. 'Rheological nonlinearity and flow instability in the deforming bed mechanism of ice stream motion', *J. Geophys. Res.*, **96**, 16,585–16,595.

Sharp, M. J., Gemmel, J. C., and Tison, J. L. 1989. 'Structure and stability of the former subglacial drainage system of the Glacier de Tsanfleuron, Switzerland', *Earth Surf. Process. Landforms*, **14**, 119–134.

Sharp, M. J., Richards, K. S., Willis, I. C., Arnold, N., Nienow, P., Lawson, W., and Tison, J. 1993. 'Geometry, bed topography and drainage system structure of the Haut Glacier D'Arolla, Switzerland', *Earth Surf. Process. Landforms*, **18**, 557–571.

Shreve, R. L. 1972. 'Movement of water in glaciers', *J. Glaciol.*, **11**, 205–214.

Walder, J. S. 1986. 'Hydraulics of subglacial cavities', *J. Glaciol.*, **32**, 439–445.

Walder, J. S. and Fowler, A. 1994. 'Channelised subglacial drainage over a deformable bed', *J. Glaciol.*, **40**, 3–15.

Walder, J. S. and Hallet, B. 1979. 'Geometry of former subglacial water channels and cavities', *J. Glaciol.*, **23**, 335–346.

Weertman, J. S. 1972. 'General theory of water flow at the base of a glacier or ice sheet'. *Rev. Geophys. Space Phys.*, **10**, 287–333.

12

THE USE OF BOREHOLE VIDEO IN INVESTIGATING THE HYDROLOGY OF A TEMPERATE GLACIER

LUKE COPLAND[1]*‡, JON HARBOR[1], SHULAMIT GORDON[2] AND MARTIN SHARP[2]

[1] *Department of Earth and Atmospheric Sciences, Purdue University, West Lafayette, IN 47907, USA and* [2] *Department of Earth and Atmospheric Sciences, University of Alberta, Edmonton, Alberta T6G 2E3, Canada*

ABSTRACT

A GeoVision Micro℗ colour video camera was used to investigate the internal structure of 11 boreholes at Haut Glacier d'Arolla, Switzerland. The boreholes were distributed across a half-section of the glacier, with closest spacing towards the glacier margin. The boreholes were used to investigate the hydrology of the glacier through automatic monitoring of borehole water level and electrical conductivity (EC) at the glacier bed. EC profiling was undertaken in several boreholes to determine the existence of water quality stratification. Temporal variations in EC stratification were used to infer borehole water sources and patterns of water circulation. Borehole video was used to confirm the conclusions made from these indirect sources of evidence, and to provide an independent source of information on the structure and hydrology of this temperate valley glacier. The video showed variations in water turbidity, englacial channels and voids, conditions at the glacier bed and down-borehole changes in ice structure. Based on the video observations, englacial channels accounted for approximately 0·1% of the vertical ice thickness, and englacial voids for approximately 0·4%. Overall, the video images provided useful qualitative and semi-quantitative data that reinforce interpretations of a range of physical and chemical parameters measured in boreholes.

INTRODUCTION

Recent advances in miniature video camera technology have allowed the observation of previously inaccessible locations such as the interior of pipelines, sewers and water wells (e.g. Westinghouse Savannah River Company, 1989). Previous glaciological applications of miniature video cameras have focused on down-borehole changes in ice structure (Harper and Humphrey, 1995), conditions at the glacier bed (Koerner *et al.*, 1981; Pohjola, 1993) and the occurrence of englacial voids (Pohjola, 1994). This paper describes video observations made in boreholes at Haut Glacier d'Arolla, Switzerland, and the information that video provides about temperate glacier hydrology.

Knowledge of glacier hydrology is important for understanding the relationship between water pressure and glacier sliding, the role of water in glacier surging and the chemistry of glacial meltwaters. Based on theoretical analyses by Röthlisberger (1972), Shreve (1972) and Weertman (1972), glacier drainage systems can be broadly classified as 'distributed' or 'channelized'. In a distributed system, water is transported over large areas of a glacier bed at low velocity, while in a channelized system water is transported at high velocity in a small number of channels. Distributed glacier drainage may occur as a thin film of water at the ice-bed interface (Weertman, 1969, 1972, 1986), as a series of linked cavities in the lee of bedrock bumps (Lliboutry, 1969; Walder, 1986; Kamb, 1987), through permeable sediment beneath a glacier (Boulton, 1974; Clarke, 1987) or through a network of broad shallow 'canals' above and within till (Walder and Fowler, 1994). Major drainage channels may be incised upwards into ice (Röthlisberger, 1972), downwards into bedrock or till (Nye, 1973) or be wholly englacial (Shreve, 1972).

---

* Author to whom correspondence should be addressed.
‡ Present address: Department of Earth and Atmospheric Sciences, University of Alberta, Edmonton, Alberta T6G 2E3, Canada.

Most conclusions about the configuration of temperate glacier drainage have been based on indirect sources of evidence such as dye-tracing results (Seaberg et al., 1988; Willis et al., 1990; Fountain, 1993; Sharp et al., 1993), meltwater chemistry (Raiswell, 1984; Tranter et al., 1993), mapping of recently deglaciated bedrock (Walder and Hallet, 1979; Sharp et al., 1989) and borehole water level fluctuations (Iken and Bindschadler, 1986; Fountain, 1994; Murray and Clarke, 1995; Waddington and Clarke, 1995). More recently, continuous in situ monitoring of EC and turbidity at the glacier bed (Stone et al., 1993; Hubbard et al., 1995) and EC profiling and artificial salt tracing (Gordon et al., in press), have provided data that demonstrate how individual open boreholes are plumbed into a glacier's drainage system. These techniques allow the identification of water inputs and outputs from a combination of supraglacial, englacial and subglacial sources, and their relative contribution to borehole water level fluctuations. Profiling techniques can be used to determine how water circulates within a borehole, and thus facilitate interpretation of in situ EC and turbidity records. In analysing such records, borehole video can be used to determine whether the signals are from water flowing past the base of a borehole (i.e. a true basal signal) or from water flowing out of the base of a borehole as a result of an englacial or supraglacial input higher in the borehole. Only when the source of these signals has been determined can the in situ data be used to make broader inferences about the drainage system structure of a temperate glacier.

Borehole video can also be used as an independent tool to verify and refine interpretations based on other borehole measurements. For example, the locations of englacial channels and voids are commonly indicated by variations in borehole water (e.g. if a borehole drains and refills during drilling it is assumed that an englacial void of finite volume has been intercepted), by electrical conductivity profiling (e.g. intrusion of dilute water into a column of high EC water suggests an englacial input) and by salt tracing (e.g. sources of inflow/outflow can be identified in boreholes that have no natural EC stratification). Borehole video enables physical identification of englacial channels and voids (both above and below the water level), and supraglacial, englacial and subglacial water inputs. In addition, borehole video can be used to catalogue the frequency and morphology of englacial channels and voids, and to evaluate relationships between their location and the structure of the enclosing ice. Borehole video also allows direct observation of the character and morphology of the glacier bed, and information on whether a glacier rests on 'soft' till or 'hard' bedrock (Koerner et al., 1981; Pohjola, 1993; Harper and Humphrey, 1995). This is important to know because the character of the glacier bed exerts a strong control on the basal drainage of a temperate glacier. In turn, the routing, residence time and storage characteristics of meltwater, together with the water pressure associated with a given meltwater input, are controlled by the structure of the glacier drainage system (Sharp, 1991).

In this paper we focus on the utility of video as an aid in interpreting the indirect information provided by other borehole-based studies. Borehole video allows independent testing of conclusions based on borehole measurements by providing direct observation of conditions within and at the bed of a glacier. It produces a continuous image that can be recorded for later analysis, and real-time viewing that allows the camera operator to focus on areas of interest as they are encountered.

FIELD SITE AND METHODOLOGY

Haut Glacier d'Arolla is a temperate valley glacier at the head of the Val d'Hérens in Valais, Switzerland (45°58′N, 7°32′E) (Figure 1). The glacier ranges in elevation from approximately 2560 to 3500 m, faces predominantly north and is about 4·5 km long. Recent work has focused on a network of boreholes drilled with high pressure hot water towards the eastern margin of the glacier, approximately 1·5 km from the terminus (Hubbard et al., 1995; Lamb et al., 1995; Tranter et al., in press; Copland et al., in press a,b). Drilling has been to the glacier bed, and a maximum depth of 142 m has been reached. The drilling area was chosen on the basis of hydrological measurements that predicted the existence of a major subglacial channel beneath this part of the glacier (Sharp et al., 1993; Hubbard et al., 1995).

A total of 25 boreholes were drilled to the base of Haut Glacier d'Arolla in July and August 1995 (Figure 2). Eleven of these were investigated with a commercially available borehole video camera system supplied by Colog Inc., and manufactured by Marks Products Inc. The GeoVision Micro® system

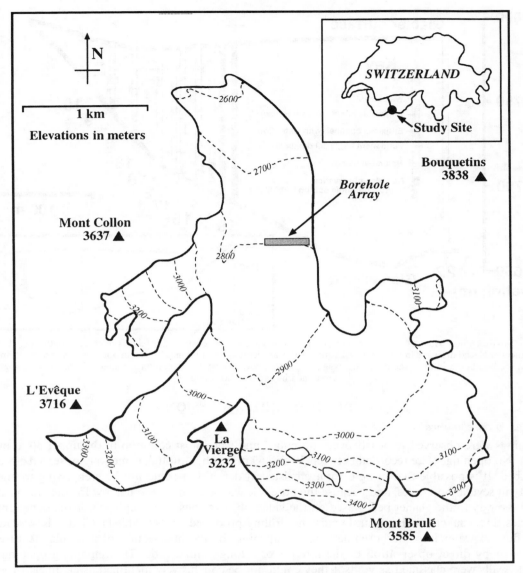

Figure 1. Haut Glacier d'Arolla. The borehole array was located towards the eastern margin of the glacier, approximately 1.5 km from the terminus

consisted of a miniature colour video camera housed in a 3 cm wide by 30 cm long waterproof stainless steel container, 230 m of cable on a hand-operated winch, a small combined colour monitor/VCR, a microphone, several side-looking mirrors and a mounting tripod (Figure 3). An in-front lighting attachment, consisting of a small bulb on the end of a thin rod, allowed exploration of boreholes more than 3 cm in diameter. Alternatively, a 7·6 cm diameter ring lighting attachment of four high-intensity bulbs could be mounted around the end of the camera. A side-looking mirror could also be attached to give a 360° view of the side of the borehole when the camera was rotated. An on-screen display showed the depth below the ice surface to an accuracy of ±1%, and this was used as the basis for all depth measurements referred to in this paper. The borehole video images were recorded on standard small-sized video cassettes for later viewing and analysis.

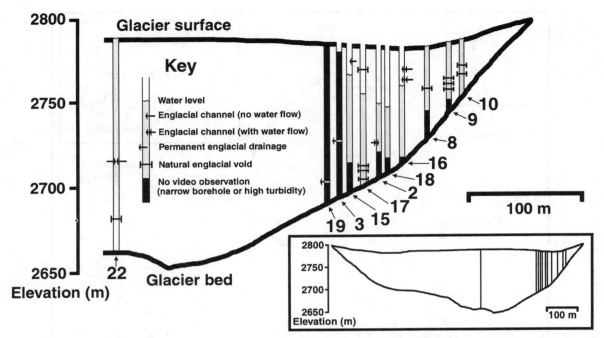

Figure 2. Cross-section of Haut Glacier d'Arolla showing location and characteristics of boreholes referred to in the text. The location of permanent englacial drainage in boreholes 95/3 and 95/19 was identified by a permanent fall in borehole water level during drilling. Note that the borehole widths in the figure are exaggerated; true borehole widths were approximately 10 cm. Basal topography was determined by radio–echo sounding

RESULTS AND DISCUSSION

Openings in borehole walls

Openings were observed in nearly every borehole, and were most common in the upper 30 m of the ice column. No openings were recorded between 30 and 59 m depth, although some were observed closer to the glacier bed. It is possible that more openings existed at depth, but could not be seen owing to high water turbidity in several boreholes, and because the borehole was sometimes too narrow for the video camera to pass all the way to the glacier bed. Based on the video observations, the openings were classified into three categories; englacial channels, natural voids and drilling-produced voids (Table I). Channels were identified by the flow of water into the borehole from an opening, by the intersection of a tubular feature by the borehole or by direct observation of the interior of a longitudinal tube. The remaining openings in the borehole walls were classified as voids if they could not be conclusively identified as channels. A void was defined as natural if it could not be linked to drilling features on the borehole wall, and had an uneven shape or interior structure. A void was defined as drilling-produced if it had a well-rounded shape, a relatively large horizontal or vertical extent compared to its depth into the ice or if it could clearly be linked to drilling features on the borehole wall. These drilling-produced voids are likely to have resulted from the melting of the borehole wall when the drill tip remained in one location for an extended period of time. The drilling process is also likely to have enlarged natural voids, although it was difficult to determine which ones and by how much. The size of an opening, as with all other features, was estimated with reference to the width of the mirror on the in-front lighting attachment (approximately 2·5 cm) or the diameter of the borehole (approximately 10 cm).

Englacial channels

Five of the nineteen openings were classified as englacial channels. The most distinctive was observed entering borehole 95/15 at a depth of 10·9 m (Figure 4). The channel opening, which was located 7·0 m above the water level in the borehole, was approximately 15 cm high and 4 cm wide, and had a thin crack

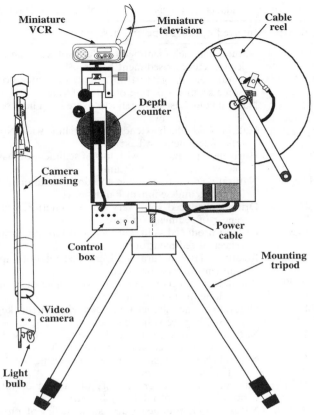

Figure 3. The GeoVision Micro® borehole video camera system (reproduced with permission from Marks Products, Inc.)

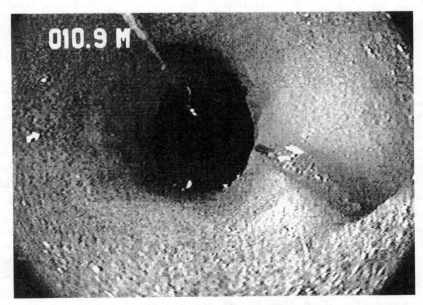

Figure 4. Video image of the englacial channel observed entering borehole 95/15 at a depth of 10·9 m (lower right hand corner of image). Borehole is approximately 10 cm in diameter

Table I. Description of the englacial channels and voids observed in Haut Glacier d'Arolla

Borehole	Depth (m)	Size (cm)*	Inferred opening type and description
95/2	59·7	2,2,?	Natural channel, identified by plume of water entering borehole. No relation to ice structure or sediment bands
95/8	26·5	60,7,?	Natural void, large vertical extent, no water flow, partially closed crevasse? Some sediment resting on base of void
95/9	22·2	7,7,5	Natural void, base of rough ice covered in a little sediment. Ice foliations present less than 5 cm away
	25·6	7,6,5	Natural void which extends into borehole wall. No relation to ice structure, but lots of sediment on base
	27·4	2,7,2	Horizontal indentation with lots of sediment on base. Probably a sediment layer that was enlarged by drilling
	29·2	6,8,8	Natural void off to one side and down from main borehole. Close (< 10 cm) to foliations and sediment bands in ice
	29·8	20,10,5	Drilling-produced void, very smooth interior, increase in size with depth, thin veneer of sediment on base
	30·5	20,10,8	Drilling-produced void, very smooth interior, increase in size with depth, thin veneer of sediment on base
95/10	14·3	14,10,7	Relatively large natural void, marked relationship to vertical ice foliation, no sediment on base or in ice
	19·8	50,20,20	Borehole intersects large natural void, uneven in shape, some sediment on base, but none in surrounding ice
95/15	10·9	15,4, >30	Natural channel, longer than seen with mirror, keyhole shape in cross-section, surrounding ice is homogeneous
95/16	14·9	6,6,?	Natural channel, identified by water entering borehole above water level, intersects near-vertical ice foliation
	20·7	8,6,?	Natural channel, identified by water entering borehole above water level, near to vertical ice foliation
95/17	14·6	20,8,?	Natural void, interior consists of mosaic of interconnecting ice fragments, occurs in area of relatively blue ice
	83·8	7,7,3	Natural void, rounded sides, some sediment on base, but no relation to ice structure or sediment frozen in ice
	86·0	8,7,3	Drilling-enlarged void, very rounded sides, sediment band visible on opposite side of borehole
	91·8	6,6,5	Natural void, rounded, lots of sediment on base, close to sediment band in ice NB: There are many indentations in the side of borehole 95/17 between 83.8 m and the bed at 93.0 m — almost all of these occur in relation to sediment bands frozen into the ice
95/22	74·1	12,10,?	Natural channel crosses borehole, openings on both sides, no relation to ice structure, but lots of sediment in ice
	109·8	14,8,5	Drilling-enlarged void, rounded sides, lots of sediment covering base and in surrounding ice

* Size of englacial channel/void (respectively): vertical size of opening, horizontal size of opening, horizontal depth into ice (? = undetermined)

4 cm long leading down from its base. The tubular channel extended further back into the ice than could be seen with the side-looking mirror (30 cm), had no detectable water flow through it and was keyhole in shape (Figure 5). It occurred in an area of homogeneous bubbly white ice, and bore no relationship to variations in ice structure seen in other parts of the borehole. The absence of water flow and the fact that the opening intersected only one side of the borehole suggests that this feature was an abandoned englacial channel isolated from the presently active drainage system. The keyhole shape of the channel is similar to that predicted by Shreve (1972) and Sugden and John (1976) for the upper part of a glacier in a zone of fluctuating water level. Using a model based on an analogy with tunnels in karst, Shreve (1972) argued that high water levels would produce an englacial channel that is circular in shape. The channel would be deepened by preferential melting of the channel base during periods of low water flow. The final result would be an englacial channel with a 'keyhole-shaped' cross-section.

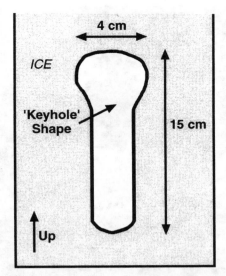

Figure 5. Sketch of the 'keyhole' shape of the englacial channel observed entering borehole 95/15 at a depth of 10.9 m

The best example of an active englacial channel below the water level was observed in borehole 95/2 at a depth of 59·7 m. It was identified by a plume of turbid water entering the water column from a circular opening in the borehole wall. The opening was approximately 2 cm high by 2 cm wide, and showed no relation to the surrounding ice structure. The plume of turbid water reached up and down the borehole from the channel opening, and resulted in a zone of turbid water within an area of generally low turbidity. This channel did not continue on the opposite side of the borehole.

Active englacial channels above the water level were observed at depths of 14·9 and 20·7 m in borehole 95/16. The water level in this borehole was at a depth of 23·4 m. The englacial channels were identified by the flow of water out of an opening and into the borehole at both locations, although conclusive identification was difficult. This is because supraglacial water running down the borehole wall may have given the same impression as water flowing out of a real opening if it ran along the inside of a depression. The discharge of water from both openings seemed to be more than was running down the borehole walls, however, which suggests that they were real englacial channels.

The only englacial channel that completely crossed a borehole was observed at a depth of 74·1 m in borehole 95/22. Circular openings approximately 12 cm high and 10 cm wide were observed on both sides of the borehole, and sediment could be seen frozen into the surrounding ice and resting on the base of the openings. Water movement below the water level was investigated by watching for the deflection of suspended sediment particles in the water column, and the movement of a piece of red thread attached to the end of the ring lighting system. With the exception of borehole 95/2, there was no detectable water flow from any englacial channel observed below the borehole water level. This does not necessarily mean that water flow from most englacial channels never occurs, however, as discussed later.

In addition to the englacial channels discussed, several others have been indicated in the past at Haut Glacier d'Arolla. Evidence has come from a sudden and permanent drop in borehole water level during drilling, and from the occasional sound of rushing water within boreholes. Permanent englacial drainage occurred twice during drilling in 1995, at depths of 60 and 85 m in boreholes 95/3 and 95/19, respectively, although these boreholes could not be accessed with the video camera. Englacial channels have also been observed in other glaciers. Raymond and Harrison (1975) identified them in ice cores from the Blue Glacier, Washington. Using borehole video, Pohjola (1994) observed them in Storglaciären, Sweden, while Harper and Humphrey (1995) observed them in Worthington Glacier, Alaska. By comparing the vertical size of the observed englacial channels with the total length of observed boreholes, it is estimated that englacial channels account for approximately 0·1% of the vertical ice thickness in the study area. However, the 1995

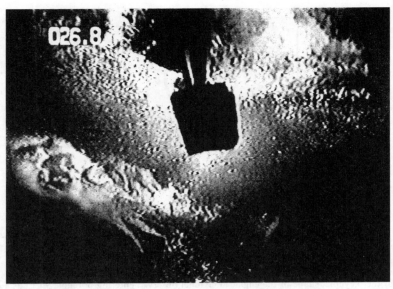

Figure 6. Video image of the large longitudinal void observed in borehole 95/8 at a depth of 26·8 m. The centre of the image is the view directly down the open void, while the upper and lower edges of the image are the ice at the edge of the void. Void is approximately 12 cm wide

drilling programme was focused on the only area of the glacier where previous more widespread drilling had identified englacial channels. Thus 0·1% is probably an overestimate for the glacier as a whole. Overall, the video observations suggest that the small englacial channels intersected by the boreholes do not drain a large proportion of the meltwater from Haut Glacier d'Arolla.

Natural and drilling-produced voids

The largest and most distinct englacial void was observed intersecting borehole 95/8 at a depth of 26·8 m (Figure 6). It was approximately 60 cm high, 12 cm wide and undefined in length. The borehole water level remained at the glacier surface during the drilling of this borehole, except for a temporary fall to the depth of the void when the void was first intersected. This indicates that the void was probably air-filled prior to drilling, finite in volume and unconnected to any active crevasse or drainage pathway. This interpretation was reinforced by the lack of detectable water flow within it. There are many small surface crevasses in the region around borehole 95/8, and the elongate shape of the void suggests it was an old crevasse that had become closed to the atmosphere.

The voids in the other boreholes were smaller in size, not detected during drilling and most appeared to be natural in origin. For example, the void observed in borehole 95/17 at a depth of 14·6 m was approximately 20 cm high, and unique because its interior consisted of a mosaic of interconnecting ice fragments. This suggests that the void may have contained partially frozen water before it was intersected by the borehole. Two of the observed 19 openings did appear to have a drilling-produced origin, however, owing to their rounded shape and close relationship to the location of the borehole. By comparing the vertical height of the observed englacial voids with the total length of observed boreholes, natural englacial voids accounted for approximately 0·4% of the ice thickness.

Particular attention was paid to the relationship between openings and surrounding ice structure and debris content during video inspection of the boreholes. As shown in Table I, most of the englacial channels and voids occurred in association with blue-ice inclusions and/or debris bands, although a few occurred in areas of apparently homogeneous ice. Sediment was also observed resting on the base of several openings, which probably originated from the intersection of local debris bands during drilling, or the settling of material in the water column after it was disturbed from the glacier bed by the drill.

The observed relation between blue-ice inclusions and englacial openings at Haut Glacier d'Arolla is similar to that described by Pohjola (1994). From borehole video observations in Storglaciären, Sweden, Pohjola identified a close association between blue-ice inclusions and englacial voids, and argued that the origin and development of these features was coupled. The burial and horizontal movement of water-filled crevasses in the accumulation areas of Storglaciären was identified as an important process in the development of blue-ice inclusions. The contact between blue-ice inclusions and firn was thought to act as a weakness along which water could flow, and a way in which englacial channels and voids could form. It is possible that a similar process occurs at Haut Glacier d'Arolla as several crevasses are present in the upper part of the glacier, although there is no evidence that any of these are ever water filled. The fact that not all openings were coupled with blue-ice inclusions also suggests that other processes are important.

The observed relation between debris bands and englacial openings at Haut Glacier d'Arolla suggests that debris may play a role in the formation of englacial channels and voids. The debris bands probably originated from the shearing of ice formerly in contact with the glacier bed (Copland *et al.*, in press a), the incorporation of avalanche and rock fall material from the slopes surrounding the glacier or the burial of patches of dirt and wind-blown dust on the glacier surface by new snowfall at the end of the summer melt season. The debris bands may provide a weakness in the ice which is exploited by meltwater to form englacial channels and voids. Alternatively, the debris bands may be relatively resistant features in the ice that restrict downward water movement and force the localization of water flow along their upper surface. The formation of englacial channels and voids may occur if the localization of meltwater continues over time and results in the melting of ice.

Water quality

Of the eleven boreholes filmed, seven had accompanying EC profiling data. EC profiles consisted of point measurements of EC at 5 m intervals from just below the borehole water level to the glacier bed. Marked spatial variations in borehole water quality were observed, superimposed upon a general increase in water turbidity with depth. Alternating zones of relatively clear and turbid water were also occasionally encountered within a single borehole. After combining the EC data with the video logs, three types of relationship between EC and turbidity stratification were identified.

(i) No turbidity or EC stratification. The video log of borehole 95/17 showed that the entire water column was composed of clear, low turbidity water. An EC profile taken 45 minutes prior to recording showed that the EC of the entire water column was <2 μS cm^{-1}. In this case clear water was characterized by low EC values. The sources of water suggested by these EC values are either from the drill water from surface meltwater streams, or supraglacial or englacial inputs. At the time of filming, this borehole was full and no water level fluctuations were observed, thus implying that it was unconnected to the basal drainage system. Borehole 95/15 also displayed no turbidity or EC stratification.

(ii) Concordant turbidity and EC stratification. The video log of borehole 95/9 showed that the water column was clear from the glacier surface to 8·8 m depth, below which it became progressively more turbid until the bed was reached. The EC profile showed that from 0–10 m below the glacier surface the EC of the water was <10 μS cm^{-1}, and below 10 m depth was 10–30 μS cm^{-1}. Consequently, there was a good correspondence between the water quality boundary seen in the video and the stratification identified from the EC profile. Clear water was characterized by low EC values, while more turbid water was associated with higher EC values. As no water level changes in this borehole were registered, stratification of the water column was likely to represent supraglacial water from drilling and melt overlying drill fluid which had churned up the bed and then acquired solute from the suspended sediment. Boreholes 95/8 and 95/2 also displayed concordant turbidity and EC stratification.

(iii) Discordant turbidity and EC stratification. The video log of borehole 95/18 showed that turbidity of the water was uniformly high along the length of the water column. EC profiling carried out two hours after filming showed that there was a boundary at 65 m below the glacier surface, where water of <10 μS cm^{-1} overlay water of 18–25 μS cm^{-1}. In this case the water quality seen in the video did not correspond to the EC stratification. A likely explanation for this phenomenon is that the borehole was reamed at the bed (the drill was lowered down the existing hole to widen it and free a stuck probe) prior to filming and

profiling. This probably disturbed unconsolidated sediment from the bed and distributed it throughout the water column. It appears that the EC stratification was then able to restabilize to the level that was recorded in a profile on the previous day, before the turbidity of the water returned to its natural state. Borehole water sampling in 1993 also produced some samples that displayed high EC, clear water, and other samples that displayed low EC, turbid waters.

Video identification of englacial channels inferred from water quality data

Borehole 95/2 was 96·5 m deep and drained to 60 m below the glacier surface two days after drilling. The water level remained constant at this level unless reaming or the input of supraglacial meltwater caused a temporary rise in water level. During one of these temporary rises in water level, EC profiling showed that below 60 m the water had stable EC values of $>10~\mu S~cm^{-1}$, while above 60 m the water had EC values of $<10~\mu S~cm^{-1}$. This suggests that there was an englacial output at 60 m because the dilute water above drained to this level overnight, while the EC stratification below 60 m remained stable.

As a result of these EC profiles, attention was paid to the area around 60 m when this borehole was viewed with the video camera. The video log showed that the water column was clear until 56·7 m, when wisps of turbid water were encountered. At 59·7 m an englacial channel was observed that appeared to be feeding the borehole with a plume of turbid water. In addition, the borehole displayed a distinct water quality split with one side of the borehole filled with turbid water, and the other side with clear water. Beyond 60·6 m the water column was clear again, until 62·5 m where more turbid water was encountered (Figure 7). It seems likely that the plume of turbid water was the result of turbid water returning to the borehole after it was forced into the englacial channel by reaming two hours prior to video observation. There are few sources of turbidity 40 m above the glacier bed, and reaming probably disturbed the turbid layers below.

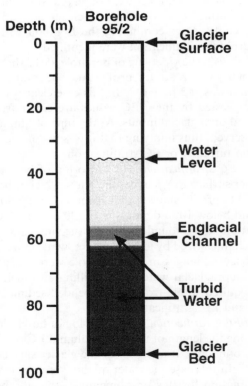

Figure 7. The patterns of water turbidity observed in borehole 95/2. Note the zone of relatively turbid water where the englacial channel enters the borehole

Further information about the water circulation and 'plumbing' of boreholes at Haut Glacier d'Arolla was provided by borehole 95/16. This borehole was drilled to a depth of 55·2 m, and drained to 24 m below the glacier surface 30 minutes after drilling had stopped. Subsequent daily EC profiles showed that the water level remained at a depth of 24 m for the following two weeks of observation, and that the water column consisted of $<3\ \mu S\ cm^{-1}$ water throughout. The video log confirmed this water level and showed that the water column was clear. The video log also showed the presence of two active englacial channels above the water level at 10·9 and 14·9 m, and the input of supraglacial water down the sides of the borehole walls.

The observations in borehole 95/16 pose an interesting question. How does the water level stay constant throughout the day when there is such an obvious input of supraglacial and englacial water into the borehole? It appears that the water output adjusts continually to match the input, so that the two are always equal. As the water column was not stratified, the output was likely to be basal as no other englacial channels were observed in this borehole; however, the adjustment process remains to be explained in detail. Had this borehole not been inspected by video, the interpretation from EC profiling would have been that the borehole connected to an englacial channel at a depth of 24 m after drilling, or to a subglacial void with the capacity to lower the borehole water level by the observed amount, and then remained passive. Now the interpretation is that the borehole water level is regulated in some way by basal water flow.

Bed conditions

The glacier bed was observed in boreholes 95/10, 95/17 and 95/22 at depths of 32·0, 93·0 and 131·7 m, respectively. The bed could not be seen in the other boreholes owing to high turbidity, or because the camera was too wide to pass all the way down the borehole. The glacier bed in borehole 95/10 consisted of subrounded and subangular clasts in a matrix of unconsolidated fine sediment (Figure 8). The bed in borehole 95/17 consisted predominantly of a relatively large, thin rock protruding vertically up into the borehole. Surrounding the rock was a layer of fine sediment. In borehole 95/22 the glacier bed consisted of fine sediment, contained no clasts or other large material and was directly in contact with the ice at the bottom of the borehole.

The character of the glacier bed observed with the video camera at Haut Glacier d'Arolla is similar to the till seen in areas recently exposed by glacier retreat, as well as in cavities at the glacier margin. It supports the

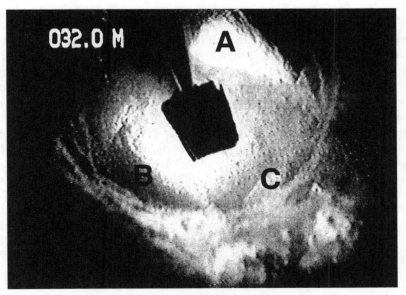

Figure 8. Video image of the glacier bed observed in borehole 95/10 at a depth of 32·0 m. Borehole is approximately 12 cm in diameter. Marker A is a clast projecting out into the borehole a few centimetres above the bed, B is a clast resting on the base of the borehole and C is a clast resting on the base of the borehole and partially frozen into the surrounding ice

conclusions of Hubbard *et al.* (1995) that Haut Glacier d'Arolla lies at least partly on an unconsolidated bed. In addition, the video logs confirm that boreholes really do reach the glacier bed. This is important as it confirms that automatic sensors are located at the glacier bed, although it must be remembered that the conditions measured by the sensors are not necessarily basal (e.g. input of supraglacial water may cause low EC outflow at the base of a borehole).

Particular attention was paid to the movement of suspended sediment particles near the glacier bed in the boreholes in which the bed could be seen. As with the englacial channels observed below the water level, water flow was not detected, although flow may still have occurred for the following reasons.

1. Water velocities may have been too low to be detected. For example, a water level rise of 20 m per hour (which was high by 1995 standards) in a 5 cm radius hole would require a flux of only 0.00004 m^3 s^{-1}, and a vertical velocity of 5 mm s^{-1}, if driven by basal inflow.

2. All basal water flow may have been through the unconsolidated sediment observed beneath the glacier, rather than at the glacier bed.

3. Basal water flow probably introduces high turbidity water into the base of boreholes, therefore precluding video observations in the boreholes where water flow is most likely.

4. Most video observations were in the afternoon, which may have been a period of relatively little englacial water movement. Salt trace studies in 1993 indicate that the direction of the hydraulic gradient between boreholes and englacial channels can change during a day. It appears that water can enter englacial channels from boreholes as the water level rises in the late morning, and then be returned to the boreholes from englacial channels in the early evening as the water level drops. We speculate that the initial water level rise is driven by both supraglacial and subglacial inputs. As a result, the borehole becomes overpressured relative to the natural pressure in the drainage system. This dams flows in the englacial conduit. Over time, the overpressure is reduced and eventually eliminated by basal outflow. This can occur once the hydraulic gradient between the borehole and channel starts to fall as discharge stabilizes in the afternoon. As the overpressure is reduced, flow recommences in the englacial conduit and this starts to feed the borehole. Consequently, water flow from englacial channels intersected by boreholes may be intermittent, and would not necessarily be expected during the afternoon.

SUMMARY AND CONCLUSIONS

Borehole video provides real-time viewing of the interior and bed of Haut Glacier d'Arolla, and can generate a continuous record of water clarity. It provides important additional information that greatly aids interpretations based on indirect sources of evidence such as EC profiling and water level variations, and allows observation of englacial features such as channels and voids that are difficult to document otherwise. By comparing the vertical height of the observed englacial channels and voids with the total length of observed boreholes, it is estimated that englacial channels account for approximately 0.1% of the vertical ice thickness, and natural voids for approximately 0.4%. Many of the englacial channels and voids are coupled with blue-ice inclusions or debris bands in the ice. This suggests that some of the englacial openings are linked to the closure of crevasses in the upper parts of the glacier, or the diversion of water flow along layers of sediment frozen into the ice. Variations in the shape of the englacial channels correlate with predictions by Shreve (1972), based on an analogy with tunnels in karst. Where the openings of a channel could be observed they were keyhole-shaped above the borehole water level, and circular below it. Finally, video observations of the glacier bed have supported inferences from borehole water level records (Hubbard *et al.*, 1995) that Haut Glacier d'Arolla lies on a bed composed at least partly of till.

HIGHLIGHTS VIDEO

A 23-minute composite video tape of the 'highlights' of the Haut Glacier d'Arolla video recordings has been produced for educational and research use. Please contact Jon Harbor to obtain a copy of this video.

ACKNOWLEDGEMENTS

This research was funded by US National Science Foundation grants OPP-9321350 and OPP-94963450, the Geological Society of America and the Canadian Natural Sciences and Engineering Research Council. The field assistance of Bryn Hubbard, Marie Minner, Simon Cross, David Gaselee and all other 1995 Arolla Glaciology Project members is greatly appreciated.

REFERENCES

Boulton, G. S. 1974. 'Processes and patterns of glacier erosion', in Coates, D.R. (Ed.), *Glacial Geomorphology*. State University of New York, Binghamton, pp. 41–87.

Clarke, G. K. C. 1987. 'Subglacial till: a physical framework for its properties and processes', *J. Geophys. Res.* **92**(B9), 9023–9036.

Copland, L., Harbor, J., Gordon, S., and Sharp, M. in press a. 'Borehole video observation of englacial and basal ice conditions in a temperate valley glacier', *Ann. Glaciol.* **24**.

Copland, L., Harbor, J., Minner, M., and Sharp, M. in press b. 'The use of borehole inclinometry in determining basal sliding and internal deformation at Haut Glacier d'Arolla, Switzerland', *Ann. Glaciol.* **24**.

Fountain, A. G. 1993. 'Geometry and flow conditions of subglacial water at South Cascade Glacier, Washington State, USA; an analysis of tracer injections', *J. Glaciol.* **39**, 143–156.

Fountain, A. G. 1994. 'Borehole water-level variations and implications for the subglacial hydraulics of South Cascade Glacier, Washington State, U.S.A.', *J. Glaciol.* **40**, 293–304.

Gordon, S., Sharp, M., Hubbard, B., Willis, I., Smart, C. C., and Ketterling, B. In press. 'Seasonal reorganisation of subglacial drainage inferred from borehole measurements', *Hydrol. Process.*

Harper, J. T. and Humphrey, N. F. 1995. 'Borehole video analysis of a temperate glacier's englacial and subglacial structure: implications for glacier flow models', *Geology*, **23**, 901–904.

Hubbard, B. P., Sharp, M. J., Willis, I. C., Nielsen, M. K., and Smart, C. C. 1995. 'Borehole water-level variations and the structure of the subglacial hydrological system of Haut Glacier d'Arolla, Valais, Switzerland', *J. Glaciol.* **41**, 572–583.

Iken, A. and Bindschadler, R. A. 1986. 'Combined measurements of subglacial water pressure and surface velocity of the Findelengletscher, Switzerland. Conclusions about drainage system and sliding mechanism', *J. Glaciol.* **32**, 101–119.

Kamb, B. 1987. 'Glacier surge mechanism based on linked cavity configuration of the basal water conduit system', *J. Geophys. Res.* **92**(B9), 9083–9100.

Koerner, R. M., Fisher, D. A., and Parnandi, M. 1981. 'Bore-hole video and photographic cameras', *Ann. Glaciol.* **2**, 34–38.

Lamb, H., Tranter, M., Brown, G. H., Gordon, S., Hubbard, B., Nielsen, M., Sharp, M., Smart, C. C., and Willis, I. C. 1995. 'The composition of meltwaters sampled from boreholes at the Haut Glacier d'Arolla, Switzerland', *Int. Assoc. Hydrol. Sci. Pub.* **228**, 395–403.

Lliboutry, L. 1969. 'Contribution à la théorie des ondes glaciaires', *Can. J. Earth Sci.* **6**, 943–953.

Murray, T. and Clarke, G. K. C. 1995. 'Black box modelling of the subglacial water system', *J. Geophys. Res.* **100**(B7), 10231–10245.

Nye, J. F. 1973. 'Water at the bed of a glacier', *Int. Assoc. Hydrol. Sci. Publ.* **95**, 189–194.

Pohjola, V. A. 1993. 'TV-video observations of bed and basal sliding on Storglaciären, Sweden', *J. Glaciol.* **39**, 111–118.

Pohjola, V. A. 1994. 'TV-video observations of englacial voids in Storglaciären, Sweden', *J. Glaciol.* **40**, 231–240.

Raiswell, R. 1984. 'Chemical models of solute acquisition in glacier melt waters', *J. Glaciol.* **30**, 49–57.

Raymond, C. F. and Harrison, W. D. 1975. 'Some observations on the behaviour of the liquid and gas phases in temperate glacier ice', *J. Glaciol.* **14**, 213–233.

Röthlisberger, H. 1972. 'Water pressure in intra- and subglacial channels', *J. Glaciol.* **11**, 177–203.

Seaberg, S. Z., Seaberg, J. Z., Hooke, R. LeB., and Wiberg, D. W. 1988. 'Character of the englacial and subglacial drainage system in the lower part of the ablation area of Storglaciären, Sweden, as revealed by dye-trace studies', *J. Glaciol.* **34**, 217–227.

Sharp, M. 1991. 'Hydrological inferences from meltwater quality data: the unfulfilled potential', *British Hydrol. Soc. Third National Hydrology Symposium*, 5·1–5·6.

Sharp, M., Gemmell, J. C., and Tison, J.-L. 1989. 'Structure and stability of the former subglacial drainage system of Glacier de Tsanfleuron, Switzerland', *Earth Surf. Process. Landf.* **14**, 119–134.

Sharp, M., Richards, K., Willis, I., Arnold, N., Nienow, P., Lawson, W., and Tison, J.-L. 1993. 'Geometry, bed topography and drainage system structure of the Haut Glacier d'Arolla, Switzerland', *Earth Surf. Process. Landf.* **18**, 557–571.

Shreve, R. L. 1972. 'Movement of water in glaciers', *J. Glaciol.* **11**, 205–214.

Stone, D. B., Clarke, G. K. C., and Blake, E. W. 1993. 'Subglacial measurement of turbidity and electrical conductivity', *J. Glaciol.* **39**, 415–420.

Sugden, D. E. and John, B. S. 1976. *Glaciers and Landscape*. Edward Arnold Ltd., London. p. 290.

Tranter, M., Brown, G., Raiswell, R., Sharp, M., and Gurnell, A. 1993. 'A conceptual model of solute acquisition by alpine glacier meltwaters', *J. Glaciol.* **39**, 573–581.

Tranter, M., Sharp, M. J., Brown, G. H., Willis, I. C., Hubbard, B. P., Nielsen, M. K., Smart, C. C., Gordon, S., Tully, M., and Lamb, H. R. 1997. 'Variability in the chemical composition of *in situ* subglacial meltwaters', *Hydrol. Process.* **11**, 59–77.

Waddington, B. S. and Clarke, G. K. C. 1995. 'Hydraulic properties of subglacial sediment determined from the mechanical response of water-filled boreholes', *J. Glaciol.* **41**, 112–124.

Walder, J. S. 1986. 'Hydraulics of subglacial cavities', *J. Glaciol.* **32**, 439–445.

Walder, J. S. and Fowler, A. 1994. 'Channelized subglacial drainage over a deformable bed', *J. Glaciol.* **40**, 3–15.

Walder, J. and Hallet, B. 1979. 'Geometry of former subglacial water channels and cavities', *J. Glaciol.* **23**, 335–346.

Weertman, J. 1969. 'Water lubrication mechanism of glacier surges', *Can. J. Earth Sci.* **6**, 929–942.

Weertman, J. 1972. 'General theory of water flow at the base of a glacier or ice sheet', *Rev. Geophys. Space Phys.* **10**, 287–333.

Weertman, J. 1986. 'Basal water and high-pressure basal ice', *J. Glaciol.* **32**, 455–463.

Westinghouse Savannah River Company. 1989. 'Demonstration of innovative monitoring technologies at the Savannah River integrated demonstration site', *Report for US Department of Energy Contract DE-AC09-89SR18035*. Westinghouse Savannah River Company, Aiken, South Carolina.

Willis, I. C., Sharp, M., and Richards, K. 1990. 'Configuration of the drainage system of Mitdalsbreen, Norway, as indicated by dye-tracing experiments', *J. Glaciol.* **36**, 89–101.

13

IN SITU MEASUREMENTS OF BASAL WATER QUALITY AND PRESSURE AS AN INDICATOR OF THE CHARACTER OF SUBGLACIAL DRAINAGE SYSTEMS

DAN B. STONE

Institute of Arctic and Alpine Research, University of Colorado, Boulder, CO 80309-0450, USA

AND

GARRY K. C. CLARKE

Department of Geophysics and Astronomy, University of British Columbia, Vancouver, British Columbia V6T 1Z4, Canada

ABSTRACT

Continuous subglacial measurements of turbidity and electrical conductivity — two indicators of basal water quality — can be used to help characterize subglacial drainage systems. These indicators of water quality yield information that complements that provided by water pressure measurements. Quantitative attributes of subglacial drainage systems, such as water velocity and subglacial residence time, as well as qualitative behaviour — for example, spatial and temporal variations in system morphology — can be deduced using water quality measurements. Interpretation is complicated by the many potential influences on turbidity and electrical conductivity, but when these complications are appreciated a richer interpretation results. To demonstrate the utility of basal water quality measurements, observations from Trapridge Glacier, Yukon Territory, Canada were examined. The data reveal complex behaviour of the drainage system, but constraints imposed by basal water quality measurements help to clarify the nature of the subglacial flow system. The measurement and interpretation methods described and demonstrated are applicable to other glaciers. As such, they should prove useful for characterizing different subglacial drainage configurations and behaviours, thereby improving our general understanding of the hydrology and dynamics of wet-based glaciers.

INTRODUCTION

It is widely accepted that the movement and storage of basal water plays an important part in the dynamics of temperate glaciers, as well as warm-based polar glaciers, ice sheets and ice streams. Nevertheless, our understanding of subglacial drainage systems remains incomplete; network morphologies are not reliably known and the spatial and temporal variability of subglacial drainage structures is only partially disclosed by existing field data. Advances in hot-water drilling technology have now allowed access to numerous glacier beds, providing direct evidence that many — if not all — glaciers are underlain by a sediment layer of variable thickness and areal extent. In the light of this, there is a growing appreciation of the important involvement of subglacial deposits in the formation and stability of basal drainage systems.

Measurements of basal water pressure have been central to many glaciological investigations (Mathews, 1964; Iken, 1972; Hodge, 1976; Engelhardt, 1978; Hodge, 1979; Hantz and Lliboutry, 1983; Kamb *et al.*, 1985; Iken and Bindschadler, 1986; Hooke *et al.*, 1989; Englhardt *et al*; 1990b; Fountain, 1994). In comparison with water pressure, however, subglacial measurements of other properties of basal water are rare: Engelhardt *et al.* (1990a) measured the electrical conductivity of water beneath Ice Stream B in

West Antarctica as part of a salt injection experiment; Blake (1992) measured DC resistivity and natural electrical potentials at the bed of Trapridge Glacier, Yukon Territory, Canada; movement of subglacial water, 'indicated by a cloud of suspended sediments', has been recorded by TV–video observations at the bed of Stoglaciären, Sweden (Pohjola, 1993).

In this paper, we show how direct measurements of the turbidity and electrical conductivity of basal water can be used to help characterize subglacial drainage systems. These properties are related to the presence of suspended solids and the concentration of dissolved ions, respectively, and are important water quality indicators for almost any purpose. In the present context, we will use the term 'basal water quality' to refer collectively to these properties of subglacial water. Combined measurements of basal water quality and pressure are complementary — together they provide a valuable approach to monitoring subglacial hydraulic conditions.

DIRECT MEASUREMENTS OF BASAL WATER PROPERTIES

In situ measurements of basal water quality require direct access to the glacier bed. We install water pressure, turbidity, and electrical conductivity sensors at the base of boreholes that have been drilled to the glacier bed using a high-pressure hot-water system. The sensors are connected by wires to Campbell CR10 dataloggers at the glacier surface.

For some glaciers, interpretation of measurements made at the bottoms of boreholes can be complicated; factors such as water inflow at the top of a borehole, the intersection of a borehole with an englacial drainage network and vertical motions of the water column within a borehole confound measurement of actual basal properties (Sharp *et al.*, 1993). We present data collected on Trapridge Glacier (61°14'N, 140°20'W). The geometry and sub-polar thermal regime of that glacier make it well suited for direct measurements of basal water properties. Trapridge Glacier is thin (depth ≈ 80 m) and mainly comprises ice at subfreezing temperatures (as low as $-6°C$ near the glacier surface), with temperate ice occurring only near the bed away from the glacier margins (Clarke and Blake, 1991). These conditions create a nearly ideal situation for measuring basal water properties, as boreholes are rapidly sealed by water freezing and englacial drainage structures are rarely encountered.

Subglacial sensors will be subjected to mechanical destruction if they are placed in direct contact with the glacier bed and, eventually, all sensors will fail due to wire breakage as ice deformation stretches the wires beyond their strain limits. To minimize costs, we use durable inexpensive turbidity and electrical conductivity sensors designed specifically for use beneath glaciers (Stone *et al.*, 1993). Figure 1 shows turbidity and conductivity sensors prepared for subglacial deployment; the essential aspects of the sensors and their calibrations are given in the following sections.

Turbidity sensors

The turbidity sensors that we use contain a miniature light source and two infra-red photodetectors. Light from the source passes through a water sample flow path and is compared with light that travels through an unobstructed reference path inside the sensor. Larger concentrations of suspended sediment in the water flow path reduce the number of photons reaching the sample detector, giving rise to increases in measured turbidity.

An accurate calibration relating turbidity to suspended sediment concentration is unrealizable because it requires a detailed knowledge of suspension properties at all times (Hach *et al.*, 1990). For subaerial streams, an approximate calibration can be obtained by collecting samples of the suspension while turbidity readings are being made, then analysing the samples to determine sediment concentration. Unfortunately, this procedure cannot be applied to subglacial water; it is problematic to collect a sufficient number of samples from the glacier bed and any samples that are collected will almost certainly be altered by the time they reach the glacier surface. For these reasons, we do not use *in situ* turbidity measurements to estimate the suspended-sediment concentration of subglacial water.

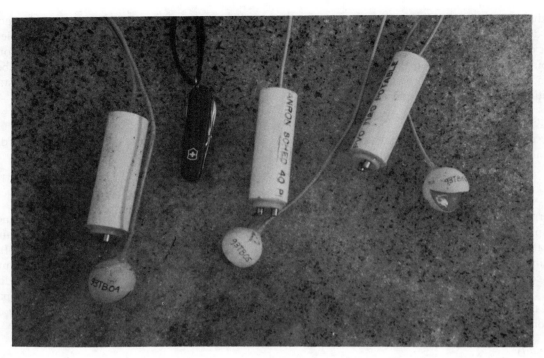

Figure 1. Pairs of turbidity and electrical conductivity sensors prepared for subglacial deployment. The turbidity sensors have been cast inside table tennis balls, out of which wedges have been cut. The conductivity sensors are housed in short cylinders of plastic conduit.

The expression

$$\tau = -\frac{1}{L}\ln\left(\frac{V}{aV_0}\right) \tag{1}$$

defines our usage of the term 'turbidity' [Stone *et al.*, 1993, their equation (5)], where τ is the turbidity, L is the light path length through the suspension, V and V_0 are measured voltages from the sensor's sample and reference circuits, respectively, and a is a positive constant. With this definition, turbidity sensor calibration simply involves measuring the light path length L and determining a value for the constant a. (For a given sensor, the value of a can be determined by measuring the voltages from both detectors when the sensor is in 'clear' water; in this instance $\tau \approx 0$ and $a = V/V_0$.) According to Equation (1), turbidity has dimensions $[L^{-1}]$ and increasing turbidity is expressed as a reduction in the effective travel distance of light through the suspension.

Electrical conductivity sensors

The electrical conductivity sensor consists of two stainless-steel electrodes housed in a non-conductive nylon cylinder (Figure 1). By applying an excitation voltage across the sensor electrodes in an AC half-bridge configuration, the conductance of subglacial water can be measured. The AC half-bridge measurement procedure is a standard function on Campbell CR10 dataloggers (Campbell Scientific, 1993: 13–15).

A standard reference solution is required for conductivity sensor calibrations. The reference solution is used to obtain a cell constant value for each sensor. The cell constant K_c is the constant of proportionality between conductivity σ and conductance G

$$\sigma = K_c G \tag{2}$$

Cell constants depend on sensor geometry and have dimensions $[L^{-1}]$. For simple geometrics, K_c can be determined analytically if the inter-electrode distance and electrode surface areas are known. However, conductivity sensors with simple geometries, such as parallel plates or concentric cylinders, tend to accumulate debris between the electrodes. The sensors that we use have been designed to avoid this problem, but their design precludes analytical determination of cell constant values. Instead, we determine cell constants empirically by submerging the electrodes in a standard solution and measuring the resistance R between them. As $G = 1/R$, these resistances can be multiplied by the known conductivity σ_s of the standard solution to obtain the cell constants

$$K_c = R\sigma_s \tag{3}$$

Cell constants obtained by this procedure are combined with measured conductances according to Equation (2) to give the conductivity of subglacial water.

UNDERSTANDING CHANGES IN BASAL WATER QUALITY

Interpreting the character of subglacial drainage systems from *in situ* water quality measurements is complicated by the many different factors that influence basal water quality. Thus in developing an interpretation it is important to consider all potential influences on turbidity and electrical conductivity. In some instances, a single process (e.g. mixing of different waters) might fully explain fluctuations in both turbidity and conductivity. In many instances, however, the two properties will be affected independently by different and, often, multiple agents. To simplify the following discussion, we will describe processes that separately lead to variations in turbidity and electrical conductivity, bearing in mind that it is usually some combination of these processes working in concert which gives rise to the measured signal.

Turbidity

Turbidity results from the fluvial entrainment and transport of sediment particles. At a particular location variations in turbidity may be due to fluctuations in water velocity, which can cause changes in the subglacial activation of sediment. Water flow at the ice–bed interface exerts a shear stress on any sediment that is present there and the stress increases with increasing flow velocity. When a sufficient shear stress is exerted, particles can be dislodged; those that are small enough will then be transported as suspended sediment. In this situation, turbidity variations directly reflect changes in basal water velocity.

Sediment disturbances at a particular location can also lead to variations in the measured turbidity. Such disturbances might be created, for instance, during the drilling of a borehole when the drill tip reaches the bed and water is flushed from the bottom of the hole. Another possible mechanism of turbidity generation is the direct mechanical action of rapidly sliding ice (Humphrey *et al.*, 1986). If the location of a disturbance is known and if the sediment is transported with sufficient turbulence to maintain suspension, a pulse of suspended sediment can serve as a subglacial tracer. Along similar lines, reservoirs of 'clear' water draining subglacially might be indicated by decreases in turbidity.

Turbidity can also fluctuate in response to basal water pressure variations that cause drainage channel migration or slumping from channel margins (Collins, 1979; Walder and Fowler, 1994). High water pressure within a channel can lead to bed separation, thereby exposing new areas of the bed to turbulent water flow. Alternatively, if the water pressure within a channel falls below the pore pressure in the surrounding sediment walls, then the hydraulic gradient drives water from the sediment pores into the channel, a situation that can generate suspendible sediment through the mechanisms of piping or slope failure.

Electrical conductivity

The electrical conductivity of a solution arises due to the presence and mobility of dissolved ions. As the ion concentration increases, the electrical conductivity of the solution also increases. However, temperature, ionic composition, ionic mobility and other properties also affect the conductivity of solutions. In natural waters, these properties vary widely and interact in complicated ways (Hem, 1985), preventing accurate determination of ion concentration from electrical conductivity measurements. Nevertheless, conductivity measurements do provide an *indication* of total dissolved solids and have proved useful in monitoring solute concentration in proglacial meltwaters (Fenn, 1987).

In subglacial environments, ionic enrichment occurs when dilute water contacts subglacial bedrock or sediment — including suspended sediment, debris-laden basal ice and basal deposits of drift — and when mixed with solute-rich waters from subglacial sources (Collins, 1977; 1981; Tranter *et al.*, 1993). Furthermore, ion concentration increases with increasing contact time between water and basal sediment (Brown *et al.*, 1994) an especially important process on glaciers having an abundance of readily soluble material (e.g. carbonates) at their beds. This effect is well known in hydrology: the conductivity of storm runoff or groundwater depends on the amount of time spent in contact with mineral sources (Pilgrim *et al.*, 1979; Hem, 1985). Such behaviour suggests that electrical conductivity measurements can be used to infer the relative residence times of subglacial water, as 'old' water will tend to be more mineralized than 'new' water.

In addition to the time of contact with subglacial bedrock or sediment, basal water can also become enriched in solute ions when ice forms; as ice crystals grow, the solute concentration of the remaining meltwater is increased. Both enrichment processes — contact with rock or sediment and ice-crystal growth — would be associated with electrical conductivity signals that monotonically increase until conditions of chemical equilibrium are achieved. Hence variations from this behaviour might indicate the influence of competing effects (e.g. dilution with fresh water).

Conductivity measurements can fluctuate due to the mixing of waters with different electrical properties. This process is evidenced by diurnal changes in conductivity associated with the daily input of fresh surface meltwater. When solutes are introduced by a subglacial source, such as a groundwater spring, changes in conductivity can also be caused by variations in basal water pressure. For instance, increasing basal water pressure decreases the hydraulic gradient driving groundwater flow toward the ice–bed contact, thereby reducing the input of solute ions and also the conductivity signal.

INTERPRETATION OF BASAL WATER QUALITY MEASUREMENTS

In light of the preceding discussion, *in situ* measurements of the turbidity and electrical conductivity of subglacial water can provide direct indications of (1) basal water velocity, (2) subglacial water provenance and (3) changes in drainage system morphology. These aspects, in turn, are useful for characterizing subglacial drainage systems. To estimate the rate and direction of water flow at the bed, subglacial arrays of turbidity and electrical conductivity sensors can be deployed. (Separation distances between sensors of $\approx 10\,\text{m}$ have been effective on Trapridge Glacier; appropriate distances on other glaciers will vary, depending primarily on the degree of hydraulic interconnectivity at the bed.) As we have mentioned, when boreholes connect with the subglacial drainage system, nearby sensors sometimes register turbidity and conductivity pulses. Conductivity sensors can also be used in borehole to borehole tracer tests, using ordinary table salt as the tracer.

The spatial distribution of basal water quality characteristics can be identified from measurements of turbidity and electrical conductivity at different subglacial locations. These characteristics place constraints on the residence time of basal water and on possible drainage system configuration. For example, areas that are efficient in evacuating surface meltwater input would be indicated by low electrical conductivity and variable turbidity. Alternatively, areas having slow water velocities, such as the cavity portions of a linked-cavity system, would correspond to reduced turbidity and increased conductivity signals relative to regions of faster transit.

Changes in basal water quality reflect variations in the sources and routings of subglacial water. Thus turbidity and conductivity measurements can also be used to monitor changes in the configuration of subglacial drainage systems. For instance, the collapse of unstable sediment-walled drainage channels would give rise to rapid and erratic fluctuations in turbidity. In some instances, migration of drainage channels might lead to the development of water flow passageways in the vicinity of a sensor, or vice versa. Such behaviour would be revealed by distinct changes in the nature of turbidity and conductivity signals — for instance, the onset of diurnal oscillations.

CHARACTERIZATION OF SUBGLACIAL DRAINAGE SYSTEMS: EXAMPLES FROM TRAPRIDGE GLACIER

In this section we show how measurements of the turbidity and electrical conductivity of basal water can be used to help characterize subglacial drainage systems. Although our discussion is focused on observations of turbidity and conductivity, basal water pressure data are included in the following examples, both to emphasize the added significance of water quality measurements and to clarify interpretations. The data were collected at two-minute sampling intervals and have been only minimally processed. (In addition to applying calibration information, turbidity and conductivity records have been slightly smoothed using an 11-point Gaussian filter.) Linear interpolation has been used for plotting between data points.

On Trapridge Glacier, the 'normal' behaviour of pressure and water quality signals is both spatially or temporally variable. At any given time, subglacial sensors in one location may record strong diurnal oscillations, while perhaps 30–40 m away another set of sensors will show irregular, large amplitude excursions that lack a diurnal component; at yet another location, pressure and water quality measurements may shown only minimal variations over an interval of several weeks. In some instances, an abrupt change in the measured signal coincides with the initiation of diurnal oscillations; in other instances, well-developed diurnal oscillations suddenly stop or gradually disappear over a few days. Such non-uniform behaviour indicates instability of subglacial drainage structures and hints at the dynamic complexity of the basal environment. The following examples are, by themselves, insufficient to fully describe the drainage system beneath Trapridge Glacier and are not intended to do so. Instead, these examples demonstrate the general usage of pressure and water quality measurements in the identification of characteristic properties and behavioural patterns of subglacial drainage systems.

Basal water velocity

The 70th borehole drilled during the 1989 field season (designated 89H70) connected with the subglacial hydraulic system, as was indicated by sudden water drainage from the borehole when the drill reached the bed. According to drilling records, connection occurred at 1555 h on 28 July. The connection of 89H70 was recorded by water pressure, turbidity and electrical conductivity sensors already in place at the glacier bed. Figure 2 shows data from three of these sensors, one of each type. The sensors, which had been installed more than one week earlier in different boreholes, were separated from 89H70 by the following straight line distances at the glacier bed: pressure sensor, 28·8 m; turbidity sensor, 31·8 m; electrical conductivity sensor, 13·1 m (Figure 2). The subglacial positions of sensors, determined from surveying and borehole inclinometry, have an estimated uncertainty of ±0·3 m.

Figure 2a shows water pressure, turbidity and electrical conductivity measurements over a four-day interval during which borehole 89H70 was drilled. Two small spikes are visible in the turbidity record (labelled **b** and **c**); data from these periods are plotted on expanded time scales in Figures 2b and 2c. The expanded data in Figure 2b show a sudden pressure increase on 28 July, which occurred shortly after 1554 h and peaked about four minutes later. The pressure rise began just when the drill reached the bed in 89H70, roughly 30 m away. It is our interpretation that this pressure disturbance was caused by the sudden opening of a water-filled borehole and that water subsequently drained from the borehole into the subglacial drainage network. A few minute after the pressure peaked, the electrical conductivity of subglacial water began to decrease, reaching a minimum at 1614 h. This was followed by a turbidity pulse, the maximum of which occurred at 1632 h, 37 minutes after the connection.

In addition to the pressure response, it is likely that the turbidity and electrical conductivity events shown in Figure 2b were also associated with the connection of borehole 89H70. The turbidity pulse would have been generated by the increase in basal water velocity near the base of the borehole as it drained. The shape of the pulse — smooth and fairly symmetrical — suggests that the sediment was derived from a brief disturbance at a single location and that the suspension had travelled for some distance before reaching the sensor. Because sediment activation is favoured in regions having the greatest water velocity, it is reasonable to assume that entrainment occurred near the base of the borehole. The pulse-like decrease in electrical conductivity would have resulted from the dilution of basal water with fresh water that drained from the

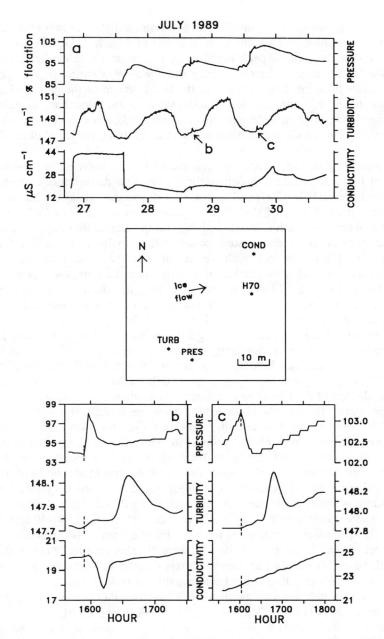

Figure 2. (a) Subglacial water pressure, turbidity and electrical conductivity recorded over a four-day interval in July 1989. The disturbances indicated in the turbidity record were related to the drilling and subsequent reopening of a nearby borehole. Data from these periods are shown (using the same units) on expanded time-scales in (b) and (c), where broken vertical lines mark disturbance times. The relative positions of sensors and the borehole are shown in the map at the centre of the figure. Note that the timing between pressure and turbidity events is similar in both (b) and (c), but the pressure and conductivity responses differ: in (b) pressure increased abruptly, whereas in (c) a sudden drop in pressure was recorded; a distinct conductivity excursion is shown in (b) and in (c) there is no apparent fluctuation in the measured conductivity

borehole. These responses, and the spatial distribution of sensors about 89H70 (Figure 2), suggest that water flow was directed radially away from the borehole base during this particular connection.

If we assume that water drained instantaneously from 89H70 at 1555 h and that sediment was stirred up at the borehole base at the same time, we can estimate an upper bound for the subglacial speed of water from the timings of events shown in Figure 2b and the known distances involved. Using the time of the electrical conductivity minimum, we obtain a basal speed of $41 \, \mathrm{m \, h^{-1}}$ for fresh water that was released from the borehole. Using the time of peak turbidity, the transport rate of the suspension was $52 \, \mathrm{m \, h^{-1}}$. For comparison dye tracer tests performed during the 1982–1983 surge of Variegated Glacier, Alaska yielded an average water flow speed of $72 \, \mathrm{m \, h^{-1}}$ in the basal system of the glacier (Kamb *et al.*, 1985); injections of dye in boreholes on South Cascade Glacier, Washington State, have revealed average subglacial flow speeds that range between 40 and $432 \, \mathrm{m \, h^{-1}}$ (Fountain, 1993); a borehole to borehole salt tracer test on Ice Stream B, West Antarctica, indicated a basal water velocity under the ice of about $30 \, \mathrm{m \, h^{-1}}$ (Engelhardt *et al.*, 1990a).

Figure 2c shows another pressure disturbance, this time a sudden decrease in basal water pressure, beginning at 1602 h on 29 July and lasting approximately 12 min. This disturbance is probably associated with the reopening of borehole 89H70, which had frozen shut overnight. Although the exact time of reconnection was not recorded, the redrilling of 89H70 took place about this time and all other fluctuations in the pressure record are well matched with different activities on the glacier that day. Note that on the afternoon of 29 July, the measured basal water pressure exceeded they hydrostatic ice flotation pressure, and the inferred reopening of 89H70 results in a slight release of pressure at the location of the sensor. For this to occur, water must have flowed from the top of 89H70 when the borehole was reopened, as the section of borehole above the ice plug would have been full of water while drilling. This cannot be confirmed from field notes; however, several times during the 1989 field season other boreholes in the vicinity were observed to have muddy water flowing out of them after they were drilled.

The ice pressure P_i at the base of a glacier having a thickness h_i is $P_i = \rho_i g h_i$ where ρ_i is the density of ice and g is the acceleration due to gravity. Similarly, the water pressure P_w at the base of a borehole filled with water to a height h_w is $P_w = \rho_w g h_w$ where ρ_w is the density of water. If the borehole is filled with water to the glacier surface then $h_w = h_i$ and, assuming $\rho_i / \rho_w = 0.9$, we have $P_w = (\rho_i / 0.09) g h_i \approx 1.11 P_i$. Thus, if water did flow from the top of 89H70 when it was reopened, the water pressure at the bottom of the borehole must have been greater than 111% of the ice flotation pressure. In this instance, a minimum value of the hydraulic gradient between 89H70 and the pressure sensor location can be estimated from survey, inclinometry and pressure measurements as follows: the elevation of the glacier surface z_2 at the top of 89H70 represents a minimum bound on the hydraulic head at that location, as the actual head value must exceed z_2 for water to flow from the top of the borehole; the hydraulic head at the pressure sensor location is $z_1 + \varphi_1$, where z_1 is the elevation of the sensor and φ_1 is the height of a water column corresponding to a measured pressure P_1. If Δx is the distance along the glacier bed between these two locations then the hydraulic gradient is simply $J = (z_2 - z_1 - \varphi_1)/\Delta x$. At the time 89H70 was reopened, this calculation yields $J \approx 0.13$, with water flow directed from the borehole towards the pressure sensor — obliquely upstream with respect to the direction of ice flow (Figure 2). This seemingly unusual water flow direction might reflect the influence of frozen bed at the glacier terminus, which prevents basal water from exiting at the frontal margin and causes ponding behind a thermally controlled dam.

A turbidity pulse, very much like that recorded the previous day, followed the pressure disturbance, peaking at 1648 h on 29 July (Figure 2c). On this occasion, however, there was no fluctuation in the measured conductivity of subglacial water. Assuming that the water jet from the drill mobilized basal sediment, the turbidity and conductivity signals can be understood, in this instance, in terms of the sensor locations (Figure 2) and the deduced hydraulic gradient: the pressure and turbidity sensors were located near each other and downstream from 89H70, as implied by the gradient direction, whereas the conductivity sensor was located on the other side of 89H70 in the upstream direction. Thus we recorded a turbidity pulse with no accompanying fluctuation in conductivity because mobilized sediment and fresh water input from the drill were carried by basal flow towards the turbidity sensor, not in the direction of the conductivity sensor. The difference between these responses and those recorded the previous day during the initial connection are, in this explanation, due to temporal changes in the gradient direction; whereas basal water flow was directed radially away from the base of the borehole during the initial connection on 28 July, it was driven in a south westerly direction during the reopening of 89H70 on the following day.

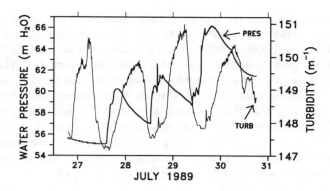

Figure 3. Data that were shown in Figure 2a, this time overlaid and plotted on expanded vertical scales. Subglacial water pressure is the thick line and turbidity is the thin line. In this figure, water pressure has been expressed as the equivalent height of a water column above the sensor; the local ice flotation pressure is roughly equal to a 63 m water column. Diurnal oscillations in both pressure and turbidity are evident, with turbidity peaks lagging pressure peaks by about 16 h. Note that turbidity continues to increase — reaching daily maximum — during periods of declining basal water pressure

If sediment was entrained during the reopening of 89H70 at 1602 h, when the basal pressure suddenly decreased, then the time of peak turbidity gives a transport rate of approximately 42 m h^{-1} for the suspension. Using this speed as a proxy for specific discharge q and a minimum hydraulic gradient of $J = 0.13$, we can estimate an upper bound for the hydraulic conductivity K of the drainage system [We have shown elsewhere (Stone and Clarke, 1993) that the hydraulic conductivity of a distributed subglacial drainage system can be defined for both laminar and turbulent flow either in a porous medium or in a very thin, sheet-like layer.] Assuming that the flow regime is turbulent and that standard empirical formulae which depend on hydraulic gradient to the one-half power apply [i.e. $q = -K \operatorname{sgn}(J)|J|^{1/2}$, where $\operatorname{sgn}(x)$ is the algebraic sign function], we obtain an estimate of 0.03 m s^{-1} for the maximum hydraulic conductivity of the drainage system in this region.

Besides the information that can be obtained from drilling-related disturbances, turbidity measurements can also provide evidence of natural variations in basal water velocity. In particular, because sediment activation is directly related to water velocity, daily periods of minimum and maximum velocity can be identified from diurnal oscillations in the turbidity of basal water. We illustrate this point in Figure 3 by overlaying the pressure and turbidity data of Figure 2a; the drilling disturbances on 28 and 29 July, which we have been discussing, are obviously superimposed on larger scale diurnal fluctuations.

The data in Figure 3 show that subglacial water pressure rises abruptly around noon, presumably when surface meltwater begins to reach the bed, whereas turbidity begins to increase a couple of hours later and continues to rise for several hours after the pressure has peaked. A number of different mechanisms could give rise to the phase lag between these two signals. One possibility is that the lag reflects the passage of sediment-rich water which was entrained earlier in the day during peak pressure and water velocity at upstream locations. This explanation is problematic, however, because it suggests that diurnal peaks in water velocity do not activate sediment in the study area; instead, it requires preferential erosion of basal sediment from 'softer' upstream sites and the presence of more cohesive deposits, which are not readily mobilized by water flow, along the flow paths and in the vicinity of the sensors. Such immobility is not characteristic of the bed material in the study area — even minor disturbances in boreholes, such as the collection of a subglacial water sample, are typically sufficient to activate basal sediment — and there are no observations which establish particular subglacial regions as sediment sources. Thus it seems more likely that suspended sediment is derived *locally* in response to an increase in water velocity, rather than being transported from an upstream source. In this instance, the data imply that diurnal pressure and velocity peaks do not coincide; basal water velocity is minimum in the early afternoon and peak velocity is achieved in the early to mid-morning, during periods of declining subglacial water pressure.

Another possibility is that conditions regulating flow in subglacial passageways might fluctuate daily in response to changes in pressure and water throughput. Consider the following: if the cross-sectional areas and hydraulic properties of drainage passageways were unchanged throughout a daily cycle, declining pressure would indicate lowering of the water surface of an upstream reservoir. However, lowering the water surface of such a reservoir would reduce the hydraulic gradient driving flow through the system, resulting in a decrease in the basal water velocity and a reduction in turbidity. The fact that turbidity continues to increase during periods of decreasing pressure suggests, instead, that subglacial passageways — including outlet portals — are bing enlarged at these times. Furthermore, the enlargement process involves, to some extent, the erosion and transport of basal sediment; it is not due solely to the melting of ice surfaces. Periods of decreasing turbidity correspond to baseflow recession; drainage passageways shrink at these times, presumably due to the inward creep of sediment while the basal water pressure is low. Because the constricted passageways cannot immediately accommodate the mid-day arrival of surface meltwater, basal water pressure increases abruptly and the cycle is repeated.

Subglacial water provenance and drainage system reorganization

During the 1990 field season, a dramatic and sudden rearrangement of the basal drainage system took place. The event spontaneously occurred just before midnight on 22 July and was recorded by subglacial sensors located throughout a central part of the glacier. The timing of subglacial events was not uniform throughout the region in which the sensors were located. Figure 4 shows that pressure sensors in different locations responded in similar ways, but at different times. In general, effects of the event appears to have propagated obliquely down-glacier, away from an icefall, in a south-easterly direction. Before the event, it is likely that only two of the pressure sensors shown in Figure 4 were well connected with each other: P1 and P2. Taking elevation differences into account, records from these two sensors indicate that basal water was 'ponded' in this region for much of the field season. Distinct incongruities between other pressure records suggest that, before the event, nearby sensors were hydraulically isolated from each other. Passage of the event temporarily broke down both the ponded conditions in the vicinity of P1 and P2 and the hydraulic barriers that isolated other nearby sensors.

At a particular location, the event onset was signalled by an initial decrease in water pressure (Figure 4). The preliminary reductions in basal water pressure were followed by rapid increases to super-flotation

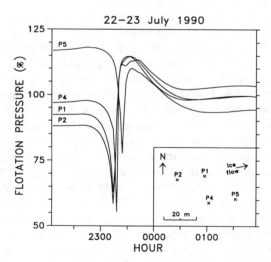

Figure 4. Fluctuations in subglacial water pressure recorded by four sensors during the 1990 'event'. The sensors were located in a central part of the glacier, approximately 500 m above the terminus; the relative positions of the sensors are shown in the inset. Arrival of the event was signalled at a particular location by a preliminary decrease in basal water pressure; this was immediately followed by a sudden increase to super-flotation pressures. From the timings of the pressure drops, it is apparent that effects of the event propagated obliquely down-glacier in a south-easterly direction. For several days before the event, pressure records from sensors P1 and P2 showed nearly identical fluctuations, indicating that these two sensors were in strong hydraulic communication. The other two sensors showed dissimilar behaviours — both between each other and between P1 and P2 — suggesting that they were in hydraulic isolation

values, which then levelled off to near-flotation pressures. Kamb and Engelhardt (1987) reported a similar pattern of water pressure variations during the 1978–1981 mini-surges of Variegated Glacier. The preliminary pressure drops during the mini-surges of Variegated Glacier were attributed to the enlargements of cavities ('basal cavitation') containing subglacial water: 'A preliminary (ice) velocity increase shortly before the pressure wave arrives is caused by the forward shove that the main accelerated mass exerts on the ice ahead of it, and the resulting preliminary basal cavitation causes the drop in water pressure shortly before the pressure wave arrives' (Kamb and Engelhardt, 1987: 27). In the case of Trapridge Glacier, the 1990 event was probably not associated with mini-surge activity. Basal water quality measurements support this view and provide evidence for an alternate explanation of the pressure fluctuations, which we now describe.

Figure 5 shows typical signals from subglacial water pressure, turbidity and electrical conductivity sensors during the event. The pressure and turbidity data in this figure were collected using two sensors which had been installed in the same borehole (the pressure sensor is P4 in Figure 4); conductivity data were obtained from a sensor that was located approximately 20 m away (near P1 in Figure 4). This conductivity sensor was originally installed in 1989 and had been in place at the glacier bed for a full year. (The data shown in Figure 2 were also obtained using this sensor.) At the beginning of the 1990 field season, the lead wires to this conductivity sensor were uncovered at the glacier surface and the sensor was reattached to a datalogger.

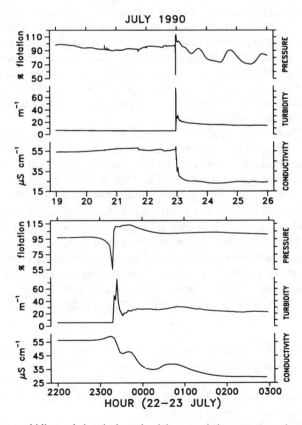

Figure 5. Subglacial water pressure, turbidity and electrical conductivity recorded over a seven-day period in July 1990 (upper panel). The large excursions just before midnight on 22 July are associated with the 1990 'event', during which a major reorganization of subglacial drainage passageways took place. Passage of the event is shown on expanded time-scales in the lower graphs. Note that there was no significant variation in turbidity nor conductivity accompanying the preliminary pressure drop; the turbidity increase and coinciding conductivity decrease began precisely when the pressure suddenly rose to super-flotation values

For several days before the 1990 event, turbidity and electrical conductivity measurements showed little variability and there were no indications of diurnal fluctuations in basal water pressure (Figure 5, top). A large turbidity spike was recorded during the event, and over the following days the turbidity of subglacial water gradually diminished. Measurements of electrical conductivity showed a substantial decrease during the event. The expanded data in the lower half of Figure 5 show that the turbidity and conductivity fluctuations lag the initial pressure response; turbidity increased and conductivity began to decrease precisely when the water pressure suddenly increased, not during the preliminary pressure drop. The fact that the preliminary pressure drop occurred without any accompanying variations in turbidity and conductivity suggest that ice motion preceded the hydrological disturbance at a given location. (The nature of the pressure signals preclude the possibility that the event was initiated by a rupture downstream from the location of the sensors; pressure increases would not be expected following a simple downstream rupture.) Furthermore, the decrease in electrical conductivity indicates the presence of fresh meltwater at the bed, whereas the coinciding abrupt increase in turbidity suggests that subglacial sediments were stirred up locally when hydraulic barriers ruptured and water velocity suddenly increased.

Based on these observations, we infer that the 1990 event was initiated by a sudden influx of surface melt-water, probably from the draining of a water-filled crevasse. (An icefall to the north-west of the study area is known to contain many water-filled crevasses.) The meltwater source supplied an enlarging body of basal water and maintained it at super-flotation pressures. Excessive pressures over a sizable region of the bed caused hydraulic lifting of the glacier, thereby reducing the ice overburden stress at surrounding locations. Hence the preliminary pressure drops resulted from decreases in the local ice pressure, not from basal cavitation. The subsequent increases to super-flotation pressures suddenly occurred when hydraulic barriers separating the expanding water body from individual sensors were breached.

This interpretation is also consistent with the observation that turbidity began increasing immediately as the pressure increased: strong hydraulic gradients, inducing water flow, would have developed as soon as connections were made between the water body and individual water pockets surrounding the sensors. Thus water flow and turbidity increases would be expected only after hydraulic barriers were breached, not during the initial stages of decreasing pressure.

Our interpretation of the 1990 event suggests that a sudden oversupply of meltwater to the bed resulted in a coupled mechanical and hydrological response. Combined basal water quality and pressure measurements indicate that the response involved: (1) the reorganization of subglacial drainage passageways; (2) sediment mobilization; (3) subglacial transfer of surface meltwater; and (4) mechanical jacking of the glacier in advance of a spreading boundary between hydraulically well connected and poorly connected regions. Initiation of water flow at a particular location coincided with the opening of downstream connections. The event terminated when the advancing hydraulic boundary established an open outlet at the glacier terminus. (This is confirmed by the sudden appearance of dyed basal water in the proglacial stream.) The opening of a terminal outlet enhanced water flow through newly connected basal regions and created a temporarily widespread drainage network, which is evidenced by the subsequent development of diurnal pressure fluctuations. The newly created drainage system collapsed as the surface and stored water supplies diminished, and after a few days sensors again became hydraulically isolated.

Summary of characteristics

Based on the two examples that we have considered, we can summarize the characteristic aspects and behavioural patterns of the drainage system beneath Trapridge Glacier.

1. Areas of the bed at least up to 30 m apart can be hydraulically well connected with each other, although hydraulically isolated regions 10–20 m apart can also exist.
2. The subglacial hydraulic conditions of a particular region — isolated or well connected to the basal drainage system — vary with time in response to changes in water pressure, water throughput and mechanical loading.
3. Subglacial water flow speeds through well connected regions are probably on the order of 40–$50 \, \mathrm{m \, h^{-1}}$.

4. The effective hydraulic conductivity of a well established drainage network could be as large as $0.03 \, \text{m} \, \text{s}^{-1}$.
5. Basal water velocity is maximum in the early to mid-morning and is minimum in the early afternoon.
6. During the summer melt season, subglacial drainage passageways can cycle, on a daily basis, between periods of enlargement — when hydraulic resistance is reduced by the erosion and transport of basal sediment — and shrinkage. Passageways presumably grow smaller due to the inward creep of sediment during periods of low basal water pressure.
7. Sudden oversupplies of meltwater to the bed can cause reorganization of subglacial water flow paths, leading to the creation of efficient, yet unstable, drainage networks. Large increases in turbidity accompanying such oversupply events indicate that mobilization of basal sediment can be involved in the reorganization of drainage passageways.

CONCLUSIONS

Direct measurements of basal water quality offer a potentially rich store of information which is unavailable from pressure data alone. Characteristic aspects of subglacial drainage systems — basal water velocity, subglacial water residence time, spatial and temporal variations in drainage system morphologies — can be deduced using basal water quality measurements.

Examples demonstrating the usage of such measurements have revealed complex behaviour of the drainage system at the bed of Trapridge Glacier; the added constraints imposed by basal water quality data are key ingredients in identifying the nature of the water flow system beneath this glacier. The measurement and interpretation methods that we have described and demonstrated are applicable to other glaciers. As such, they should prove useful for characterizing different subglacial drainage configurations and behaviours, thereby improving general understanding of the hydrology and dynamics of wet-based glaciers.

ACKNOWLEDGEMENTS

This research was funded by the Natural Sciences and Engineering Research Council of Canada and by the US National Science Foundation, grant DPP92-24244. The field work was carried out in Kluane National Park with permission from Parks Canada and the Yukon Territory Government. Logistical support for the field programme was provided by Kluane Lake Research Station, owned and operated by the Arctic Institute of North America, University of Calgary. David Bahr and an anonymous reviewer made helpful comments on the manuscript.

REFERENCES

Blake, E. W. 1992. 'The deforming bed beneath a surge-type glacier: measurement of mechanical and electrical properties', *PhD Thesis*, University of British Columbia.
Brown, G. H., Sharp, M. J., Tranter, M., Gurnell, A. M., and Neinow, P. N. 1994. 'The impact of post-mixing chemical reactions on the major ion chemistry of bulk meltwaters draining the Haut Glacier D'Arolla, Valais, Switzerland', *Hydrol. Process.*, **8**, 465–480.
Campbell Scientific 1993. *CR10 Measurement and Control Module Operator's Manual* Campbell Scientific, Logan. 192 pp.
Clarke, G. K C. and Blake, E. W. 1991. 'Geometric and thermal evolution of a surge-type glacier in its quiescent state: Trapridge Glacier, Yukon Territory, Canada, 1969–89', *J. Glaciol.*, **37**, 158–169.
Collins, D. N. 1977. 'Hydrology of an alpine glacier as indicated by the chemical composition of meltwater', *Z. Gletscherk. Glazialgeol.*, **13**, 219–238.
Collins, D. N. 1979. 'Sediment concentration in melt waters as an indicator of erosion processes beneath an alpine glacier', *J. Glaciol.*, **23**, 247–257.
Collins, D. N. 1981. 'Seasonal variation of solute concentration in melt waters draining from an alpine glacier', *Ann. Glaciol.*, **2**, 11–16.
Engelhardt, H. 1978, 'Water in glaciers: observations and theory of the behaviour of water levels in boreholes', *Z. Gletscherk. Glazialgeol.*, **14**, 35–60.
Engelhardt, H., Humphrey, N., and Kamb, B. 1990a. 'Borehole geophysical observations on Ice Stream B, Antarctica', *Antarct. J. US.*, **25**, 80–82.
Engelhardt, H., Humphrey, N., Kamb, B., and Fahnestock, M. 1990b. 'Physical conditions at the base of a fast moving Antarctic ice stream', *Science*, **248**, 57–59.

Fenn, C. R. 1987. 'Electrical conductivity' in Gurnell, A. M. and Clark, M. J. (Eds), *Glacio-Fluvial Sediment Transfer: an Alpine Perspective*. Wiley, Chichester. pp. 377–414.

Fountain, A. G. 1993. 'Geometry and flow conditions of subglacial water at South Cascade Glacier, Washington State, U.S.A.; an analysis of tracer injections', *J. Glaciol.*, **39**, 143–156.

Fountain, A. G. 1994. 'Borehole water-level variations and implications for the subglacial hydraulics of South Cascade Glacer, Washington State, U.S.A.', *J. Glaciol.*, **40**, 293–304.

Hach, C. C., Vanous, R. D., and Heer, J. M. 1990. 'Understanding turbidity measurement', *Technical Information Series — Booklet No. 11*. Hach, Loveland. 11 pp.

Hantz, D. and Lliboutry, L. 1983. 'Waterways, ice permeability at depth, and water pressures at glacier d'Argentière, French Alps', *J. Glaciol.*, **29**, 227–239.

Hem, J. D. 1985. 'Study and interpretation of the chemical characteristics of natural water', *US Geol. Surv. Water Supply Pap.*, **2254**, 263 pp.

Hodge, S. M. 1976. 'Direct measurement of basal water pressures: a pilot study', *J. Glaciol.*, **16**, 205–218.

Hodge, S. M. 1979. 'Direct measurement of basal water pressures: progress and problems', *J. Glaciol.*, **23**, 309–319.

Hooke, R. L., Calla, P., Holmlund, P., Nilssson, M., and Stroeven, A. 1989. 'A 3 year record of seasonal variations in surface velocity, Storglaciären, Sweden', *J. Glaciol.*, **35**, 235–247.

Humphrey, N., Raymond, C., and Harrison, W. 1986. 'Discharges of turbid water during mini-surges of Variegated Glacier, Alaska, U.S.A.', *J. Glaciol.*, **32**, 195–207.

Iken, A. 1972. 'Measurements of water pressure in moulins as part of a movement study of the White Glacier, Axel Heiberg Island, Northwest Territories, Canada', *J. Glaciol.*, **11**, 53–58.

Iken A. and Bindschadler, R. A. 1986. 'Combined measurements of subglacial water pressure and surface velocity of Findelengletscher, Switzerland: conclusions about drainage systems and sliding mechanism', *J. Glaciol.*, **32**, 101–119.

Kamb, B. and Engelhardt, H. 1987. 'Waves of accelerated motion in a glacier approaching surge: the mini-surges of Variegated Glacier, Alaska, U.S.A.', *J. Glaciol.*, **33**, 27–46.

Kamb, B., Raymond, C. F., Harrison, W. D., Engelhardt, H., Echelmeyer, K. A., Humphrey, N., Brugman, M. M., and Pfeffer, T. 1985. 'Glacier surge mechanisms: 1982–83 surge of Variegated Glacier, Alaska', *Science*, **227**, 469–479.

Mathews, W. H. 1964. 'Water pressure under a glacier', *J. Glaciol.*, **5**, 235–240.

Pilgrim, D. H., Huff, D. D., and Steele, T. D. 1979. 'Use of specific conductance and contact time relations for separating flow components in storm runoff', *Wat. Resour. Res.*, **15**, 329–339.

Pohjola, V. A. 1993. 'TV-video observations of bed and basal sliding on Storglaciären, Sweden', *J. Glaciol.*, **39**, 111–118.

Sharp, M., Willis, I., Hubbard, B., Nielsen, M., Brown, G., Tranter, M., and Smart, C. 1993. 'Borehole water quality profiling: explaining water level variations in boreholes' [abstract], *International Workshop on Glacier Hydrology, 8–10 September 1993, Cambridge, UK*.

Stone, D. B. and Clarke, G. K. C. 1993. 'Estimation of subglacial hydraulic properties from induced changes in basal water pressure: a theoretical framework for borehole response tests', *J. Glaciol.*, **39**, 327–340.

Stone, D. B., Clarke, G. K. C., and Blake, E. W. 1993. 'Subglacial measurement of turbidity and electrical conductivity', *J. Glaciol.*, **39**, 415–420.

Tranter, M., Brown, G., Raiswell, R., Sharp, M., and Gurnell, A. 1993. 'A conceptual model of solute acquisition by Alpine glacial meltwaters', *J. Glaciol.*, **39**, 573–581.

Walder, J. S. and Fowler, A. 1994. 'Channelised subglacial drainage over a deformable bed', *J. Glaciol.*, **40**, 3–15.

14

VARIABILITY IN THE CHEMICAL COMPOSITION OF *IN SITU* SUBGLACIAL MELTWATERS

M. TRANTER,[1] M. J. SHARP,[2] G. H. BROWN,[3] I. C. WILLIS,[4] B. P. HUBBARD,[3]
M. K. NIELSEN,[4] C. C. SMART,[5] S. GORDON,[1,2] M. TULLEY,[4]
AND H. R. LAMB[1]

[1]*Department of Geography, University of Bristol, Bristol BS8 1SS, UK*
[2]*Department of Earth and Atmospheric Sciences, University of Alberta, Edmonton, Canada T6G 2E3*
[3]*Centre of Glaciology, Institute for Earth Studies, The University of Wales, Aberystwyth SY23 3DB, UK*
[4]*Department of Geography, University of Cambridge, Cambridge CB2 3EN, UK*
[5]*Department of Geography, University of Western Ontario, London, Ontario, Canada N6A 5C2*

ABSTRACT

Meltwaters collected from boreholes drilled to the base of the Haut Glacier d'Arolla, Switzerland have chemical compositions that can be classified into three main groups. The first group is dilute, whereas the second group is similar to, though generally less concentrated in major ions, than contemporaneous bulk glacial runoff. The third group is more concentrated than any observed bulk runoff, including periods of flow recession. Waters of the first group are believed to represent supraglacial meltwater and ice melted during drilling. Limited solutes may be derived from interactions with debris in the borehole. The spatial pattern of borehole water levels and borehole water column stratification, combined with the chemical composition of the different groups, suggest that the second group represent samples of subglacial waters that exchange with channel water on a diurnal basis, and that the third group represent samples of water draining through a 'distributed' subglacial hydraulic system. High NO_3^- concentrations in the third group suggest that snowmelt may provide a significant proportion of the waters and that the residence time of the waters at the bed in this particular section of the distributed system is of the order of a few months. The high NO_3^- concentrations also suggest that some snowmelt is routed along different subglacial flowpaths to those used by icemelt. The average SO_4^{2-} : $(HCO_3^- + SO_4^{2-})$ ratio of the third group of meltwaters is 0·3, suggesting that sulphide oxidation and carbonate dissolution (which gives rise to a ratio of 0·5) cannot provide all the HCO_3^- to solution. Hence, carbonate hydrolysis may be occurring before sulphide oxidation, or there may be subglacial sources of CO_2, perhaps arising from microbial oxidation of organic C in bedrock, air bubbles in glacier ice or pockets of air trapped in subglacial cavities. The channel marginal zone is identified as an area that may influence the composition of bulk meltwater during periods of recession flow and low diurnal discharge regimes.

INTRODUCTION

The chemical composition of subglacial meltwaters has hitherto been determined by indirect methods rather than from *in situ* measurements. In particular, the association between discharge and either electrical conductivity (EC) (Collins, 1978) or SO_4^{2-} (Tranter and Raiswell, 1991) in bulk glacial runoff has been used to establish the quasi-constant composition of two types of subglacial waters that are believed to have different flow paths through subglacial drainage systems. Assigning quasi-constant compositions to subglacial waters is fraught with uncertainty (Sharp *et al.*, 1995). Hot water drilling has allowed access to the glacier bed for some time (Hodge, 1976; Englehardt, 1978; Iken and Binschadler, 1986) and the design of new sampling and monitoring equipment enables direct sampling of *in situ* subglacial waters (Blake and Clarke, 1991; Stone *et al.*, 1993; Sharp *et al.*, 1993a; Nielsen, Unpublished data; Lamb *et al.*, 1995). To date, there has been no thorough investigation into the EC or chemistry of waters in subglacial drainage systems, although basal waters with EC in the range of ~15–75 μS/cm have been reported from Trapridge Glacier, Yukon Territory, Canada (Stone *et al.*, 1993; Stone and Clark, in press). These waters exhibit variability in turbidity and conductivity on the time-scale of hours. This paper is one of a series that will document an integrated, multi-disciplinary study of the subglacial hydrology of the Haut Glacier d'Arolla by means of borehole monitoring, sampling and experimentation. Here our intent is to present the results of studies of the EC and the chemistry of *in situ* subglacial waters. Our aim is to demonstrate that subglacial hydrochemistry is more complex than hitherto appreciated from studies of bulk runoff and that there are strong links between hydrological flow paths and geochemical weathering environments at the bed.

The chemical composition of glacial meltwaters is dependent on both the source of the melt and the flow paths that the melt follows. In particular, subglacial flow paths are important controls on meltwater chemistry (Tranter *et al.*, 1993). Subglacial drainage systems often consist of two different components (e.g. Humphrey *et al.*, 1986; Fountain, 1994; Humphrey and Raymond, 1995). The subglacial drainage system beneath the Haut Glacier d'Arolla is also believed to consist of two components, a 'distributed' subglacial hydraulic (or slow transit) system and a channelized (or rapid transit) system (Nienow *et al.*, in press). Analysis of discharge–concentration associations in bulk glacial runoff (Tranter *et al.*, 1993), in combination with the results of dye tracing experiments (Nienow *et al.*, in press), suggests the following model of solute acquisition in subglacial environments. A 'subglacial hydraulic' distributed system underlies snow-covered ice. Relatively long residence times, of the order of tens of hours, and high concentrations of freshly comminuted debris in areas occupied by the distributed drainage system promote solute acquisition that is dominated by carbonate dissolution and sulphide oxidation. Waters drain from the distributed system into the channelized drainage system, and are termed the delayed flow component of bulk runoff. The channelized drainage system is also fed by icemelt, routed rapidly to the bed from supraglacial streams via crevasses and moulins (Sharp *et al.*, 1993b). This rapidly routed icemelt is termed the quickflow component of bulk runoff. Reactive sulphides are largely exhausted by chemical weathering in the distributed system, hence further solute acquisition in the channelized system is via carbonation reactions. These so-called post-mixing reactions provide the bulk of the solute found in dilute runoff at the height of the ablation season (Brown *et al.*, 1994a), when transit times through the channelized system are of the order of hours.

Hubbard *et al.* (1995) suggest that major subglacial channels are flanked by a channel marginal zone that may be up to 70 m wide. The channel marginal zone is believed to consist of a vertically confined layer of sediment whose hydraulic conductivity approaches that of fine sand near the channel margin and glacial till near the transition zone to the distributed system. The channel marginal zone exchanges water on a diurnal basis with the channel, storing water during the ascending limb of the hydrograph and releasing water during the descending limb. A chemical signature for chemical weathering reactions in this zone has not yet been identified, although the influence of the zone is likely to diminish over the ablation season as fines are eluted from the zone.

If the above hydrochemical model is correct, we may anticipate that at least three types of subglacial meltwater chemistry exist. First, there may be dilute waters corresponding to snowmelt and icemelt that have yet to interact with debris. Second, there may be channelized flow with a chemical composition resembling that of bulk runoff and, third, there may be concentrated waters corresponding to those draining through the distributed system. Finally, there may be waters in the channel marginal zone that may contain characteristics of channel water and chemical weathering in the zone. It is unlikely that these three or four water types exist as discrete types, as the progressive chemical evolution of the subglacial meltwaters depends on factors such as the water to rock ratio, access to reactive minerals, the duration of water–rock contact and the availability of protons for chemical weathering (Raiswell, 1984; Brown *et al.*, 1994a; in press; Tranter *et al.*, in press). These factors are likely to vary within each of the hydrological reservoirs and flow paths identified here and are likely to fluctuate on a diurnal and/or seasonal basis as the different flow paths evolve as the ablation season progresses.

STUDY SITE

The Haut Glacier d'Arolla is located at the head of Val d'Hérens, Switzerland and has an area of ~6·3 km^2.

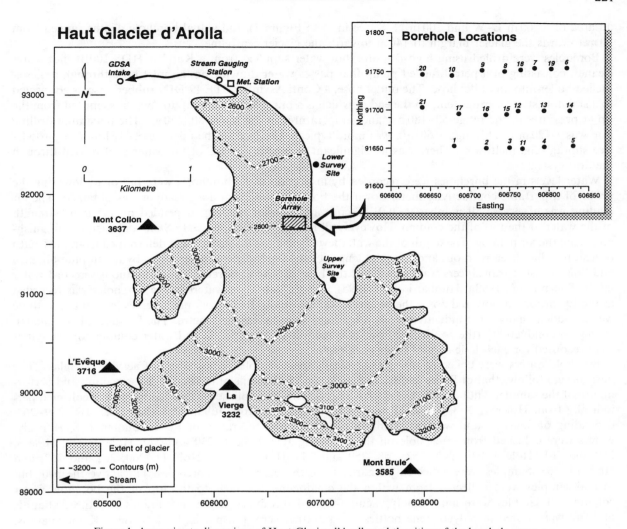

Figure 1. Approximate dimensions of Haut Glacier d'Arolla and the siting of the borehole array

The main ice tongue descends to the snout at 2560 m and is fed from a compound basin of maximum elevation 3838 m (see Figure 1). The glacier has been in retreat during recent years (by 720 m during 1967–1989). The glacier is probably warm-based and has a maximum ice thickness of 180 m (Sharp *et al.*, 1993b). The catchment is underlain by schists and gneisses of the Arolla series and by amphibolite. The bedrock contains trace amounts of geochemically reactive minerals such as carbonates and sulphides (Brown 1991). Albite, anorthite, microcline and sanidine are the main feldspars and diopside, enstatite and spodumene the main pyroxenes. Actinolite (an amphibole), muscovite (a mica), cordierite, haematite, hydrobasaluminite, magnetite, quartz and talc are also found (D. Webb, pers. comm., 1995).

METHODS

Hot water drilling was undertaken during August and September 1992 on the eastern side of the main body of the glacier at an altitude of ~2800 m (see Figure 1). The site was chosen on the basis of a theoretical reconstruction of the subglacial drainage system that predicted a major subglacial channel in this vicinity (Sharp *et al.*, 1993b). About 20 boreholes were drilled to the bed in an array of 3 × E–W transects, covering an area of ~200 m E–W and 100 m N–S (see inset to Figure 1). Individual borehole depths ranged from 50 m towards the glacier margin to 140 m towards the glacier centreline.

Boreholes were drilled using high-pressure, hot water supplied from a Kärcher HDS 100BE hot water cleaner, consisting of a petrol-fuelled pump that passes water drawn from an inlet hose through a diesel-fuelled boiler into an outlet hose. The outlet hose, a Conti Asymflex 3TE 12 DIN rubber hose, is connected to a stainless-steel steering stem, on the end of which is a brass drill head and tip. Water is emitted from the tip at pressures in the range 50–180 bar and temperatures in the range 80–150°C. The maximum drilling rate was ~120 m/h, although ~60 m/h was more typical. The feed to the inlet was > 15 l/min of particle-free supraglacial meltwater. There was no significant change in the EC of the water as it passed through the drill system.

Water levels in the boreholes were recorded by inspection if the boreholes were full or by lowering the probe of a portable conductivity meter down the borehole until a non-air reading was obtained. The air reading was a very small number ($\approx 0.01 \mu$S/cm), and non-air readings depended on the ionic strength of the water at the top of the column. However, all non-air values were $> 1 \mu$S/cm and easily distinguishable from the air reading. The depth of the water level below the ice surface was determined from the length of lead that had been paid out and the total length of the borehole was determined by depth plumbing. In addition, pressure transducers were located at the base of several boreholes to continuously record water levels (Sharp *et al.*, 1993a; Hubbard *et al.*, 1995). The EC profile of waters in the boreholes was recorded as the EC probe was lowered down the borehole water column. Readings were taken every 5 m unless there was a sudden change in conductivity, when readings were taken each metre. The location of the sudden change in conductivity (the condocline) was also recorded. Water levels and water column conductivities were recorded for each hole several times each day.

Borehole waters were collected using a chemically inert sampler, constructed of acrylic Plexiglass. The basic design follows that of Blake and Clarke (1991) and full details of the design, construction and performance of the sampler will be published elsewhere (Nielsen, Unpublished data). Samples were collected periodically from Holes 1, 2, 3, 4, 10, 11 and 14 (see Figure 1) from Julian Day (hereafter, JD) 239–263, depending on logistics and weather conditions, and covered the range of EC encountered. Specifically, waters were collected from each hole on the following JDs: Hole 1, 239 and 242; Hole 2, 239; Hole 3, 260 and 261; Hole 4, 239, 242, 252, 258, 259 and 263; Hole 10, 263; Hole 11, 260, 261 and 263; and Hole 14, 263. Samples were collected primarily from the base of the boreholes, 1 m above the bed, but several samples were collected from within the condocline and near to the top of the borehole water column. All samples were immediately vacuum-filtered through 0.45 μm membranes and three aliquots of 30 ml were stored in polyethylene bottles. Two bottles were transported back to the UK for analysis; one was used for cation and total dissolved silicon (hereafter, Si) analysis and another was used for anion and alkalinity determinations. The third was stored for the determination of pH in the field laboratory up to 4 h later. In addition, an unfiltered, 30-ml aliquot was stored in an HDPE bottle for determination of O_2 up to 4 h later using an Orion 97-08 oxygen electrode connected to an Orion model 290A pH/ISE meter (Brown *et al.*, 1994b).

From JD 232–264 1992, bulk runoff was sampled twice daily at 10·00 and 17·00 h, when discharge approximated the daily minimum and maximum respectively, ~ 100 m from the portal (see Figure 1). Samples were collected and treated as described elsewhere (Gurnell *et al.*, 1994).

Discharge was continuously monitored from May to September at a weir operated by Grande Dixence SA, about 1 km from the portal (see Figure 1). The precision and accuracy of these structures is $\approx 5\%$. The weir was operational from JD 132 to 252 only because of engineering maintenance. A second gauging station was established ≈ 850 m from the portal and run from JD 232 to 264. The output from a Druck PDCR850 pressure transducer and an in-house rod-cylinder EC probe (measured across an AC half-bridge) was logged by a Campbell Scientific CR-10 micro-logger. This station and the Grand Dixence weir were cross-calibrated for the period JD 244–252 and the calibration equation was applied to the transducer output to produce the hydrograph for the period JD 232–264. The resulting hydrograph is shown in Figure 2.

Water samples were collected from the main supraglacial streams draining the eastern side of the glacier tongue and near the drill site from JD 233 to 259. Samples were collected on most days.

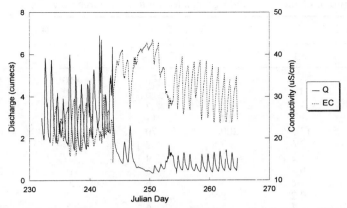

Figure 2. Discharge and electrical conductivity of bulk runoff, Julian Days 232–264 1994

Snow was sampled from snow pits dug near to the glacier snout during November 1992 and January 1993 (45 samples in all) and fresh snowfall was collected at the village of Arolla from JD 298, 1992 to JD 85, 1993 (27 samples in all). Sampling and sample treatment were similar to those used by Tranter *et al.* (1986). Although these samples were collected after the main field campaign, they provide illustrative values for the range of concentrations of ions found in snow on the glacier.

The major cations (Ca^{2+}, Mg^{2+}, Na^+, K^+) were determined by atomic adsorption spectrometry (AAS) using an air–acetylene flame. Spectrochemical buffers [$La(NO_3)_3$ and $CsCl$] were added to overcome chemical interferences and ionization, respectively. Detection limits were $\approx 1\,\mu equiv./l$, and the precision was $\pm 5\%$ at concentrations $> 50\mu equiv./l$. However, at concentrations below $10\,\mu equiv./l$ precision is at best $\pm 30\%$ and at $1\,\mu equiv./l$ it is $\pm 100\%$. Major anions were determined by ion chromatography on a Dionex 2000i or 4000i instrument. Detection limits were $\approx 0\cdot 1\,\mu equiv./l$, and the precision of determinations was $\pm 3\%$ at concentrations $> 50\,\mu equiv./l$. Similar deterioration in precision occurs at lower concentrations, as described for the above cations.

Si was determined by flow injection analysis using an FIAstar 5010 system, consisting of a FIAstar 5023 spectrophotometer, V100 injector and Chemifold III. Samples of 1 ml were injected into a stream of ammonium molybdate, oxalic acid and acidified stannous chloride. The amount of molybdate-reactive Si was determined spectrophotometrically at a wavelength of 695 nm. The detection limit was $1\,\mu mol/l$ and precision was $\approx \pm 5\%$ at concentrations $> 10\mu mol/l$.

Alkalinity (HCO_3^-) was determined by titration to endpoint $4\cdot 5$ using $0\cdot 01$ M HCl and BDH $4\cdot 5$ indicator. The precision of measurement is dependent on the titre volume and ranges from $\pm 1\%$ at higher values to $\pm 50\%$ at low values. The HCO_3^- concentration was determined from the alkalinity after correcting for the H^+ needed to acidify the combined sample and titre volume to pH $4\cdot 5$.

The pH was determined using a Ross combination research grade electrode connected to an Orion 290A meter, calibrated using Orion Low Ionic Strength Buffers of pH $4\cdot 00$ and $6\cdot 97$. The method of measurement was modified for field conditions from that of McQuaker *et al.* (1983). The precision of measurement was $\pm 0\cdot 2$ pH units.

Charge balance errors (CBE), expressed as percentages, were calculated as follows

$$CBE = \frac{\left(\sum^+ - \sum^-\right)}{\left(\sum^+ + \sum^-\right)} \times 100\% \tag{1}$$

where $\sum^+$ is the sum of cation equivalents and $\sum^-$ is the sum of anion equivalents. The mean CBE values were: bulk runoff, $+4\%$; borehole waters, mode 1, -9%; mode 2, -2%; mode 3, -4%. Definitions of the modes are given later. The CBE values are larger and more variable for dilute solutions because of the analytical uncertainty at low concentrations. The CBE values are not presented for the snow and supraglacial samples for this reason.

Table I. Chemical composition of the three main groups (or modes) of borehole waters

| | Mode 1 | | Mode 2 | | Mode 3 | | Bulk runoff |
	$\bar{x}$	sd	$\bar{x}$	sd	$\bar{x}$	sd	Range
Ca^{2+}	22	25	190	46	810	50	250–560
Mg^{2+}	3·4	4·0	17	5	90	7	26–110
Na^+	3·9	1·9	11	7	15	1	11–55
K^+	1·8	1·5	14	13	16	1	8·6–36
HCO_3^-	27	13	200	70	680	40	220–360
Cl^-	4·0	2·6	2·8	2·4	4·7	1·3	0·5–14
NO_3^-	0·8	1·2	2·5	4·1	27	1	0·0–8·5
SO_4^{2-}	8·0	10	35	16	310	22	46–220
Si	3·6	1·8	15	6	25	4	7·1–27
pH	6·4	0·2	8·9	1·2	7·5	0·2	7·0–8·9
O_2	6·9	1·1	7·2	0·2	7·7	1·1	3·4–8·7
$p(CO_2)$	−3·4	0·3	−4·2	1·3	−3·0	0·1	−2·8 to 4·9
EC_0	(1·1)		12		64		16–43*
n	11		12		23		66

$\bar{x}$ and sd, mean and standard deviation, respectively.
* these are measured, and not calculated, EC.
EC_0 is a derived value [see Equation (2)] and is not accurate for mode 1 waters.
Units are μequiv./l, except for Si (μmol/l), pH, O_2 (ppm) and EC (μS/cm).

The logarithm of the partial pressure of CO_2 with which the solution is in apparent equilibrium, hereafter $p(CO_2)$, is defined and calculated as

$$p(CO_2) = \log_{10}\{(HCO_3^-)(H^+)/(K_H^* K_1)\} = \log_{10}\{(HCO_3^-)\} - pH + 1·7 \tag{2}$$

where units of HCO_3^- are μmol/l, $K_H = 10^{-1.12}$ mol atm^{-1} and $K_1 = 10^{-6.58}$ mol l^{-1} at 0°C (Garrels and Christ, 1965).

RESULTS

We first describe the chemical composition of bulk runoff, then document the chemical composition of the

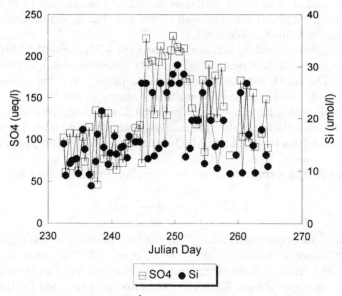

Figure 3. Variation of SO_4^{2-} and Si concentrations in bulk runoff

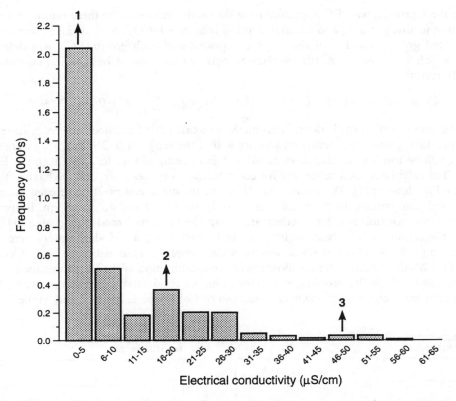

Figure 4. Histogram of the observed electrical conductivity of borehole waters measured during borehole profiling

different borehole waters encountered during sampling and conclude with an account of where these different waters are found in individual borehole water columns and throughout the borehole array.

Figure 2 shows the discharge and conductivity of the bulk runoff during the sampling period. High discharges (up to $7 \, m^3/s$) were recorded from JD 232–244, with large diurnal amplitudes of up to $4 \, m^3/s$. A snowfall and associated sub-zero temperatures during JD 244–245 brought about a period of recession flow lasting until JD 249, with a minimum recorded discharge of $\sim 0.5 m^3/s$. Thereafter the discharge increased and weak diurnal discharge cycles recommenced. As is common at other Alpine glaciers, the EC and discharge are inversely associated on both diurnal and seasonal timescales (Fenn, 1987; Röthlisberger and Lang, 1987). The range of EC recorded was from $\sim 15 \, \mu S/cm$ at maximum discharge to $\sim 43 \, \mu S/cm$ during the period of recession flow (see Table I).

Figure 3 shows the variation in the concentration of two relatively conservative species (see below), SO_4^{2-} and Si, in bulk runoff. The concentrations of both species have an inverse diurnal association with discharge, and both increase during the period of recession flow after JD 245. It is evident that SO_4^{2-} is enriched relative to Si in runoff during and following the recession flow. The range of concentrations encountered is 46–220 μequiv./l for SO_4^{2-} and 10–30 μmol/l for Si.

Figure 4 shows a histogram of EC measurements made during the EC profiling of the boreholes. Approximately 4000 measurements are included. There are three populations of EC with modes at 0–10, 11–30 and 46–55 μS/cm (hereafter designated Modes 1, 2 and 3, respectively). The waters with the lowest EC (0–10 μS/cm) have similar EC values to supraglacial runoff (3–4 μS/cm). The waters of ntermediate EC (11–30 μS/cm) were turbid, as revealed by borehole water sampling, and overlap with the EC of bulk glacial runoff (16–43 μS/cm). The waters of highest EC (46–55 μS/cm), which are largely free of suspended sediment, are more concentrated than the bulk runoff during the period of recession flow.

Table I shows the mean chemical composition of waters sampled from the boreholes. The waters have been split into three groups, with EC approximating the modal values of the three groups described earlier. The major cation is always Ca^{2+}, and the major anion is always HCO_3^-. SO_4^{2-} is also a major anion in the most concentrated group. The EC of the mean composition of each group, EC_0, was determined from Equation (1), which is derived from the association between the sum of base cation equivalents and the EC of the bulk runoff

$$Log_{10}(EC) = -1\cdot73(\pm0\cdot05) + 1\cdot19(\pm0\cdot07)log_{10}\left(\sum{}^+\right) \quad r^2 = 0\cdot84(n = 54) \tag{3}$$

where $\sum{}^+$ is the sum of cation equivalents. This method of calculating EC does not necessitate temperature correction of the data, as all water temperatures are within the range of 0–$2°C$ (we are aware that the EC_0 of dilute water will be too low, as the effect of pH < 6 may be to add up to $2\,\mu S/cm$ to the EC_0 of supra-glacial runoff. The calculated EC_0 values are for comparative purposes only, and we have left our low EC_0 data uncorrected for lower pH). The values calculated for the mean composition of each group are $1\cdot1$, 13 and $64\,\mu S/cm$. This compares with the modal ranges of 0–10, 11–30 and 46–$55\,\mu S/cm$. It appears that the sampled Mode 3 composition is more concentrated than the measured modal EC value (64 cf. 46–$55\,\mu S/cm$). However, the conductivity of basal water in Hole 4, from which all Mode 3 waters were sampled, was typically in the range 50–$60\,\mu S/cm$ at times near to water sampling. Thus either we have overestimated the conductivity of the Mode 3 mean water composition or cross-calibration of the EC probes used for bulk runoff EC monitoring and borehole EC profiling is in error at higher conductivities. In summary, we are confident that we have presented representative chemical compositions of Mode 1, Mode 2 and Mode 3 waters.

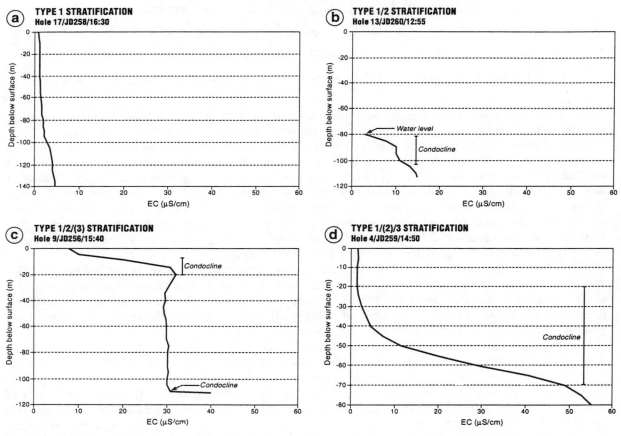

Figure 5. Examples of the types of conductivity stratification found in the borehole water columns. (a) Type 1 stratification; (b) Type 1/2 stratification; (c) Type 1/2/(3) stratification; and (d) Type 1/(2)/3 stratification

Table II. Mean percentage of each mode measured within individual boreholes and the stratification type of each borehole

Hole	% Mode 1	% Mode 2	% Mode 3	Stratification
1	15	81	3·9	1/2/(3)
2	100			1
3	70	30		1/2
4	48	19	33	1/(2)/3
5	59	37	3·6	1/2/(3)
6	100			1
7	70	30		1/2
8	100			1
9	6·0	89	5·4	1/2/(3)
10	79	21		1/2
11	58	43		1/2
12	95	5·3		1/2
13	61	39		1/2
14	92	7·6		1/2
15	79	3·0	19	1/(2)/3
16	99	1·3		1
17	100			1
18	80	20		1/2
19	63	38		1/2
20	100			1

The stratification of waters in each hole conformed to one of the following four patterns (see Figure 5):

1. Type 1 stratification denotes a borehole water column containing Mode 1 (i.e. very dilute) waters only.
2. Type1/2 stratification denotes a borehole water column in which Mode 1 water overlies Mode 2 water. The junction between the waters is marked by a rapid change in conductivity with depth, which we refer to as the condocline.
3. Type 1/2/(3) stratification denotes a borehole water column in which Mode 1 water usually overlies Mode 2 water, but where Mode 3 water may enter the bottom of the holes occasionally and lie beneath the Mode 2 water.
4. Type 1/(2)/3 stratification denotes a borehole water column in which Mode 1 water overlies Mode 3 water, with water of intermediate composition (often approximating the composition of Mode 2 water) found in the condocline.

It is evident from Figure 5 that the condocline is usually > 10 m thick and that its boundaries are gradational. It is also evident that Mode 3 waters need to be sampled at the maximum depth that is practical to obtain representative samples of these waters.

Table II shows that of the 20 holes routinely monitored, six contained Type 1 and nine contained Type1/2 stratification. Three exhibit Type 1/2/(3) stratification and the remaining two contained Type 1/(2)/3 stratification.

DISCUSSION

The chemical composition of waters draining through Alpine glaciers is dependent on the flow paths they follow (Tranter *et al.*, 1993; in press). Therefore it is necessary to give a brief account of the hydrology of the boreholes and their association with the subglacial drainage system before discussing the chemistry of the borehole waters. Fuller accounts of borehole hydrology will be published elsewhere (Hubbard *et al.*, 1995; Willis *et al.*, Unpublished data).

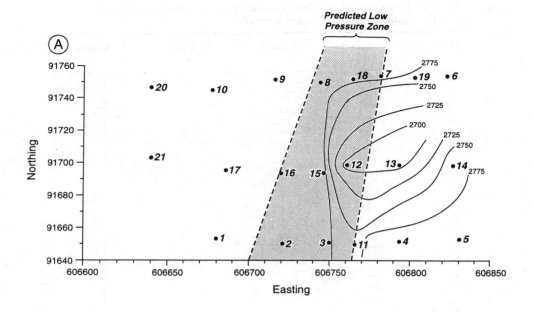

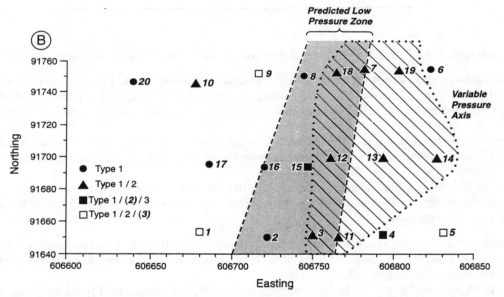

Figure 6. Location of the low pressure axis as determined from the piezometric surface (expressed as metres above sea level) in bore-holes and the predicted location of the low pressure axis (Sharp *et al.*, 1993a). The location of the low pressure axis is also the location of maximum diurnal variations in water levels and is therefore called the variable pressure axis (Hubbard *et al.*, 1995). (b) Distribution of borehole stratification in relation to the position of the variable pressure axis

Borehole 'connections' to the subglacial hydrological system

Boreholes may fill from the top (supraglacial input), the sides (englacial input; Pohjola, 1994) or from the bottom (subglacial input). They may drain from the bottom, the sides and the top (if the subglacial water pressure is greater than the hydraulic head of the borehole water column). Borehole water levels indicate, to some extent, the degree to which a hole has connected with the dynamic or variable portion of the sub-glacial hydrological system (Hodge, 1976; Engelhardt, 1978; Fountain, 1994; Waddington and Clarke,

1995). Connections are suggested by low-fluctuating water levels. Falling water levels indicate that outflow from the holes is greater than the inflow and a reasonable assumption is that much of the water is draining from the bottom of the hole into the subglacial hydrological system (although we caution that drainage can also occur from the sizes of boreholes into englacial storage or flowpaths; Pohjola, 1994). Holes 1, 3, 4, 11, 12, 13 and 14 have low and/or fluctuating water levels. The remaining holes were stable at water pressures greater than the local overburden, which is taken to indicate relative isolation from the subglacial hydrological system. The piezometric surface reconstructed from the manual water level measurements for 12·00 h, JD 260 (see Figure 6) clearly indicates a low pressure axis that is also subject to diurnal variations in water level (Hubbard et al., 1995). This variable pressure axis (VPA) is located close to the predicted subglacial channel (Sharp et al., 1993b). The extent of the VPA changes during the ablation season (Willis et al., Unpublished data). Some holes (e.g. 15 and 18) took a relatively long time to connect, whereas others showed intermittent connection and disconnection (e.g. 4 and 14).

Because none of the boreholes drained completely, it is not believed that a connection was made with a major subglacial channel. Rather, it is believed that a marginal zone between the channel and the distributed system has been tapped. This channel marginal zone may consist of a vertically confined sediment layer that is flooded with water from channels during rising discharge and partially drains during falling discharge. It is subjected to repeated diurnal elution of fines (Hubbard et al., 1995). The consequence of the diurnal flooding and elution is that reactive minerals and surface sites are depleted in the marginal zone as the ablation season progresses. These zones may therefore become more geochemically inert as the number of flooding/elution events increases.

Distribution of borehole stratification

Given the existence of the VPA, the stratification of the borehole water columns is easily explained. Figure 6 shows the distribution of stratification types. A clear pattern emerges. Holes with Type 1/2 stratification are strongly clustered in the VPA, wheres holes with Type 1/(2)/3 stratification are generally found along the margin. Those with Type 1/2/(3) stratification are usually sited slightly further away. Holes with Type 1 stratification are 'unconnected' and are generally located at greater distance from the VPA. The occurrence of Mode 2 waters at the base of boreholes in the VPA, and the fact that these waters have similar conductivities to the bulk runoff, strongly suggests that the VPA is a region affected by the presence of a major subglacial channel. Away from the VPA, the higher and more stable water levels suggest the presence of a distributed drainage system or the relative hydrological isolation of these areas of the bed (Hubbard et al., 1995; Stone and Clarke, in press).

Borehole water chemistry

The results presented here are among the first to describe the chemical composition and the variability of subglacial waters (see also Lamb et al., 1995). Two issues will be explored; how each mode acquires solute and how representative are the modes of waters constituting the bulk runoff.

Mode 1 water chemistry. Mode 1 waters fill 'unconnected' holes and cap the water columns of 'connected' holes. They are the most common waters found in the boreholes, given our sampling regime (see Figure 4). Mode 1 waters were sampled from Holes 2, 3, 10, 11 and 14. The most dilute waters were found at the top of the borehole water columns (see Figure 5a). Field observations show that supraglacial runoff drains into boreholes, either directly or via englacial conduits. In addition, supraglacial water used to drill the holes and ice melted during drilling may remain in unconnected holes. All water sources are dilute. There may be some initial turbidity in the water column associated with the drilling (e.g. suspending debris from the bed or melting out debris from basal ice layers) and this suspended sediment may interact with the dilute water to liberate solute. However, the magnitude of solute release is likely to be limited because there is no sustainable source of protons to fuel chemical weathering reactions (Raiswell, 1984).

Table III gives illustrative values for the mean composition of supraglacial runoff derived largely from icemelt. The dominant cation is Ca^{2+}, and there are low (~ 1 μequiv./l) concentrations of the major acid anions (Cl^-, NO_3^- and SO_4^{2-}). The concentrations of these latter ions are usually higher in snow (see Table III). Comparison of the composition of Mode 1 waters (see Table I) with the composition of

Table III. Chemical composition of supraglacial runoff and snow

	Supraglacial runoff		Snow	
	$\bar{x}$	sd	$\bar{x}$	sd
Ca^{2+}	4·9	1·5	8·4	8·3
Mg^{2+}	0·3	0·2	2·7	4·3
Na^+	0·2	0·5	1·3	1·8
K^+	0·2	0·4	0·0	0·0
HCO_3^-	4·2	3·3	*	
Cl^-	1·2	1·3	3·8	3·6
NO_3^-	0·4	1·3	6·9	9·4
SO_4^{2-}	1·3	2·7	6·6	7·2
Si	$< 1^\dagger$		$< 1^\dagger$	
O_2	5·3	0·8	*	
pH	6·0	0·2	5·4	0·4
$p(CO_2)$	−4·0	0·4	*	
EC_0		(0·4)		(0·4)

$\bar{x}$, mean; sd standard deviation.
* Unavailable.
† Only four samples were analysed and the concentrations were at or below the detection limit.
Units are μequiv./l, except for Si (μmol/l), pH, O_2, (ppm) and EC (μS/cm). Ec_0 is a derived value [see Equation (1)] that is not accurate for dilute waters.
The values are not true representations of the mean composition of all waters sampled, as not all samples were analysed for all species. The number of analyses range from: supraglacial runoff, $HCO_3^- = 16$ and $SO_4^{2-} = 57$; snow, $Ca^{2+} = 26$ and $SO_4^{2-} = 70$. Hence the apparent excess of negative charge arises because the mean of the cations and the anions were drawn from populations of different size.

the supraglacial runoff (see Table III) shows that some of the Mode 1 waters are enriched in base cations and HCO_3^-. Where glacial flour is suspended from the bed or melted out of ice, the waters appear to acquire some solute (mainly Ca^{2+} and HCO_3^-) from carbonation reactions such as those illustrated by Equations (4) and (5).

$$CaCO_3(s) + CO_2(aq) + H_2O(aq) \rightleftharpoons Ca^{2+}(aq) + 2HCO_3^-(aq) \qquad (4)$$

calcite

$$CaAl_2Si_2O_8(s) + 2CO_2(aq) + 2H_2O(aq) \rightleftharpoons Ca^{2+}(aq) + 2HCO_3^-(aq) + H_2Al_2Si_2O_8(s) \qquad (5)$$

anorthite (Ca-feldspar)   weathered feldspar surfaces

It is also possible that diffusion of ions from Mode 2 and Mode 3 waters lower in the hole enriches Mode 1 waters. The calculation shown in the Appendix suggests that this is unlikely. It is more likely that the conductivity profiles shown in Figure 5 are affected by turbulent mixing associated with any rise or fall of borehole water levels or the transit of the sampler and EC probe up and down the borehole. Turbulent mixing probably controls the dimensions of the condocline and results in some dilution of Mode 2 and Mode 3 waters in the vicinity of the condocline.

Mode 2 water chemistry. Mode 2 waters are found at the base of the borehole water columns in the VPA, and are also found between Mode 1 and Mode 3 waters in boreholes with Type 1/(2)/3 stratification. We refer to the former waters as Mode 2A waters and the latter as Mode 2B waters.

Mode 2A waters were sampled from Holes 1 and 3. Mode 2A waters are driven into the base of boreholes during periods of increased water pressure in the channel, when waters flood the channel marginal zone (Hubbard *et al.*, 1995). They are therefore likely to have a chemical composition that is similar to that of bulk runoff, minus any solute derived from post-mixing reactions that may occur between the sampling

Table IV. Chemical composition of mode 2A and 2B waters

	Mode 2A		Mode 2B	
	$\bar{x}$	sd	$\bar{x}$	sd
Ca^{2+}	210	1·5	150	17
Mg^{2+}	16	0·2	20	4
Na^+	13	0·5	7·0	3·1
K^+	17	0·4	5·7	1·9
HCO_3^-	220	30	130	3
Cl^-	2·6	1·3	3·1	2·3
NO_3^-	1·9	1·3	3·8	5·4
SO_4^{2-}	31	2·7	41	12
Si	15	4	11	6
pH	8·9	0·2	6·7	0*
O_2	7·4	1·1	6·5	0*
$p(CO_2)$	−4·6	1·2	−2·9	0*
EC_0	14		9·2	
n	7		3	

$\bar{x}$ mean; sd, standard deviation
* One value only.
Units are μequiv./l, except for Si (μmol/l), pH, O_2 (ppm) and EC (μS/cm).

point and the glacier terminus. They may also carry a fingerprint of any chemical weathering that occurs in the channel marginal zone. Table IV shows that the composition of mode 2A waters is similar to that of dilute bulk runoff (see Table I). The mean concentration of the major ions, Ca^{2+}, Mg^{2+}, HCO_3^- and SO_4^{2-}, is lower or equal to the most dilute bulk meltwater sampled, but the mean concentration of the other minor ions lies inside the range of concentrations found in the bulk runoff. The dominant ions are Ca^{2+} and HCO_3^-, and the ratio of SO_4^{2-} to (SO_4^{2-} + HCO_3^-), hereafter the S-ratio, is relatively low (0·11). The mean $p(CO_2)$ is −4·6, so the waters are undersaturated with respect to atmospheric CO_2 (the $p\,CO_2$ at the glacier snout is −3·6). This type of water chemistry is similar to that inferred for subglacial channelised flow by Tranter et al. (1993) and by laboratory dissolution experiments (Brown et al., 1994a). It is interesting to note that the ratio of Na^+ or K^+ to Mg^{2+} is higher in mode 2A waters than in the bulk runoff, suggesting that aluminosilicates such as feldspars are contributing relatively more solute than ferro-magnesium minerals or carbonates.

It is dangerous to be too categorical about the interpretation of a small number of samples from only two boreholes, but possible explanations for the difference between the compositions of Mode 2A and bulk runoff follow. Firstly, the Mode 2A waters will acquire further solute en route to the portal through post-mixing reactions (Brown et al., 1994a). Transit times for waters in subglacial channels from beneath the borehole array to the snout are of the order of an hour (Nienow et al., in press), over which time chemical weathering of glacial flour can liberate solute even in waters already containing solute (Tranter et al., 1989; Brown et al., 1994a). These so-called 'post mixing reactions' are unlikely to supply SO_4^{2-} to solution, however. Secondly, Sharp et al. (1993b) predict that a number of major or arterial subglacial channels contribute to the bulk runoff. If each of these different channels has Mode 2A water of a slightly different composition, dependent on the mineralogy of the bedrock, for example (Lamb et al., 1995), there need not necessarily be a similarity between the Mode 2A waters sampled from our borehole array and the bulk runoff. Finally, Mode 2A waters may also be derived from sidewall stream contributions from small cirque glaciers on the eastern wall of the catchment. Further differences between the composition of the Mode 2A waters and the bulk runoff are explored below.

Figure 4 shows that there is a broad spectrum of Mode 2B water compositions, that are most likely formed by the turbulent mixing of Mode 1 and 3 waters during rising and falling water levels or the transit of probes or samplers. Mode 2B water was sampled from the condocline in Hole 4. We believe that the broad correspondence of the conductivities and composition of the 2A and 2B waters (see Table IV) is fortuitous and therefore that Mode 2B waters have little intrinsic value in constraining the nature of the subglacial drainage system.

Mode 3 water chemistry. Mode 3 waters were samples from Hole 4. They are the most concentrated waters sampled during the field season. The EC of bulk runoff approaches the lower range of the EC of Mode 3 waters during the period of recession flow, when waters draining from the distributed system make most contribution to bulk runoff. We believe that Mode 3 waters are representative of those flowing in the distributed drainage system because they are found in holes that showed relatively little diurnal fluctuation in water level or in the position and shape of the condocline, and because their water chemistries match those of delayed flow inferred from bulk runoff chemistry (Tranter *et al.*, 1993). It may seem strange that of the holes that were routinely profiled for EC, only five of the 20 contained Mode 3 water (see Table II). However, very low hydraulic transmissivities at the base of certain boreholes prevent their water levels from equilibrating with local subglacial water pressures. Water levels may remain high-standing in such boreholes, forcing the small but continuous input of surface-derived Mode 1 waters down and out through their bases. Mode 3 waters may therefore never be sampled at the base of such boreholes, even though these waters may occur locally.

Table I shows that Mode 3 waters are dominated by Ca^{2+}, HCO_3^- and SO_4^{2-}. The mean S-ratio is relatively high (0·31). The mean $p(CO_2)$ is $-3·0$, showing that the waters are supersaturated with respect to atmospheric CO_2. This type of water chemistry can be derived from coupled sulphide oxidation and carbonate dissolution, illustrated by Equation (5), believed to be a dominant chemical weathering reaction in the distributed system (Tranter *et al.*, 1993).

$$4FeS_2(s) + 16CaCO_3(s) + 15O_2(aq) + 14H_2O(aq)$$

pyrite   calcite

$$\rightleftharpoons \tag{6}$$

$$16Ca^{2+}(aq) + 16HCO_3^-(aq) + 8SO_4^{2-}(aq) + 4Fe(OH)_3(s)$$

Mode 3 waters contain high concentrations of all major base cations, SO_4^{2-} and HCO_3^-, which must be the result of the chemical weathering of rock flour in the distributed drainage system. This assertion is supported by the relatively high Si concentrations found in Mode 3 waters. Silica is liberated from the relatively slow dissolution of aluminosilicate–silicate–SiO_2 minerals (Rimstidt and Barnes, 1980; Lerman, 1979) in rock flour, illustrated in Equations (7) and (8), and the distributed drainage system provides the combination of sufficient reactive rock and rock to water contact times of sufficient duration of significant chemical weathering of these minerals to occur. An additional environment in which Si may be liberated in significant concentrations, the channel marginal zone, is discussed in the following

$$SiO_2(s) + 2H_2O(aq) \rightleftharpoons H_4SiO_4(aq) \tag{7}$$

amorphous   silica disolved silicon, silicic acid or 'Si'

$$Mg_2SiO_4(s) + 4H^+(aq) \rightleftharpoons 2Mg^{2+}(aq) + H_4SiO_4(aq) \tag{8}$$

fosterite

The average S-ratio in the borehole waters is 0·31. This has implications for mechanisms of subglacial chemical weathering. The coupling of sulphide oxidation and carbonate dissolution should give rise to an S-ratio of 0·5 (where units of concentration are expressed in equivalents), as shown in Equation (6). The excess HCO_3^- found in Mode 3 waters may arise from the hydrolysis of $CaCO_3$ (Fairchild *et al.*, 1994), but this rapidly produces a solution of high pH (and low pCO_2), in which further carbonate dissolution quickly stops if there is no additional source of protons. Hence, hydrolysis may occur initially, and thereafter sulphide oxidation and carbonate dissolution dominate (Fairchild *et al.*, 1994). Alternatively, a subglacial source of CO_2 is required to dissolve either carbonates or aluminosilicate–silicate minerals, as illustrated by Equations (4) and (5). The subglacial source of CO_2 might be one or a combination of the following: (a) pockets of air trapped in subglacial cavities; (b) gas bubbles in glacier ice; (c) microbial oxidation of organic carbon in the freshly comminuted bedrock; or (d) possible (but improbable) metamorphic CO_2 exhalations from the bedrock (Holland, 1978). Currently, we cannot distinguish between these potential CO_2 sources.

The mean NO_3^- concentration (27 μequiv./l) of the Mode 3 waters is unusually high (see Table I). The mean concentration is at least an order of magnitude greater than found in supraglacial runoff (0·4 μequiv./l see Table III) and a factor of $\sim$ 4 greater than those of average snow (6·9 μequiv./l; see Table III). By contrast, Cl^- concentrations in Mode 3 waters are only a factor of 1·2 greater than those of average snow. The greater relative concentration of NO_3^- with respect to Cl^- in Mode 3 waters might arise from preferential elution of NO_3^- from the snowpack (Tranter, 1991). The concentration of NO_3^- in the bulk runoff (0·0–8·5 μequiv./l) is at least a factor of three lower than in Mode 3 waters and approaches Mode 3 concentrations only during the onset of the ablation season, as a result of fractionation of solute from the snowpack into the first meltwaters (Tranter, 1991; Tranter *et al.*, 1994). If snowmelt is the source of NO_3^- found in these Mode 3 waters, then the snowmelt has been in long-term storage at the glacier bed. An estimate of this longer term storage would be of the order of months, as the ablation season commences during May and Mode 3 samples were collected during September. This implies that some snowmelt may be routed along different subglacial flow paths than icemelt. Other possible sources of NO_3^- to Mode 3 waters are rainfall, partial freezing of a former dilute water, brine drainage from the glacier ice or microbial N_2 fixation. None of these sources is presently thought likely.

Correspondence of the chemical composition of Mode 1, 2A and 3 waters with bulk runoff

The chemical composition of bulk runoff integrates the different compositions of a spectrum of waters (Collins, 1978), including supraglacial runoff, drainage from the distributed system and the output from several subglacial channels and channel marginal zones. Post-mixing reactions may add solute to waters

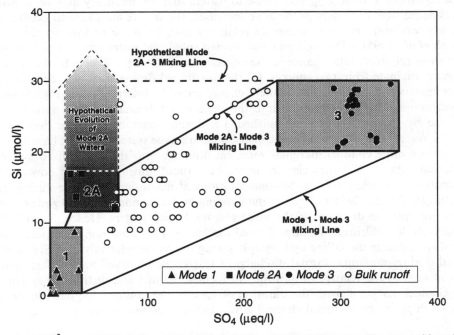

Figure 7. Scatterplot of SO_4^{2-} versus Si for bulk runoff and Mode 1, 2A and 3 borehole waters. The composition of the bulk runoff is largely confined within the mixing zone of Mode 1 and Mode 3 waters (lower solid line), Mode 1 and Mode 2A waters (solid line on the left) and the hypothetical mixing line (upper broken line) which defines the mixing of hypothetically evolved Mode 2A waters and Mode 3 waters

(e.g. Ca^{2+}, Mg^{2+} and HCO_3^-; Brown *et al.*, in press; Sharp *et al.*, 1995). The variety of possible water sources and the possibility that they may not mix conservatively creates difficulties in deciding how representative the borehole waters are of those comprising the bulk runoff. We choose to examine the correspondence of two dissolved species, Si and SO_4^{2-}, that are believed to mix quasi-conservatively when distributed drainage system waters enter the channelized system and are diluted by icemelt. We believe that Si is quasi-conservative because the dissolution kinetics of Si from alumino-silicate, silicate and SiO_2 minerals are slow (Rimstidt and Barnes, 1980; Lerman, 1979). We believe that SO_4^{2-} behaves quasi-conservatively because glacial flour that has already been transported through the channelized drainage system does not liberate significant amounts of SO_4^{2-} during laboratory dissolution experiments (Brown *et al.*, 1994a).

Figure 7 shows a plot of Si versus SO_4^{2-} for the bulk runoff, Mode 1, Mode 2A and Mode 3 waters. It is clear that there is a broad linear association between the two species in bulk runoff. Mode 1 and 3 waters lie at either end of the linear cluster of bulk runoff data points, suggesting that variable dilution of Mode 3 waters with Mode 1 waters could give rise to waters approximating the composition of some of the bulk runoff. Approximately 50% of the bulk runoff samples lie in the vicinity of this mixing line, whereas the other 50% could arise from the variable mixing of Mode 2A waters and Mode 3 (denoted by the Mode 2A–Mode 3 mixing line).

Although we believe that the waters mix quasi-conservatively for SO_4^{2-} and Si in the channelized drainage system, it is unlikely that the end member compositions are constant, neither throughout the ablation season nor on a diurnal time scale. We illustrate this assertion by focusing our attention on the bulk runoff that is relatively enriched in Si. Waters with Si > 21 μmol/l were sampled during the period of recession flow and at minimum discharge during the end of the sampling season when bulk runoff was low (see Figures 2 and 3). During these periods of low discharge, waters of relatively long residence time in the channel marginal zone must drain into the channelized drainage system. Water to rock ratios in the channel marginal zone are believed to be high (Hubbard *et al.*, 1995) and diurnal flushing of the zone must exhaust reactive minerals and surface sites during the course of the ablation season. Hence the dominant chemical weathering reactions that occur in this zone are the dissolution of SiO_2, silicate and alumino-silicate mineral lattices [see Equations (7) and (8)], giving rise to waters that are relatively high in Si with respect to SO_4^{2-}. As the residence time of waters in the zone increases, the waters are likely to become increasingly enriched in Si, whereas SO_4^{2-} is likely to remain relatively low. Evidence to support this assertion may be found in Lamb *et al.* (1995), where certain concentrated subglacial waters show a wide range of Si concentrations, but show relatively little change in the concentration of SO_4^{2-}. Waters further into the channel marginal zone may originate from two sources: (a) the distributed drainage system, characterized by high SO_4^{2-} and Si concentrations; and (b) from channels, characterized by intermediate to high Si and intermediate SO_4^{2-} concentrations. These waters drain into the channelized drainage system only if the diurnal recharge of the zone by channel water is diminished, i.e. during times of recession flow or low diurnal discharge regimes. Waters that originated in the distributed drainage system and that have been stored temporarily in the more remote channel marginal zone, on drainage into the channelized system, may give rise to the observed high Si–SO_4^{-2} associations in the bulk runoff during recession flow and low diurnal discharge. By contrast, the relatively low Si concentrations at maximum discharge during the period of low diurnal discharge (JD 253–265) may reflect the flooding of the channel marginal zone during the rising hydrograph that prevents the drainage of the more remote, Si-rich waters. Hence, Si and SO_4^{2-} concentrations are relatively low during maximum diurnal discharge. The more remote waters emerge for the channel marginal zone during the falling hydrograph, giving rise to the relatively high Si and SO_4^{2-} concentrations observed during minimum diurnal discharge. Clearly, the observed Si and SO_4^{2-} concentrations depend on factors such as the relative proportions of former channel and distributed system waters that drain from the channel marginal zone, the duration of rock–water contact, the particle size distribution and reactive mineralogy in the marginal channel zone, and the relative proportion of quickflow in bulk runoff.

The model we present here of the hydrochemistry of bulk runoff and the subglacial drainage system is radically different to that proposed by earlier workers (e.g. Collins, 1979). We do not believe that bulk runoff consists of two components; rather there are at least three. We do not believe that the end-member compositions are constant; rather they may vary on diurnal time-scales and are controlled by the kinetics of geochemical reactions and the hydrological flow paths in operation within a range of subglacial environments.

CONCLUSIONS

This study presents an analysis of borehole water chemistry that we believe is representative of *in situ* subglacial drainage waters. We believe that future studies of meltwater hydrochemistry and subglacial chemical weathering need to include similar, but extended, sampling, so that the integrated chemical signals observed in bulk runoff throughout the ablation season can be properly interpreted.

Three broad types of water were identified in the boreholes from EC profiling and chemical analysis. Dilute, Mode 1 waters are believed to be supraglacial runoff that may have been modified by limited chemical weathering of debris melted out of the ice during drilling or sediment suspended from the bed by the hot water jet. Mode 2A waters are believed to represent that in the channel marginal zone, closely associated with active subglacial channels. They have a major ion composition that is more dilute than contemporaneous bulk runoff, and Ca^{2+} and HCO_3^- are the dominant ions. They have intermediate Si concentrations, a high Si to SO_4^{2-} ratio, a low S-ratio and a low $p(CO_2)$. By contrast, Mode 3 waters are more concentrated than contemporaneous runoff. The dominant ions are Ca^{2+}, HCO_3^- and SO_4^{2-}, concentrations of Si, the S-ratio and the $p(CO_2)$ is relatively high. These waters are believed to represent those flowing through a distributed drainage system.

The chemical composition of subglacial drainage waters from the Haut Glacier d'Arolla is broadly in line with that predicted from studies of bulk glacial runoff (Tranter *et al.*, 1993), but there are aspects which were not anticipated. These throw light on subglacial water routing and subglacial chemical weathering mechanisms. The high NO_3^- concentration of Mode 3 waters suggests that concentrated snowmelt feeds a distributed drainage system and that this snowmelt is routed in a different manner to icemelt. Storage of waters in the distributed drainage system may be of the order of months, longer than indicated by dye tracing studies (Nienow *et al.*, in press). The high S-ratio of the concentrated subglacial drainage waters suggests that either carbonate hydrolysis is occurring before sulphide oxidation (Fairchild *et al.*, 1994) or a subglacial source of CO_2 must be present to augment the chemical weathering fuelled by sulphide oxidation. The source may be from CO_2 in bubbles in glacier ice, englacial voids or subglacial microbial activity. The composition of many waters at or near the earth's surface are modified by microbial activity. It may well be that microbial activity has a greater impact on subglacial chemical weathering than hitherto appreciated (Tranter *et al.*, 1994). Finally, the channel marginal zone may be a significant source of solute during periods of flow recession or low diurnal discharge regimes.

ACKNOWLEDGEMENTS

This work was funded by NERC Grant No. GR4/8114, a NERC Fellowship No. GT5/F/91/AAPS/13 (GHB) and NERC Studentship GT4/93/113 (HRL) and GT4/91/GS/27 (MJT). Y. Bams, P. Bournusson and Grande Dixence SA (especially J.-M. Bovin and M. Beytresson) are gratefully thanked for their logistical support. Neil Young's 'Change your mind' and 'Cocaine eyes' were a constant source of tranquility during the frequent revision of this manuscript.

REFERENCES

Blake, E. W. and Clarke, G. K. C. 1991. 'Subglacial water and sediment samplers. *J. Glaciol.*, **37**, 188–190.
Brown, G. H. 1991. 'Solute provenance and transport pathways in Alpine glaciers', *Unpublished PhD Thesis*, Univ. Southampton.
Brown, G. H., Tranter, M., Sharp, M. J., Davies, T. D., and Tsiouris, S. 1994b. 'Dissolved oxygen variations in Alpine glacial meltwaters', *Earth Surf. Process. Landforms*, **19**, 247–253.
Brown, G. H., Sharp, M. J., Tranter, M., Gurnell, A. M., and Nienow, P. N. 1994a. 'The impact of post-mixing chemical reactions on the major ion chemistry of bulk meltwaters draining the Haut Glacier d'Arolla, Valais, Switzerland', *Hydro. Process.*, **8**, 465–480.
Brown, G. H., Tranter, M., and Sharp, M. J. 1996. 'Experimental investigations of the weathering of suspended sediment by Alpine glacial meltwater', *Hydrol. Process.*, **10**, 579–597.

Collins, D. N. 1978. 'Hydrology of an alpine glacier as indicated by the chemical composition of meltwater', *Z. Gletscherk. Glazialgeo.*, **13**, 219–238.

Collins, D. N. 11979. 'Quantitative determination of the subglacial hydrology of two alpine glaciers', *J. Glaciol.*, **23**, 347–361.

Englehardt, H. 1978. 'Waters in glaciers: observations and theory of the behaviour of water levels in boreholes', *Z. Gletscherk. Glaziolgeol.*, **14**, 35–60.

Fairchild, I. J., Bradby, L., Sharp, M., and Tison, J. -L. 1994. 'Hydrochemistry of carbonate terrains in Alpine glacial settings', *Earth Surf. Process. Landforms*, **19**, 33–54.

Fenn, C. R. 1987. 'Electrical conductivity', in Gurnell, A. M. and Clark, M. J. (Eds), *Glacio-fluvial Sediment Transfer*. Wiley, Chichester. pp. 377–414.

Fountain, A. G. 1994. 'Borehole water level variations and implications for the subglacial hydraulics of South Cascade Glacier, Washington State, U.S.A.', *J. Glaciol.*, **40**, 293–304.

Garrels, R. M. and Christ, C. L. 1965. *Solutions, Minerals and Equilibria*. Freeman Cooper, San Fransisco.

Gurnell, A. M., Brown, G. H., and Tranter, M. 1994. 'A sampling strategy to describe the temporal hydrochemical characteristics of an Alpine proglacial stream', *Hydrol. Process.*, **8**, 1–25.

Hodge, S. M. 1976. 'Direct measurements of basal water pressures: a pilot study', *J. Glaciol.*, **16**, 205–218.

Holland, H. D. 1978. *The chemistry of Atmosphere and Oceans*. Wiley-Interscience, New York.

Hubbard, B. P., Sharp, M. J., Willis, I. C., Nielsen, M. K., and Smart, C. C. 1995. 'Borehole water-pressure variations and the structure of the subglacial hydrological system of the Haut Glacier d'Arolla, Valais, Switzerland', *J. Glaciol.*, **41**, 572–583.

Humphrey, N. F. and Raymond, C. F. 1995. 'Hydrology, erosion and sediment production in a surging glacier: Variegated Glacier, Alaska, 1982–83', *J. Glaciol.*, **40**, 539–552.

Humphrey, N. F., Raymond, C., and Harrison, W. 1986. 'Discharge of turbid water during mini-surges of Variegated Glacier, Alaska, U.S.A.', *J. Glaciol.*, **32**, 195–207.

Iken, A. and Bindschadler, R. A. 1986. 'Combined measurements of subglacial water pressure and surface velocity of Findelengletcher, Switzerland: conclusions about drainage system and sliding mechanism', *J. Glaciol.*, **32**, 101–119.

Lamb, H. R., Tranter, M., Brown, G. H., Hubbard, B. P., Sharp, M. J., Smart, C. C., Willis, I. C., and Nielsen, M. K. 1995. 'The composition of subglacial meltwaters sampler from boreholes at the Haut Glacier d'Arolla, Switzerland', *IAHS Publ.*, **228**, 395–403.

Lerman, A. 1979. *Geochemical Processes. Water and Sediment Environments*. Wiley, New York.

McQuaker, N. R., Kluckner, P. D., and Sandberg, D. K. 1993. 'Chemical analysis of acidic precipitation: pH and acidity determinations', *Environ. Sci. Technol.*, **17**, 431–439.

Nienow, P., Sharp, M. J., Willis, I. C., and Richards, K. S. 'Dye tracer investigations at the Haut Glacier d'Arolla, Switzerland: seasonal changes in the morphology of the subglacial drainage system', *Earth Surf. Process. Landforms*, in press.

Pohjola, V. A. 1994. 'TV-video observations of englacial voids in Storglaciären, Sweden', *J. Glaciol.*, **40**, 231–240.

Rimstidt, J. D. and Barnes, H. L. 1980. 'The kinetics of silica–water reactions', *Geochim. Cosmochim. Acta*, **44**, 1683–1699.

Röthlisberger, H. and Lang, H. 1987. 'Glacier hydrology', in Gurnell, A. M. and Clark, M. J. (Eds), *Glacio-fluvial Sediment Transfer*. Wiley, Chichester. pp. 207–284.

Sharp, M., Willis, I. C., Hubbard, B., Nielsen, M., Brown, G. H., Tranter, M., and Smart, C. C. 1993a. 'Water storage, drainage evolution and water quality in Alpine glacial environments', *Interim Report on NERC Grant No. GR3/8114*.

Sharp, M., Richards, K., Willis, I., Arnold, N., Nienow, P., Lawson, W., and Tison, J. -L. 1993b/ 'Geometry, bed topography and drainage system structure of the Haut Glacier d'Arolla, Switzerland', *Earth Surf. Process. Landforms*, **18**, 557–571.

Sharp, M.J, Brown, G. H., Tranter, M., Willis, I. C., and Hubbard, B. P. 1995. 'Some comments on the use of chemically-based mixing models in glacier hydrology', *J. Glaciol.*, **41**, 241–246.

Stone, D. B. and Clarke, G. K. C. 'In situ measurements of basal-water quality and pressure as an indicator of the character of subglacial drainage systems', *Hydrol. Process.*, In press.

Stone, D. B., Clarke, G. K. C., and Blake, E. W. 1993. 'Subglacial turbidity and electrical conductivity', *J. Glaciol.*, **39**, 415–420.

Tranter, M. 1991. 'Controls on the composition of snowmelt', in Davies, T. D., Tranter, M., and Jones, H. G. (Eds), *Seasonal Snowpacks. Processes of Compositional Change. NATO ASI Series*, **G28**, 241–271.

Tranter, M. and Raiswell, R. 1991. 'The composition of the englacial and subglacial components in bulk meltwaters draining the Gornergletscher,' *J. Glaciol.*, **37**, 59–66.

Tranter, M., Brimblecombe, P., Davies, T. D., Vincent, C. E., Abrahams, P. W., and Blackwood, I. 1986. 'The chemical composition of snowpack, snowfall and meltwater in the Scottish Highlands—evidence for preferential elution', *Atmos. Environ.*, **20**, 517–525.

Tranter, M., Raiswell, R., and Mills, R. A. 1989. 'The geochemistry of Alpine glacial meltwaters', In Miles, D. L. (Ed.), *Proceedings, 6th International Symposium on Water–Rock Interaction, Malvern, 1989*. pp. 687–690.

Tranter, M., Brown, G. H., Raiswell, R., Sharp, M. J., and Gurnell, A. M. 1993. 'A conceptual model of solute acquisition by Alpine glacial meltwaters', *J. Glaciol.*, **39**, 573–581.

Tranter, M., Brown, G. H., Hodson, A. J., Gurnell, A. M., and Sharp, M. J. 1994. 'Nitrate variations in glacial runoff from Alpine and sub-Arctic glaciers', *IAHS Publ.*, **223**, 299–311.

Tranter, M., Brown, G. H., Hodson, A. J., and Gurnell, A. M. 1996. 'Hydrochemistry as an indicator of the nature of subglacial drainage system structure: a comparison of Arctic and Alpine environments', *Hydrol. Process.*, **10**, 541–556.

Waddington, B. S. and Clarke, G. K. C. 1995. 'Hydraulic properties of subglacial sediment determined from the mechanical response of water-filled boreholes', *J. Glaciol.*, **41**, 112–124.

APPENDIX

The following calculation, which maximizes the impact of diffusion, shows that diffusion is unlikely to be a significant factor in defining the observed chemical compositions or the chemical (EC) profiles observed in the boreholes. Fickian diffusion of ions is proportional to the diffusion coefficient, D (cm^2/s), of the ion of interest and the concentration gradient, dC/dz (g/cm^4). The flux of the ion of interest, F (g/cm^{2}s), is defined as (Lerman, 1979)

$$F = D\frac{dC}{dz} \qquad (A1)$$

We choose Ca^{2+} as an example to illustrate the probable impact of diffusion. The diffusion coefficient of Ca^{2+} at $0°C$ is $3·73 \times 10^{-6}$ cm^2/s (Lerman, 1979). The difference in the mean Ca^{2+} concentration of Mode 1 and 3 waters provides a likely maximum concentration difference, namely 790 μequiv./l (see Table I). Let us assume that the thickness of the condocline is 1 m (examination of Figure 5 shows that this is an order of magnitude too small). Hence the maximum likely concentration gradient is set at 7·9 μequiv./l per cm (0·16 μg/cm^4). The diffusional flux across this layer is $\approx 0·6 \times 10^{-6}$ μg/cm^2/s. Over the course of 30 days, the amount of solute transferred under steady state conditions would be $\sim 1·5 \mu$g/cm^2. If there were only 10 m of water column above the condocline (so that 1 l of water overlies each cm^2), the increase in Ca^{2+} would be 0·1 μequiv./l. Given that concentration gradients observed in the boreholes are at least one order of magnitude lower than given above, and that the depth of Mode 1 water above the condocline is usually tens of metres, we feel it unlikely that significant amounts of solute diffuse from Mode 3 into Mode 1 waters.

15

SEASONAL REORGANIZATION OF SUBGLACIAL DRAINAGE INFERRED FROM MEASUREMENTS IN BOREHOLES

SHULAMIT GORDON,[1] MARTIN SHARP,[1,*] BRYN HUBBARD,[2] CHRIS SMART,[3] BRAD KETTERLING[3] AND IAN WILLIS[4]

[1]Department of Earth and Atmospheric Sciences, University of Alberta, Edmonton, Alberta, T6G 2E3, Canada
[2]Centre for Glaciology, Institute of Earth Studies, University of Wales, Aberystwyth, Dyfed, SY23 3DB, UK
[3]Department of Geography, University of Western Ontario, London, Ontario, N6A 5C2, Canada
[4]Department of Geography, University of Cambridge, Downing Place, Cambridge, CB2 3EN, UK

ABSTRACT

The effect of the formation of a major subglacial drainage channel on the behaviour of the subglacial drainage system of Haut Glacier d'Arolla, Switzerland, was investigated using measurements of borehole water level and the electrical conductivity and turbidity of basal meltwaters. Electrical conductivity profiles were also measured within borehole water columns to identify the water sources driving water level changes, and to determine patterns of water circulation in boreholes. Prior to channel formation, boreholes showed idiosyncratic and poorly coordinated behaviour. Diurnal water level fluctuations were small and driven by supraglacial/englacial water inputs, even when boreholes were connected to a subglacial drainage system. This system appeared to consist of hydraulically impermeable patches interspersed with storage spaces, and transmitted a very low water flux. Drainage reorganization, which occurred around 31 July, 1993, in response to rapidly rising meltwater and rainfall inputs, seems to have involved the creation of a connection between an incipient channel and a well-established channelized system located further down-glacier. Once a major channel existed within the area of the borehole array, borehole water level fluctuations were forced by discharge-related changes in channel water pressure, although a diversity of responses was observed. These included (i) synchronous, (ii) damped and lagged, (iii) inverse, and (iv) alternating inverse/lagged responses. Synchronous responses occurred in boreholes connected directly to the channel, while damped and lagged responses occurred in boreholes connected to it by a more resistive drainage system. Pressure variations within the channel resulted in diurnal transfer of mechanical support for the ice overburden between connected and unconnected areas of the bed, producing inverse and alternating patterns of water level response.

INTRODUCTION

Since 1989, we have conducted a detailed study of the hydrology and dynamics of Haut Glacier d'Arolla, Switzerland (Richards *et al.*, 1996). Our aim has been to determine the structure of the subglacial drainage system and its evolution over an annual cycle (Nienow, 1993; Sharp *et al.*, 1993; Hubbard *et al.*, 1995), to identify the influence of subglacial drainage conditions on meltwater quality (Tranter *et al.*, 1993; Brown *et al.*, 1994; Lamb *et al.*, 1995; Tranter *et al.*, 1997), and to understand how the hydrological behaviour of the

* Correspondence to: Martin Sharp, Department of Earth and Atmospheric Sciences, University of Alberta, Edmonton, Alberta, T6G 2E3, Canada.

Contract grant sponsor: NERC.
Contract grant number: GR3/8114.

glacier influences its flow dynamics (Harbor *et al.*, 1997). Since subglacial water pressure is known to exert a major influence on glacier flow mechanics (Iken, 1981; Bindschadler, 1983; Iken and Bindschadler, 1986; Boulton and Hindmarsh, 1987), it is particularly important to understand the relationship between water pressure and seasonal changes in the configuration of the subglacial drainage system. In this paper, we present measurements of borehole water level (WL) fluctuations over the period 22 July to 21 August 1993. We utilize *in situ* measurements of water quality [electrical conductivity (hereafter EC) and turbidity (hereafter τ)] from the base of the boreholes, together with results from EC profiling of borehole water columns, to interpret these observations in terms of seasonal changes in the configuration and behaviour of the subglacial drainage system. The results allow us to evaluate and refine a model of seasonal drainage evolution based on interpretations of the results of dye tracing experiments (Nienow, 1993; Richards *et al.*, 1996).

FIELD SITE AND METHODS

Haut Glacier d'Arolla is located at the head of the Val d'Hérens, Valais, Switzerland (Figure 1). It has an area of 6·3 km^2, spans an elevation range of 2560 to *c.* 3500 m a.s.l., and is believed to be warm based. Observations in subglacial cavities (Hubbard, 1992) and by borehole video (Copland *et al.*, 1997), together with interpretations of borehole water level measurements (Hubbard *et al.*, 1995), suggest that the glacier is underlain, at least in part, by unconsolidated sediments. During the summer of 1993, hourly meteorological data were collected at an automatic weather station located on the glacier, and meltwater discharge was monitored hourly at a hydroelectric intake structure located approximately 1 km from the snout (Figure 1). Some of the resulting time-series are plotted in Figure 2.

Twenty four boreholes were drilled in an area 250 m by 50 m close to the eastern margin of the glacier, some 1·5 km from the glacier snout (Figure 1). Average borehole spacing was about 25 m. The drill site location was chosen because a major channel was predicted to exist in this area during the melt season (Sharp *et al.*, 1993). Boreholes were drilled using high pressure, hot water, and varied in depth from 23 m (H38) to 142 m (H46). Sensors similar to those described by Stone *et al.* (1993) and Stone and Clarke (1996) were placed a few decimetres above the bed in 12 holes (H27, H29, H30, H34, H35, H36, H37, H40a and b, H41, H42, H43) to measure WL, EC and τ. Measurements were logged as means of 30 readings every 10 minutes using Campbell Scientific CR-10 dataloggers. For the purposes of this paper, the records were resampled at hourly intervals. Hubbard *et al.* (1995) provide details of the drilling and sensor calibration procedures.

EC is presented as measured in water at *c.* 0°C, and τ as relative turbidity:

$$\tau = (\tau[c] - \tau[s])/\tau[c] \tag{1}$$

where $\tau[c]$ and $\tau[s]$ are voltage signals recorded in clear water and sample, respectively. Since voltage drops as turbidity increases, clear water has $\tau \approx 0$ and opaque water has $\tau \approx 1$. Hubbard *et al.* (1995) give an approximate calibration of turbidity in terms of suspended sediment concentration. Measurements of EC and τ made during periods when a borehole was dry have been excised from the records.

Measurements of borehole WL are used to characterize the behaviour of the subglacial drainage system, and calculations of the difference in elevation of the water surface between adjacent pairs of boreholes (head difference; HD) are used to infer the likely pattern of water flow at the glacier bed. Although we presume that borehole WL is related in some way to the water pressure in the subglacial drainage system, it is likely that the relationship is neither simple, consistent between boreholes, nor constant over time (Engelhardt, 1978). Instead, it will be mediated by the magnitude and timing of supraglacial/englacial water inputs to each borehole, and by the nature of the hydraulic link between each borehole base and the subglacial drainage system. We therefore consciously use the term 'borehole WL' to describe our measurements, rather than the term 'basal water pressure'. At times when a borehole was dry, the elevation of the underlying bed was used in HD calculations since this determines the hydraulic potential at the base of the hole. Hydraulic gradients between adjacent boreholes were not calculated for two reasons: (a) the true position of the base of each

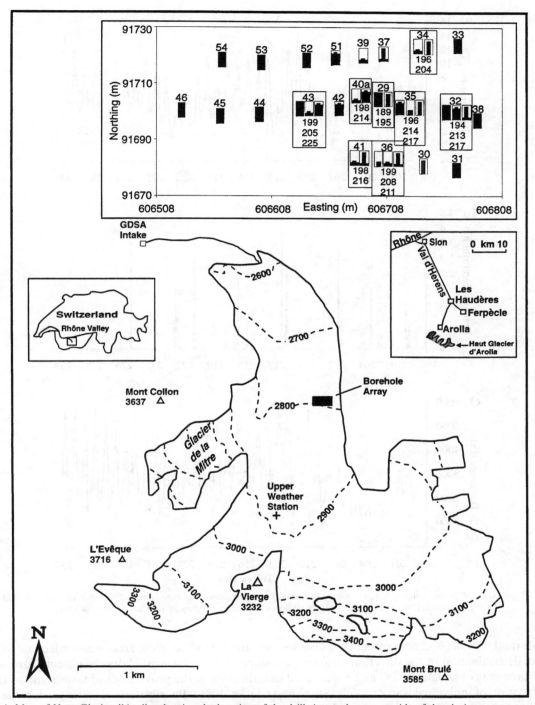

Figure 1. Map of Haut Glacier d'Arolla, showing the location of the drill site on the eastern side of the glacier tongue some 1500 m from the snout. Inset shows the 1993 borehole array. Symbols denote the status of the borehole water level (see text) which is classified as: unconnected ■; HIGH–LOW ■; LOW–LOW ■; LOW–HIGH ■; DRY–LOW ■; DRY–HIGH ■. Where several symbols are plotted for a single hole, this indicates a change in status over time. The number below the symbol denotes the Julian day on which the hole achieved the status indicated by the symbol

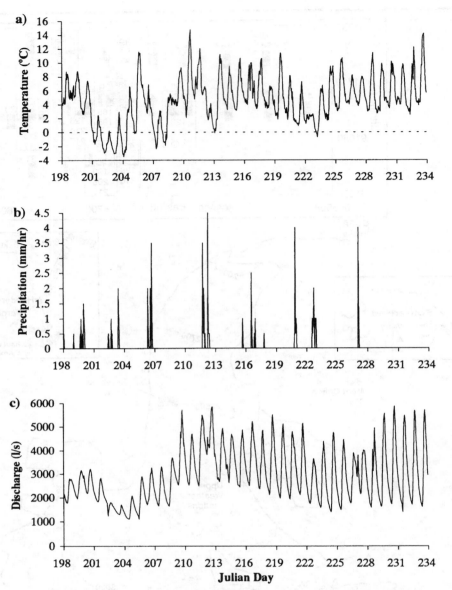

Figure 2. Time-series of (a) air temperature, (b) precipitation and (c) meltwater discharge recorded at Haut Glacier d'Arolla during the study period. Locations of the weather station and stream gauging station are shown on Figure 1

borehole (and thus the borehole spacing) was not known, and (b) the borehole array was inadequate to allow accurate determination of the direction of maximum slope of the subglacial hydraulic potential surface.

Measurements of subglacial EC and τ were used as indicators of the provenance of waters and as tracers for the motion of individual water parcels (Stone and Clarke, 1996). Basally derived water tends to be more mineralized and more turbid than supraglacially/englacially derived water. However, *in situ* water quality data must be interpreted with caution because measurements were made in open boreholes. In such boreholes, pressure transducers record changes in WL rather than changes in subglacial water pressure (as would be the case in a borehole sealed near its base). WL changes require fluxes of water into and out of the borehole. These can be supplied by any combination of basal, englacial and supraglacial sources, and complex patterns of water circulation can therefore develop within boreholes. Waters of different provenance can be mixed within the borehole, and individual parcels of water may leave the borehole by pathways different from those by which they entered it. When water drains from the base of a borehole (which can occur even when WL is rising), *in situ* sensors record the character of the water leaving the borehole, rather than that of water draining through the local subglacial system.

To determine the water sources that contributed to WL changes in individual boreholes, and the patterns of circulation that developed within them, the technique of borehole EC profiling was developed. This involves repeatedly measuring vertical EC profiles within a borehole (at 5 m depth intervals), and using temporal changes in water quality stratification (Tranter *et al.*, 1997) as a tracer for circulation within the borehole (Ketterling, 1995). When described at 5 m resolution, the stratification within a borehole is not significantly altered by diffusion or turbulent mixing, or by the action of repeatedly raising and lowering the profiling instrument through the water column. If used regularly through a melt season, the EC profiling technique allows the identification of changes in the stratification and pattern of water circulation within a borehole. These suggest changes in the way in which a borehole is connected to the glacier drainage system, and can be used to assist interpretation of changes taking place within that system.

The analysis presented in this paper is based on data from seven boreholes: H29, H34, H35, H36, H40a, H41 and H42 (Figure 1). H34 and H41 have continuous *in situ* WL, EC and τ records spanning the entire observation period [Julian days (JD 203–233)], and they also have good EC profiling coverage. WL measurements alone are analysed from H36, H40a and H42 because insufficient profiling data are available to allow reliable interpretation of the *in situ* water quality measurements. WL measurements for H40a are available only for the period JD 203–220. The WL record for H35 is continuous, but water quality measurements from this hole cannot be used since the base of the hole is believed to have been blind (see below). Records for H29 are patchy, but in the later part of the observation period WL records from this hole are indistinguishable from those from H35 (Hubbard *et al.*, 1995). Records from H29 are therefore used instead of those from H35 whenever possible to allow examination of water quality behaviour.

THE DISTRIBUTION OF CONNECTED AND UNCONNECTED HOLES

All boreholes were initially classified as 'connected' or 'unconnected' to the glacier drainage system. In connected holes, WL fell during or after drilling and subsequently fluctuated. Unconnected holes remained full of water after drilling. Of the 24 holes drilled in 1993, nine (H31, H33, H38, H44, H45, H46, H52, H53, H54) were unconnected. Five of the connected holes (H27, H30, H40a, H41 and H42) initially connected to an englacial drainage system during the course of drilling, while two (H26 and H36) connected when the drill reached the bed. The other eight connected sometime after drilling was completed. Three of the initially englacially connected holes (H30, H40a and H41) subsequently developed basal connections (Gordon, 1996).

Connected holes were further classified on the basis of the height of their daily minimum water level and amplitude of diurnal water level fluctuation, using a modified version of the scheme proposed by Smart (1996). Daily minimum water levels were termed DRY (hole drains completely), LOW (<50% of ice overburden pressure) or HIGH (>50% of ice overburden pressure). Diurnal amplitudes were taken as the difference between daily minimum and maximum water levels, and were termed ZERO (no fluctuation), LOW (fluctuation <50% of ice overburden pressure) or HIGH (fluctuation >50% of ice overburden pressure). Individual holes were thus classified as DRY–HIGH and so on.

Figure 1 shows the distribution of hole types within the borehole array, and indicates where the classification of an individual hole changed over time. At the end of the observation period, there was a clear spatial pattern to the distribution of borehole types. A 120 m wide, NE–SW trending swathe of connected holes cut through the centre of the borehole array, defining what Hubbard *et al.* (1995) termed a variable pressure axis (VPA). Within the connected area, there was an outward transition from a central core zone of DRY–HIGH holes to holes with higher base WLs and lower amplitude diurnal WL fluctuations, and finally to

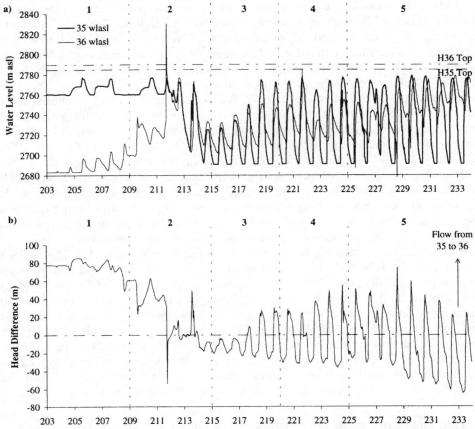

Figure 3. (a) Water level records from H35 and H36. (b) H35–H36 head difference series. The subdivision of the monitoring period into subperiods is shown, along with the elevations of the top and base of each borehole

unconnected holes. This pattern suggests that there was a major drainage pathway in the centre of the connected area which drew down water pressures in surrounding areas of the bed, and which was the source of a discharge-related diurnal pressure forcing (cf. Hantz and Lliboutry, 1983; Fountain, 1994; Hubbard *et al.*, 1995; Alley, 1996). Henceforth, we refer to this pathway as the 'major drainage path', while the term 'subglacial drainage system' is used in a more general sense to include both the path and hydraulic links between it and adjacent areas of the glacier bed.

Eight holes (H29, H32, H34, H35, H36, H40a, H41 and H43) changed classification during the observation period. With the exception of H40a and H43, all of these holes showed increases in the amplitude of diurnal WL fluctuations over time. Between JD 204 and 217, there was a trend towards a decrease in base WL and/or an increase in the amplitude of diurnal WL fluctuations in boreholes located towards the centre of the array. By contrast, H40a evolved from a LOW–LOW to a HIGH–LOW hole on JD 214, while H43 connected on JD 205 and exhibited LOW–LOW behaviour until JD 225, when its base WL rose to HIGH. There is therefore evidence for a change in the pressure regime within the connected area of the bed over the course of the season, and for disconnection of formerly connected holes. As outlined in more detail below, most of the observed changes in borehole behaviour occurred within a seven day window in late July and early August.

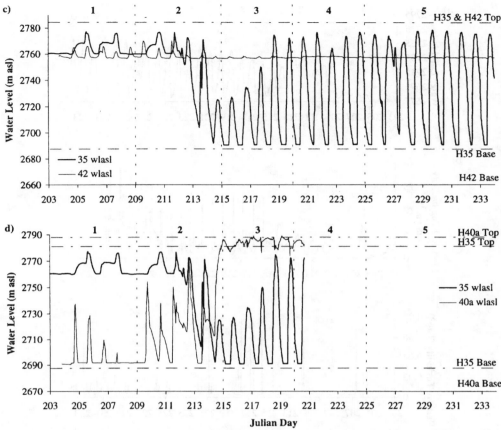

Figure 3. (c) Water level records from H35 and H42. (d) Water level records from H35 and H40a. The subdivision of the monitoring period into subperiods is shown, along with the elevations of the top and base of each borehole

WATER LEVEL, HEAD DIFFERENCE AND WATER QUALITY TIME-SERIES

Over the study period, *in situ* records of WL and HD from the seven boreholes displayed significant changes in background levels, and in the frequency, amplitude and relative timing of fluctuations (Figures 3–5). The EC and τ records from H34 and H41 showed major changes which can be correlated with changes in WL and HD records (Figures 4 and 5). Based on these changes, the records were split into five subperiods (JD 203–208; 209–214; 215–219; 220–224; 225–233), which are used as a framework for describing the major changes in borehole behaviour that occurred during the study period. In this discussion, we assume that water with an EC of $< 10\ \mu S\ cm^{-1}$ is derived directly from surface melt (either by supraglacial/englacial runoff or as drill fluid), while more concentrated water has become mineralized as a result of contact with subglacial sediments. This assumption is justified on the basis of measurements of the EC and chemical composition of supraglacial and bulk runoff, and of waters sampled from the base of boreholes (Lamb *et al.*, 1995; Tranter *et al.*, 1997).

Period 1 (22–27 July)

All seven boreholes were 'connected' to the glacier drainage system. Most holes displayed weak, low amplitude diurnal WL cycles (Figure 6), although H35 showed a two-day WL cycle (Figure 3). There was no consistency to the order in which holes reached their diurnal peak WL (Figure 7). HD records suggest that

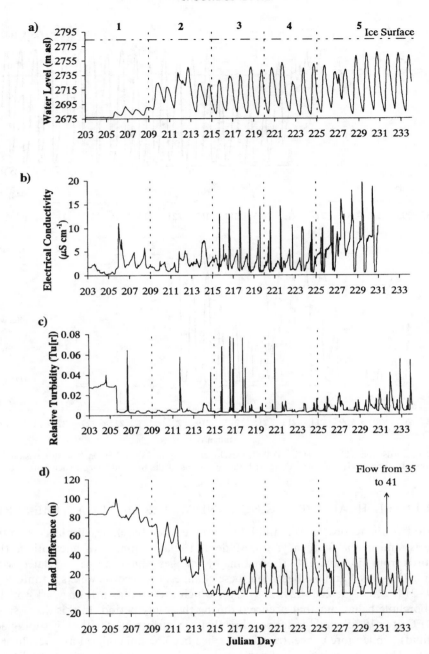

Figure 4. (a) Water level record from H41, summer 1993, showing the subdivision of the monitoring period into subperiods. (b) *In situ* electrical conductivity record from the base of H41. (c) *In situ* turbidity record from the base of H41. (d) H35–H41 head difference series

basal water flow was always directed away from H35 (Figures 3b, 4d and 5d), but this impression may be misleading since H35 had probably not developed a *basal* connection to the glacier drainage system. Its WL may therefore have overestimated the true subglacial water pressure (see below). The main feature of the τ records was a peak related to a rain storm on JD 206, although there was a weak diurnal cycle in H41

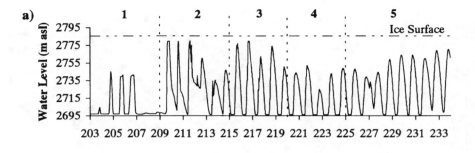

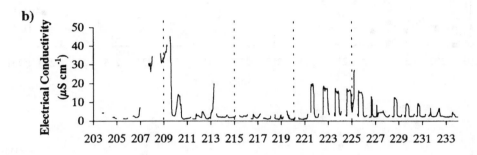

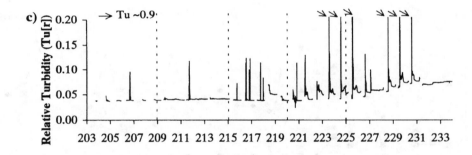

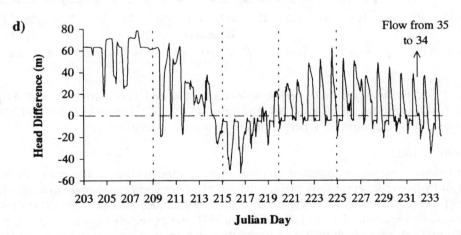

Figure 5. (a) Water level record from H34, summer 1993, showing the subdivision of the monitoring period into subperiods. (b) *In situ* electrical conductivity record from the base of H34. (c) *In situ* turbidity record from the base of H34. (d) H35–H34 head difference series

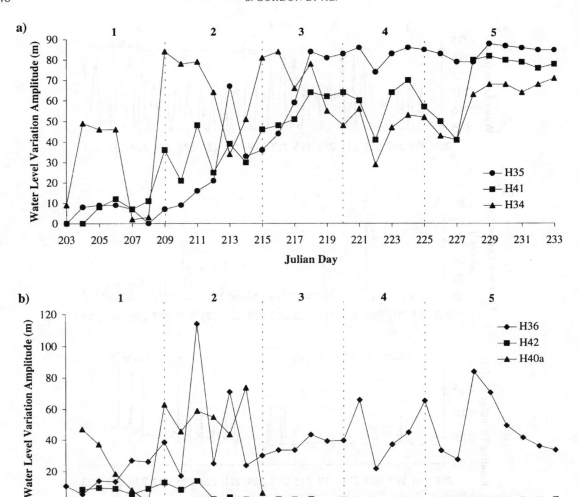

Figure 6. Amplitude of the diurnal water level cycle in (a) H34, H35, H41 and (b) H36, H40a and H42 as a function of Julian day in summer 1993

(Figures 4c and 5c). EC was generally very low in both H34 and H41, although a diurnal cycle was recognizable in H41 (Figure 4b). In H34, EC rose to around 45 μS cm^{-1} between JD 207 and 209 (Figure 5b) as the hole became almost dry during a period of very low air temperatures.

Period 2 (28 July–2 August)

The transient summer snow-line migrated up-glacier through the drill site at the end of Period 1. During Period 2, a major hydrological event took place, which resulted in significant changes in borehole behaviour. This event was preceded by three days of generally increasing surface melt (JD 209–211; Figure 2) which produced higher daily minimum WLs in H34, H36 and H41 and extremely high diurnal WL peaks in H34 and H40a (Figures 3–5). These high melt days were followed by a major rain storm between JD 211·71 and JD 212·25, when 14 mm of rain were recorded at the glacier weather station (Figure 2). Water inputs from this storm produced τ spikes in H34 and H41 (but no obvious change in *in situ* EC) (Figures 4 and 5), and a

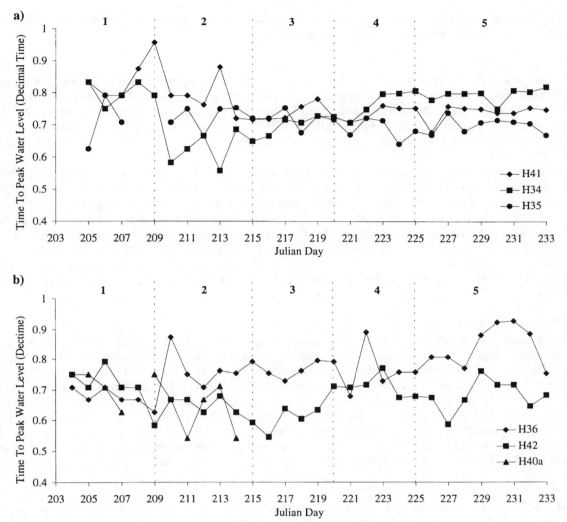

Figure 7. Time (in decimal fractions of a day) at which daily peak water level was reached in (a) H34, H35, H41 and (b) H36, H40a and H42, summer 1993

short period when WLs suggested super-flotation pressures in H36 (Figure 3) (super-flotation pressures could be recorded in H36 because it froze shut after drilling). These overnight inputs prevented WLs in H34, H36, H40a and H41 falling to their normal base levels between JD 211 and 212, so these holes recorded a sustained period of high water levels (Figures 3–5). During this period, H35 connected to the subglacial drainage system, while H40a became disconnected from it (Figure 3).

During the connection event (JD 212–214), the WL in H35 fell by 70 m (Figure 3). The daily minimum WL dropped in three stages to successively lower base levels of 2750, 2705 and 2690 m a.s.l. Clear diurnal water level peaks separated these successive base levels. Following the connection event, a diurnal WL cycle replaced the two-day WL cycle in H35 and more consistent diurnal WL cycles developed in H34, H36 and H41 (Figures 3–5). Amplitudes of diurnal WL cycles were greater than in Period 1 (20–50 m in H41, 30–80 m in H34 and H35, 30–70 m in H36; Figure 6), and the diurnal WL peak was reached first by H34, followed by H35, H41 and H36 (except on JD 214) (Figure 7). The connection of H35 caused a dramatic reduction in apparent HD between H35 and H41, H34 and H36, and there were short-lived reversals in the direction of H35–H34 and H35–H36 HD (Figures 3–5).

Period 3 (3–7 August)

Diurnal WL cycles in H34, H35, H36 and H41 increased further in amplitude (Figure 6). The order in which peak diurnal WL was reached changed over the period (Figure 7), with H35 leading H41, H34 and H36. WL behaviour in H42 and H40a continued to contrast with that observed in the other holes. WL fluctuations in H42 were barely perceptible, while H40a WL remained close to the surface, and showed small diurnal fluctuations (Figure 3). By the end of the period, there were clear diurnal HD cycles between H35 and H41, H34 and H36 (Figures 3–5).

During this period, there was a major change in the EC signal in H41. A double peaked cycle developed with an initial, smaller and more attenuated peak of ~ 5 μS cm^{-1}, which was followed by a larger, sharper peak of $10–15$ μS cm^{-1}. This suggests that basal water was detected regularly at the base of H41 during Period 3 (Figure 4). Between JD 215 and 217, major τ peaks were superimposed upon the diurnal τ cycle in H41, and these coincided with times when high EC waters were recorded at the base of the hole (Figure 4). This may suggest that erosional enlargement of a hydraulic link between H41 and the major drainage path was taking place. Similar τ events were also recorded in H34, although the *in situ* EC in this hole remained very low (Figure 5). Thus, the *in situ* water quality records for Period 3 clearly suggest a significant basal influence.

Period 4 (8–12 August)

The diurnal WL amplitude in all holes observed was relatively stable (except on JD 222, when air temperatures were low) (Figures 2 and 6). The order in which boreholes reached their diurnal peak WL was also stable (Figure 7). H35 registered the largest WL fluctuations, and was the first hole to reach its diurnal peak WL. Strong diurnal cycles in HD occurred between H35 and H41, H34 and H36 (Figures 3–5), and the direction of HD between H35 and H34 and H36 reversed on a diurnal basis. τ peaks became less frequent in H41, but larger and more frequent in H34 (Figures 4 and 5). A clear diurnal EC signal developed in H34, suggesting that basal water was regularly present at the base of this hole (Figure 5b). These changes in EC and τ behaviour in H34 were associated with a major change in the sign of H35–H34 HD, which was directed predominantly away from H34 in Period 3, but predominantly towards it in Period 4 (Figure 5d).

Period 5 (13–21 August)

WL and HD behaviour were very similar to that in Period 4, with a few exceptions. In H36, the base WL rose from 30 to 70 m above the bed, rapid drainage events at the end of the falling limb of the diurnal WL curve began on JD 224 and continued throughout the period, and a secondary WL peak emerged on JD 229 (Figure 3a). As outlined below, these trends indicate that H36 was becoming progressively less well connected to the subglacial drainage system.

In H34, τ peaks continued to occur on a regular basis, while in H41 a double-peaked diurnal τ cycle developed. The two peaks were coincident with peaks in H35/H41 HD during the rising and falling limbs of the diurnal WL cycle (Figures 4c). In H34 the major τ peaks were superimposed on a general rise in background τ, suggesting continued instability of the link between this borehole and the major drainage path. The diurnal EC cycle in H34 became less marked as the peak daily H35/H34 HD declined progressively over time (Figure 5). In H41, by contrast, the two EC peaks that first appeared in Period 3 merged into one, and EC increased during both the daily peak and the preceding attenuated shoulder (Figure 4).

CROSS-CORRELATION ANALYSES

The results presented above suggest that a major subglacial drainage reorganization occurred during Period 2. After this event, borehole WL fluctuations were very different in character from those earlier in the season, and there was clear evidence of a basal influence on basal water quality. In an attempt to quantify the changes that occurred during and after this event, cross-correlation analyses were carried out for each subperiod between the different WL series, and between these series and air temperature. Maximum cross-correlation coefficients and associated lags were used to express the degree of coherence between different data series. To emphasize diurnal cycles in the records, hourly data series were detrended prior to analysis by calculating a 25-point centred running mean, subtracting this from the raw data, dividing the data into the relevant subperiods and then re-expressing them as standardized residuals from the running mean.

The results suggest that H34, H35, H36 and H41 behaved in a similar fashion throughout the study period, while H40a and H42 behaved rather differently. Holes in the first group showed a generally increasing correlation between WL and air temperature over time, with WL eventually lagging air temperature by 3–5 hours (Table I). Correlations between pairs of WL records from holes in this group increased substantially through the study period, with a particularly marked increase occurring between Periods 2 and 3 (Table II). During Period 5, maximum correlation coefficients for pairs of WL records from boreholes in this group exceeded 0·86. Lags between records ranged from 1–4 hours at the start of the period, but the records became essentially synchronous during Period 3 when they lagged air temperature by 3–4 hours. By Period 5, a systematic pattern of lead–lag relationships developed, with H35 WL lagging air temperature by 3 hours

Table I. Maximum cross-correlation coefficients and associated lags (hours) between temperature and borehole water level records for each of the five subperiods of the 1993 melt season. See text for definition of subperiods. Positive lags indicate that the air temperature series leads the water level series

	Period				
	1	2	3	4	5
T/H35	0·41 (−1)	0·65 (5)	0·83 (3)	0·90 (3)	0·91 (3)
T/H41	0·60 (6)	0·81 (6)	0·91 (4)	0·90 (4)	0·93 (4)
T/H34	0·58 (2)	0·72 (3)	0·91 (4)	0·91 (4)	0·94 (5)
T/H36	0·63 (1)	0·61 (1)	0·88 (4)	0·82 (4)	0·82 (5)
T/H40a	0·61 (0)	0·72 (−9)	−0·63 (1)		
T/H42	0·73 (1)	0·55 (−1)	−0·64 (−13)	−0·47 (6)	0·64 (2)

Table II. Maximum cross-correlation coefficients and associated lags (hours) between pairs of water level records for each of the five subperiods of the 1993 melt season. See text for definition of subperiods. Positive lags indicate that the first series leads the second

	Period				
	1	2	3	4	5
H35/H41	0·64 (4)	0·77 (1)	0·97 (0)	0·97 (1)	0·97 (1)
H35/H34	0·47 (−1)	0·52 (−1)	0·89 (0)	0·95 (1)	0·96 (2)
H35/H36	0·66 (−1)	0·70 (0)	0·96 (0)	0·89 (1)	0·89 (1)
H41/H34	0·60 (−4)	0·79 (−2)	0·95 (−1)	0·97 (0)	0·97 (1)
H41/H36	0·86 (−4)	0·80 (−1)	0·99 (0)	0·93 (0)	0·90 (1)
H34/H36	0·44 (−1)	0·55 (3)	0·94 (0)	0·92 (0)	0·86 (1)
H36/H42	0·82 (0)	0·55 (−5)	0·75 (−3)	0·66 (0)	0·73 (−2)
H35/H42	0·59 (−1)	0·21 (−5)	0·67 (−3)	0·59 (2)	0·73 (1)
H41/H42	0·74 (−4)	0·57 (−6)	0·72 (−3)	0·61 (1)	0·80 (−1)
H34/H42	0·71 (−1)	0·65 (−3)	0·80 (−3)	0·56 (0)	0·68 (−2)
H35/H40a	0·36 (−2)	0·62 (−1)	−0·58 (−1)		
H41/H40a	0·40 (−4)	0·85 (−3)	−0·55 (−2)		
H34/H40a	0·79 (−2)	0·84 (0)	−0·60 (−2)		
H36/H40a	0·44 (0)	0·62 (−2)	0·57 (−16)		
H42/H40a	0·72 (−1)	0·55 (3)	0·47 (12)		

and leading H36 and H41 WL by 1 hour and H34 WL by 2 hours. These results confirm the shift towards a more coherent pattern of diurnal water level fluctuations within this group of boreholes after Period 2. Although correlation coefficients between WL in H36 and the other holes in the group remained high after Period 2, they did decrease slightly after Period 3, confirming the suggestion that H36 was becoming less well connected to the main drainage path over time.

The results of cross-correlation analysis emphasize that H42 and H40a behaved rather differently from the other boreholes studied. H42 WL was always positively correlated with both air temperature and other WL records, but the correlations were always weaker than for holes in the first group (always <0.82). There was no consistent increase in the degree of correlation between WLs in H42 and the other boreholes after Period 2, and WL fluctuations were not synchronized with those in the other holes during Period 3. By Period 5, however, H42 WL lagged that in H35 by only 1 hour, suggesting that it was then functioning in a similar way to holes in the first group.

During Periods 1 and 2, H40a WL was positively correlated with WL in the other holes (especially with H34 and H41 in Period 2). After the hole became disconnected at the end of Period 2, however, its WL was anticorrelated at small lags with WL in all other holes, and with air temperature. This confirms the inverse relationship between water pressure fluctuations in connected and unconnected areas of the bed found by Murray and Clarke (1995) at Trapridge Glacier, Yukon Territory.

FORCING–RESPONSE PLOTS

Forcing–response plots (F–R plots) (Murray and Clarke, 1995) provide a useful means of investigating phase lags between pairs of borehole WL records, and they facilitate the identification of non-linearities in the behaviour of the glacier drainage system. For the purposes of this analysis, the record from the hole that was first to reach its diurnal WL peak was designated as the forcing record, and the other as the response record. This implicitly assumes that boreholes that directly penetrate the main drainage path will reach their daily WL peak before holes that are more remote from it. It also assumes that diurnal pressure variations within this drainage system provide the forcing for the WL responses recorded in boreholes. Whilst this approach is consistent with that adopted by Murray and Clarke (1995), it must be remembered that open boreholes may also be forced by direct water inputs from the glacier surface, and that there may be times when this is the dominant forcing. It is therefore important to understand the source of the forcing driving water level variations in each borehole. This issue is addressed below using the results of borehole EC profiling.

To allow comparison of WL behaviours at the beginning and end of the observation period, F–R plots were constructed for JD 206 (Period 1) and JD 229 (Period 5) (Figures 8 and 9). Since WL records for H40a ceased before JD 229, F–R plots for this hole were constructed for JD 218. For JD 218 and 229, the WL record from H35 was used as the forcing record, since this was consistently the first record to peak and thus most likely to represent pressure variations in the major drainage path.

Julian day 206

F–R plots for JD 206 (Figure 8) show a clear phase lag between all pairs of WL records, as demonstrated by open hysteresis loops. The rectangular shapes of the plots for H42/H40a, H34/H35 and H36/H35 suggest that WL fluctuations in these holes were essentially independent of each other, whereas the positive slopes of plots for H36/H41, H42/H41 and H40a/H35 imply some relationship between WL variations in these holes. The form of these plots is different from any of those described by Murray and Clarke (1995) from Trapridge Glacier. Whilst it is clear that WLs in all holes were responding to a forcing associated with diurnal runoff variations, the response was different in every hole. This implies either that the detailed nature of the forcing differed between holes (as might be the case if the forcing was surface runoff into each borehole) and/or that, if the forcing originated subglacially within the major drainage path, all holes were separated by areas of bed that offered differential resistance to water flow. This issue will be considered further below.

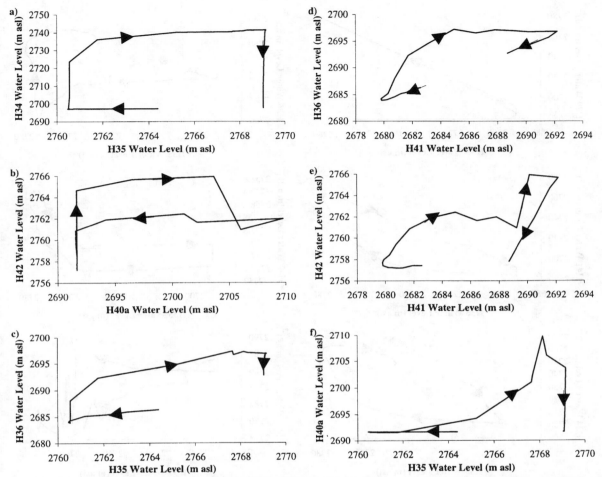

Figure 8. Forcing–response (F–R) plots for Julian day 206 (25 July), 1993. (a) H34 versus H35. (b) H42 versus H40a. (c) H36 versus H35. (d) H36 versus H41. (e) H42 versus H41. (f) H40a versus H35

Julian days 218 and 229

A variety of responses to H35 forcing were observed on JD218 and 229.

(i) Direct response. The H29/H35 plot (Figure 9a) shows no phase lag, and demonstrates that the records are essentially identical. This behaviour is comparable with that reported by Murray and Clarke (1995) for a pair of boreholes that intersected the same water body. It indicates minimal resistance to flow between the two boreholes.

(ii) Continuous, lagged response. F–R plots for H41/H35 and H34/H35 (Figure 9b and c) show a positive slope and open hysteresis loops, indicating a phase lag between the records. This suggests that WL variations in H41 and H34 were a differentially lagged response to a common forcing. Lagged response implies that H41 and H34 were separated from the major drainage path by areas of bed that resisted water flow, but it appears that the magnitude of the resistance varied between the rising and falling limbs of the diurnal WL cycle. While large head differences were maintained during the rising limb of the cycle, indicating significant hydraulic resistance, HD dropped to values near zero during the falling limb (Figures 4 and 5). This suggests

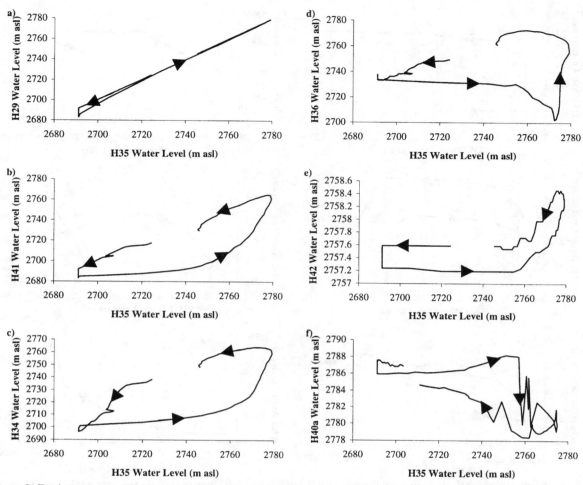

Figure 9. Forcing–response (F–R) plots for Julian days 229 (17 August) and 218 (6 August), 1993. (a) JD 229, H29 versus H35. (b) JD 229, H41 versus H35. (c) JD 229, H34 versus H35. (d) JD 229, H36 versus H35. (e) JD 229, H42 versus H35. (f) JD 218, H40a versus H35

either that events during the rising limb of the cycle (such as erosion of underlying sediments, melt of overlying ice or uplift of the glacier sole) reduced the hydraulic resistance of the connection, or that the resistance was directionally asymmetric.

The slope of the rising part of the F–R plots increased as H35 WL increased, suggesting greater sensitivity to H35 forcing at higher pressures. Two factors might contribute to this pressure-dependent sensitivity. (a) The hydraulic gradient between the major drainage path and adjacent areas of the bed increased with the pressure in that system. A higher hydraulic gradient would encourage more rapid water flow, and this in turn would increase the rate of melt of overlying ice and the potential of the flow to erode underlying sediments. Both of these effects would favour more rapid enlargement of the link between a borehole and the main drainage path, and would act to reduce the hydraulic resistance along the link. (b) Physical uplift of the glacier by hydraulic jacking effects at the glacier bed may be a non-linear function of subglacial water pressure (Iken, 1981). Opening of links by this process, which would also reduce the hydraulic resistance between borehole and major drainage path, may thus be more effective at higher pressures.

Murray and Clarke (1995) did not identify lagged behaviour of this sort at Trapridge Glacier (perhaps because their experiment was conducted in an area much smaller than ours), but it would be expected, for instance, if a diurnal pressure wave propagated laterally across the bed from a source in a linear drainage element such as a subglacial channel (Hubbard *et al.*, 1995; Alley, 1996). Wave propagation could occur through either a permeable subglacial sediment, a water film or a network of cavities and/or microchannels at the ice-bed interface.

(iii) Intermittent, lagged response. The H42/H35 plot shows that H42 WL responded to forcing from H35 only after H35 WL rose above H42 WL at a threshold value of around 2757 m a.s.l. (Figure 9e). This implies that a connection between H42 and the drainage system in the vicinity of H35 was established on an intermittent basis. When it existed, H42 WL responded in a lagged manner to H35 forcing. Drilling records indicate that an englacial conduit was intercepted at an elevation of 2757 m a.s.l. in H42. If this conduit eventually connected to the major drainage path, the daily WL rise in H42 may have resulted from water backing up within this conduit and into the borehole as water pressure rose within the major drainage path.

(iv) Inverse response with premonitory WL rise. The H40a/H35 plot for JD 218 shows a generally inverse relationship (Figure 9f). A marked fall in H40a WL occurred when H35 WL rose to 2757 m a.s.l. (this value is coincidental as it was different on other days), but this fall was preceded by a slight WL rise in H40a. Murray and Clarke (1995, their Figure 6b) recorded similar behaviour between a connected and unconnected borehole at Trapridge Glacier, and interpreted it as a result of the transfer of mechanical support of the ice overburden between connected and unconnected areas of the bed as water pressure within the connected areas fluctuated. The premonitory WL rise may thus reflect transient loading of the bed around H40a as areas closer to H35 were unloaded, while the subsequent WL fall suggests that the zone of diurnal unloading expanded to include H40a. However, if H35 WL is taken as a measure of pressure in the connected system, unloading of the unconnected system began while this pressure was significantly below that required to float the overlying ice. H40a WL rose again as H35 WL fell, and an increasing proportion of ice overburden pressure was again supported by unconnected areas of the bed.

The form of the F–R plot shows that the response of the unconnected system to forcing from the connected system is more sensitive when the pressure in the connected system is high. Murray and Clarke (1995) argued that this behaviour was linked to the influence of ambient pressure on the sensitivity of changes in the porosity of subglacial sediments to an incremental change in pressure. However, although the diurnal change in H35 WL ($\sim$ 80 m) was larger than the forcing recorded by Murray and Clarke ($\sim$ 50 m), the resultant WL change in the unconnected system was much smaller (10 m as opposed to 40 m).

(v) Alternating response. When H35 WL passed 2757 m a.s.l. (this threshold value is fortuitous because it increased progressively through Period 5), H36 WL initially fell, then rose rapidly as H35 WL approached its daily maximum value, and finally peaked after H35 WL had begun to fall (Figure 9d). This behaviour suggests that H36 alternated between being connected and unconnected to the major drainage path (Murray and Clarke, 1995). We interpret the F–R plot as follows. The initial WL fall in H36 represents unloading of the area of bed around H36 as water pressure in the major drainage path rose. The subsequent WL rise indicates expansion of the connected area to include H36. The response to H35 forcing was, however, lagged, indicating that there was resistance to flow between the major drainage path and H36. A secondary WL peak preceding the drainage event became a prominent feature of the H36 WL record after JD 229 (Figure 3). Since this secondary peak occurred while H35 WL was still falling it probably resulted from reloading of the bed around H36 as pressure in the connected region decreased. It should therefore be distinguished from the premonitory rise recorded in H40a on JD 218, which occurred while H35 WL was rising.

If the rise in pressure at the onset of the secondary peak is taken as the transition to unconnected behaviour, and the rise at the end of the drainage event as the transition back to connected behaviour, it is apparent that the switch to connected behaviour always occurred at a higher value of H35 WL than the switch to unconnected behaviour (Figure 3a). It is also clear that transitions between the connected and unconnected states did not occur when water pressures in the two systems were matched. H35 WL exceeded H36 WL at the transition to connected behaviour, and the reverse was true at the transition to unconnected behaviour. These observations imply that it was easier to exit the connected state than to enter it. The switch to connected behaviour was tripped by a combination of rapidly rising H35 WL and rapidly falling H36 WL, which produced a very large head difference (up to 70 m over a horizontal distance of ~ 25 m) between connected and unconnected regions of the bed. As outlined under (ii) above, this creates conditions favourable for opening of links between the areas of high and low pressure. The plot of H35/H36 HD (Figure 3b) shows a diurnal alternation between large positive and negative values, indicating that the hydraulic barrier that developed while H36 was unconnected offered considerable resistance to flow in both directions.

BOREHOLE ELECTRICAL CONDUCTIVITY PROFILING

The above discussion of WL, HD and *in situ* EC and τ records suggests a major change in borehole/drainage system behaviour between Periods 2 and 3. To understand the factors behind this change better, EC profiling results were used to identify changes in the sources of water driving WL fluctuations, and to reconstruct patterns of water circulation within boreholes. The discussion focuses on differences in profiling results between Periods 1/2 and 4/5, and on the process by which H35 became connected to the subglacial drainage system.

Periods 1 and 2

Prior to the marked drop between JD 212 and 214, H35 WL varied with a two-day cycle. WL increased, from a base level 74 m above the base of the borehole (a.b.), by 8 m on the first day of each cycle, and by a further 9 m on the second, before falling rapidly back to 74 m a.b. (Figure 10). Profiling on JD 210 showed a 50 m thick plug of water with EC > 2 μS cm^{-1} at the base of the borehole, the thickness of which remained unchanged as WL varied. The WL changes were thus accommodated by variations in the thickness of a cap layer with EC < 2 μS cm^{-1}. This layer was presumably fed by supraglacial or englacial inflow, and drained by episodic outflow through an englacial conduit located 74 m a.b. Since borehole WL consistently rose to

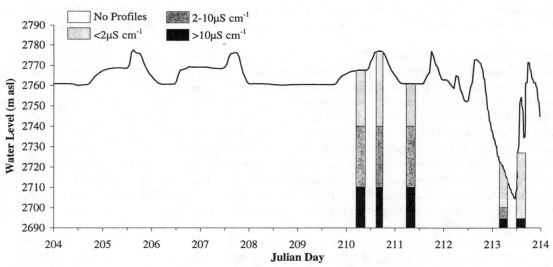

Figure 10. Water level and electrical conductivity profiling results from H35 for the period Julian day 204–214 (23 July–2 August), 1993

91 m a.b. before englacial drainage occurred, some threshold pressure was apparently required to initiate outflow. Outflow ceased when borehole WL was drawn down to the level of the englacial conduit. This behaviour is suggestive of the making and breaking of a siphon within the englacial conduit, as described from Bondhusbreen, Norway, by Hooke *et al.* (1985). Clearly, there was no consistent connection between H35 and the subglacial drainage system during this period, so records of HD calculated from H35 WLs (Figures 3–5) do not provide a true indication of hydraulic gradients within the subglacial system at this time.

Minimum WLs in both H34 and H41 were at or just above the glacier bed, indicating that drainage occurred from the base of each hole. However, the water columns in both holes always consisted entirely of water with EC < 1 μS cm^{-1}. WL variations were therefore presumably driven by supraglacial/englacial inputs. Nevertheless, EC and τ records from the base of both boreholes did show some changes. The minor variations observed may reflect variable solute acquisition arising from changing contact times between water and subglacial sediments, or disturbance of the bed by outflowing water. Larger changes, such as those in H34 between JD 207 and 210, suggest the presence of basal water at the bottom of the boreholes (Figure 5b). Such water was never observed to rise into the boreholes, however, and it was most evident when holes were nearly dry and there was therefore limited basal outflow of supraglacially or englacially derived waters.

Both H34 and H41 connected basally to a drainage system that was able to absorb the volume of water introduced to the boreholes during drilling. These boreholes filled during the day as a result of water inputs from supraglacial/englacial sources, and they drained when these inputs ceased. The fall in *in situ* EC as WL rose suggests that there was basal outflow even at times of rising WL, implying that WL rose because supraglacial/englacial inputs exceeded the drainage capacity of the subglacial system. In these circumstances, the boreholes were probably over-pressured by the daily melt input, and measured WL does not provide a true record of basal water pressure. Basal water detected by *in situ* sensors was never able to rise up the boreholes, suggesting that the natural water flux across these areas of the bed at this stage of the season was very low.

The connection of H35 to the subglacial drainage system

Profiling data suggest that the connection of H35 to the subglacial drainage system occurred via an englacial conduit located close to the bed. A profile measured at 09:59 on JD 213 showed a 5 m thick layer of water with EC of ~ 20 μS cm^{-1} at the base of the borehole (Figure 10). This layer was apparently a remnant of a 20 m thick layer which had been present prior to JD 212, suggesting that the drainage event of that day had occurred by a pathway located no more than 5 m above the bed. At 09:59 on JD 213, this layer was overlain by 5 m of 2–10 μS cm^{-1} water and 10 m of <2 μS cm^{-1} water. By 12:27 on the same day, by which time the WL was rising again having passed through a short-lived minimum, the cap consisted of 30 m of <2 μS cm^{-1} water. This suggests that the WL rise was driven by supraglacial/englacial inputs, and that these had forced the water of intervening EC out of the borehole walls, leaving the 5 m- thick high EC layer undisturbed at the base of the hole. This implies that the drainage of H35 occurred via an englacial connection located 5 m above the glacier bed. By JD 215, the WL record appears to show that H35 drained completely overnight (Figure 3a), but the stable overnight WL readings in fact record a WL 5 m above the pressure sensor. Thus, values of HD calculated for periods when H35 WL was at its diurnal minimum may be in error by as much as ± 5 m. The EC record from H35 shows a steady increase over the observation period, consistent with solute acquisition by weathering of subglacial sediment (Brown *et al.*, 1996). It thus seems that the *in situ* sensors were located near the bottom of a blind hole below the level of an active englacial connection which fixed the height of the daily minimum WL in the hole.

Periods 4 and 5

Periods 4 and 5 were characterized by clear, high amplitude diurnal WL cycles in H29, H34, H35, H36 and H41. The amplitude of these cycles was greatest in H29 and H35, and the sequence in which WL peaked was stable, with H29 and H35 leading the other boreholes. Records for JD 224 are used to illustrate borehole behaviour during this time interval, and measurements from H29 are substituted for those from H35 since water quality data are available for this hole.

In H29, the diurnal WL peak was associated with large head differences directed away from H29 and with low values of *in situ* EC and τ (Figure 11). It is not clear whether these dilute waters entered the hole from the base or the surface, but they were clearly a product of the new daily melt cycle. EC and τ peaked during the final stages of the falling limb of the WL cycle. At this time, H29–H41 HD was close to zero, and H29–H34 HD was directed towards H29. This suggests that the *in situ* sensors were detecting drainage of subglacial waters past the hole at a time of minimal dilute outflow from its base.

While H41 WL was rising, H29–H41 HD was always directed towards H41. The initial WL rise was fed by supraglacial/englacial inputs, resulting in a dilute water column and a fall in *in situ* EC (Figure 11c). Basal water with high EC and τ entered the borehole after JD 224·58, rising to a height of 40 m a.b. By JD 224·63, this high EC water column had been split in two by more dilute water, leaving a 10 m thick layer of water with EC > 10 μS cm^{-1} m a.b. and continued high *in situ* EC. Subsequently, *in situ* EC fell to < 2 μS cm^{-1} by JD 224·67, at which time there was no high EC water in the borehole. This behaviour suggests that H41 intersected an englacial conduit ~ 30 m a.b. This conduit fed dilute water into the borehole and split the rising column of high EC water. Initiation of this input triggered basal outflow, driving high EC water from the base of the borehole and reducing *in situ* EC. Subsequently, continued high supraglacial inputs drove the upper layer of high EC water out of the englacial conduit, so that by JD 224·67 there was no high EC water in the hole. As there was no borehole source of high EC water, the overnight rise in *in situ* EC and τ must indicate subglacial water flow past the hole. As the borehole WL fell and the rate of dilute outflow from the borehole base decreased, more concentrated basal water was diluted to a diminishing degree, allowing *in situ* EC to rise.

The *in situ* record shows that H34 began to be influenced by the inflow of basal water during Period 4. Throughout the daily WL peak, H29–H34 HD was directed towards H34 (Figure 12). Low *in situ* EC values as the WL started to rise suggest that the rise was driven by supraglacial/englacial inputs. At JD 224·56, however, water with high τ and EC of ~ 16 μS cm^{-1} entered the base of the hole and rose to 40 m a.b. This water was overlain by a cap of more dilute water which thickened slightly during the day, suggesting continued supraglacial/englacial inputs to the borehole. However, 80% of the observed WL rise was fed by basally derived water. *In situ* EC fell as the WL fell, reflecting the downward motion of the EC stratification within the borehole. Very low values of *in situ* EC before the borehole went dry represent the remnants of the dilute cap. The timing of *in situ* EC changes suggests that the stratification fell more rapidly than the WL, and that basal outflow must have been initiated before WL started to fall. This part of the subglacial drainage system drained completely overnight.

The most important conclusion to be derived from the results of EC profiling is that there was a significant change in the source of waters driving WL fluctuations between the start and end of the study period. In Periods 1 and 2, water level fluctuations were driven by supraglacial and englacial inputs, and there was minimal evidence for the presence of true basal water at the bottom of the boreholes. In Periods 4 and 5, however, *in situ* sensors clearly recorded the regular presence of high EC basal water, and profiling revealed that this water made a significant contribution to borehole WL fluctuations. Nevertheless, supraglacial and englacial inputs to boreholes continued and may have resulted in temporary over-pressurization of boreholes, with the result that basal outflow was initiated while WL was still rising.

CHANGES IN THE CHARACTER OF THE SUBGLACIAL DRAINAGE SYSTEM

The results presented above indicate that a major change in the hydrology of the drill site area occurred between JD 210 and 215, 1993. In the following discussion, we attempt to identify the causes of this change, and to characterize the drainage systems that existed before and after it took place.

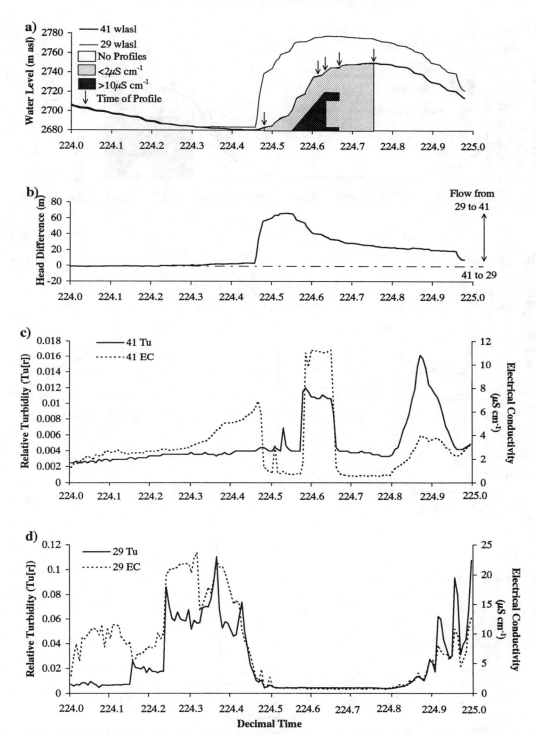

Figure 11. (a) Water level and electrical conductivity profiling results from H41 on JD 224 (12 August), 1993. (b) Difference in hydraulic head between H29 and H41 on JD 224, 1993. (c) *In situ* electrical conductivity and turbidity records from the base of H41, JD 224, 1993. (d) *In situ* electrical conductivity and turbidity records from the base of H29, JD 224, 1993

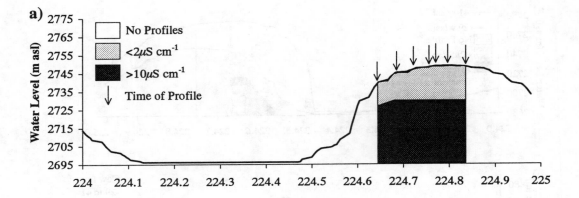

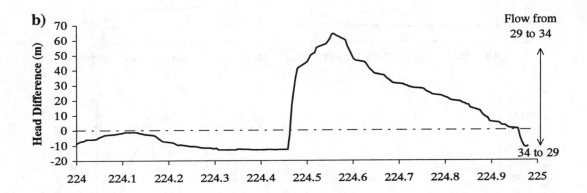

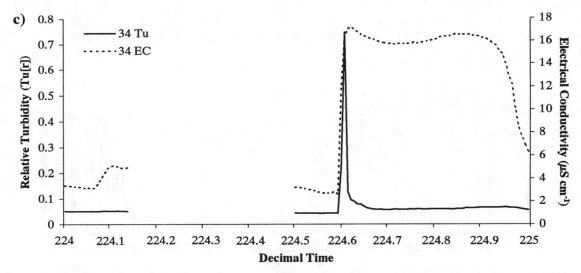

Figure 12. (a) Water level and electrical conductivity profiling results from H34 on JD 224 (6 August), 1993. (b) Difference in hydraulic head between H29 and H34 on JD 224, 1993. (c) *In situ* electrical conductivity and turbidity records from the base of H34, JD 224, 1993.

The subglacial drainage system before Julian day 210

Since many of the boreholes drilled connected to the glacier drainage system before the events of JD 210–215 (either during drilling or shortly afterwards), it is apparent that the area of the bed described as the VPA was hydrologically active before these events took place. However, *in situ* EC and τ sensors detected mineralized and/or turbid basal water at the glacier bed only following rain storms and during a prolonged period of low water levels associated with cold weather. EC profiling confirmed that WL fluctuations were driven entirely by dilute englacial and/or supraglacial inputs since basal water never rose into the boreholes. WL responses to diurnal changes in these inputs were relatively weak and poorly correlated between boreholes. Four of the boreholes studied (H34, H36, H40a and H41) drained completely overnight when surface inputs ceased. H35 and H42 connected englacially to the glacier drainage system, but apparently overlay hydrologically inactive patches of glacier bed within the connected region.

From these observations we conclude:

(i)   Prior to JD 210, WL variations were more a response to diurnal variations in direct supraglacial/englacial inputs to boreholes than to a discharge-related diurnal pressure signal originating within the subglacial drainage system. This indicates either the absence of a subglacial drainage system connected to surface water inputs, the existence of sufficient storage volume within such a system to accommodate changes in discharge with no significant change in water pressure, or the lack of a distinct diurnal rhythm to subglacial meltwater discharge at this stage of the melt season.

(ii) There is no evidence for significant fluxes of mineralized basal water across those areas of the bed intersected by our boreholes. If such fluxes occurred they must therefore have been restricted to quite small areas of the bed.

(iii) During periods of high water inputs to boreholes with a basal connection to the subglacial drainage system, borehole WLs rose. When inputs ceased at night, the holes drained completely. This implies that, in the absence of water inputs via boreholes, some areas of the bed were at atmospheric pressure. This suggests that drainage may have occurred through a network of interconnected cavities opened by glacier sliding in the lee of bedrock bumps. In the absence of large water fluxes, such a system could be expected to operate at low water pressure (Walder, 1986; Kamb, 1987). When borehole inputs were active, however, borehole WLs rose, indicating that the cavity system had insufficient free storage volume to accommodate these inputs.

(iv) If, as suggested above, subglacial storage volume was limited, it seems unlikely that there were large diurnal discharge variations within the natural subglacial drainage system at this stage of the season. This could be the case if the system was fed by a source (such as basal melt or seepage from the supraglacial snowpack) that generated only minor diurnal variations in water flux. Although diurnal WL variations in boreholes indicate that there were diurnal variations in the rate of surface runoff at this stage of the season, small changes in input can translate into relatively large changes in WL in narrow boreholes connected to a hydraulically resistive drainage system (Engelhardt, 1978). Under these circumstances, slight differences in borehole dimensions, in the transmissivity of the systems to which each borehole was connected and in the water input history to which each hole was subjected could produce significantly different WL responses. So long as the mean flow resistance in the subglacial drainage system was high, the head differences generated by such variable WL responses would not readily be evened out by basal water flow.

(v)   H35 and H42 did not develop basal connections at this stage of the season. Presumably, they intersected areas of the bed that were isolated from the interconnected cavity network. The upstream sides of bedrock bumps and patches of till overlying bedrock are, possible areas where such hydrological isolation could occur.

We therefore conclude that, at the start of the study period, the drainage system over much of the connected area of the glacier bed consisted of a hydraulically resistive network of interconnected cavities separated by hydrologically isolated areas. Under conditions of low water flux, this system operated at low ambient pressure. If a more transmissive drainage system fed by surface inputs existed at this time, it was spatially restricted and probably operated at near-constant pressure under conditions of limited discharge variability.

The drainage reorganization event of Julian days 210–215

Three days of increasing surface melt, followed by a major rain storm apparently provided the trigger for a major change in the hydrology of the drill site area. During this period, H34, H36, H40a and H41 experienced a sustained period of higher than normal water levels, while H35 developed a near-basal connection to the glacier drainage system and H40a eventually became disconnected from it. Following this event, borehole WL behaviour in all the holes studied was very different from that in the earlier period and there was clear evidence for basal forcing of WL fluctuations.

As outlined above, H35 connected to the subglacial drainage system via an englacial conduit located 5 m above the glacier bed. The connection event lasted three days, and diurnal WL cycles were superimposed on the 70 m fall in WL. These diurnal cycles supplanted the two-day WL cycle that had occurred prior to connection. The onset of the connection event in H35 on JD 211 coincided with an interval of extremely high water levels in H34 and H36. Synchronous with these water level events, turbidity peaks were recorded in H34, H40a, H41 and H42. The disconnection of H40a began on JD 213 and was complete by JD 215.

The connection event in H35 was rather different in character from borehole connection events recorded elsewhere. Extremely rapid water level drops which occurred when a borehole intersected the subglacial drainage system during drilling were recorded by Hodge (1976) and Stone and Clarke (1993), while drops lasting a few hours, which resulted from the progressive enlargement of hydraulic links between an over-pressured borehole and the subglacial drainage system, were described by Engelhardt (1978) and Fountain (1994). The event in H35 was thus unusual in its duration and also because it involved formation of a new englacial connection. Given that the event began at a time when the pressure transducer in H36 recorded a transient water level indicative of substantial artesian pressure, and that the head difference between H35 and H36 fell to near-zero for the first time immediately after this event (Figure 3b), we suggest that the connection was initiated by hydrofracturing of the basal ice. The initial connection cannot have provided an effective drainage route since the WL fall in H35 was rapidly reversed by the diurnal melt cycle on JD 212. The most dramatic WL fall occurred between JD 212 and 213, but this too was reversed by the melt cycle of JD 213, so that the WL fall was not completed until JD 214. This suggests that enlargement of the initial fracture by ice melt to form an efficient connection to the subglacial drainage system was a relatively slow process.

The turbidity events that occurred during the connection of H35 indicate simultaneous disturbance of the glacier bed. Since EC records show no major perturbations during this period, it seems likely that the turbidity events resulted from mechanical disturbance of the bed by the glacier, rather than from enhanced subglacial water flow. Such disturbance could have been induced by either increased glacier sliding or deformation of subglacial sediments. If increased sliding was involved, it may have resulted in the growth of lee-side cavities, creating new void space beneath the glacier and helping to form the pathways by which basal water flowed across the bed after JD 215. Frequent turbidity events after JD 215 may have been the result of erosional enlargement of these pathways by subglacial water flow. The disconnection of H40a from the subglacial drainage system was probably linked to these events. Increased sliding could have advected the base of this borehole into a hydrologically isolated area of the bed such as those on which H35 and H42 were apparently located before JD 210, while mechanical disturbance of subglacial sediments could have blocked the hydraulic links between the base of the borehole and the subglacial drainage system.

The subglacial drainage system after Julian day 215

After JD 215, the subglacial drainage system seems to have been much more complex than it was before JD 210. F–R plots allow the identification of at least three elements to the system, and provide evidence that some parts of the bed changed hydrological status on a diurnal basis. Within the connected area of the bed, there was clearly a highly transmissive flow system which could not support large hydraulic gradients, and also a more resistive system which was connected either continuously or intermittently to the transmissive system. The transmissive system experienced large, discharge-related diurnal pressure fluctuations which, together with direct inputs from the glacier surface, acted as a forcing for water level variations in boreholes located within the more resistive system. With increasing distance from boreholes connected to the transmissive system, diurnal water level fluctuations became increasingly damped and lagged behind those within the transmissive system. Head differences between the two systems oscillated on a diurnal time-scale, but were generally much higher at times of rising pressure than at times when pressure was falling. This suggests that flow resistance within the resistive system may have been directionally asymmetric (cf. Alley, 1996, p. 655).

The transmissive system was probably a major subglacial channel, but the nature of the resistive system is more difficult to ascertain. This system obviously included links between the bases of some boreholes and the transmissive system, but it is not clear whether these links developed solely because of the creation of the boreholes, or whether they formed part of a natural drainage system. However, although the character of the links may have been significantly altered by drainage of the water introduced into the boreholes during drilling, the fact that not all boreholes connected suggests that the resistive system is not entirely a borehole-induced artefact. It is unclear whether the links within the resistive system consisted of microchannels, interconnected cavities, a macroporous horizon overlying a till surface or the pore space within permeable sediments.

Within the unconnected regime, water levels fluctuated inversely, but non-linearly, with those in the transmissive system. This behaviour reflects cyclic transfer of mechanical support for ice overburden between connected and unconnected areas as water pressure fluctuated within the transmissive system. Over some areas of the bed, WL fluctuations alternated between being inverse with and lagged behind those in the transmissive system regime on a diurnal time-scale. This suggests that the extent of the connected area of the bed varied diurnally. A link to the connected regime was established when pressure in the transmissive system was rising, and broken again as it fell. When the link was sealed, borehole WLs in these intermittently connected areas varied inversely with those in the transmissive system. When it was open, they lagged slightly behind those in the transmissive system. There was, however, a difference in the height of the transmissive system WLs at which the link opened and closed, indicating that it was much harder for an unconnected area to become connected than vice versa. Opening of the link was probably accomplished by a combination of glacier uplift in response to rising water pressures, and erosion or ice melt by water flowing between the connected and unconnected areas of the bed under very high hydraulic gradients.

Thus the major feature that distinguishes the drainage system after JD 215 from that before JD 210 is the existence of a transmissive element, probably a subglacial channel, which acted as the source of a strong diurnal water pressure forcing. This forcing had three main effects which were not apparent in the earlier period: (i) it initiated spatially propagating pressure waves, which travelled through the resistive system producing damped and systematically lagged water level fluctuations in boreholes connected to this system; (ii) it involved water pressure fluctuations, which were large enough to produce load transfer between connected and unconnected areas of the bed, thus inducing the inverse and alternating modes of water level fluctuation; (iii) it produced large head differences between the transmissive and resistive systems and thus induced the flow of mineralized basal water across much larger areas of the bed than in the early period. The key question to be addressed concerning the transmissive system is whether it actually developed during the hydrological event of JD 210–215, or whether it simply became easier to detect after this event because of a change in discharge regime that produced a more variable pressure regime within it. This question is addressed in detail below.

SUBGLACIAL CHANNEL FORMATION

Apart from the borehole measurements described above, the best evidence for subglacial drainage evolution at Haut Glacier d'Arolla comes from dye-tracing studies conducted in 1990 and 1991. Nienow (1993) found that when repeated dye injections were made at individual moulins, the dye travel time from moulin to glacier snout decreased progressively over the melt season, until it eventually reached a stable minimum value. This value increased in magnitude and was reached later in the season at moulins located further up-glacier. The decrease in travel time began before the transient summer snow-line passed each moulin, but the minimum value was not reached until up to five days after the snow-line had passed. Nienow concluded that by the time the minimum value was reached, a system of major subglacial channels connected the moulin to the glacier snout. Before this date, the channel system was growing headward towards the moulin, but a section of inefficient distributed drainage system lay between the moulin and the channel system. Nienow argued that channel formation was stimulated by changes in the magnitude and diurnal variability of water inputs to moulins which occurred following exposure of an impermeable, low-albedo ice surface in the drainage catchment upstream of the moulin.

To determine whether the events of JD 210–215 could be linked to a change in water input regime to moulins located 250 m up-glacier from the drill site, the runoff hydrograph into these moulins was simulated using a model developed by Arnold *et al.* (in press). This model simulates surface melt within the catchment feeding the moulins using an energy balance approach (Arnold *et al.*, 1996) and routes this melt to the moulins using an algorithm that simulates vertical percolation through a snowpack, horizontal flow through a saturated layer at the base of the snowpack and overland flow across bare ice surfaces (Arnold *et al.*, in press). The results (Figure 13) show a major change in runoff regime after JD 210, some 4–5 days after the snow-line passed through the drill site. Between JD 210 and 215, both the daily minimum and maximum runoff values increased significantly above peak values recorded earlier in the season. After JD 215, however, the daily minimum runoff decreased rapidly, eventually reaching zero, while the daily maximum increased

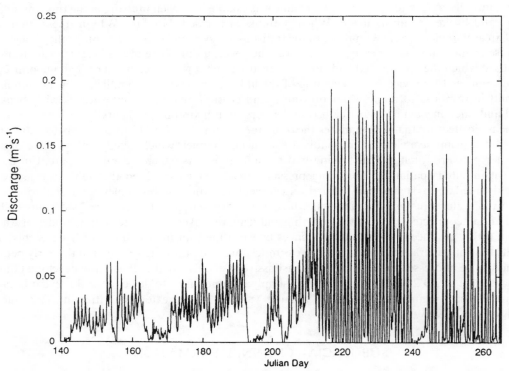

Figure 13. Simulated time-series of meltwater inputs to the group of moulins located immediately up-glacier of the drill site during the 1993 melt season

rapidly. These changes resulted from the removal of the snowpack from the catchment feeding the moulins, and are entirely consistent with those which Nienow (1993) hypothesized were responsible for subglacial channel formation in 1990 and 1991. We therefore suggest that channel formation was initiated by the rise in both maximum and mean diurnal water inputs to upstream moulins between JD 210 and 215. The heavy rainstorm of JD 211–212 may also have provided a stimulus to channel formation, since the basal connection in H35 developed immediately after this event. The onset of strong diurnal pressure variations within the subglacial drainage system was, however, linked to the initiation of a highly variable water input regime after JD 215. We note that, had a subglacial conduit existed before JD 210, the modelled change in moulin water inputs could have produced a change in the pressure regime within such a conduit to one similar to that observed after JD 215 (Arnold *et al.*, unpubl.). Nevertheless, in the absence of clear evidence for the existence of such a conduit before JD 210, we prefer the interpretation that it formed between JD 210 and 215.

If the inferred changes in the configuration of the subglacial drainage system within the drill site area were initiated by changes in the discharge regime into moulins located up-glacier from the drill site, it seems likely that, at the local scale, channel growth occurred in a down-glacier direction. However, as the snow-line retreated up-glacier during the melt season, it exposed glacier ice within the catchments of moulins located progressively further from the glacier snout. Thus, the process of channel formation occurred later in the season further up-glacier, giving the impression that the channel system as a whole developed by headward growth (Nienow, 1993). Individual channel segments grew by local melt-induced enlargement and down-glacier extension, and eventually linked up with established channel systems located further down-glacier. When the connection to the established channel system was made, free drainage became possible in the area between the feeder moulins and this system. Water stored within this area could then be drained, and water pressure could oscillate freely in response to the diurnal discharge regime. Clarke (1996, his Figure 9) simulated behaviour of this sort using a lumped element model of a subglacial drainage circuit.

CONCLUSIONS

Borehole measurements have clarified our understanding of subglacial drainage system evolution at Haut Glacier d'Arolla. Major changes in borehole behaviour resulted from the development of a subglacial channel during the period JD 210–215, 1993. Prior to channel formation, borehole WL fluctuations were driven by direct inputs of surface water to individual boreholes and were low in amplitude and poorly correlated. Basal water pressures and subglacial water fluxes across connected areas of the glacier bed are inferred to have been low. Channel formation was promoted by rapidly rising melt and rainfall-induced discharges which resulted in a period of high subglacial water pressures. It was associated with a marked change in the diurnal discharge regime of the stream draining into moulins located immediately up-glacier from the drill site area. This change resulted from the removal of the snowpack from the catchment feeding these moulins and involved a significant increase in the amplitude of diurnal discharge fluctuations. Following channel formation, borehole WLs were driven in part by discharge-related fluctuations in channel water pressure, though the character of the fluctuations was not uniform. Holes connected directly to the channel system showed identical WL fluctuations, while holes connected to it via a more resistive drainage system showed damped and systematically lagged fluctuations. Holes in unconnected areas of the bed showed inverse WL fluctuations, while holes in intermediate areas showed an alternating response which included elements of the unconnected and lagged responses. The inverse and alternating behaviours provide evidence for diurnal transfer of mechanical support for the ice overburden between connected and unconnected areas of the bed as channel water pressures fluctuated, and for diurnal changes in the extent of the connected regime. Once the channel existed, large diurnal variations in head difference developed between it and the adjacent resistive drainage system. These induced lateral flow of mineralized basal water through the resistive system, and this in turn entrained suspended sediment.

ACKNOWLEDGEMENTS

This work was supported by grants from the UK Natural Environment Research Council (GR3/8114), the Royal Society and the Natural Sciences and Engineering Research Council of Canada (to C.S. and M.S.). We thank Michael Nielsen for the design and construction of field equipment, and Alun Hubbard, Keith Carr and Matthew Tully for field assistance. Martyn Tranter and Peter Nienow provided constructive reviews of an earlier draft. Yvonne Bams, Patricia and Basile Bournissen provided logistical support in Arolla, where M. Beytreyson of Grande Dixence S. A. also gave us invaluable assistance. We are indebted to Garry Clarke for stimulating and encouraging our interest in subglacial hydrology.

REFERENCES

Alley, R. B. 1996. 'Towards a hydrologic model for computerised ice-sheet simulations', *Hydrol. Process.*, **10**, 649–660.

Arnold, N. S., Willis, I. C., Sharp, M., Richards, K. S., and Lawson, W. 1996. 'A distributed surface energy balance model for a small valley glacier. I. Development and testing for the Haut Glacier d'Arolla, Valais, Switzerland', *J. Glaciol.*, **42**, 77–89.

Arnold, N. S., Richards, K. S., Willis, I. C., and Sharp, M. In press. 'Initial results from a semi-distributed, physically-based model of glacier hydrology', *Hydrol. Process.*, in press.

Bindschadler, R. A. 1983. 'The importance of pressurised subglacial water in separation and sliding at the glacier bed', *J. Glaciol.*, **29**, 3–19.

Brown, G. H., Sharp, M., Tranter, M., Nienow, P., and Gurnell, A. M. 1994. 'The impact of post-mixing chemical reactions on the major ion chemistry of bulk meltwaters draining the Haut Glacier d'Arolla, Valais, Switzerland', *Hydrol. Process.*, **8**, 465–480.

Brown, G. H., Tranter, M., and Sharp, M. J. 1996. 'Experimental investigations of the weathering of suspended sediment by alpine glacial meltwater', *Hydrol. Process.*, **10**, 579–598.

Boulton, G. S. and Hindmarsh, R. C. A. 1987. 'Sediment deformation beneath glaciers: rheology and geological consequences', *J. Geophys. Res.* **92** (B9), 8903–8911.

Clarke, G. K. C. 1996. 'Lumped element analysis of subglacial hydraulic circuits', *J. Geophys. Res.*, **101** (B8), 17547–17559.

Copland, L., Harbor, J., Gordon, S., and Sharp, M. 1997. 'The use of borehole video in the investigation of the hydrology of a temperate glacier', *Hydrol. Process.*, **11**, 211–224.

Engelhardt, H. F. 1978. 'Water in glaciers: observations and theory of the behaviour of water levels in boreholes', *Z. Gletscher. Glazialgeol.*, **14**, 35–60.

Fountain, A. G. 1994. 'Borehole water level variations and implications for the subglacial hydraulics of South Cascade Glacier, Washington State, USA', *J. Glaciol.*, **40**, 293–304.

Gordon, S. 1996. 'Borehole-based investigations of subglacial hydrology', *MSc Thesis*, University of Alberta, 142 pp.

Hantz, D. and Lliboutry, L. 1983. 'Waterways, ice permeability at depth, and water pressures at Glacier d'Argentière, French Alps', *J. Glaciol.*, **29**, 227–239.

Harbor, J. M., Sharp, M., Copland, L., Hubbard, B., Nienow, P., and Mair, D. 1997. 'The influence of subglacial drainage conditions on the distribution of velocity within a glacier cross-section', *Geology*, **25**, 739–742.

Hodge, S. M. 1976. 'Direct measurement of basal water pressure: a pilot study', *J. Glaciol.*, **16**, 205–218.

Hooke, R. LeB., Wold, B. and Hagen, J.-O. 1985. 'Subglacial hydrology and sediment transport at Bondhusbreen, south-west Norway', *Geol. Soc. Am. Bull.*, **96**, 388–397.

Hubbard, B. P. 1992. 'Basal ice facies and their formation in the western Alps', *PhD Thesis*, University of Cambridge. 258 pp.

Hubbard, B., Sharp, M., Willis, I. C., Nielsen, M., and Smart, C. C. 1995. 'Borehole water level variations and the structure of the subglacial hydrological system of Haut Glacier d'Arolla, Switzerland', *J. Glaciol.* **41**, 572–583.

Iken, A. 1981. 'The effect of the subglacial water pressure on the sliding velocity of a glacier in an idealised numerical model', *J. Glaciol.*, **27**, 407–421.

Iken, A. and Bindschadler, R. A. 1986. 'Combined measurements of subglacial water pressure and surface velocity of Findelengletscher, Switzerland: conclusions about drainage system and sliding mechanism', *J. Glaciol.*, **32**, 101–119.

Kamb, W. B. 1987. 'Glacier surge mechanism based on linked cavity configuration of the basal water conduit system', *J. Geophy. Res.*, **92**, (B9), 9083–9100.

Ketterling, B. 1995. 'Electrical conductivity of waters in glacier boreholes', *MSc Thesis*, University of Western Ontario. 156 pp.

Lamb, H., Tranter, M., Brown, G. H., Gordon, S., Hubbard, B., Nielsen, M., Sharp, M., Smart, C. C., and Wallis, I. C. 1995. 'The composition of meltwaters sampled from boreholes at the Haut Glacier d'Arolla, Switzerland', *IAHS Publ.*, **228**, 395–403.

Murray, T. and Clarke, G. K. C. 1995. 'Black-box modelling of the subglacial water system', *J. Geophys. Res.*, **100** (B7), 10231–10245.

Nienow, P. W. 1993. 'Dye-tracer investigations of glacier hydrological systems', *PhD Thesis*, University of Cambridge. 337 pp.

Richards, K. S., Sharp, M., Arnold, N., Gurnell, A. M., Clark, M., Tranter, M., Nienow, P., Brown, G. H., Willis, I. C., and Lawson, W. 1996. 'An integrated approach to modelling hydrology and water quality in glacierized catchments', *Hydrol. Process.*, **10**, 479–508.

Sharp, M., Richards, K. S., Willis, I. C., Arnold, N., Nienow, P., and Tison, J.-L. 1993. 'Geometry, bed topography and drainage system structure of the Haut Glacier d'Arolla, Switzerland', *Earth Surf. Process. Landf.*, **18**, 557–571.

Smart, C. C. Statistical evaluation of boreholes as indicators of basal drainage systems', *Hydrol. Process.*, **10**, 599–614.

Stone, D. B. and Clarke. G. K. C. 1993. 'Estimation of subglacial hydraulic properties from induced changes in basal water pressure: a theoretical framework for borehole response tests', *J. Glaciol.*, **39**, 327–340.

Stone, D. B. and Clarke, G. K. C. 1996. 'In situ measurements of basal water quality and pressure as an indicator of the character of subglacial drainage systems', *Hydrol. Process.*, **10**, 615–628.

Stone, D. B., Clarke, G. K. C., and Blake, E. W. 1993. 'Subglacial measurement of turbidity and electrical conductivity', *J. Glaciol.*, **39**, 415–420.

Tranter, M., Brown, G. H., Raiswell, R., Sharp, M., and Gurnell, A. M. 1993. 'A conceptual model of solute acquisition by alpine glacial meltwaters', *J. Glaciol.*, **39**, 573–581.

Tranter, M., Sharp, M., Willis, I. C., Brown, G. H., Hubbard, B., Nielsen, M., Smart, C. C., Gordon, S., Lamb, H., and Tully, M. 1997. 'Variability in the chemical composition of *in situ* subglacial meltwaters', *Hydrol. Process.*, **11**, 59–77.

Walder, J. S. 1986. 'Hydraulics of subglacial cavities', *J. Glaciol.*, **32**, 439–445.

16

AN INTEGRATED APPROACH TO MODELLING HYDROLOGY AND WATER QUALITY IN GLACIERIZED CATCHMENTS

KEITH RICHARDS

Department of Geography, University of Cambridge, Cambridge, UK

MARTIN SHARP

Department of Geography, University of Alberta, Alberta, Canada

NEIL ARNOLD*

Department of Geography University of Cambridge, Cambridge, UK

ANGELA GURNELL

Department of Geography, University of Birmingham, Birmingham, UK

MICHAEL CLARK

Department of Geography, University of Southampton, Southampton, UK

MARTIN TRANTER

Department of Geography, University of Bristol, Bristol, UK

PETER NIENOW

Department of Geography, University of Edinburgh, Edinburgh, UK

GILES BROWN

Department of Earth Studies, University of Wales, Aberystwth, UK

IAN WILLIS

Department of Geography, University of Cambridge, Cambridge, UK

AND

WENDY LAWSON

Department of Geography, University of Auckland, Auckland, New Zealand

ABSTRACT

The results are summarized of an integrated investigation of glacier geometry, ablation patterns, water balance, melt-water routing, hydrochemistry and suspended sediment yield. The ultimate objective is to evaluate the assumptions of lumped, two-component mixing models as descriptors of glacier hydrology, and to develop a semi-distributed physically based model as an alternative. The results of the study demonstrate that a reconstruction of probable subglacial drainage alignments can be achieved through a combination of terrain modelling based on estimated potential surface and dye tracing experiments. Recession curve analysis, evidence of the seasonal instability of the englacial and subglacial electrical conductivities assumed in a mixing model, evidence of the non-conservative behaviour of water chemistry in the presence of suspended sediment, and evidence of the seasonal evolution of the subglacial drainage system based on dye tracing all indicate that an alternative to a lumped, static model of the hydrology is necessary.

* Now at Department of Geophysics and Astronomy, University of British Columbia, Vancouver, B.C., Canada

The alternative presented in this paper is based on the combination of an energy balance model for surface melt which operates on an hourly time step and accounts for the changing spatial distribution of melt through the day as shading patterns change, and routing procedures that transfer surface melt to moulins on the basis of glacier surface gradients, then route water through reconstructed conduit systems using a hydraulic sewer-flow routing procedure.

INTRODUCTION

This paper reviews some aspects of the interpretation and modelling of glacier hydrology by drawing on a research project undertaken at the Haut Glacier d'Arolla, Valais, Switzerland since 1989. This research project was initiated in the light of several perceived scientific imperatives. These included the need to adopt an integrated approach to the study of glacier hydrology and water quality; the need to break away from the limited approach of interpreting glacial hydrology from the records of a restricted number of outflow water quality parameters (e.g. specific conductivity); the need to abandon the simplified and restrictive assumptions of lumped, two-component (englacial/subglacial) models of glacier hydrology; and the need to explore in the context of glacier hydrology the potential of the physically based semi-distributed modelling strategies now common in conventional hydrology.

Three specific objectives of the research project emerged from these considerations. The first was to develop and calibrate such a physically based model of glacier hydrology to simulate water routing and residence time in relation to distributed meltwater inputs. The second objective was to evaluate the predictions of this model against those of solute-based flow separation methods used to distinguish rapidly and slowly routed components, and to undertake a more rigorous analysis of glacial hydrochemistry in relation to routing and residence time. The results of these first two objectives were then to be used to develop multivariate rating and time series approaches to the analysis of suspended sediment data which would be sensitive to the influence of sediment supply and delivery. This third objective, involving the interpretation of suspended sediment data, was considered qualitatively, through the establishment of a sound basis for defining stationary series suspended sediment data after the analysis of hydrological and hydrochemical data. The need for an integrated and iterative approach is exemplified by the fact that the hydrochemical data can themselves only be interpreted in the context of information on suspended sediment concentrations.

PROGRAMME OF FIELD INVESTIGATIONS

The $11.7 \, \text{km}^2$ basin of the $6.3 \, \text{km}^2$ Haut Glacier d'Arolla (Figure 1) was chosen as the field area for the project. The glacier is a classic tongue extending down-valley from a basin-shaped accumulation area and is set in a topographically clearly defined catchment. The glacier rests on a predominantly till-covered bed. Data were available from a nearby meteorological station and flow in the outlet melt stream was monitored by Grand Dixence SA (GDSA). These data were augmented by several field seasons of research conducted by the Department of Geography at Southampton University. The data considered in this paper are from two long field seasons (June–September) undertaken there in 1989 and 1990, and a shorter one in 1991. These provided the background data necessary for a fully integrated study of hydrology and water quality. To satisfy the discussed objectives, four specific requirements were identified: (i) to establish the character of the drainage system and its evolution over time; (ii) to determine the parameters required to model meltwater production processes, water inputs to the subglacial drainage system and water routing characteristics; (iii) to evaluate the water storage behaviour of the glacier; and (iv) to describe water quality variations and explain how these may relate to drainage evolution and water storage.

Given these requirements, the basic data collected are indicated in Tables I–III and typical data series for one season are presented in Figure 2.

The meteorological data listed in Table I were required as input to a distributed model of surface melt

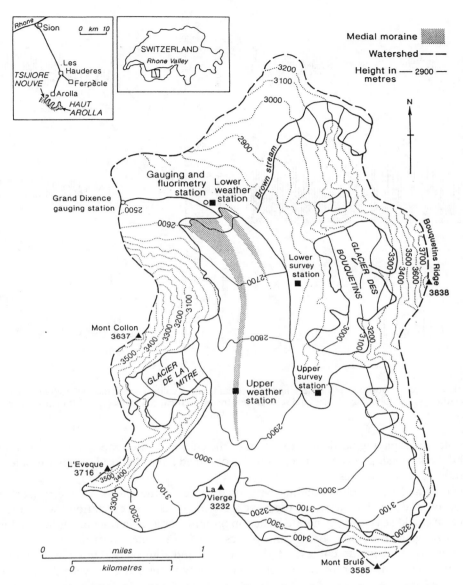

Figure 1. Map of the Haut Glacier d'Arolla and its catchment, showing locations of gauging, fluorimetry, weather and survey stations. UWS and LWS are upper and lower weather stations

developed during the course of the project (see section on Surface Melt and Water Balance). A meteorological station was established at a site roughly 300 m from the glacier terminus at 2560 m and this was augmented in 1990 by a second at 2885 m. Daily ablation monitoring at a series of centreline and cross-section stakes (Table III) provided a check on melt predictions by the model by giving the integrated daily melt at different locations. Discharge monitoring was by pressure transducer measurements of stage at a gauging station established at c. 2550 m on the main meltwater stream. These were calibrated by stage–discharge relationships and checked using GDSA records (the melt stream drains into a hydroelectric power intake operated by GDSA). The discharge record was used to estimate the water balance and to calculate chemical and suspended sediment loadings. Turbidity and conductivity were also monitored at the gauging station (Table II). A pump sampler was installed at the gauging station to calibrate the turbidity record. Gurnell *et al.* (1992b) have assessed the reliability and representativeness of this combination of turbidity

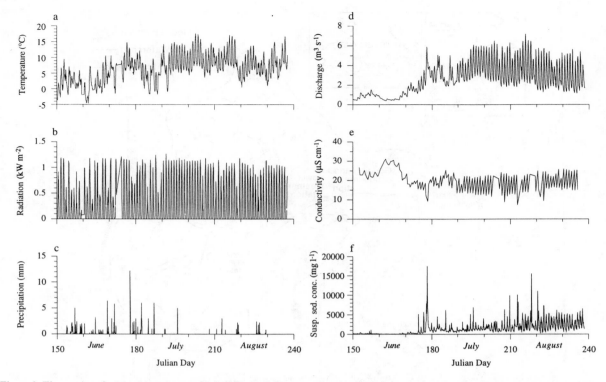

Figure 2. Time series of selected hourly meteorological, hydrological and water quality data for the 1990 melt season. (a) Temperature; (b) shortwave radiation; (c) precipitation; (d) stream discharge; (e) stream electrical conductivity; and (f) stream suspended sediment concentration

and pump sampler data, paying particular attention to the choice of filter paper used to obtain the calibration data, and the relationship between concentrations obtained from the fixed sampler intake and section-integrated USDH-48 hand samples.

Ground survey and radio-echo sounding were used to define the glacier geometry (surface and subglacial topography, and ice thickness; see the next section). Fixed survey stations were established and re-used in successive field seasons to generate basic topographic data. Dye tracing studies enabled the identification of two subglacial drainage basins and routes and travel times from moulins to the glacier terminus, and are interpreted in the context of the seasonal and diurnal water storage changes. Finally, water quality sampling, ranging from continuous conductivity monitoring to twice-daily (and sometimes

Table I. Field data collection programme: weather station

	1989 Season		1990 Season	
	Sensed	Averaged	Sensed	Averaged
Precipitation	Accumulated	10 min	Accumulated	10 min
Incident radiation	30 min	30 min	30 s	10 min
Relative humidity	30 min	30 min	10 min	10 min
Air temperature	1 min	30 min	30 s	10 min
Wind speed	1 min	30 min	30 s	10 min
Wind direction	1 min	30 min	30 s	10 min

Delta-T automatic weather station. In 1989 and 1990 a weather station was established at 2550 m below the glacier terminus; in 1990 a second weather station was established at 2850 m on the western medial moraine.

Table II. Field data collection programme: river gauging station

	1989 Season		1990 Season	
	Sensed	Averaged	Sensed	Averaged
Stage	10 s	10 min	10 s	10 min
Conductivity	30 s	10 min	30 s	10 min
Water temperature	30 s	10 min	30 s	10 min
Turbidity	30 s	10 min	30 s	10 min
Suspended sediment	2 h	2 h	2 h	2 h

Campbell data logger, Druck pressure transducer, PHOX conductivity meter, Partech turbidity meter, ISCO pump sampler and Solapak solar panels. In 1989 and 1990 the gauging station was established at c. 2550 m in a section where all melt streams had combined into a single channel. The gauging station was also the location for fluorimetry using a Turner fluorimeter.

hourly) sampling for a wide range of determinants (pH, pCO_2, $SI_{calcite}$, cations, anions), was undertaken to characterize the hydrochemistry and subglacial weathering environment to assist in the interpretation of subglacial flow pathways and storages. In addition to sampling at the gauging station, this programme included extensive sampling and analysis of supraglacial waters. This paper only reports the interpretation of a subset of the total accumulated data gathered during the project.

GLACIER GEOMETRY AND DRAINAGE RECONSTRUCTION

The geometry of the Haut Glacier d'Arolla, which ranges in elevation from 2560 to 3400 m, was established using a combination of surface topographic survey and radio-echo sounding. The survey was conducted by theodolite and electronic distance meter (EDM) (1989) and geodimeter total station (1990) from two fixed survey stations overlooking the upper basin and the main lower tongue (at 3018·6 and 2772·3 m), and from roving surveys in areas obscured from those vantage points. Over 820 data points were surveyed. The bed topography was established by radio-echo sounding using a 1–10 MHz monopulse radar along a series of 19 cross-sections and three long profiles. Ice thicknesses were established at 242 points. The resulting irregular pattern of elevation data was converted using the UNIRAS bi-linear interpolation routines (UNIRAS, 1990) to provide a digital elevation model (DEM) of bed and surface topographies, which are shown in Figure 3. Three bedrock basins appear to exist, the lowest of which seems to be overdeepened. The other bedrock hollows appear high in the western tributary glacier and below the bedrock step down-glacier from the main eastern upper firn basin.

In addition to the field-surveyed heights and ice thicknesses, elevations of the surrounding watershed surface were obtained from the Swiss 1:25 000 topographic sheet (by scanning the mapped contours) and the detailed glacier DEM was located within a more generalized DEM of the whole valley. This allowed the development of a distributed surface energy balance and melt model (Arnold *et al.*, in press) capable of dealing with temporally varying patterns of topographic shading on the glacier surface.

The spatially distributed meltwater calculated by this model then forms the input for a model of subglacial drainage. The route taken by this water is dependent on the hydraulic potential beneath the glacier.

Table III. Field data collection programme: ablation survey

	1989 Season		1990 Season	
	No. of sites	Measured	No. of sites	Measured
Ablation	15 stakes	Daily	20 stakes	Daily
Albedo			20 sites	Weekly

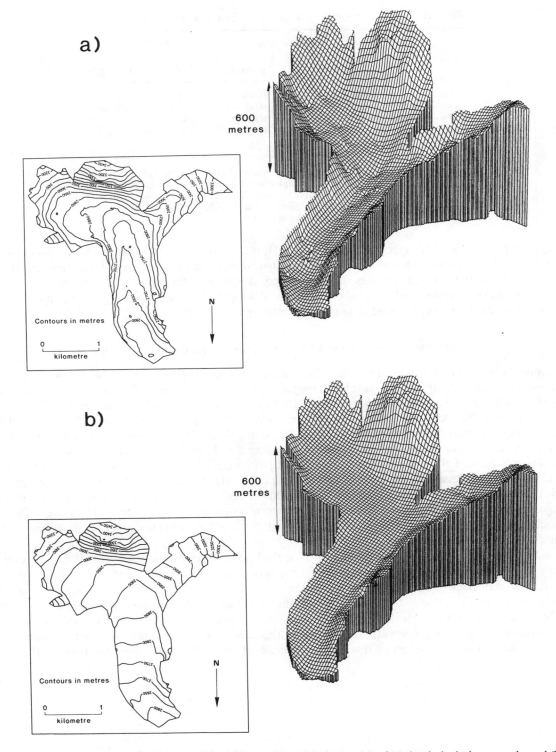

Figure 3. Three-dimensional and contour representations of the digital elevation models of (a) the glacier bed topography and (b) the surface topography of the Haut Glacier d'Arolla

Using Shreve's (1972) expression for the hydraulic potential (Φ)

$$\Phi = \rho_w g H_b + \rho_i g (H_s - H_b) \tag{1}$$

where the heights H are of the bed (subscript b) and surface (s), the densities ρ are for water (w) and ice (i), and g is the gravitational constant, the elevation of a potential surface was calculated for each cell in the DEM. By deriving cell slope magnitudes and directions the cumulative area (the total number) of up-slope cells whose melt production passes through each cell can be calculated using algorithms similar to those suggested by Zevenbergen and Thorne (1987). The 'up-glacier contributing area' on the potential surface is then mapped (Figure 4) and this provides an initial estimate of the probable location of subglacial drainage conduits (Sharp *et al.*, 1993). The figure shows two main axes of drainage, on the west and east sides of the main tongue of the glacier which drain the western and eastern upper basins, respectively.

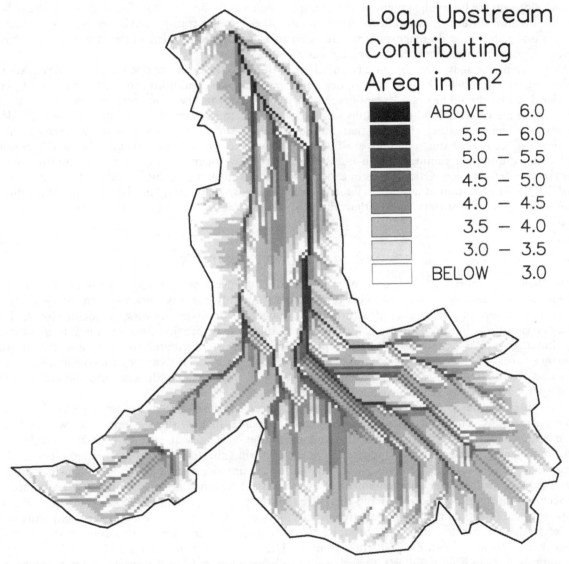

Figure 4. Map of the up-glacier contributing area on the hydraulic potential surface, illustrating the means of identifying the likely subglacial drainage configuration. The DEM is based on a 20 m grid and the key defines classes of the common logarithm of contributing drainage area

A similar reconstruction was made for drainage on the bedrock surface. This differed in predicting only a single channel beneath the main glacier tongue and in predicting two channels in the eastern firn basin, the more northerly of which is more remote from the valley walls than that suggested by the potential surface. The reconstruction from the potential surface appears to match moulin and conduit locations rather better, both as observed visually and as interpreted from dye tracing experiments. Different algorithms for routing area (a surrogate for flow) downslope were investigated, such as assuming that the area is passed to only the lowest adjacent cell, or to all lower cells proportionally to gradient. Differences between these methods were less pronounced than those between the potential surface and bedrock surface drainage patterns. In addition, sensitivity of the map of 'up-glacier contributing area' to the grid resolution (10, 20 and 40 m) was assessed; the 10 m grid was considered too detailed for the data and the 40 m grid unduly generalized.

The moulins selected for dye tracing experiments were surveyed and mapped and were found to lie along the lines of the conduits predicted from the potential surface. This in part reflects the dominant effect of surface slope on the hydraulic potential, which implies that valleys on the ice surface tend to lie above those on the potential surface. It also reflects the closure of other moulins transported by ice flow away from the surface and subsurface drainage systems, but also to some degree the deliberate choice for dye tracing studies of moulins with evidence of running water.

An apparent anomaly arises because calculations of the expected extent of open and surcharged conduits (following Hooke, 1984) suggest that under summer discharge conditions, conduits beneath the Haut Glacier d'Arolla should be open, except immediately upstream from the lower overdeepened basin. If these calculations are correct, conduit locations would be expected to be determined by the form of the bedrock topographic surface rather than by that of the potential surface. This apparent discrepancy can be explained if, as is probable, conduits are full when they develop, and subsequently the total amount of wall melting during a summer season is insufficient to allow them to migrate any significant distance from their initial location. Calculations of expected melt rates suggest that this is likely to be the case (Sharp *et al.*, 1993). The location of moulins, as noted earlier, may also help to explain the locations of conduits in lows on the potential surface, rather than the bedrock surface, as they would pin channels to particular locations.

DYE TRACING EXPERIMENTS

In the 1989, 1990 and 1991 field seasons over 500 dye tracing injections were made to moulins and crevasses whose locations had been mapped. This programme (probably the most intensive on any glacier) was designed to determine the spatial structure of the subglacial drainage network, to identify diurnal and seasonal variations in travel time and drainage structure and to provide data for the interpretation of the drainage system hydraulics. Such interpretation therefore contributes to an assessment of water residence times, of the open or closed system nature of the subglacial weathering environment in terms of access to O_2 and CO_2 sources, and of likely access to supplies of sediment, and therefore informs both hydrochemical and sediment transport studies.

In all three seasons rhodamine B was used as the tracer, whereas in 1990 fluorescein was also used so simultaneous injections from different moulins could be made. In 1989 the dye emergence was monitored by manual sampling and fluorimetry at the gauging station, and subsequently by continuous siphon flow and data logging. When the dye appeared at the monitoring site (downstream of the confluence of the five outlet streams shown in Figure 5), the five outlet streams were sampled to identify which of them were carrying the dye. Over a large number of experiments it was then possible to identify the connections between moulins and outlet streams and therefore the catchments draining to the outlet streams. Figure 5 shows the catchments of the five outlet streams and illustrates that the north-east (lower right) flank of the glacier forms a separate catchment. Broadly, the conclusions on the spatial structure of the subglacial network are consistent with the prediction from the DEM of the potential surface (Figure 4). The typical dye return curves in Figure 6 illustrate both a rapid response and a delayed response from, respectively, a moulin well connected to the conduit system and one that is poorly connected and feeding a more distributed form of subglacial drainage. Dye returns from the 'connected' moulins suggest average velocities, of

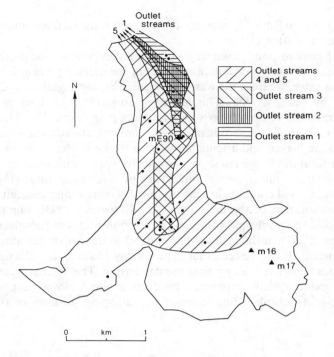

Figure 5. Catchments of the outlet streams as determined by dye tracing experiments carried out in 1989 from the moulins as mapped (moulins 16, 17 and E90 are marked; see Figures 6 and 7)

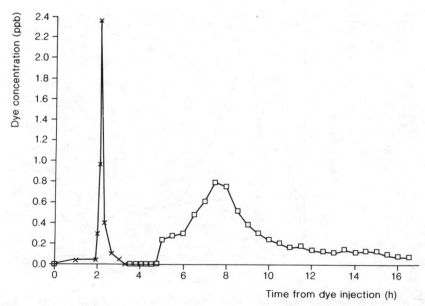

Figure 6. Typical dye return curves from moulin 16 (connected to the conduit network; plotted as crosses) and moulin 17 (unconnected; plotted as squares)

$0.5\,\mathrm{m\,s^{-1}}$ (ranging from 0.3 to $0.8\,\mathrm{m\,s^{-1}}$), whereas the average velocity from 'unconnected' moulins is an order of magnitude lower (c. $0.01\,\mathrm{m\,s^{-1}}$).

Diurnal variations in the dye response from single injection points reflect the pattern of insolation receipt; the eastern moulins have significantly longer travel times in the morning when the glacier surface remains shaded by the Bouquetins ridge, whereas the western moulins display slightly longer travel times in the late afternoon when shaded by Mont Collon. This gives rise to strong hysteresis in the velocity–discharge relationships shown by multiple returns from single moulins on a given day. This behaviour reflects the fact that inputs to individual moulins have diurnal cycles similar to the radiation (and melt) cycles, whereas the total discharge cycles are lagged and attenuated relative to these. Hence the local discharge falls in the late afternoon, while the total discharge rises. As the velocity varies with local discharge, this also falls in the late afternoon for western moulins. Travel times vary by a factor of three (Figure 7), suggesting that much of the travel path is in small conduits linking moulins to more major conduits. The seasonal changes in travel time reflect the evolution of the drainage system (Nienow, 1993). Figure 8 illustrates the 1990 experimental data, as a plot of travel time contours in the coordinates of 'up-glacier distance to injection site' and 'date of experiment'. When the contour lies parallel to the Julian day axis, the travel time from a particular location is constant. It can be seen that between day 160 and day 210 the contours are inclined, indicating that travel times from particular points are decreasing. This is interpreted as the result of head-ward extension of the conduit system, expanding headwards into a slower responding distributed basal drainage system up-glacier; the broken line suggests the changing position of the head of the conduit

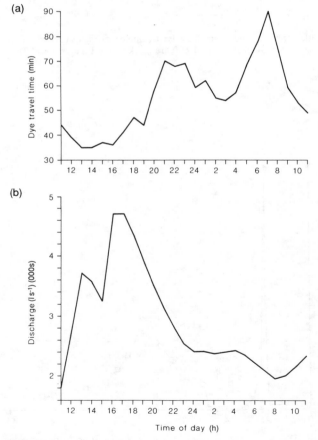

Figure 7. (a) Variation in dye travel time for experiments conducted from moulin E90 (see Figure 5) at hourly intervals on 16–17 August 1990. The diurnal hydrograph is shown in (b). The reduction in travel time at 2.00 am is attributable to a rainfall event which halted the normal fall in discharge at this time of the day

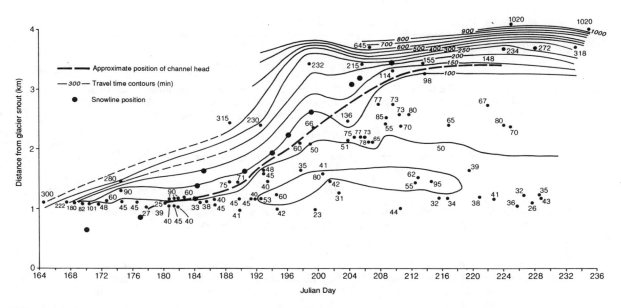

Figure 8. Time–space variation in dye travel time during the 1990 melt season based on injections carried out between 11.00 and 17.00 h. The seasonal growth of a rapid transit drainage system in the lower 3 km of the glacier is apparent and the approximate position of the conduit head and the transient snowline are also indicated

network over time. Also plotted on the diagram is the position of the transient snow line, which suggests a close association between its up-glacier retreat and the development of subglacial channelized drainage. This is likely to reflect the effect of the increasing amplitude of the diurnal meltwater recharge cycle, which serves to open the conduits, and which occurs once snow cover has been removed from the glacier surface. This is in turn due to the lower albedo of ice surfaces imposing larger diurnal peak volumes of water, as well as to the removal of the reservoir effect of the snowpack which attenuates and delays surface runoff.

SURFACE MELT AND WATER BALANCE

The surface energy balance model (Arnold *et al.*, in press) predicts surface ablation energy as the sum of short wave solar radiation, turbulent heat flux and long wave radiation, following Braithwaite and Olesen (1990). Table IV identifies the relationships and parameters used in deriving the surface energy balance. The parameters were measured at the lower meteorological station and values in each cell of the DEM are then calculated using a standard lapse rate and elevation–pressure relationship. For the calculation of short wave radiation, cell slope angles and aspects are derived from the DEM by examining the eight surrounding cells. Topographic shading is determined by calculating solar elevation and azimuth at the time of concern, and 'walking' away from the surface along the solar beam until a ground elevation higher than the solar beam is encountered. For a sunny cell, the radiometer measurement is then adjusted to account for the cell slope and aspect effects. If a cell is shaded, it receives only diffuse radiation. Figure 9 shows three-dimensional views of the DEM with shading calculated within the model; these confirm the conclusions about diurnal variations in dye travel times related to shading and melt production.

   Cell albedos are calculated using relationships proposed by Oerlemans (1993), with an exponential relationship describing the albedo of snow as a function of snow depth, accumulated melt and background albedo. The albedo is further adjusted given the addition of fresh snow during the melt season (assumed to occur when the cell temperature is below 1°C at a time when precipitation is recorded at the meteorological station). The model initial condition is a snow depth distribution over the glacier, based on measured snow depths at the start of the melt season; it then accounts for the melt thinning of old snow and the gain and loss of new snow until ice is exposed. All of the associated albedo changes are calculated accordingly.

Table IV. Equations and parameters in the melt model

Basic equation

$$ABL = SHF + LHF + SWR + LWR$$

where ABL is ablation, SHF and LHF are sensible and turbulent heat fluxes, and SWR and LWR are short and long wave radiation fluxes (all expressed in equivalent ablation units of mm water per unit time). Components of the energy balance are

$$SHF = K_s PTV$$

where K_s is given below (and includes L), P is atmospheric pressure (Pa), T is air temperature (°C) and V is wind speed (m s^{-1}).

$$LHF = K_1 \delta e V$$

where K_1 is given below (and includes L) and δe is the difference between the vapour pressure of the air and the saturation vapour pressure at the glacier surface (Pa).

$$SWR = (1 - \alpha) Q_i / L$$

where α is albedo, Q_i is solar radiation received at the surface (W m^{-2}) and L is the latent heat of fusion of water (J kg^{-1}). The radiation measured by the radiometer at the meteorological station (Q) is converted to the radiation received in a cell on the glacier surface (Q_i) by

$$Q_i = (Q / \sin S)[\sin S \cos Z + \cos S \sin Z \cos(A - Y)]$$

where S is the angle of the sun above the horizon, A is the solar azimuth, and Z and Y are the cell slope angle and aspect derived from the DEM. The albedo α is given by

$$\alpha = \alpha_{sn} - (\alpha_{sn} - \alpha_b) e^{-5d} - a_3 M$$

where α_{sn} is the standard albedo of old snow, d is snow depth (m of water equivalent), M is accumulated melt, and α_b is the background albedo profile. This is given by

$$\alpha_b = a_1 \arctan [(h - E + 300)/200] + a_2$$

in which E is the equilibrium line elevation (m) and h is height. After new snowfall, the albedo is redefined as α_{ns}, defined by

$$\alpha = \alpha_{sn'} - (\alpha_{sn'} - \alpha) e^{-5p/20} - a_4 M$$

where $\alpha_{sn'}$ is the standard albedo of fresh snow of depth p. The parameters a_1 to a_4 are empirically based on field albedo data.

$$LWR = (I_{in} - I_{out})/L$$

where I_{in} and I_{out} are incoming and outgoing long wave radiation (W m^{-2}). I_{out} for a glacier surface at °C is given below and

$$I_{in} = \varepsilon \sigma T_a^4$$

where ε is the effective emmissivity of the sky and σ the Stefan–Boltzmann constant.

Parameters used in the model

Sensible heat scalars			
(Ice)	K_s	$6 \cdot 34 \times 10^{-6}$	m kg^{-1} K^{-1} s^2
(Snow)	K_s	$4 \cdot 42 \times 10^{-6}$	m kg^{-1} K^{-1} s^{-2}
Latent heat scalars			
(Ice, condensation)	K_1	$9 \cdot 83 \times 10^{-3}$	m kg^{-1} s^2
(Ice, evaporation)	K_1	$11 \cdot 14 \times 10^{-3}$	m kg^{-1} s^2
(Snow, condensation)	K_1	$6 \cdot 86 \times 10^{-3}$	m kg^{-1} s^2
(Snow, evaporation)	K_1	$7 \cdot 77 \times 10^{-3}$	m kg^{-1} s^2
(Water, fusion)	L	$3 \cdot 34 \times 10^5$	J kg^{-1}

Table IV. Continued

Equilibrium line elevation	E	3000	m
Snow albedo	α_{sn}	0·75	
Mean catchment albedo	α_m	0·4	
Stefan-Boltzmann constant	σ	$5 \cdot 70 \times 10^{-8}$	$W\,m^{-1}\,K^{-4}$
Black-body radiation from ice	I_{out}	316	$W\,m^{-2}$
Cloud constant	k	0·26	
Lapse rate		0·0065	$°C\,m^{-1}$

Figure 10 compares model albedo estimates for two dates with different snow distributions and Figure 11 illustrates the predicted spatial pattern of the melt for these days. This figure shows an up-glacier decrease in melt related to altitude, a cross-glacier trend which shows the effects of west-side shading by Mont Collon, and the difference between Figure 11a and 11b shows the effects on melt rates of the seasonal change in albedo as ice is exposed. To test the model the hourly calculated melt values were cumulated over 24-hour periods to allow comparison with the daily field measurements of ablation. The results for two of the ablation stake sites are shown in Figure 12 and show generally good correspondence.

The glacier water balance for 1990, calculated as the difference between the modelled melt values over the glacier surface (plus any precipitation) and the measured stream discharge at the gauging station, is shown in Figure 13. In the early part of the melt season (to approximately day 180), the distributed system is recharged by snowmelt, which varies only weakly over diurnal time-scales, and water storage is fairly constant. Variations do occur, however, and seem to be linked to changes in the weather. For instance, the sunny weather (and associated higher melt rates) from days 151 to 156 result in water storage within the glacier; this period is followed by a cold, wet period in which the water balance is negative. As melting of snow continues, exposure of a low albedo ice surface results in greater amounts of recharge and stronger diurnal variations which induce transient water pressure peaks which destabilize the distributed system and allow conduit growth. At this point (in late June–early July, at about day 180), the water balance switches from approximately constant to negative (from no storage to net loss) and there is a contemporaneous pronounced rise in suspended sediment concentrations suggestive of a restructuring of subglacial drainage. Thus the water balance becomes negative during episodes of rapid snowline retreat and conduit growth, when the subglacial water reservoir drains as water stored in cavities is tapped and efficiently evacuated by growing conduits. This results from the rearrangement of the piezometric surface as conduits form and locally steep pressure gradients develop from cavities to conduits. This process is currently being observed in pressure transducer records from boreholes drilled along the axis of the main drainage line of the glacier. Once conduit growth stops, the water balance again becomes positive; presumably net drainage only occurs again when meltwater inputs reduce and eventually cease at the end of the melt season.

INVESTIGATIONS OF HYDROCHEMISTRY AND SUSPEND SEDIMENT

Studies of water quality were undertaken both to provide a basis for the critical assessment of existing models of glacier hydrology and to provide data which would constrain interpretation of the hydrological processes occurring at the Haut Glacier d'Arolla. Sharp (1991) has stressed the considerable potential for interpretation of glacier hydrology based on detailed hydrochemical studies and criticized both the inappropriate choice of determinands for measurement (conductivity or base cations) and the weak bases for interpretation in much existing research (mixing models, rating curves and time series models). The hydrochemical data collected in the 1989 and 1990 field seasons were evaluated initially by Gurnell *et al.* (1994) to confirm the validity of a hierarchical sampling strategy involving (i) automatic monitoring of specific conductivity, (ii) twice-daily (high and low flow) sampling for the analysis of major cations and anions and (iii) intermittent hourly sampling through diurnal cycles. The use of conductivity as a surrogate for total dissolved solids is well established, but it is now evident that to draw inferences about the hydrological system and its associated weathering environments from an understanding of the dynamics of meltwater chemistry

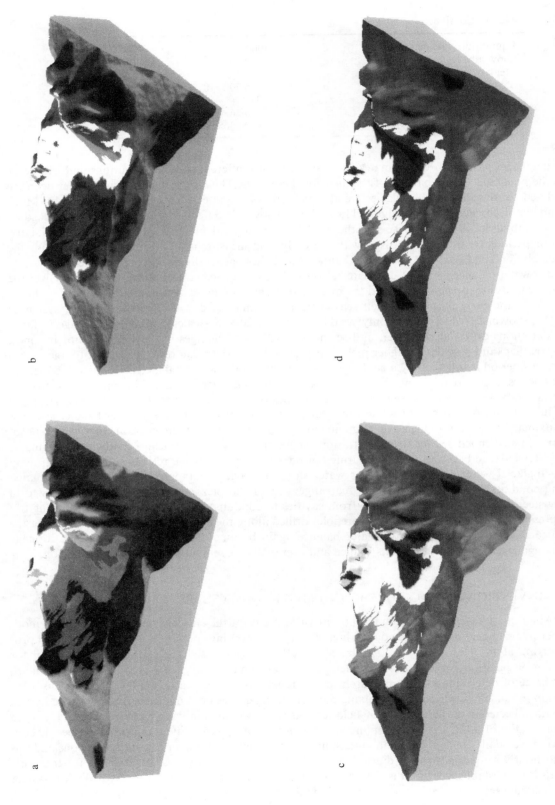

Figure 9. Three-dimensional computer images of the Haut Glacier d'Arolla, looking SSE towards Mont Brulé, with topographic shading as calculated in the melt model for 21 June. (a) 0700; (b) 0900; (c) 1700; and (d) 1900 h

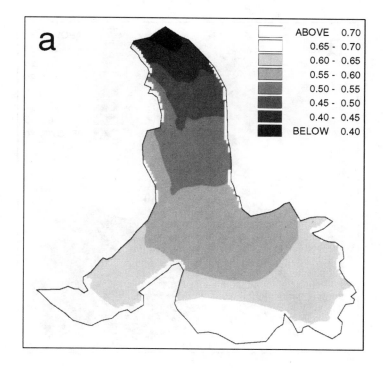

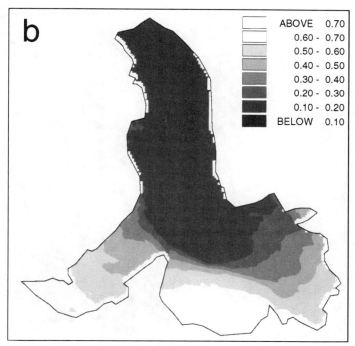

Figure 10. Modelled albedo patterns on (a) 30 May and (b) 15 August 1990

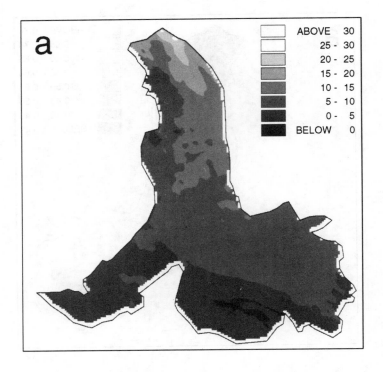

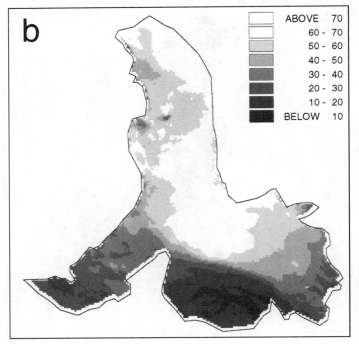

Figure 11. Predicted spatial pattern of melt for the two dates whose predicted albedo is shown in Figure 10. Units are mm day^{-1}

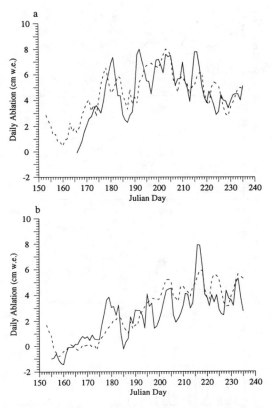

Figure 12. Modelled and measured daily ablation at ablation stakes (a) B-90 (close to the glacier snout) and (b) N-90 (close to the head of the main basin). Broken line, modelled values; solid line, measured values

it is essential to acquire more detailed cation and anion data. However, this is very demanding of field and laboratory time. The evaluation of the sampling strategy adopted during the 1989 field season confirmed that a hierarchical scheme is an acceptable basis for interpretation of the detailed hydrochemistry, with regression, Box–Jenkins ARIMA or transfer function models being used to interpolate hourly concentrations on days when only two samples are taken. This rigorous sampling and analysis of a wide range of determinands has allowed a detailed reassessment and redefinition of the underlying assumptions of the two-component mixing model frequently used to represent glacier hydrology (e.g. Collins, 1979).

The first major issue is that it is arbitrary to assume that there are only two forms of glacial drainage, which are typically believed to represent englacial and subglacial waters, though there is some confusion about this terminology in published work. Here we will use 'subglacial to refer to the component routed slowly through the glacier (usually near the bed) and 'englacial' to refer to the component routed quickly through the glacier (often away from the bed). Secondly, it is evident that most of the chemical parameters used to characterize the englacial and subglacial components are variable and non-conservative. Brown and Tranter (1990) show that it is impossible to define fixed C_e and C_s parameters (where C is the concentration of a determinand and subscripts e, s and b denote englacial, subglacial or bulk water). These have usually been obtained from the lowest and highest values observed in bulk waters through the melt season. However, inverse linear associations often exist between C_b and Q_b (where Q is water discharge) over the rising and falling limbs of individual diurnal discharge hydrographs and take the form

$$C_b = mQ_b + k \tag{2}$$

where m and k are constants. Conservative mixing of the englacial and subglacial components requires that

$$Q_e/Q_b = (C_s - C_b)/(C_s - C_e) \tag{3}$$

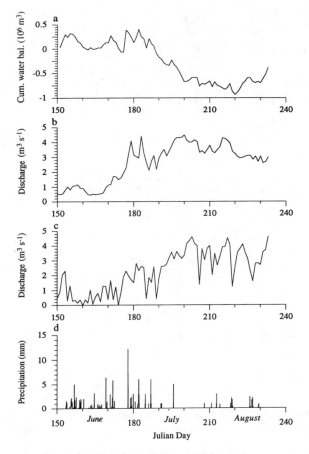

Figure 13. Cumulative water balance for summer 1990. (a) Water balance; (b) measured river discharge; (c) modelled surface melt; and (d) measured precipitation adjusted for altitude variation

Substitution of C_b and rearranging implies that

$$Q_e/Q_b = (-mQ_b + C_s - k)/(C_s - C_e) \qquad (4)$$

If C_s and C_e are assumed to be constant, the bulk discharge is therefore a simple linear measure of the mass fraction of the englacial component in the bulk water (Q_e/Q_b). The value of C_s is defined when the mass fraction of the englacial component is zero. Given this condition, it follows that

$$Q_b = (C_s - k)/m \qquad (5)$$

Because m is negative, C_s cannot be greater than k (because Q_b cannot be negative). Constant values of C_s and C_e, together with the assumption of only two components, imply that C_s cannot be less than k, and hence it follows that $C_s = k$ (Brown and Tranter, 1990). C_s can therefore be estimated by extrapolation of the relationship between C_b and Q_b. However, the value of C_s ($= k$) for calcium determined from single diurnal discharge cycles during the 1989 season varies from 310 in June to 390 μ equiv. 1^{-1} Ca in July. The concentration of the subglacial component is neither constant nor equivalent to the maximum recorded value in the bulk water (470 μ equiv. 1^{-1}).

The second key point, as demonstrated by Sharp *et al.* (1995), is that very few of the conventionally available determinands are conservative during the mixing process. In the case of Haut Glacier d'Arolla the behaviour of SO_4^{2-} alone approximates to that of a chemically conservative quality parameter for hydrograph separation in a mixing model. Tranter *et al.* (1991) have exploited this parameter in mixing model flow separations, defining a seasonally varying subglacial flow sulphate concentration using the linear extrapolation method outlined above, and assuming a sulphate concentration of zero in ice melt-derived englacial flow. The results again show that separation of the discharge in the Haut Arolla meltwater stream for typical June, July and August diurnal hydrographs in 1989 and 1990 results in very different englacial and subglacial flow proportions from those predicted by a conventional constant composition mixing model (Figure 14). This difference is significant for subsequent interpretation of the seasonal evolution of the glacial drainage system because the different proportions of englacial and subglacial flow predicted by the two methods carry implications for the interpretation of the configuration of the drainage system.

However, in relation to the non-conservative behaviour of water chemistry, Sharp (1991) has emphasized that surface exchange of base cations for protons in solution is rapid and largely irreversible during initial water–sediment interaction and results in an increase of pH and reduction of pCO_2. Proton supply by dissociation of H_2CO_3 may restore the pCO_2 to equilibrium with the atmosphere, in an 'open' system, but failure to achieve equilibrium (either because of a lack of access to atmospheric CO_2 or because of a kinetic effect when CO_2 diffusion is slower than proton consumption) occurs in a 'closed' system (Raiswell, 1984). It is often assumed that conduit flow involves short residence times and that solute acquisition is limited by comparison to that occurring in the distributed system; however, the large, fresh specific surface area of glacially derived suspended sediment may result in the rapid consumption of protons and the pCO_2

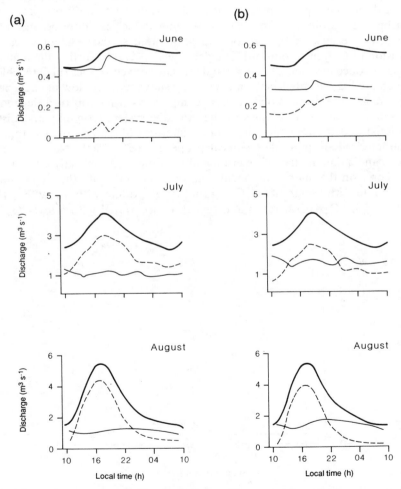

Figure 14. Separation of the bulk meltwater hydrograph (upper solid line) into quickflow (broken line) and delayed flow (lower solid line) using (a) the constant component method of Collins (1978) and (b) the variable composition method as described in the text (Tranter et al., 1991). Note that the difference between the two methods decreases as the melt season progresses

of meltwaters may therefore respond to variations in suspended sediment concentration. Furthermore, whether or not the conduits are surcharged will determine access to atmospheric CO_2. Additional evidence of the 'open' or 'closed' nature of subglacial drainage is provided by NO_3^- concentrations which may drop to zero for prolonged periods in bulk waters, even though nitrate is still present in supraglacial snowmelt. This may be explained by reduction of NO_3^-, possibly coupled with pyrite oxidation, which appears to occur under suboxic conditions in water-filled cavities (Tranter *et al.*, 1994).

This consideration of water–sediment interaction proves to be critical to hydrograph separation based on a two-component mixing model. Comparing hydrograph separations using SO_4^{2-} (being the most conservative parameter) with those using conductivity, Mg^{2+}, Na^+, HCO_3^-, Ca^{2+} and K^+, it is possible to identify this as the rank order of the degree of conservatism of these solute species. Plotting the difference between estimates of subglacial flow using separations based (a) on SO_4^{2-} and (b) on a second determinand (e.g. electrical conductivity) results in a clear positive association with the contemporaneous suspended sediment concentration (Figure 15), which in turn suggests that a significant proportion of the solute load of bulk meltwaters is derived from post-mixing reaction with suspended sediment. This proportion may reach 70% under high discharge conditions in August (Brown *et al.*, 1994).

Further hydrochemical research questions are concerned with the process of solute acquisition and the characterization of glacial waters (Tranter *et al.*, 1993). It is possible to distinguish subglacial water (which is solute-rich and exhibits high pCO_2 because of its slow transit time and the predominance of sulphide oxidation and carbonate dissolution) from englacial water (which is dilute because of rapid transit time and exhibits low pCO_2 because of the predominance of rapid surface exchange reactions together with slow gaseous diffusion of CO_2 into solution). Subglacial waters can be further distinguished from englacial waters on the basis of concentrations of aerosol and sea salt derived species (nitrate, chloride) leached from the snowpack which are high in subglacial waters (fed by snowmelt) and low in englacial waters (fed by ice melt). Bulk waters reflect the proportions of these components. which in turn depend on the amount of meltwater input, the structure and volumes of the drainage system (conduits and distributed systems in changing proportions) and the residence times and quantities of water stored within these components. If waters are characterized by multiple chemical properties (EC, Ca^{2+}, Na^+, K^+, Mg^{2+}, SO_4^{2-}, NO_3^-, HCO_3^-, SO_4/HCO_3) and a data matrix of determinand by sample time is analysed by principal components methods, sample scores on the main components can be used to classify the waters into four basic types. These are: type 0, dilute with a high SO_4/HCO_3 ratio; type 1, dilute with a low SO_4/HCO_3 ratio; type 2, concentrated with a high SO_4/HCO_3 ratio; and type 3, concentrated with a low SO_4/HCO_3 ratio. These

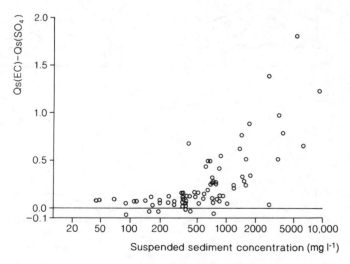

Figure 15. Plot of the difference between estimates of delayed flow discharge derived from hydrograph separations based on the methods using variable SO_4^{2-} concentrations and variable EC, and the contemporaneous suspended sediment concentration. Note that a threshold sediment concentration occurs above which post-mixing solute acquisition affects the results of flow separation

four water types are considered to reflect distinctive sources: type 0 is unmodified supraglacial water; type 1 is supraglacial water modified by limited contact with suspended sediment within the conduit system; type 2 is water supplied by snowmelt and passing relatively slowly through the subglacial distributed drainage system, and type 3 is essentially type 2 water mixed with either type 0 or type 1 water, and which has had some contact with suspended sediment during its passage through the conduit system. Figure 16 shows that the results of this analysis are consistent between years in identifying diurnal fluctuations between two end-member classes, which differ during the meltwater season as the drainage system evolves. In particular, during 1990 (Figure 16b) the early melt season is dominated by types 2 (at low flow) and 3 (at high flow) waters, but as subglacial channels begin to develop, type 0 waters replace type 3 at high flow. Towards the end of the season, type 1 waters replace type 0 at high flow as the suspended sediment concentrations increase at high discharges, and type 3 waters replace type 2 at low discharges, indicating that icemelt inputs remain dominant even at night. This hydrochemical variation provides evidence for a subdivision of the meltwater season consistent with that obtained from dye tracing and hydrometeorological and water balance data (see previous two sections).

One objective of the Arolla research project was to identify ways of incorporating the influence of sediment supply and delivery into multivariate rating and time series analyses of suspended sediment data. This is far from straightforward in terms of identifying numerical parameters to represent these influences. However, the recognition of distinct phases in the evolution of the interdependent glacier melt–meltwater drainage–meltwater chemistry–suspended sediment system enables the establishment of separate simple regression and time series models for statistically stationary data obtained within each phase. A qualitative description of five phases of the melt season is summarized in Table V and uses information about the

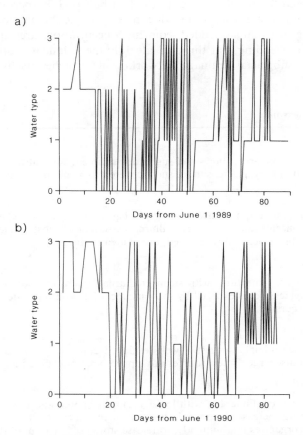

Figure 16. Seasonal variation in water type draining from the Haut Glacier d'Arolla in (a) 1989 and (b) 1990. Water types are as defined by principle components analysis of meltwater chemistry data and are defined in the text

changing pattern of melt, the position of the snowline, the drainage network structure as inferred from dye tracing studies (conduit extent and rate of headward growth), and both mean levels and diurnal variations in the hydrochemistry and suspended sediment concentrations.

Given the phases of the melt season defined in Table V, and as Gurnell *et al.* (1992a) have noted, it is possible to derive distinct univariate Box–Jenkins ARIMA models and bivariate transfer function models to describe hydro-meteorological data in each phase. Table VI identifies examples of cross-correlation analyses which show that the air temperature series leads the ($\log_{10}$ transformed) discharge series by 19 h in the first period but by 2 h in period 4, reflecting the changing efficiency of transfer of melt energy into the diurnal hydrograph via a low albedo ice surface and an expanded conduit system. Discharge variation lags conductivity by 5 h in period 1 and by 1 h in period 4, whereas it leads turbidity in period 1 but coincides with turbidity variation in period 4. This demonstrates that any attempt to describe the interdependence among hydro-meteorological variables by models based on whole season data is likely to be invalid and will lead to incorrect inferences concerning the associated processes because of non-stationary model structure and parameter values. However, qualitatively, a multivariate basis does exist for subdividing the melt season into physically meaningful sub-periods, rather than using arbitrary statistical procedures for the identification of statistical stationarity.

One alternative to time series modelling of seasonal variations in suspended sediment concentration after such a subdivision of the melt season is illustrated by Clifford *et al.* (1995). They focus on the whole season and interpret the residuals from a simple bivariate power function concentration–discharge rating relationship. The residuals from this function vary systematically through the melt season. The residuals have very low amplitude variations in June, then drift to increasingly negative and more variable values (i.e. lower concentrations than expected for the discharge) in the first half of July (period 2 in Table V), recover in late July (period 3) and remain consistently positive in August (i.e. higher concentrations than expected for the discharge). This suggests that suspended sediment concentrations are supply-limited during the period when conduits are still extending, but through late July the sediment sources are being tapped by an extending and integrating subglacial drainage network which reaches its limit in August, after which

Table V. Phases of the melt season

Period 1 (early to mid-June)

    Low air temperatures; low and invariant flows and suspended sediment concentrations; high solute concentrations; recession after period of melt in May; glacier surface snow-covered; no conduits, distributed drainage over glacier bed

Period 2a (mid-to late June)

    Gradual increase in discharge and appearance of clear diurnal discharge cycles reflecting increased levels of melt energy; decreasing solute concentrations, but larger diurnal variation of most solute species; suspended sediment concentrations relatively stable; glacier snow-covered; distributed drainage

Period 2b (early to mid July)

    Higher suspended sediment concentrations with stronger diurnal cycles; low solute concentrations; higher mean discharges as lower albedo glacier ice is exposed by snowline recession, continued development of diurnal discharge cycles as conduit growth occurs

Period 3 (mid July to early August)

    Higher peak flows and lower minima suggest a well-developed conduit system; unstable solute and suspended sediment relationships with discharge as new sources are tapped; area of exposed ice increasing

Period 4 (early August onwards)

    Strong diurnal discharge cycles continue, but with lower mean, maximum and minimum discharges; variable melt especially when snowfall events increase albedo; a decline and stepped increase of sediment concentration may occur if the subglacial drainage reorganizes intermittently, to release stored sediment until the supply becomes exhausted; maximum up-glacier retreat of snowline separating exposed glacier ice from the previous winter's snow cover

---

Table VI. Bivariate time series models for seasonal sub-periods, after first differencing (lagged to best match position). Based on Gurnell *et al.* (1992a)

Time period	Input series	Output series	Lag (h)	Residual MS
1	δtemp	δlog$Q$	+19	0·000
	δlog$Q$	δcond	−5	0·04
	δlog$Q$	δlogSSC	+17	0·01
4	δtemp	δlog$Q$	+2	0·001
	δlog$Q$	δcond	−1	1·16
	δlog$Q$	δlogSSC	0	0·01

SSC, Suspended sediment concentration.

time concentrations remain high despite the gradual decline in discharge. Approaching the suspended sediment data in this way provides confirmation of the general subdivision of the melt season shown in Table V and illustrates the value of an integrated analysis in which such confirmatory evidence can be provided through the monitoring of interdependent variables.

LUMPED HYDROLOGICAL MODELS

Having obtained a rigorous description of the seasonally changing character of the glacier, its surface snow cover, its drainage system and its hydrology and water quality, it is desirable to attempt to model its meltwater runoff processes. The two-component mixing model commonly used to interpret glacier hydrology is equivalent to assuming a two-reservoir lumped model. One approach to the identification of reservoir storages and the glacier hydrology which differs from the conventional mixing model identification based on outflow chemistry involves an empirical analysis of diurnal hydrograph characteristics (Gurnell, 1993). Such an analysis reveals that a two-component structure may be an inadequate representation of the hydrology of the Haut Glacier d'Arolla. This analysis is based on fitting exponentials to recession curves of the form

$$Q_t = Q_{0 \cdot e}^{(-(t-t_0)/K)} \qquad (6)$$

where Q_0 is the initial discharge at time t_0, Q_t is discharge at time t and K is a reservoir recession coefficient (Figure 17). Recession curves often contain more than one linear component when plotted on log-linear graphs and these components can be interpreted as the recessions from different reservoirs. The 1989 data suggest three distinct reservoirs and longer term data suggest that a fourth 'very slow' reservoir may also exist. Approximate average reservoir constants are 12, 27 and 72 h for the three most frequently observed recessions, although some dependency on the level of initial discharge suggests a non-linearity of reservoir behaviour.

Another test of the validity of a lumped reservoir-based representation of the hydrology is to perform numerical experiments using a variety of simple reservoir models. Such experiments can be undertaken simply using a spreadsheet-based model, and were undertaken as part of the Arolla study in the form of simulation modelling of flow patterns assuming both linear and non-linear reservoirs. For example, a linear case assumes

$$Q_t V_t / K \qquad (7)$$

where Q_t and V_t are the discharge and reservoir volume at time t and K is the reservoir storage constant. If the predicted output diurnal hydrographs for different storage constant (K) values are described by (a) a dimensionless amplitude and (b) the lag between the input melt cycle chosen to drive the simulation and the output discharge, these descriptive parameters can be related to the reservoir characteristics. For a single reservoir, as the reservoir coefficient increases, the experiments show both that the discharge amplitude decreases as a proportion of the recharge amplitude and that the lag increases. For a two-reservoir

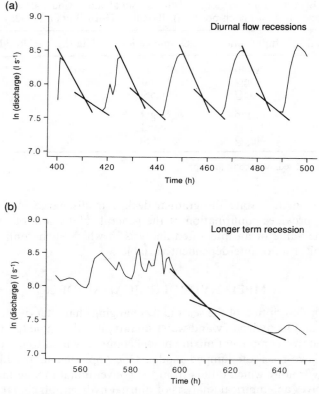

Figure 17. Examples of semi-logarithmic recession curves for diurnal and longer term discharge variations, illustrating the separation of flow components draining from different reservoirs

system it is evident that the controls of the output amplitude and lag are multivariate, relating to both the reservoir coefficient and the recharge amplitude. This implies that inference based on fitting a two-component mixing model of the outflow chemistry is under-determined and liable to be physically erroneous, although the output form may be most closely controlled by the behaviour of the faster reservoir.

This conclusion is reinforced by inspection of the dimensionless discharge amplitude and the time of peak discharge in the observed diurnal hydrograph parameters for the 1990 Arolla record. The dye tracing results, and calculations of the status of conduits as open channel or surcharged, together suggest that the melt season can be subdivided into four main periods as in Table V: (i) early to mid-June, a distributed system over the whole glacier; (ii) mid-June to mid-July, snowline retreats across the lower closed basin, and a surcharged conduit system develops in the lower glacier; (iii) mid-July to early August, further snowline retreat and associated headward conduit migration resulting in an open conduit system feeding the lower surcharged conduit and (iv) early August onwards, conduit growth no longer keeps pace with snowline retreat and a residual distributed system remains in the upper firn basin.

These results are consistent with those based on hydrochemical data (Figure 16), and only differ from the conclusions summarised in Table V in that periods 2a and 2b are combined. Figure 18 shows that the hydrograph parameters adjust between these four periods and simulation experiments have been conducted to simulate these changes. Period (i) is best represented as a single reservoir with $K = 22$ h. To simulate the sharp reduction in lag between periods (i) and (ii) it is necessary to introduce a second, faster reservoir, as suggested earlier. As the relative proportions of the glacier bed drained by distributed and conduit systems change from periods (ii) to (iii) and then to (iv), the reservoir constants must be altered so that the observed outflow hydrograph characteristics can be matched. Thus these experiments confirm the hydrochemical and recession curve investigations in illustrating the dangers of assuming a constant form of two-component

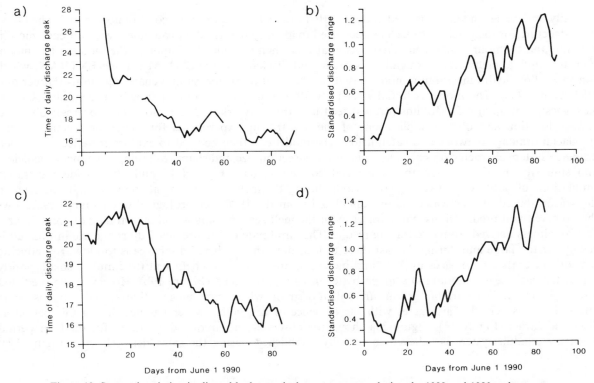

Figure 18. Seasonal variation in diurnal hydrograph shape parameters during the 1989 and 1990 melt seasons

system and imply that a model with a more distributed and dynamic structure is required to describe glacier hydrology. Significantly, the form of the outflow hydrograph through the meltwater season is very similar for 1989 and 1991, but the timing of changes varies in a way which is consistent with the timing of snowline retreat.

DISTRIBUTED HYDROLOGICAL MODEL

A more sophisticated model must accommodate spatially distributed melt and the seasonal changes in water storage and flow routing. Initially for this project a model was considered that was based on an empirical multiple regression model of daily melt and a two dimensional single-conduit model of water routing using a finite difference implementation of the equations for unsteady, non-uniform flow in subglacial conduits presented by Spring and Hutter (1981). Although this model successfully identified the two-zone structure of the subglacial drainage system (upper basin distributed system and lower tongue conduit), it could not deal with the emerging dye tracing evidence of two sub-basins and three major conduits. As the field evidence indicated the obvious three dimensionality of the drainage system, further model development was undertaken to couple the numerical simulation of melt production and flow routing to the three-dimensional DEM format of the glacier and basin geometry data.

This has involved combining (i) the physically based distributed melt model (described earlier and in Arnold *et al.*, in press) running on an hourly time step and predicting melt generation in 20 m grid cells distributed over the glacier surface and the consequent meltwater inputs to particular known crevasse or moulin locations; with (ii) the sewer-flow element of the US EPA storm water management model (SWMM; Huber and Dickinson, 1988). This latter component provides a physically-based hydraulic routing procedure for a network of conduits (sewers) receiving inputs from a system of moulins (drains). The result is the first full distributed and physically based modelling structure for glacier hydrology.

The melt model predictions appear consistent with the empirical evidence of ablation when cumulated to the daily time-scale, and are thus used at the hourly time-scale as input to the conduit flow model. The DEM data structure for the glacier surface has been used to identify the tributary catchment area for each moulin location and the hourly melt generated over that area then provides the input to the moulin. The model used to simulate meltwater throughflow, storage and discharge is the SWMM routine EXTRAN, which is a sewer-flow hydraulic routing model based on the solution of the Saint Venant equations (Huber and Dickinson, 1988). The model is initially set up with a drainage network of subglacial conduits (sewers) as suggested by Figure 5. Moulins (drains) feed into this network at points dependent on their observed surface location and on the implications of the dye tracing experiments for the reconstruction of the subglacial drainage network and catchments. The model can then be used to explore the effects of varied drainage system characteristics such as conduit cross-section size and shape, Manning roughness coefficient and sinuosity in the unobservable subglacial conduit system. Figures 19 and 20 provide illustrative simulations of outflow for two 10-day periods during the melt season, one at the start and one later in the season, based on meltwater inflows calculated from the 1990 meteorological data. Figure 19a shows the observed and predicted discharge for early in the melt season using an initial estimate of conduit sizes and roughness derived from dye tracing results. The amplitude of the predicted diurnal cycle in this run is clearly far too large and during the last three to four days the predicted discharge is too low. By reducing the conduit cross-section sizes by 75% and increasing the roughness coefficient (to simulate small, poorly connected early season conduits) a much better fit can be obtained (Figure 19b). However, the predicted discharges in the last part of the run are still too small, which may suggest that stored water is being released in this period. This agrees with the evidence of a negative water balance during this period as discussed earlier (Figure 13). Figure 20a shows the observed and predicted discharge for the later period when the conduit system is believed to be fully established. Using the same network as in Figure 19a,

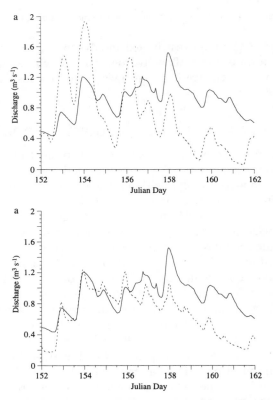

Figure 19. Early season hydrograph simulated using EXTRAN and the network structure based on Figure 5: (a) with 'standard' conduit sizes (see text); and (b) with smaller, rougher conduits. Broken line, modelled values; solid line, measured values

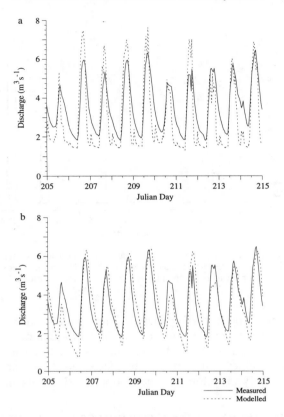

Figure 20. Late season hydrograph simulated using EXTRAN and the network structure based on Figure 5: (a) with 'standard' conduit sizes (see text); and (b) with rougher conduits and slower snowpack percolation. Broken line, modelled values; and solid line, measured values

the fit is clearly much better, although again the amplitude of the diurnal cycle is rather too large and the recession limb of each cycle rather too steep. Again, increasing the conduit roughness, and in this instance showing the percolation velocity of water through the remaining snowpack, greatly improves the match (Figure 20b).

Other experiments have been performed, including simulating the transfer of water to a distributed subglacial drainage system by inserting a side-spilling weir at each moulin location which extracts water at peak flow and diverts it to a 100 m long, 20 m wide, dead-end conduit which is shallow (0·05 m) and rough (Manning's $n = 0·25$) and which simulates the possible transfer under high pressure of water from a conduit to a distributed system. This modification again produces a diurnal hydrograph behaviour which is realistic, though only with rough main conduits, and implies that at least two classes of subglacial storage can be simulated: the major conduits themselves and a distributed system which stores water temporarily during periods of high subglacial water pressure before releasing it back to the main (conduit) drainage system when water pressures fall. These will have a distinctive recession character, as implied by the analyses outlined in the preceding section. Conduit dimensions can be varied to examine the sensitivity of model predictions to changing conduit size. The dynamic storage created when the flow backs up in narrow conduits is insufficient in total volume to dampen the outflow hydrograph significantly and moulin (drain) overflow can result. This is, however, only considered an acceptable prediction in certain locations where water has been observed in the field spouting from crevasses or moulins, or where moulin water pressure records (Figure 21) indicate the occurrence of such overspill events — observations that help to guide the choice of conduit diameters. EXTRAN output indicates when and where conduits become surcharged (i.e. where closed conduit flow occurs) and predicts the height of the hydraulic gradeline. Therefore there

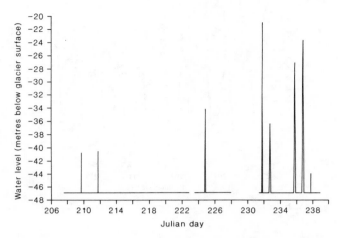

Figure 21. Water pressures measured over a 30 day period by a pressure transducer lowered to a depth of 47 m into a moulin in the lower part of the main accumulation basin. Pressures are averaged over 10 minutes and transient peak pressures when water was observed to reach the glacier surface are not represented

remains much glaciological potential in further experimental simulation modelling, which can be carried out now that a distributed model structure exists in which spatio-temporal variations of meltwater input can be routed through a glacier's hypothesized plumbing system. Furthermore, the wealth of empirical data available for the glacier enables constraints to be placed on the general range of parameter values, when model flow predictions imply variations in other properties that are consistent with the field evidence.

SUMMARY AND FUTURE RESEARCH NEEDS

In summary, this integrated investigation of glacier geometry, hydrology, hydrochemistry and sediment load has provided several valuable insights into the workings of a system which is notoriously difficult to observe. A variety of disparate methods has been combined to illuminate the process in which a seasonally retreating snowline alters the volume and spatial pattern of meltwater input to a drainage system that is consequently destabilized and whose reorganization during a period of net water storage results in changes in the subglacial weathering environment both as a result of altered access to CO_2 and increased access to and mobilization of readily weathered suspended sediment. Water residence times also change as the system evolves. The early season distributed system is fed by snowmelt and hence drains water with a higher content of anions derived from sea salt and acid aerosols than the channelized waters which are fed later in the season by melt derived from ice that has been leached of such species during the firnification process. Both empirical studies of recession curves and routing studies using a new physically based and spatially distributed model of meltwater production and routing suggest that it may be inappropriate to assume that glacier hydrology can be adequately described by a two-component lumped model, particularly if the character and behaviour of the two components are assumed to be invariant through time. The hydrochemical studies reported here have shown that the characterization of the chemistry of the two component flows in the mixing calculations based on such a model is likely to be arbitrary and of doubtful physical significance, as the component chemistries are neither constant through time nor conservative, particularly in the presence of suspended sediment.

A commonly applied methodology in studies of glacial water quality has been Box–Jenkins times series modelling (Gurnell and Fenn, 1984). The results of this integrated investigation do not provide convenient numerical parameters to improve multivariate transfer functions between discharge and sediment supply 'input' variables and quality (e.g. chemical determinand or suspended sediment concentration) 'output' variables. However, they demonstrate a physical basis, in terms of changes in the structure of the glacier drainage system, for the non-stationarity that often exists in whole season time series, emphasizing the importance of only developing such black-box models for sub periods within which reliable parameter estimates can be made. There are diagnostic statistical criteria for identifying the existence of non-stationarity within data, but these do not provide information about the causes of non-stationarity and may be unreliable when applied to short runs of data. An understanding of the seasonal evolution of a glacial system (which may be replicated from year to year at a given glacier) provides a better basis for time series modelling.

Although the research conducted at the Haut Glacier d'Arolla has provided a comprehensive understanding of its hydrology, and a methodology which is more widely applicable, further research remains to be carried out. For example, a hot-water drilling programme and measurements of basal water pressures and water quality parameters are providing further insight into conditions at the glacier bed. The fixed survey stations were established to enable repeat (daily) measurements of fixed points on the glacier surface. These are providing additional data on variations in glacier motion and can be compared with predictions of the variation in water pressures made by the routing model, with measurements of basal water pressure in boreholes, and with suspended sediment data (Willis *et al.*, this issue). Further studies involving weathering experiments have simulated the chemical interaction between meltwater and suspended sediment (Brown *et al.*, this issue, Brown *et al.*, 1994), and can investigate the kinetics of dissolution processes under appropriate temperature (and possibly pressure) conditions.

Further numerical model development includes an additional routine for simulating water flow through supraglacial snow and firn to provide an early season runoff delay. Simulation of the seasonal hydrograph for a season's simulated meltwater input (with retreating snowline and associated changes in the recharge amplitude) could be undertaken with fixed conduit geometry and structure, but with conduits that progressively enlarge as a function of wall-melting by the cumulative flow transmitted. The more realistic model structure introduced in this paper (compared with lumped two-component models) thus has considerable potential for incorporating and testing a range of theoretical ideas in glacio-hydrology.

ACKNOWLEDGEMENTS

The authors acknowledge the considerable support of the UK Natural Environment Research Council (NERC) in the form of two research grants (GR3/7004 and GR3/8114), two research studentships (P.N. and G.B.) and two post-doctoral research fellowships (G.B. and P.N.). We also thank undergraduates and others too numerous to mention for invaluable assistance in the field over several field seasons, and Jenny Wyatt for preparing the diagrams for this paper.

REFERENCES

Arnold, N., Sharp, M. J., Richards, K. S., and Lawson, W. 'A distributed surface energy balance model for a small valley glacier: I. Development and testing for the Haut Glacier d'Arolla, Valais, Switzerland', *J. Glaciol.*, in press.

Braithwaite, R. J. and Olesen, O. B. 1990. 'A simple energy balance model to calculate ice ablation at the margin of the Greenland ice sheet', *J. Glaciol.* **36**, 222–228.

Brown, G. H. and Tranter, M. 1990. 'Hydrograph and chemograph separation of bulk meltwaters draining the Upper Arolla Glacier, Valais, Switzerland', *Int. Assoc. Hydrol. Sci. Publ.*, **193**, 429–437.

Brown, G. H., Sharp, M. J., Tranter, M., Gurnell, A. M., and Nienow, P. W. 1994. 'The impact of post-mixing chemical reactions on the major ion chemistry of bulk meltwaters draining the Haut Glacier d'Arolla, Valais, Switzerland', *Hydrol. Process.*, **8**, 465–480.

Clifford, N. J., Richards, K. S., Brown, R. A., and Lane, S. N. 1995. 'Scales of variation in suspended sediment concentration and turbidity in a glacial meltwater stream', *Geogr. Ann. Ser. A*, **77**, 45–65.

Collins, D. N. 1979. 'Quantitative determination of the subglacial hydrology of two Alpine glaciers', *J. Glaciol.*, **23**, 347–362.

Gurnell, A. M., 1993. 'How many reservoirs? An analysis of flow recessions from a glacier basin', *J. Glaciol.*, **39**, 409–414.

Gurnell, A. M. and Fenn, C. R. 1984. Flow separation, sediment source areas and suspended sediment transport in a pro-glacial stream', *Catena Suppl.*, **5**, 109–119.

Gurnell, A. M., Clark, M. J., and Hill, C. T. 1992a. 'Analysis and interpretation of patterns within and between hydro-climatological time series in an alpine glacier basin', *Earth Surf. Process. Landforms*, **17**, 821–839.

Gurnell, A. M., Clark, M. J., Hill, C. T., and Greenhalgh, J. 1992b. 'Reliability and representativeness of a suspended sediment concentration monitoring programme for a remote alpine proglacial river' in *Int. Assoc. Sci. Hydrol. Symp. Erosion and Sediment Transport Monitoring Programmes in River Basins, Oslo, Norway, 24–28 August 1992*. IAHS, Oxford. pp. 191–200.

Gurnell, A. M., Brown, G. H., and Tranter, M. 1994 'A sampling strategy to describe the temporal hydrochemical characteristics of an alpine proglacial stream', *Hydrol. Process.*, **8**, 1–25.

Hooke, R. leB. 1984. 'On the role of mechanical energy in maintaining subglacial water conduits at atmospheric pressure', *J. Glaciol.*, **30**, 180–187.

Huber, W. C. and Dickinson, R. E. 1988. *Storm Water Management Model, Version 4: User's Manual.* United States Environmental Protection Agency, Athens, Georgia. 406 pp.

Nienow, P. W. 1994. Dye tracer investigations of glacier hydrological systems', *Unpublished PhD Thesis*, University of Cambridge.

Oerlemans, J. 1993. 'A model for the surface balance of ice masses: part 1. Alpine glaciers', *Z. Gletscherk. Glazialgeol.*, **27/28**, 63–83.

Raiswell, R. 1984. 'Chemical models of solute acquisition in glacial meltwaters', *J. Glaciol.* , **30**, 49–57.

Sharp, M. 1991. 'Hydrological inferences from meltwater quality data: the unfulfilled potential' In *Proc. British Hydrol. Soc. National Symp., Southampton, 16–18 September 1991*, 5.1–5.8.

Sharp, M., Richards, K. S., Arnold, N., Lawson, W., Willis, I., Nienow, P., and Tison, J.-L. 1993. 'Geometry, bed topography and drainage system structure of the Haut Glacier d'Arolla, Switzerland', *Earth Surf. Process. Landforms*, **18**, 557–571.

Sharp, M., Brown, G. H., Tranter, M., Willis, I. C., and Hubbard, B. 1995. 'Some comments on the use of chemically-based mixing models in glacier hydrology', *J. Glaciol.*, **41**, 241–246.

Shreve, R. L. 1972. 'Water movement in glaciers', *J. Glaciol.*, **11**, 205–214.

Spring, U. and Hutter, K. 1981. 'Numerical studies of Jökulhlaups', *Cold Regions Sci. Technol.*, **4**, 227–244.

Tranter, M., Brown, G. H., and Raiswell, R. 1991. 'The separation of alpine glacial meltwater hydrographs into quickflow and delayed flow components' in *Proc. British Hydrol. Soc. National Symp., Southampton, 16–18 September 1991*, 3.31–3.37.

Tranter, M., Brown, G. H., Raiswell, R., Sharp, M. J., and Gurnell, A. M. 1993. 'A conceptual model of solute acquisition by alpine glacial meltwaters,' *J. Glaciol.*, **39**, 573–581.

Tranter, M., Brown, G. H., Hodson, A., Gurnell, A. M., and Sharp, M. J. 1994. 'Nitrate variations in glacial runoff from Alpine and sub-Arctic glaciers', *IAHS Publ.* , **223**, 399–311.

UNIRAS 1990. *Unimap 200 Users Manual. Version 6.* UNIRAS Ltd, Soborg. 255pp.

Zevenbergen, L. W. and Thorne, C. R. 1987. 'Quantiative analysis of land surface topography', *Earth Surf. Process. Landforms*, **12**, 47–56.

17

INITIAL RESULTS FROM A DISTRIBUTED, PHYSICALLY BASED MODEL OF GLACIER HYDROLOGY

NEIL ARNOLD,[1]* KEITH RICHARDS,[2] IAN WILLIS[2] AND MARTIN SHARP[3]

[1]*Scott Polar Research Institute, University of Cambridge, Lensfield Road, Cambridge, CB2 1ER, UK*
[2]*Department of Geography, University of Cambridge, Downing Place, Cambridge, CB2 3EN, UK*
[3]*Department of Earth and Atmospheric Sciences, University of Alberta, Edmonton, AB, T6G 2EH, Canada*

ABSTRACT

This paper describes the development and testing of a distributed, physically based model of glacier hydrology. The model is used to investigate the behaviour of the hydrological system of Haut Glacier d'Arolla, Valais, Switzerland. The model has an hourly time-step and three main components: a surface energy balance submodel, a surface flow routing submodel and a subglacial hydrology submodel. The energy balance submodel is used to calculate meltwater production over the entire glacier surface. The surface routing submodel routes meltwater over the glacier surface from where it is produced to where it either enters the subglacial hydrological system via moulins or runs off the glacier surface. The subglacial hydrology submodel calculates water flow in a network of conduits, which can evolve over the course of a melt season simulation in response to changing meltwater inputs. The main model inputs are a digital elevation model of the glacier surface and its surrounding topography, start-of-season snow depth distribution data and meteorological data. Model performance is evaluated by comparing predictions with field measurements of proglacial stream discharge, subglacial water pressure (measured in a borehole drilled to the glacier bed) and water velocities inferred from dye tracer tests. The model performs best in comparison with the measured proglacial stream discharges, but some of the substantial features of the other two records are also reproduced. In particular, the model results show the high amplitude water pressure cycles observed in the borehole in the mid-melt season and the complex velocity/discharge hysteresis cycles observed in dye tracer tests. The results show that to model outflow hydrographs from glacierized catchments effectively, it is necessary to simulate spatial and temporal variations in surface melt rates, the delaying effect of the surface snowpack and the configuration of the subglacial drainage system itself. The model's ability to predict detailed spatial and temporal patterns of subglacial water pressures and velocities should make it a valuable tool for aiding the understanding of glacier dynamics and hydrochemistry.

INTRODUCTION

Distributed, physically based modelling strategies are now a common feature of conventional hydrology (Anderson and Burt, 1990). There has, however, been no adoption of this approach to the study of glacier hydrology. Accordingly, the overall objective of the work presented in this paper was to explore the potential of developing and using a distributed, physically based model both to predict and to explain how glacier drainage systems function. Existing models of glacier hydrology tend to be largely statistical (e.g. Willis *et al.*,

---

* Correspondence to: Neil Arnold.

Contract grant sponsor: UK NERC.
Contract grant number: GR3/8971A, GR3/7004A, GR3/8114, GT4/89/AAPS/53.

1993) or lumped (e.g. Baker *et al.*, 1982) and are used for predicting output hydrographs for forecasting purposes (see Fountain and Tangborn, 1985, for a review). However, while such statistical or lumped models might provide useful predictions of proglacial stream discharges in the catchments for which they were developed, they cannot necessarily be transferred for use in other catchments. Nor can they reliably be used for explanatory purposes.

This paper reports on the initial results from a distributed, physically based model of glacier hydrology. It was developed and tested using data collected at Haut Glacier d'Arolla, Valais, Switzerland (Figure 1). The hydrology of this glacier has been studied in detail since 1989 (Sharp *et al.*, 1993; Hubbard *et al.*, 1995; Nienow *et al.*, 1996; Richards *et al.*, 1996) and so many different data sets were available for model testing. The model is used not only to predict outflow hydrographs from the glacier, but also to examine the functioning of the glacier's internal hydrological system.

MODEL DESCRIPTION

The model has three main components:

  (i) a surface energy balance submodel which calculates meltwater production across the glacier surface;

  (ii) a surface routing submodel which routes meltwater and rain water over the glacier surface from where it is produced to where it either enters the glacier hydrological system via moulins, or runs off the glacier; and

  (iii) a subglacial hydrology submodel which predicts the flow conditions in the subglacial hydrological system. The configuration of this system can change between a distributed system and a channelized system in response to changing meltwater inputs. The cross-sectional area of the conduits can also enlarge and contract in response to changes in the balance between wall melting and creep closure which accompany changes in meltwater discharge.

The model has an hourly time-step and can be used to calculate spatial and temporal variations in surface melt, supraglacial stream discharges, subglacial water velocities, discharges and pressures, as well as temporal variations in proglacial stream discharges over an entire melt season.

The surface energy balance submodel

The surface energy balance submodel is described fully by Arnold *et al.* (1996). Briefly, the model uses a 20 m × 20 m digital elevation model (DEM) of the glacier surface and the surrounding topography, knowledge of the snow distribution at the start of the summer and meteorological variables measured at a site in front of the glacier to calculate the surface energy balance of each DEM cell from the sum of net short-wave radiation flux, net long-wave radiation flux, sensible heat flux and latent heat flux. Of these components, solar radiation is treated in most detail, as it is generally acknowledged to be the greatest source of melt energy (e.g. Munro and Young, 1982; Oerlemans, 1993). In particular, the model accounts for the effects of glacier surface slope, aspect and shading by surrounding topography on solar radiation receipts, and it includes empirical relationships for the changing glacier surface albedo as the winter snow cover is removed or as new snow falls. The basic model equations are given in the Appendix. Rainfall measured at a site in front of the glacier is assumed to fall evenly across the glacier. Though rainfall rates will undoubtedly vary across the glacier, in general, the amounts of rainfall recorded are small in comparison with daily melt totals, and so this approximation should make little difference to the moulin input hydrographs. Thus, output from the surface energy balance submodel and measurements of rainfall can be used to calculate hourly variations of water input to each DEM cell.

The surface routing submodel

The surface routing submodel routes the calculated melt and rainfall over the surface of the glacier to the individual moulins that feed the subglacial drainage system. The glacier surface DEM allows the drainage

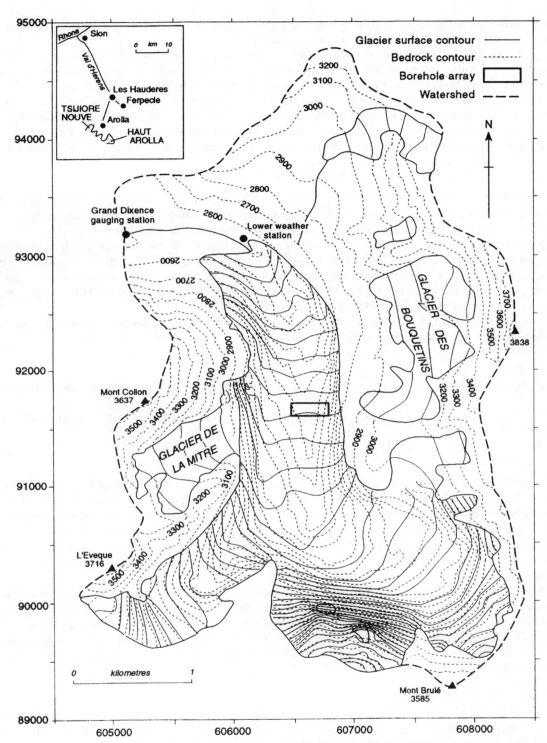

Figure 1. Location map of Haut Glacier d'Arolla, Valais, Switzerland, showing location of the stream gauging station, meteorological station and borehole drilling array

basins of individual moulins to be established, and an algorithm was developed to estimate the travel time between each DEM cell and the moulin to which it drains.

Using the glacier surface DEM, the path from any cell on the glacier surface to the appropriate cell containing the moulin was calculated by assuming that water flows from one cell to the next down the steepest slope. Thus, both the path length and the actual route taken by the water are known. This latter is important since surface conditions affecting supraglacial water velocity (particularly the presence or absence of snow) can vary spatially and temporally over the glacier as snow cover melts over a melt season.

For all cells with snow cover at a given time, the time taken for water to percolate from the surface to the base of the snowpack is calculated using Colbeck's (1978) equation for vertical flow in unsaturated snow:

$$D = \kappa_e d / (3\rho_w g/\mu)^{1/3} k^{1/3} q^{2/3}) \tag{1}$$

where D is the time taken to travel vertically through the snowpack (s), κ_e is the effective porosity of the snowpack, d is the depth of the snowpack (m), ρ_w is the density of water (kg m^{-3}), g is acceleration due to gravity (m s^{-1}), μ is the viscosity of water (Pa s), k is the permeability of snow (m^2) and q is the water flux through the snowpack, here taken to be the melt rate as calculated by the surface energy balance submodel, plus any liquid precipitation (m^3 s^{-1}). Given the difficulty of measuring κ_e and k in the field, and the changes in these parameters through time in response to melt, constant values were adopted, since modelling snowpack evolution at a micro-scale is beyond the scope of this project. The values used in this study are for 'medium grain old dry snow', from Male and Gray (1981, p. 401).

Colbeck's (1978) equation for flow in a saturated layer at the base of a snowpack is used to calculate the water flow velocity across each snow-covered cell:

$$C_s = (\rho_w g/\mu)k\theta/\kappa \tag{2}$$

where C_s is the water velocity under a snowpack (m s^{-1}), θ is the surface slope and κ is the porosity of the snow.

For each cell where the surface is ice, the water flow velocity on ice (C_i) is calculated using Manning's equation:

$$C_i = R^{2/3}\theta^{1/2}/n \tag{3}$$

where R is the hydraulic radius of the supraglacial channel (m) and n is the Manning roughness coefficient (m$^{-1/3}$ s). Because of the high roughness and the rapidly changing nature of the ice surface during the melt season, once the snow cover has gone, and reflecting the observation that most supraglacial streams tend to flow in a series of small, sub parallel, interconnected channels, a constant value for R is assumed. Given that the major factor affecting hydrograph shape is the presence or absence of snow cover in a moulin's subcatchment, variations in R have a minor effect on the hydrograph shape. All parameter values in Equations (1)–(3) are given in Table I.

From these equations, the time taken for water to cross each cell (be it snow or ice covered) can be calculated. By adding together the individual times for all cells on every path between a melt source cell (and the vertical infiltration time if the source cell is snow covered) and the relevant moulin cell, the total time taken for each hourly melt/rain increment to reach each moulin can be calculated. Thus, moulins develop distinctive input hydrograph shapes that depend not only on the shape of the basin draining into the moulin, but also on the surface conditions in the basin.

The subglacial hydrology submodel

Having calculated the input hydrographs to each moulin, the water is routed beneath the glacier via the subglacial drainage system. This is done using the EXTRAN block of the US Environmental Protection Agency Storm Water Management Model (SWMM), which simulates flow of rain water through a sewer network based on a solution of the Saint Venant equations (Roesner et al., 1988). EXTRAN simulates

Table I. Model parameter values

Parameter	Symbol	Value	Units
Hydraulic radius	R	0·035	m
Manning's roughness	n	0·05	$m^{-1/3}$ s
Water density	ρ_w	1000·0	kg m^{-3}
Ice density	ρ_i	900·0	kg m^{-3}
Gravity	g	9·81	m s^{-2}
Water viscosity	μ	$1·8 \times 10^{-3}$	Pa s
Snow permeability	k	6×10^{-9}	m^2
Snow porosity	κ	0·68	—
Snow effective porosity	κ_e	0·63	—
Friction parameter	f_r	0·25	—
Latent heat of fusion of water	L	$3·34 \times 10^5$	J kg^{-1}
Arrhenius parameter	B	$5·8 \times 10^7$	N m^{-2} s$^{1/m}$
(SI equivalent)		$6·8 \times 10^{-15}$	s^{-1} kPa^{-3}
Ice flow law exponent	m	3	—

sewage pipe systems using a network of 'drains' where water can enter (or leave) the system, connected by 'pipes', which can have individually specified shapes, sizes, lengths and roughnesses. Thus, for a glacier hydrological system, the 'drains' are used to represent moulins where surface water can enter (or overflow from) the system, and the 'pipes' are used to represent subglacial conduits whose properties can be specified individually as required. In the rest of this paper, the term 'network structure' refers to the layout of the links between the individual moulins and 'drainage configuration' refers to the nature of the links between moulins (that is, the size, shape and number of conduits between moulins).

A problem with the use of this model for glacier drainage systems is that both field evidence (Nienow, 1993) and theoretical considerations (Röthlisberger, 1972; Kamb, 1987) suggest that such systems will change over the course of a melt season in response to changing water inputs. This change can occur in two ways. First, the character of the system may change between a distributed system, in which cavities between the bottom of the glacier and its bed are linked by many small conduits, and a channelized system, consisting of a few, large conduits, during the melt season (Kamb, 1987; Nienow, 1993). Secondly, the dimensions of the conduits themselves may change in response to changes in the balance between wall melting by frictional heating from the passing water, and creep closure by ice deformation (Röthlisberger, 1972; Kamb, 1987). To model flow in subglacial drainage systems, it is necessary to simulate both these processes. In this study, two different strategies were adopted.

A distributed system was represented in one of two ways: by 'bundles' of eight small, circular, rough conduits, or by one very shallow, wide, rectangular, rough conduit. A channelized system was represented by a single circular conduit. Conduit roughnesses were taken from the literature, and the values derived by Nienow (1993). These varied from 0·05 to 0·25 $m^{-1/3}$ s. These values were linearly related to conduit area, such that a conduit with a cross-section of 5 m^2 had a roughness of 0·05, while one with a minimum cross-section of 0·05 m^2 (this minimum was set to preserve continuity) had a roughness of 0·25 $m^{-1/3}$ s. Subglacial conduit roughnesses are still the subject of debate in the literature (Nienow, 1993), and early experimentation with the model showed that the results were much more sensitive to conduit size variations than conduit roughness variations, so this simple parameterization was felt to be adequate as a first approximation.

Dye tracing studies at Haut Glacier d'Arolla indicate that a transition from a distributed to a channelized system generally coincides with the passage of the snow line over the area (Nienow, 1993). Given this, the whole melt season was simulated in some model runs using a series of shorter runs during which the drainage configuration remained constant (although individual conduit dimensions could vary in response to wall melting and conduit closure, see below), with the drainage configuration being adjusted on the appropriate dates as the snow-line passed each moulin, and using the final flow conditions of the previous run as input to the next run. 'Distributed' links were changed to 'channelized' links as the modelled snow-line passed each moulin. The possibility of a change from channelized back to distributed drainage at the end of the melt season following late summer snowfall was also investigated in some model runs.

To account for changing conduit dimensions we assume that conduits enlarge because of the release of frictional heat in the water flowing in the conduits, and close in response to ice deformation (Röthsliberger, 1972). Spring and Hutter (1981) show that the rate of conduit-wall melting (M) owing to frictional heating, expressed as the mass melted per unit length of conduit per unit time, is

$$M = [(\pi S)^{1/2} \rho_w (f_r v^3 / 4)] / L \tag{4}$$

where S is the conduit cross-section (m^2), f_r is a friction parameter, v is the water velocity in the conduit (m s^{-1}) and L is the latent heat of fusion of water (J kg^{-1}). The rate of conduit closure through ice deformation (A), expressed as the change in cross-section area per unit time, is

$$A = -(p_i - p_w)|p_i - p_w|^{m-1} 2(1/mB)^m S \tag{5}$$

where p_i and p_w are, respectively, ice overburden and water pressures (Pa), m is the exponent in Glen's flow law and B is the Arrhenius parameter in Glen's flow law (N m^{-2} s$^{1/m}$). The parameter values in Equations (4) and (5) are given in Table I. Where these variables are not physical constants, values are taken from the literature discussed, or are typical, empirically determined values (e.g. Paterson, 1994, p. 97, for B and m). The equations were solved using the water velocities and pressures calculated by EXTRAN for each conduit, and the conduit sizes were adjusted every hour.

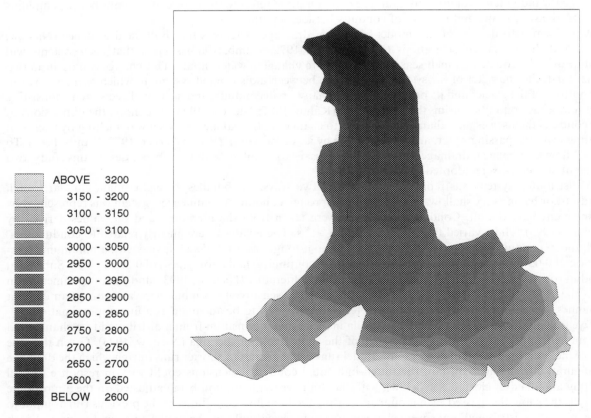

ABOVE	3200
	3150 - 3200
	3100 - 3150
	3050 - 3100
	3000 - 3050
	2950 - 3000
	2900 - 2950
	2850 - 2900
	2800 - 2850
	2750 - 2800
	2700 - 2750
	2650 - 2700
	2600 - 2650
BELOW	2600

Figure 2. 1993 Digital elevation model of Haut Glacier d'Arolla, derived from photogrammetric survey. Units in metres above sea level

The complete model was implemented on a DEC Alpha 3000/400 workstation. A whole-season run of the melt/surface routing submodels takes approximately six hours of computer time; a whole season run of the subglacial hydrology submodel takes between 20 and 60 minutes, depending on the complexity of the drainage configuration used.

DATA REQUIREMENTS

The model requires three main data sets: (i) initialization data to set the boundary conditions; (ii) meteorological data to drive the surface melt submodel and determine patterns of rainfall; (iii) field data to test the model outputs against. The model was initialized, driven and tested using data collected in 1989, 1990, 1991 and 1993.

Initialization data comprised a 1993 DEM of the glacier surface and its surrounding catchment; the initial distribution of snow thickness at the start of the 1993 melt season; the inferred network structure; and the start-of-season drainage configuration. The DEM was constructed from photogrammetrically derived data for the glacier surface supplemented by Swiss 1:25 000 map data for the surrounding areas of the catchment (Willis *et al.*, in press). The glacier surface DEM is shown in Figure 2. Initial snow thickness was measured in the field at 87 points across the glacier during the period 17–22 June 1993 (Copland, in press). These measurements were interpolated on to the DEM to produce estimates of snow thickness for each cell on 22 June (Willis *et al.*, in press). This snow thickness distribution was used in an initial model run driven by meteorological data for the period 20 May to 22 June, which calculated the amount of melt over this period.

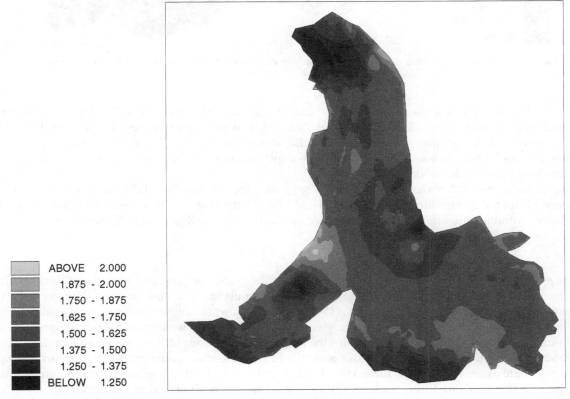

Figure 3. 1993 Snow depth for Julian day 140 (20 May). Units are metres of water equivalent depth (that is, the snow depth multiplied by the snow density)

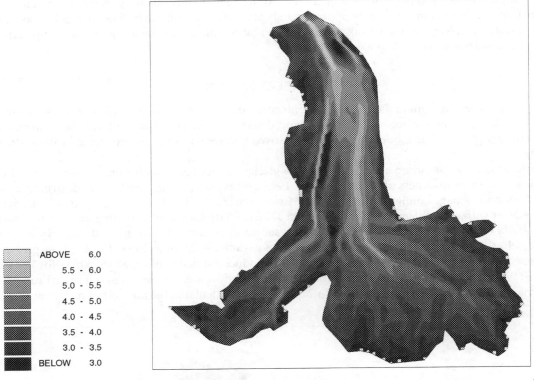

	ABOVE 6.0
	5.5 - 6.0
	5.0 - 5.5
	4.5 - 5.0
	4.0 - 4.5
	3.5 - 4.0
	3.0 - 3.5
	BELOW 3.0

Figure 4. Up-glacier contributing area on the hydraulic potential surface. Units are common logarithm of area, in m^2

As the whole glacier was snow covered on 22 June, all the calculated melt from 20 May to 22 June was assumed to be of snow. The calculated melt was then added to the snow thickness distribution derived for 22 June to provide an estimate of snow thickness distribution on 20 May (Figure 3). This distribution was used in all model runs. The back-calculation of snow depth allowed us to include early season meteorological and discharge data in the model experiments.

The drainage network structure was constructed using the 1993 surface DEM and a bedrock DEM developed from radio-echo surveys made in 1989 and 1990 (Sharp *et al.*, 1993). These two DEMs allowed the subglacial hydraulic potential (ϕ) to be calculated for each cell using Shreve's (1972) equation:

$$\phi = \rho_w gz + \rho_i gH \tag{6}$$

where ρ_i is the ice density (kg m^{-3}), z is the elevation of the glacier bed (m) and H is the ice thickness (m). The locations of subglacial conduits were inferred by cumulating the upslope area draining over this potential surface, as described in Sharp *et al.* (1993). Figure 4 shows the cumulative area, which indicates that two main drainage axes exist in the lower glacier, which drain the western and eastern upper basins. Additionally, there is a short axis which drains the lower, far-eastern side of the glacier, and a central axis on the lower glacier. Surveyed moulin locations generally lie on these main drainage axes, so the links between the moulins are assumed to follow these axes. The inferred subglacial drainage network structure is shown in Figure 5. This structure forms the basis for all of the model runs discussed in this paper.

The start-of-season drainage configuration varied between runs and is therefore described in the Experimental Design section below.

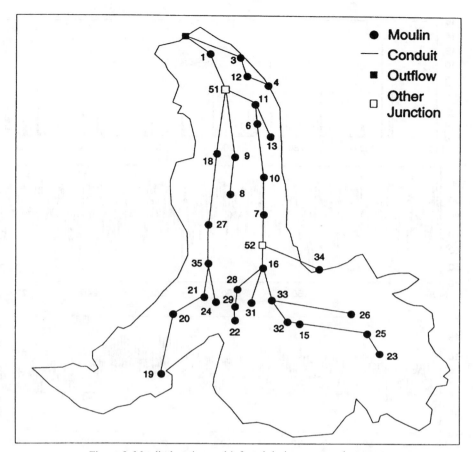

Figure 5. Moulin location and inferred drainage network structure

The meteorological data were collected in 1993 at a site in front of the glacier shown in Figure 1. The meteorological station sampled every 30 seconds, and Figure 6 shows the hourly averages of incoming short-wave radiation, air temperature, wet bulb temperature and wind speed (not shown), together with hourly totals of precipitation which were used to determine melt and rain inputs. Full details are given in Arnold *et al.* (1996).

The model results were tested by comparing them with field data. The main data set used for this purpose was hourly proglacial stream discharge variations collected in 1993 by the hydroelectric power company Grande Dixence S.A. at a site approximately 1 km from the glacier snout (Figure 1). The time taken for water to travel from the snout to this station is less than the hourly output time-step of the model, so should not introduce any significant errors. One of the advantages of distributed models over simpler statistical or lumped models, is that they make predictions about the internal system state, as well as predictions of the output variables. Thus, it is important to test distributed models wherever possible with such internal variables. In this study, two such variables are available: subglacial water pressure measurements obtained in 1993 from a borehole located close to the subglacial conduit running down the eastern half of the glacier tongue between moulins 7 and 10 (cf. Figures 1 and 5) [i.e. Borehole 35 discussed by Hubbard *et al.* (1995)], and subglacial water velocities derived from dye tracer experiments conducted from moulins during 1990 and 1991, described by Nienow (1993) and Nienow *et al.* (1996).

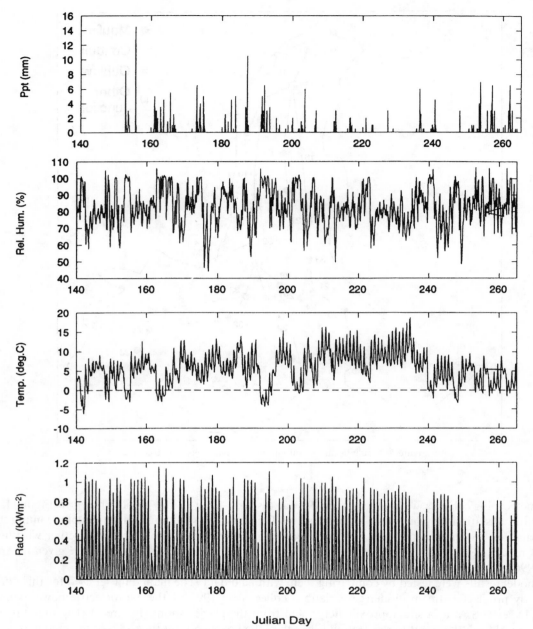

Figure 6. Meteorological data used by the surface energy balance submodel

EXPERIMENTAL DESIGN

Two main sets of experiments were performed, which are summarized in Table II. Runs S1, S2 and S3 are short period runs and were used to examine the behaviour of the subglacial hydrology submodel in isolation from the melt/surface routing submodels. In these runs, diurnally repeating moulin hydrographs were used as input to the subglacial hydrology submodel to investigate the effects of different subglacial boundary conditions (i.e. start-of-season drainage configuration and its subsequent evolution) on output hydrographs.

Table II. Summary of model runs

Run no.	Diurnal input hydrographs	Initial conduit diameter	Change between distributed and channelized	Change in conduit sizes
S1	Season mean	Large	No	No
S2	Season mean	Large	No	Yes
S3	Season mean	Small	No	Yes
L1	Variable	Large	No	Yes
L2	Variable	Small	No	Yes
L3	Variable	Small	Yes	Yes
L4	Variable	Small	Yes	Yes

The input hydrographs used were the season-long means of the diurnal input discharges to each moulin. Runs L1–L4 are long period runs and were full-model, non-steady-state simulations where the changing moulin input hydrographs calculated from the melt/surface routing submodels were used as input to the subglacial hydrology submodel.

Three progressively more realistic models of subglacial drainage were used:

(a) a fixed drainage configuration and conduit geometry (run S1);
(b) fixed drainage configuration, but with changing conduit dimensions (runs S2, S3, L1 and L2); and
(c) changing drainage configuration and conduit geometry (runs L3 and L4).

For runs S1, S2 and L1, the start-of-season conduit sizes were derived from dye tracer-based calculations of conduit dimensions by Nienow (1993). He assumed that the conduit cross-sectional area downstream of a given moulin could be calculated from the mean throughflow velocity derived from the dye tracer test and the local discharge at the dye injection site. This was calculated by multiplying the measured proglacial stream discharge by the proportion of the total area of the glacier above the injection site (calculated from the glacier hypsometry), and allowing for a linear decrease in melt rate with increasing altitude. This procedure produced results which allowed realistic reconstructions of the flow velocities derived from the dye return curves when used with realistic values of Manning's roughness ($0.05–0.1$ m$^{-1/3}$ s) and sinuosity (2–3) (Nienow, 1993). The inferred conduit dimensions used in this study were taken from those derived from dye tracer tests undertaken in early August 1990, when ice was exposed over most of the glacier, and dye injections could be conducted from most moulins. The use of tracing results from this short period created a 'snap-shot' picture of conduit dimensions. The reconstructed sizes are shown in Table III, column 4. The disadvantage of using results from this period is that it is relatively late in the melt season, meaning that much conduit evolution would have already occurred, and the reconstructed conduit sizes would be inaccurate early in the melt season.

For this reason, a second set of start-of-season conduit sizes were calculated by running the subglacial hydrology model for six months with no input discharge, using the conduit sizes from the end of run L1 as an initial condition, in order to simulate the winter closure of conduits owing to ice deformation. The sizes at the end of this run were then used as the start-of-season conduit sizes for runs S3, L2, L3 and L4. Conduits which closed completely over the winter were given a fixed minimum size (0·25 m diameter), in order to preserve the connectivity of the system, which is necessary for the model to function. The reconstructed sizes are shown in Table III, column 5.

MODEL RESULTS

Short period, subglacial boundary condition experiments

The results of run S1 (not shown) demonstrated that the initial empty condition of the drainage system was unimportant. Given constant diurnal input cycles, drainage configuration and conduit geometry, the system produced constant diurnal output hydrographs within one day.

Table III. Modelled conduit dimensions

Upstream junction	Downstream junction	Length (m)	Conduit diameter (m)	
			Large	Small
1	Outlet	199	3·0	6·0
3	Outlet	450	1·27	1·06
12	3	156	2·65	0·71
4	12	198	1·56	0·98
51	1	386	2·50	2·19
11	51	236	2·46	1·16
6	11	427	2·36	0·69
13	11	569	2·65	0·74
10	6	474	2·26	0·36
7	10	336	2·22	0·25
52	7	287	2·28	0·25
34	52	539	1·92	0·25
16	52	190	2·34	0·25
31	16	324	1·83	0·25
28	16	289	2·50	0·25
29	28	152	1·91	0·25
22	29	123	2·23	0·25
33	16	297	1·84	0·25
26	33	694	1·32	0·25
32	33	234	1·64	0·25
15	32	156	1·45	0·25
25	15	587	1·33	0·25
23	25	300	0·75	0·25
9	51	558	2·78	0·25
8	9	331	2·65	0·25
18	51	529	1·97	0·51
27	18	636	1·78	0·77
35	27	345	2·23	0·69
24	35	348	2·04	0·48
21	35	300	2·11	0·52
20	21	305	0·91	0·68
19	20	541	0·65	0·83

'Large' denotes conduit size reconstructed from late season dye tracing experiments (see text).
'Small' denotes conduit size reconstructed as above but allowing them to close over winter (see text).

Figure 7 shows the results of run S2, in which the initial drainage configuration characterized by large conduits was allowed to change in response to wall melting and conduit closure. Again, the system reached a steady diurnal hydrograph shape within one day. However, after *c.* 20 days, the peak flow discharges started to drop. This was caused by the gradual closure of the conduits during the simulation, as the simulated water pressure at the start of the model run was generally atmospheric, owing to the large conduit areas. Thus, conduit closure rates exceeded wall melting rates. After *c.* 20 days, pressurized flow started to occur during the day, when meltwater inputs were high. As the conduits were full during these periods, peak flow rates were limited. The rise in water pressure reduced the conduit closure rate, and eventually a new steady-state flow regime was reached after *c.* 5 days, characterized by pressurized flow during the day, and flow at atmospheric pressure at night, and in which daytime wall melting was balanced by nighttime closure by ice deformation.

The results of run S3 are shown in Figure 8. In this run, the initial drainage configuration, characterized by small conduits, was allowed to evolve. The results are essentially the reverse of run S2. As water pressures are initially above atmospheric pressure, peak flow discharges started to rise immediately, as the small conduits

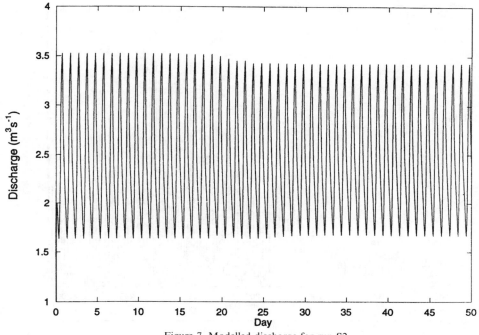

Figure 7. Modelled discharge for run S2

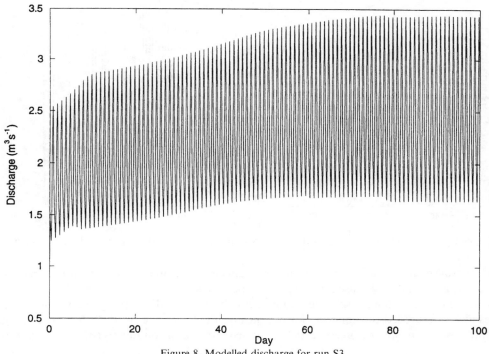

Figure 8. Modelled discharge for run S3

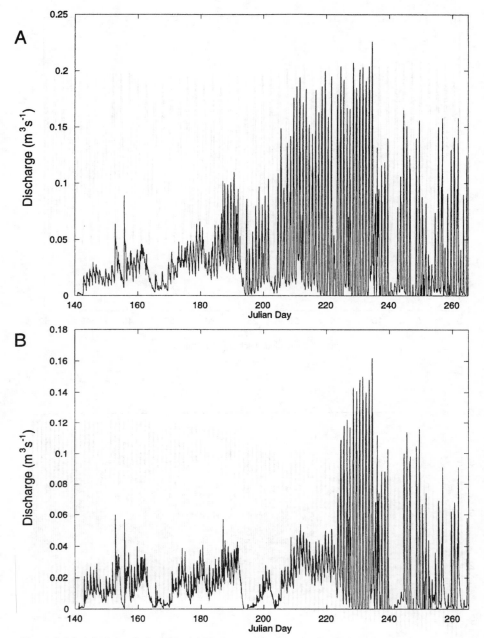

Figure 9. Modelled moulin input hydrographs. (a) Moulin 9 (lower glacier); (b) moulin 33 (upper glacier)

were opened by wall melting, and the high water pressures limited conduit closure. In this case, however, it took *c*. 80 days for a steady-state discharge regime to be reached.

For runs S2 and S3, the final discharge regime actually continued to change very slightly, as the wall melting rates always exceeded the conduit closure rates on the lower one-third of the glacier tongue. However, once flow was at atmospheric pressure, the continued increase in conduit area made very little difference to the flow regime.

Long period, full simulation experiments

Two typical moulin discharge hydrographs generated by the melt/surface routing submodels are shown in Figure 9. For the moulin on the lower glacier (moulin 9), there is a change in the hydrograph shape as the snow-line crosses the moulin at around Julian day (JD) 190 (9 July) (Figure 9a). Before this time, the peak discharges are relatively low, reflecting the low melt rates, and the diurnal range is also small, owing to the lag effect caused by the slow flow of water through the snowpack. After this time, however, the peak discharge increases, because of the generally higher melt rates on ice surfaces, and the diurnal range also increases, as water runs off much more quickly on ice. Towards the end of the melt season, the low-flow discharges drop to zero, as all the melt generated in the moulin 9 catchment runs off in less than 24 hours. For the moulin on the upper glacier (moulin 33), the change in hydrograph regime occurs later in the year at around JD 220 (8 August) (Figure 9b). These two moulins also illustrate the influence of the shape of the catchment feeding an individual moulin on its hydrograph. The moulin 9 catchment is narrow, but extends *c*. 800 m up-glacier; the moulin 33 catchment is wide, and extends just *c*. 100 m up-glacier. Thus, the transition from a 'snow' regime (small peak flow, small diurnal range) to an 'ice' regime (high peak flow and high diurnal range, often with zero low flow) occurs relatively slowly for moulin 9, since it takes 8–10 days for the snow-line to cross the catchment from bottom to top. For moulin 33, this change occurs in only 2–3 days.

The output hydrograph for run L1 is shown in Figure 10, together with the measured stream discharge. There is generally a good match between modelled and measured discharge, especially between JD 160 (9 June) and JD 220 (8 August). Before this period, there are two anomalous high flow events in the model record, which do not occur in the measured data. These are caused by heavy rain on JD 152 and 155 (1 and 4 June). Later in the melt season, however, the diurnal range of the model hydrograph is too large. After JD 240 (28 August), when there is heavy snowfall, this mismatch becomes even more apparent; in particular, peak discharges are 4–5 m^3 s^{-1} too large and minimum discharges are approximately 0·25–0·5 m^3 s^{-1} too low.

As most of the dye tracing experiments used to calculate initial conduit sizes for this run were conducted in early August (around JD 200), the good match near this period is perhaps not surprising. However, the

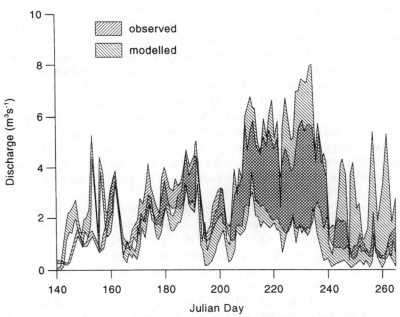

Figure 10. Modelled discharge for run L1, and measured discharge. The width of the shaded bands at any point corresponds to the diurnal range in values for that day

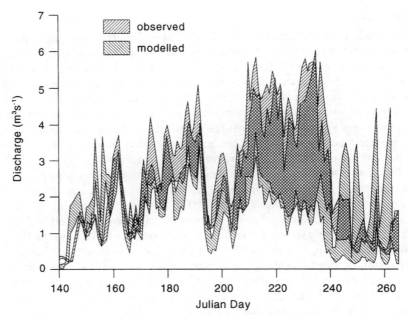

Figure 11. Modelled discharge for run L2, and measured discharge. The width of the shaded bands at any point corresponds to the diurnal range in values for that day

pressure calculations from run L1 (not shown) demonstrate that water rarely flowed above atmospheric pressure, whereas 1993 borehole water pressure records, as well as observations of water emerging from moulins and crevasses at the glacier surface in 1993, suggest that pressurized flow did occur. Together with the over prediction of peak flow discharges later in the melt season, the lack of pressurized flow suggests that the initial conduit sizes based on late-season dye tracer tests were too large.

Figure 11 shows the output hydrograph for run L2, in which the initial conduit sizes were smaller than in run L1, together with the measured discharge. It shows a much better fit between the model results and the measured data than that for run L1. The overprediction during the two rain storms near the start of the melt season is still present, but in general, the peak discharge magnitudes and the diurnal discharge range both match very well. However, for the earlier part of the season (before *c*. JD 200), the model generally under-predicts low-flow discharges. After this time, the low-flow magnitudes match rather better. After the snowfall on JD 240 (28 August), however, this generally good match again breaks down, After this time, the modelled diurnal range is again too large, owing mainly to modelled peak discharges being $2-3$ m^3 s^{-1} too high. As in run L1, modelled minimum discharges are also approximately $0.25-0.5$ m^3 s^{-1} too low after this time. The general dampening of the hydrographs between runs L1 and L2 is caused by the smaller conduits in run L2, which restrict peak flow discharges, causing water to back up in the system. This backed up water is released later in the day, raising low-flow discharges.

The next two runs attempt to model the possible evolution of the drainage system from 'distributed' (represented in run L3 as bundles of eight small circular conduits between each moulin, and in run L4 as shallow, wide, rough rectangular conduits) to 'channelized' (represented by single circular conduits) as the snow-line crossed each moulin. They also try to account for the possible reversion to 'distributed' drainage beneath part of the glacier following the late summer snowfall on JD 240 (28 August). Because of the number of moulins and conduits in the system, and the amount of data manipulation required for each run, the adjustment was made in four steps. From JD 140 (20 May) to JD 190 (9 July) all links were 'distributed'. From JD 191 to JD 213 (1 August) links below moulin 7 for the main eastern system, below 27 for the western system, and the whole of the central and far-eastern systems were 'channelized ', while all other links

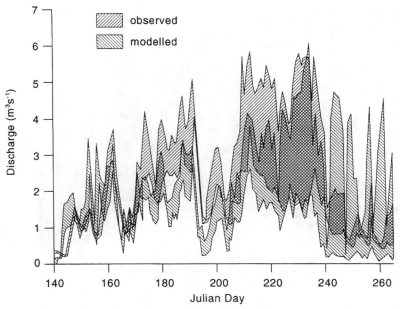

Figure 12. Modelled discharge for run L3, and measured discharge. The width of the shaded bands at any point corresponds to the diurnal range in values for that day

remained 'distributed'. From JD 214 to JD 240 (28 August), the whole system was 'channelized', except for the links immediately below moulins 23, 25 and 19. After the heavy snowfall on JD 240, all links above moulins 6, 9 and 18 reverted to 'distributed' while the lower links remained 'channelized'.

Figure 12 shows the modelled discharge hydrograph for run L3. This is very similar to the hydrograph for run L2. The main differences are even greater underestimation of low-flow discharges, by approximately 0.5 m^3 s^{-1} in the early part of the melt season, and an underestimation of peak flow discharges for rather longer, until around JD 230 (18 August). After JD 240, the model again does not capture the change in discharge regime.

Figure 13 shows the modelled discharge regime for run L4. This shows a very good match with the measured discharge before JD 190, when the entire system is distributed. The match is not as good when the partially distributed system is in place from JD 191 to 213; low-flow discharges in the model again underestimate the observed ones, though by less than in runs L2 and L3. However, after JD 213, modelled peak flow discharges are lower than observed values. After JD 240, modelled low-flow values match observed values very well, but peak values are still too high.

Table IV summarizes the performance of the four model runs in predicting outflow hydrographs, using the statistical measures proposed by Fountain and Tangborn (1985). The whole-season values compare well with

Table IV. Statistical summary of model performance: discharge

Run no.	Period (JD)	No. of data points	Coefficient of determination	Seasonal volumetric difference	Standard error	Relative error	Absolute error
L1	140–265	3000	0·414	−0·028	0·464	0·028	0·333
L2	140–265	3000	0·655	0·072	0·306	−0·072	0·240
L2	210–240	720	0·776	0·053	0·165	−0·053	0·131
L3	140–265	3000	0·433	0·208	0·457	−0·208	0·357
L4	140–265	3000	0·548	0·054	0·408	−0·053	0·298

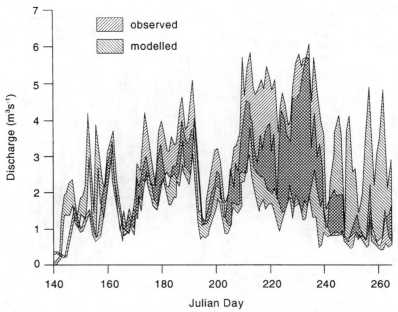

Figure 13. Modelled discharge for run L4, and measured discharge. The width of the shaded bands at any point corresponds to the diurnal range in values for that day

other glacier hydrology modelling studies in the literature (Fountain and Tangborn, 1985; Willis *et al*., 1993), especially as our study is based on 3000 hourly data points, over a four-month period. For run L2, the best model overall, the statistical measures for the period JD 210 to 240 (when the channelized drainage system is essentially fully developed) are also given. Given the channelized structure of the model drainage system, the predictive power of the model is better during this time than for the whole season. The generally positive values of the seasonal volumetric difference means the model is predicting less runoff than actually occurs.

Modelled and measured subglacial water pressures for run L2 are shown in Figure 14. As explained earlier, the measured water pressure data were obtained from a borehole that was inferred to be close to a major subglacial conduit running down the eastern half of the glacier tongue. In the model, this borehole is represented as a junction that is linked to the main conduit between moulins 7 and 10 by a short, wide, low, rectangular, rough conduit. From JD 200 to *c*. JD 218, the mismatch in the records may be a result of a poor connection between the just-drilled borehole and the subglacial conduit (Gordon *et al*., 1998). After this time, however, the borehole seems to become well connected to the main channel, and the water pressure varies diurnally from atmospheric to the ice overburden pressure. In the model, however, peak pressures are somewhat low, and the low pressures never drop to atmospheric pressure, although the modelled diurnal pressure range does increase throughout this period, until it reaches a maximum at JD 230 (6 August). From this time until JD 240 (28 August) the predicted pressures show a very good match with the measured pressures, with water pressure varying diurnally from zero at night to at or near the ice overburden pressure at peak flow discharges in the late afternoon. However, after JD 240, the match again breaks down. This is the same time at which the modelled outflow hydrograph diverges from the measured values.

Figure 15 shows the modelled subglacial water pressure for run L3. At no time does the water pressure vary between atmospheric pressure and the ice overburden pressure, as is seen in the measured record. However, after JD 240, when the modelled drainage configuration reverts back to a mostly distributed one, the modelled pressure variations match the measured values slightly better. For the early part of this period, the modelled and measured data match very well. Later on, the modelled diurnal pressure range again exceeds the measured range, although model pressures never drop to atmospheric pressure as they did in run L2.

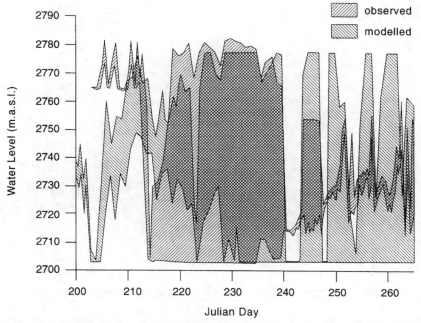

Figure 14. Modelled borehole water level elevation for run L2, and measured water level elevation. The width of the shaded bands at any point corresponds to the diurnal range in values for that day

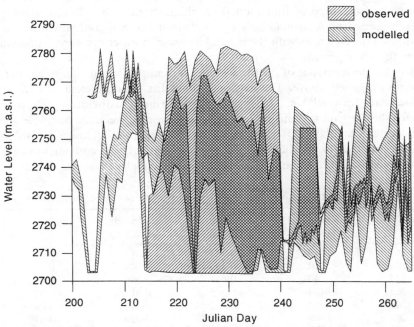

Figure 15. Modelled borehole water level elevation for run L3, and measured water level elevation. The width of the shaded bands at any point corresponds to the diurnal range in values for that day

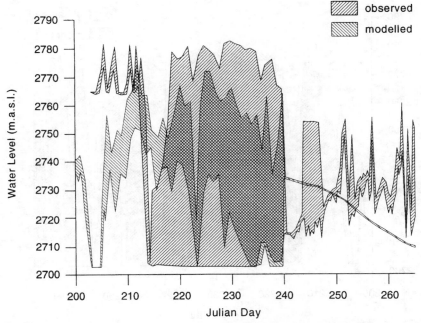

Figure 16. Modelled borehole water level elevation for run 4, and measured water level elevation. The width of the shaded bands at any point corresponds to the diurnal range in values for that day

Figure 16 shows the modelled subglacial water pressures for run L4. Before JD 240, these are very similar to those for run L3 because the drainage configuration is similar to that in run L3 and has evolved from a distributed to a channelized system over the same time period. After JD 240, however, there is a surge in modelled water pressure at the time of transition from channelized to distributed drainage in the upper glacier, after which pressures drop to around 50% of ice overburden then gradually decrease and show very small diurnal variations. This contrasts with the general increase in water pressure, with some large diurnal cycles, observed over the same period.

Table V summarizes the performance of the three model runs in predicting borehole water pressure, using the statistical measures discussed above. The negative values for the coefficient of determination for the whole period for which data are available, and for all the model runs, are obviously disappointing. However, as discussed above, for the early part of this period, the connection between the borehole and the subglacial drainage system is uncertain. Similarly, after the snowfall on JD 240, the drainage system obviously adjusts in a way that is not accounted for by the model at present. Thus, if the comparisons are limited to the period when the borehole is believed to be well connected to the subglacial drainage system, and before the

Table V. Statistical summary of model performance: borehole water level

Run no.	Period (JD)	No. of data points	Coefficient of determination	Standard error	Relative error	Absolute error
L2	203–265	1488	−0·145	0·010	−0·010	0·008
L2	215–240	600	0·329	0·009	0·001	0·007
L2	230–240	240	0·657	0·006	0·001	0·004
L3	203–265	1488	−0·245	0·019	−0·012	0·011
L3	215–240	600	0·102	0·010	−0·001	0·008
L4	203–265	1488	−0·288	0·020	−0·012	0·011
L4	215–240	600	0·107	0·010	−0·001	0·008

end-of-season drainage adjustment (i.e. JD 218–240), the coefficients rise significantly, particularly for run L2, which shows good predictive power, especially for the period JD 230–240. The low values of the error statistics for all runs are caused by the small range in water level values (c. 80 m) compared with the mean value (c. 2735 m).

Because few dye tracer experiments were conducted during the 1993 field season, the model results are compared with results from dye injections conducted at comparable stages of the 1990 and 1991 melt seasons, when the configuration of the drainage system is likely to be similar. Figure 17 shows the seasonal evolution of moulin to snout travel times calculated from 1990 and 1991 dye tracer tests for three moulins, and the equivalent modelled travel times for run L2. Once the melt season is well established, the modelled values generally agree well with the observed values. The observed early season decrease in travel time (believed to be owing to the establishment of channelized drainage) is less well marked in the model results, where only the dimensions of conduits changed.

Nienow *et al.* (1996) observed very complex diurnal velocity–discharge relationships from dye tracer experiments on Haut Glacier d'Arolla. They observed both linear and hysteretic (clockwise and anti-clockwise) velocity–discharge relationships from various moulins at various stages during the melt season. They have also found that individual moulins can give different relationships at different times in the melt season. Such behaviour is also found in the model results. Figure 18 shows four diurnal velocity–discharge cycles for run L2. Following Nienow *et al.* (1996), the 'mean' discharge (Q_m) during an experiment is plotted, where $Q_m = \{[(Q_{b1} + Q_{b2})/2) + (Q_{11} + Q_{12})/2]\}/2$. Here, Q_b is the bulk proglacial stream discharge, Q_1 is the local discharge in the conduit below the moulin concerned, and subscripts 1 and 2 refer to discharges at the start and end of the tracer test. Here, the tests are assumed to take 60 minutes. Figure 18a–c shows the changing velocity–discharge relationships for moulin 6 on the lower glacier [called moulin m2Ec in Nienow *et al.* (1996)]. This moulin shows a near-linear relationship early in the melt season, clockwise hysteresis during the period when large diurnal pressure cycles occur and a more linear relationship again late in the melt season, when water pressures are lower. Nienow *et al.* (1996) found both linear early season cycles, and clockwise mid-season cycles at this location in 1991. By contrast, moulin 24 in the upper ablation area (called moulin m5Wf in Nienow *et al.*) shows anticlockwise hysteresis on JD 230 (Figure 18d). Nienow *et al.* similarly found anticlockwise hysteresis from this location in 1991 in mid to late season. In the model, anticlockwise hysteresis also occurs early in the melt season at this location, and stops only after JD 240, when meltwater inputs drop dramatically.

INTERPRETATION AND DISCUSSION

The short period runs (S1, S2 and S3) indicate that the drainage system fills quickly at the start of the melt season, and is unlikely to be influenced by initial conditions for more than one or two days, given fixed conduit geometry. However, they indicate that the time needed for an evolving drainage system to reach a steady state (if, in fact, it ever does) is comparable to the length of a melt season. Given the marked diurnal changes in melt inputs, and the seasonal changes in the diurnal input hydrograph shapes in response to changing snow-line position and snow thickness over the course of a melt season, the subglacial drainage system probably never reaches an equilibrium state.

The long period runs (L1–L4) indicate that in order to model water outflow from a glacier effectively over a melt season, it is necessary to account for the changing patterns of melt over the course of the melt season, the changing meltwater inputs to moulins caused by changing snow cover and also the evolution of the subglacial drainage system itself. For Haut Glacier d'Arolla, the importance of these various factors seems to change over the course of a melt season. The similarity of the model output hydrographs for runs L1–L4 in the early part of the melt season (when the nature of the drainage system varies between the runs) indicates that it is supply of melt that is the most important determinant of runoff. Later in the melt season, however, the model output hydrographs diverge. This suggests that the transport capacity of the subglacial hydrological system is the most important determinant. If the system is too efficient (as in run L1), peak discharges

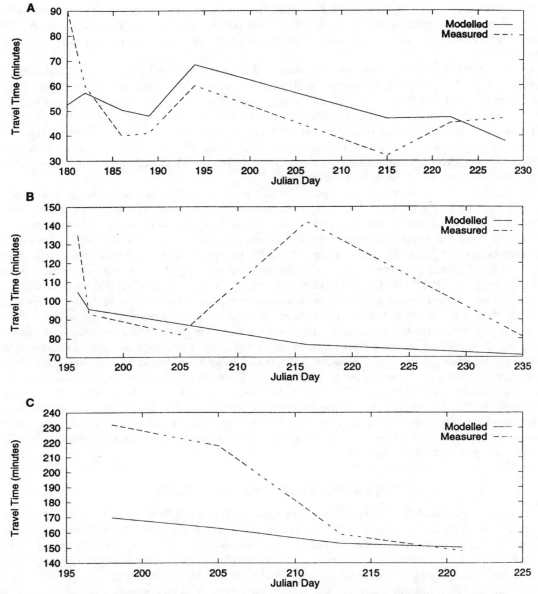

Figure 17. Seasonal variations in subglacial water travel time from moulin to snout, as modelled for 1993 and observed in dye tracer tests in 1990 and 1991. (a) Moulin 9 (1990); (b) moulin 34 (1991); (c) moulin 23 (1990)

in particular are too high. If, however, the system is too restrictive (as in runs L3 and L4, when the channelized system has had less time to enlarge than in runs L1 and L2), peak discharges are too low.

In spite of this apparent sensitivity however, there is still a wide range of possible conduit configurations (sizes and roughnesses) which will transport approximately the correct amount of water. The borehole water pressure measurements become very valuable as the conduit dimensions can be much more precisely defined. In particular, reproducing in the model the very large diurnal pressure variations observed in the field constrained the possible conduit configurations quite closely. Indeed, none of the simulations proved able to

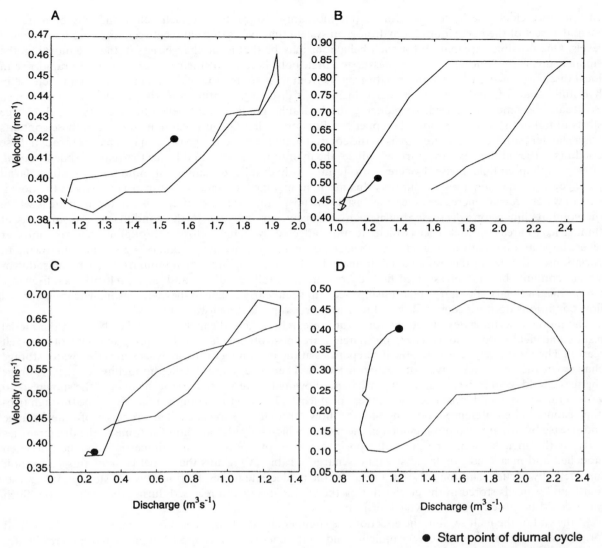

● **Start point of diurnal cycle**

Figure 18. Typical model diurnal velocity–discharge relationships. (a) Moulin 6, for JD 190 (9 July); (b) moulin 6, for JD 230 (18 August); (c) moulin 6, for JD 243 (31 August); (d) moulin 24, for JD 230 (18 August)

match the observed pressure variations for the whole period for which water pressure observations were available. While this may seem a weakness in the model, it should be borne in mind that the pressure observations were not made directly in the subglacial conduit, but in a nearby borehole. The nature of the link between the borehole and the conduit almost certainly changed over the course of the observations (Gordon *et al.*, 1998) but simulating such changes was beyond the scope of this modelling exercise. Given these uncertainties, the fact that the model matched many of the main features of the record for some 20 days is encouraging.

The fact that the complex velocity–discharge relationships observed in dye tracer tests are also reproduced by the model further increases confidence in the model. Nienow *et al.* (1996) argue that clockwise hysteresis can be explained by assuming that a moulin drains into a small tributary channel, rather than directly into an arterial channel, and that discharge variations in the two channels are out of phase. The fact that the model can produce clockwise hysteresis without this mechanism suggests that such behaviour may be a fundamental aspect of the behaviour of a hydrological system that cycles diurnally between high and low pressure levels. One possible explanation for such behaviour may be that as discharge rises in the late morning, the smaller, upstream conduits become pressurized first, both because of their size and the more rapid rise in local discharge owing to the shorter path lengths from melt source to conduit. This leads to an increase in hydraulic gradient, and hence to higher velocities. As discharge continues to rise, the lower conduits also eventually become pressurized, so reducing the hydraulic gradient, and hence the velocity. As discharge starts to fall in the late afternoon, the upper conduits drain first (again because meltwater in these conduits has a shorter path from its source areas to the conduit in question, so the peak discharge passes through these conduits earlier in the day); this further reduces the hydraulic gradient, and hence the velocity. Nienow *et al.* (1996) explain anticlockwise hysteresis in terms of hydraulic damming of moulins or sub/englacial passageways. The model results support this mechanism, since it seems to occur in areas where the downstream water pressure increases at least as fast as the upstream water pressure. This can occur in areas where the conduit dimensions decrease downstream, as occurs in areas where the conduits pass from areas of thinner ice to areas of thicker ice (which are more affected by channel closure through ice deformation), or where large amounts of water enter the subglacial drainage system downstream of a given moulin, owing to variations in the size of the supraglacial drainage basins supplying individual moulins. In such areas, downstream conduits become pressurized first, decreasing hydraulic gradients and hence velocities as discharge rises. Linear velocity–discharge relationships seem to occur early or late in the melt season, when pressurized flow conditions do not occur, owing to lower and/or less peaked meltwater inputs.

The model also indicates directions for future research, largely through mismatches between the model results and field observations. These mismatches are most obvious at the beginning and end of the melt season. The most likely reason for the discrepancies at the beginning of the melt season is that water storage changes in the snowpack have not been considered. Therefore, more detailed modelling of these storage changes would seem to be required. In particular, allowing the saturated zone at the base of the snowpack to vary in thickness would most likely solve these problems. One way to accomplish this would be through two-dimensional finite-difference modelling of water flow through the snowpack. The processes of water storage and release by firn are also not included in the model, which may also account for some of the discrepancies early in the melt season. The generally positive values of the seasonal volumetric difference between modelled and measured proglacial stream discharges (Table IV) means the model is predicting less runoff than actually occurs. This may also be a result of the inadequate representation of water storage. It suggests that liquid water is stored in the glacier during the winter/spring and released during the melt season. Such water could be stored in the snowpack or in firn.

At the end of the melt season, the effect of late summer snowfall also seems to be inadequately accounted for. Measurement of snowfall is very difficult, and so the energy balance model in particular underestimates snowfall at present. This in turn means too much meltwater continues to be produced (as the albedo of the glacier remains too low), and runs off too quickly (as snow depths, and hence flow delays, are smaller). Thus, the overprediction of peak flow discharge at the end of the melt season is at least partially a result of these problems. However, the water pressure record also undergoes marked changes at the end of the melt season after the first snowfalls. This suggests that the glacier drainage system may revert to some form of distributed system at this time.

At present, both model representations of a distributed system were in some way unsatisfactory. In particular, the representation of the system as bundles of small conduits (run L3) did not really work, though whether this would be improved by having more than eight conduits could not be tested, owing to computer memory limitations. Using wide, shallow, rough rectangular conduits (run L4) was much more successful in terms of hydrograph prediction, but during the late season the pressure records did not correspond well. This may indicate that other factors influence the configuration of the subglacial drainage system that are at present unaccounted for, in both the model and in theoretical glacier hydrology. Such factors may involve sudden partial or total blockage of conduits by ice collapse or by subglacial debris following sudden drops in conduit water pressures. Another possibility is that as subglacial discharges rise at the start of the melt season, the glacier becomes hydraulically 'jacked-up' off its bed, allowing subglacial drainage paths to open faster than would otherwise be expected, and that at the end of the melt season, when subglacial discharges fall, the glacier settles back on its bed causing faster closure of subglacial drainage paths (Iken, 1981; Iken *et al.*, 1983).

Another factor that may be very important, especially in the early and late season when the main subglacial conduit system is inefficient or poorly developed, is temporary storage of water in subglacial sediments. Hubbard *et al.* (1995) have suggested that subglacial conduits may 'leak', releasing water into subglacial sediments during the day when conduit water pressures are high. This water then flows back into the conduits at night, at times of low conduit water pressures. This process is not represented in the model at present, and may explain the general overprediction of high flows and underprediction of low flows. Having such a drainage component may allow higher low-flow values without the need for the very restrictive rectangular conduits that currently provide the best representation of a distributed system. If such conduits were slightly more efficient than at present, they may then show the weak diurnal pressure variations currently observed during the last part of the melt season, when the drainage system is believed to be distributed. Allowing for seasonal variations in storage and release of water in subglacial sediments may also account for some of the seasonal volumetric difference between modelled and measured proglacial stream discharges (Table IV).

SUMMARY AND CONCLUSIONS

In this study a physically based, semi-distributed modelling strategy was adopted to model the hydrology of a temperate valley glacier, Haut Glacier d'Arolla, Valais, Switzerland. This approach involved linking three separate submodels: (i) a surface energy balance submodel was used to calculate seasonal patterns of melt variation over the glacier surface; (ii) this was coupled to a surface routing submodel that accounts for differences between flow on snow-covered and ice surfaces; and (iii) this was then coupled to a subglacial hydrology submodel that can account for changing drainage configuration (distributed or channelized) and for the changing geometry of subglacial conduits (owing to wall melting and closure).

The model captured the essential features of both the diurnal and seasonal measured proglacial stream hydrographs. The model also produced subglacial water pressure records that were in reasonable agreement for some periods during the melt season with those obtained in the field. Reconstructed water travel times and flow velocities matched those obtained from dye tracer studies for much of the melt season.

The best matches between model output and field observations were obtained in the middle of the melt season, once the channelized system was established under most of the glacier. The biggest discrepancies occurred at the beginning of the summer, when large parts of the glacier were snow covered and underlain by a distributed drainage system, and at the end of the summer following a late season snowfall, when some form of distributed drainage system appears to have been re-established beneath at least part of the glacier.

Given the overall effectiveness of the model, specific conclusions regarding glacier hydrology can be made. To predict outflow hydrographs accurately, the delaying effect of the winter snowpack, and changes in this snow cover over the course of a melt season, have to be accounted for. However, changes in the configuration of the drainage system from distributed to channelized, and the seasonal evolution of subglacial conduits owing to wall melting and conduit closure must also be included. This is particularly true for the later parts of a melt season, when outflow hydrographs seem to be controlled largely by the transport capacity of the hydrological system, rather than by the amount of meltwater supplied to the system, which seems to control discharge earlier in the season. The complex spatial and temporal patterns of diurnal water pressure variation in the hydrological system give rise to very complex velocity–discharge histories in different areas and at different times. These patterns are very hard to predict in advance since they depend on subtle interactions between local, upstream and downstream conduit geometries and meltwater inputs. This generally supports the conclusions of Nienow *et al.* (1996), who argue that because of such complexity, using velocity–discharge relationships to reconstruct flow conditions in subglacial drainage systems is very difficult. However, the use of a model such as the one developed in this study does overcome many of these problems since the changing flow conditions encountered by a parcel of water as it moves through the subglacial hydrological system are taken into account.

However, the discrepancies between model outputs and field observations show that much work remains to be done in terms of field observations, modelling and development of theory of glacier hydrological systems. In particular, there is a need for increased understanding of the role of water storage changes within snow, firn and at the base of glaciers, especially at the beginning of the melt season. Furthermore, more research is required to investigate the possibility that processes other than conduit wall melting and closure (e.g. conduit blockage by ice collapse or sediment, or hydraulic jacking) are responsible for rapid changes in the configuration of subglacial drainage systems at both the start and end of the melt season.

Despite the remaining problems, the model should have several scientific applications. Its ability to predict spatial and temporal patterns of subglacial water pressures will make it a valuable tool for interpreting complex patterns of ice flow dynamics (where rates of glacier sliding and bed deformation are generally acknowledged to depend strongly on subglacial water pressure, e.g. Boulton and Hindmarsh, 1987; Fowler, 1987). Its ability to predict patterns of water pressures, as well as spatial and temporal variations in subglacial water velocities will also be useful for aiding the interpretation of hydrochemical variations observed in proglacial streams in terms of the processes and rates of solute acquisition (where pressurized or unpressurized flow conditions affect the availability of protons and oxidants from atmospheric gases and subglacial residence times influence the magnitude and relative importance of surface exchange and dissolution of mineral lattices, e.g. Sharp, 1991; Tranter *et al.*, 1993).

The model should also have significant practical application for water resource management in mountainous regions as a tool to predict hourly variations in glacier meltwater production, both at the present time, and, given predictions of possible climatic change, into the future (Willis and Bonvin, 1995).

ACKNOWLEDGEMENTS

This work was supported by the UK Natural Environment Research Council (Grants GR3/8971A, GR3/7004A and GR3/8114 and Studentship GT4/89/AAPS/53) and Earthwatch. The meteorological station was borrowed from the NERC equipment pool. Logistical support in the field was provided by Grande Dixence S.A., Yvonne Bams, and Patricia and Basile Bournissen. The field data used for model inputs and testing were collected by many individuals. We thank in particular Ben Brock, Luke Copland, Bryn Hubbard, Michael Nielsen and Pete Nienow. The photogrammetric DEM was provided by Grande Dixence S.A. The manuscript was completed while one author (N.A.) held an Isaak Walton Killam Memorial Postdoctoral Fellowship at the Department of Geophysics and Astronomy, University of British Columbia and while I.W. was on sabbatical leave at the Quaternary Research Center, University of Washington. We therefore thank Garry Clarke and Bernard Hallet for their support.

APPENDIX. SURFACE ENERGY BALANCE MODEL EQUATIONS

The energy balance model assumes there are four significant energy flux components at the surface of a glacier, which determine the ablation:

$$ABL = SWR + LWR + SHF + LHF$$

where ABL is ablation, SWR and LWR are short- and long-wave radiation fluxes and SHF and LHF are sensible and latent turbulent heat fluxes. All are expressed in equivalent ablation units, of mm of water per unit time. The individual energy balance components are calculated as follows. All parameter values are given in Table AI.

Table AI. Melt model parameter values

Parameter	Value	Units
K_s (ice)	$6\cdot34 \times 10^{-6}$	m kg^{-1} K^{-1} s^2
K_s (snow)	$4\cdot42 \times 10^{-6}$	m kg^{-1} K^{-1} s^2
K_1 (ice, condensation)	$9\cdot83 \times 10^{-3}$	m kg^{-1} s^2
K_1 (ice, evaporation)	$11\cdot14 \times 10^{-3}$	m kg^{-1} s^2
K_1 (snow, condensation)	$6\cdot86 \times 10^{-3}$	m kg^{-1} s^2
K_1 (snow, evaporation)	$7\cdot77 \times 10^{-3}$	m kg^{-1} s^2
L	$3\cdot34 \times 10^5$	J kg^{-1}
E	3000	m
σ	$5\cdot7 \times 10^{-8}$	W m^{-1} K^{-4}
k	$0\cdot26$	—
a_{os}	$0\cdot75$	—
a_{ns}	$0\cdot85$	—
a_1	$0\cdot137$	—
a_2	$0\cdot344$	—
a_3	$-1\cdot26$	—
a_4	$0\cdot03$	—
a_5	$-5000\cdot0$	—

For radiative fluxes:

$$SWR = (1 - \alpha)Q_i/L$$

where a is the surface albedo, Q_i is the solar radiation received at the surface (W m^{-2}), modified for slope, aspect and shading, and L is the latent heat of fusion of water (J kg^{-1}). Albedo is calculated using the relationships proposed by Oerlemans (1992, 1993), with empirical parameters derived from measurements at Haut Glacier d'Arolla. Briefly, a 'background' albedo profile is defined as:

$$\alpha_b = a_1 \arctan[(h - E + 300)/200] + a_2$$

where α_b is the background albedo, a_1 and a_2 are empirical parameters, h is the surface elevation and E is the elevation of the equilibrium line. This value is then modified by the presence of snow, and ageing of the surface:

$$\alpha = \alpha_{os} - (\alpha_{os} - \alpha_b)e^{a_3 d} - a_4 M$$

where α is the resulting surface albedo, α_{os} is a standard albedo for old snow, d is the snow depth in metres of water equivalent, M is the accumulated melt over the whole melt season, and a_3 and a_4 are empirical parameters. This value is further modified if 'new' snow is present in a given area, where new snow is assumed to be snow that fell less than three days before the time period under consideration:

$$\alpha' = \alpha_{ns} - (\alpha_{ns} - \alpha)e^{a_5 p}$$

where α' is the surface albedo if new snow is present, α_{ns} is a standard albedo for new snow, p is the depth of new snow in metres water equivalent and a_5 is an empirical parameter.

$$LWR = (I_{in} - I_{out})/L$$

where I_{in} and I_{out} are incoming and outgoing long-wave radiation (W m^{-2}). I_{out} for a glacier surface at 0°C, assuming it radiates as a black body, has a constant value of 316 W m^{-2}. I_{in} is given by:

$$I_{in} = \varepsilon\sigma T_a^4$$

where ε is the effective emissivity of the sky, σ is the Stefan–Boltzmann constant and T_a is the absolute temperature. ε is calculated as:

$$\varepsilon = (1 + kn)\varepsilon_0$$

where k is a constant related to cloudiness, n is cloud amount as a fraction and ε_0 is the clear sky emissivity, given by:

$$\varepsilon_0 = 8.733 \times 10^{-3} T_a^{0.788}$$

For turbulent fluxes:

$$SHF = K_S PTV$$

where K_S is a constant (including a correction for the latent heat of fusion of water), P is the atmospheric pressure (Pa), T is the air temperature (°C) and V is the wind speed (m s^{-1}). All these values are measured 2 m above the ice surface.

$$LHF = K_1 \delta_e V$$

where K_1 is a constant (including a correction for the latent heat of fusion of water), and δ_e is the difference between the vapour pressure of the air and the saturated vapour pressure over the glacier surface (Pa).

REFERENCES

Anderson, M. G. and Burt, T. P. (Eds), 1990. *Process Studies in Hillslope Hydrology*. John Wiley & Sons, Chichester. 539 pp.

Arnold, N., Willis, I. C., Sharp, M. J., Richards, K. S., and Lawson, W. J. 1996. 'A distributed surface energy balance model for a small valley glacier. I. Development and testing for Haut Glacier d'Arolla, Valais, Switzerland', *J. Glaciol.*, **42**, 77–89.

Baker, D., Escher-Vetter, H., Moser, H., Oerter, H., and Reinworth, O. 1982. 'A glacier discharge model based on the results of field studies of energy balance, water storage and flow', *IAHS Publ.*, **138**, 103–112.

Boulton, G. S. and Hindmarsh, R. C. A. 1987. 'Sediment deformation beneath glaciers: rheology and geological consequences', *J. Geophys. Res.*, **92(B9)**, 9059–9082.

Colbeck, S. C. 1978. 'The physical aspects of water flow through snow', in Chow, V. T. (Ed.), *Advances in Hydroscience*, Vol. 11. Academic Press, New York. pp. 165–206.

Copland, L. in press. 'The use of terrain analysis in the evaluation of snow cover over an alpine glacier', in Lane, S. N., Chandler, J. H., and Richards, K. S. (Eds), *Landform Monitoring, Modelling and Analysis*. John Wiley & Sons, Chichester.

Fountain, A. G. and Tangborn, W. V. 1985. 'Overview of contemporary techniques', *IAHS Publ.*, **149**, 27–41.

Fowler, A. C. 1987. 'Sliding with cavity formation', *J. Glaciol.*, **33**, 255–267.

Gordon, S., Sharp, M., Hubbard, B., Smart, C., Ketterling, B., and Willis, I. 1998. 'Seasonal reorganization of subglacial drainage inferred from measurements in boreholes', *Hydrol. Process.*, **12**, 105–134.

Hubbard, B. P., Sharp, M. J., Willis, I. C., Neilsen, M. K., and Smart, C. C. 1995. 'Borehole water level variations and the structure of the subglacial hydrological system of Haut Glacier d'Arolla, Valais, Switzerland', *J. Glaciol.*, **41**, 572–583.

Iken, A. 1981. 'The effect of the subglacial water pressure on the sliding velocity of a glacier in an idealised numerical model', *J. Glaciol.*, **27**, 401–427.

Iken, A., Röthlisberger, H., Flotron, A., and Haeberli, W. 1983. 'The uplift of Unteraagletscher at the beginning of the melt season — a consequence of water storage at the bed?', *J. Glaciol.*, **29**, 28–47.

Kamb, B. 1987. 'Glacier surge mechanism based on linked cavity configuration of the basal water conduit system', *J. Geophys. Res.*, **92(B9)**, 9083–9100.

Male, D. H. and Gray, D. M. 1981. 'Snow cover ablation and runoff', in Gray, D. M. and Male, D. H. (Eds), *Handbook of Snow*. Pergamon, Toronto. pp. 360–436.

Munro, D. S. and Young, G. J. 1982. 'An operational shortwave radiation model for glacier basins', *Wat. Resour. Res.*, **18(2)**, 220–230.

Nienow, P. W. 1993. 'Dye tracer investigations of glacier hydrological systems', *PhD Thesis*, University of Cambridge, Cambridge.

Nienow, P. W., Sharp, M., and Willis, I. C. 1996. 'Velocity–discharge relationships derived from dye tracer experiments in glacial meltwaters: implications for subglacial flow conditions', *Hydrol. Process.*, **10**, 1411–1426.

Oerlemans, J. 1992. 'Climate sensitivity of glaciers in Southern Norway: application of an energy balance model to Nigardsbreen, Hellstugbreen and Alfotbreen', *J. Glaciol.*, **38**, 223–232.

Oerlemans, J. 1993. 'A model for the surface balance of ice masses: part 1. Alpine glaciers', *Z. Gletscher. Glazialgeol.*, **27/28**, 63–83.

Paterson, W. S. B. 1994. *The Physics of Glaciers*, 3rd edn. Pergamon/Elsevier Science, Oxford.

Richards, K. S., Sharp, M. J., Arnold, N., Gurnell, A. M., Clarke, M. J., Tranter, M., Nienow, P. W., Brown, G. H., Willis, I. C., and Lawson, W. J. 1996. 'An integrated approach to modelling hydrology and water quality in glacierized catchments', *Hydrol. Process.*, **10**, 479–508.

Roesner, L.A., Aldrich, J. A., and Dickinson, R. E. 1988. *Storm Water Management Model User's Manual version 4; EXTRAN Addendum*. US Environmental Protection Agency, Athens, Georgia. 188 p.

Röthlisberger, H. 1972. 'Water in intra- and sub-glacial channels', *J. Glaciol.*, **11**, 177–203.

Sharp, M. J. 1991. 'Hydrological inferences from meltwater quality data: the unfulfilled potential', *Proceedings of the British Hydrological Society National Symposium, Southampton, 16–18 September 1991*, 5.1–5.8.

Sharp, M. J., Richards, K. S., Willis, I. C., Arnold, N., Nienow, P. W., Lawson, W. J., and Tison, J.-L. 1993. 'Geometry, bed topography and drainage system structure of Haut Glacier d'Arolla, Switzerland', *Earth Surf. Process. Landf.*, **18**, 557–571.

Shreve, R. L. 1972. 'Water movement in glaciers', *J. Glaciol.*, **11**, 205–214.

Spring, U. and Hutter, K. 1981. 'Numerical studies of Jökulhlaups', *Cold Reg. Sci. Technol.*, **4**, 227–244.

Tranter, M., Brown, G. H., Raiswell, R., Sharp, M., and Gurnell, A. M. 1993. 'A conceptual model of solute acquisition by alpine glacial meltwaters', *J. Glaciol.*, **39**, 573–581.

Willis, I. C. and Bonvin, J.-M. 1995. 'Climate change in mountain environments: hydrological and water resource implications', *Geography*, **80(3)**, 247–261.

Willis, I. C., Arnold, N., Bonvin, J.-M., Sharp, M., and Hubbard, B. In press. 'Mass balance and flow variations of Haut Glacier d'Arolla, Switzerland calculated using digital terrain modelling techniques', in Lane, S. N., Chandler, J. H., and Richards, K. S. (Eds), *Landform Monitoring, Modelling and Analysis*. John Wiley & Sons, Chichester.

Willis, I. C., Sharp, M. J., and Richards, K. S., 1993. 'Studies of the water balance of Midtdalsbreen, Handangerjökulen, Norway. II. Water storage and runoff prediction', *Z. Gletscher. Glazialgeol.*, **27/28**, 117–138.

18

TOWARDS A HYDROLOGICAL MODEL FOR COMPUTERIZED ICE-SHEET SIMULATIONS

RICHARD B. ALLEY

*Earth System Science Center and Department of Geosciences, The Pennsylvania State University,
University Park, PA 16802, USA*

ABSTRACT

Ice-sheet modelling typically uses grid cells 10 km or more on a side, so any hydrological and sliding model must average or parameterize processes that vary over shorter distances than this. Observations and theory suggest that basally produced water remains in a distributed, high-pressure system unless it encounters low-pressure channels fed by surface melt. Such distributed systems appear to exhibit increasing water storage, water transmission and water lubrication of sliding with increasing water pressure. A model based on these assumptions successfully simulates some aspects of the non-steady response of mountain glaciers to externally forced channel-pressure variations; it merits testing in ice-sheet modelling.

INTRODUCTION

Research in glacier hydrology ranges from the highly practical problems of finding subglacial drainages for hydropower generation, to the speculative exercise of estimating the behaviour of past ice sheets. Much of the recent progress in understanding water routing in and beneath temperate glaciers is of great value in solving problems of alpine regions; however, the increasingly complex view of glacier hydrology is not easily generalized to a scale suitable for computer modelling of ice sheets.

For whole ice-sheet models, grid cells typically are of the order of 10–100 km on a side. Any processes occurring in smaller regions must be averaged in some way to obtain their grid-scale behaviour. This 'homogenization' or 'parameterization' step is often viewed with alarm by those who understand the true complexity of the system, but is absolutely essential to large numerical experiments.

Here, a simple parameterization of basal processes is advanced. It begins with three main assumptions based on theory and observation: that basal water remains in a distributed, high-pressure system unless it encounters a low-pressure channel fed by surface melt; that water storage, water transmission and ice–bed separation by water increase with water pressure in a distributed system; and that the sliding velocity of ice increases with ice–bed separation. These assumptions allow a model to be assembled for water flow and its coupling to basal ice sliding using relations modified slightly from those already published.

Insufficient data are available from large ice sheets to allow testing of this model, so it is tested against mountain glacier data. Beneath mountain glaciers, regions between moulin-fed channels contain distributed water systems analogous to those under ice sheets, and the externally forced changes in channel pressure cause large changes in the distributed systems that are fairly well documented by observations. Finally, order-of-magnitude arguments are used to show that some terms in the model are important for mountain glaciers, but can be eliminated for most regions of ice sheets, producing a simplified basal model that can be coupled to an ice deformation model.

ASSUMPTIONS

Distributed basal water

The range of glacier water-system elements is fairly large — we can choose from R channels in ice (Röthlisberger, 1972), N channels incised into rock (Nye, 1973), linked cavities (Walder, 1986; Kamb, 1987), broad, shallow canals over unconsolidated sediments (Walder and Fowler, 1994) and films (Weertman, 1972), among others. The foremost distinction seems to be between those 'channelized' drainage systems for which the steady water pressure decreases with increasing water flux (R channels, with or without subjacent N channels) and those 'distributed' systems for which the steady water pressure increases with increasing water flux (linked cavities, canals and films). This distinction is critical because lubrication of basal sliding is believed to increase with increasing water flux in steady distributed systems, but to be independent of or to decrease with increasing water flux in steady channelized systems.

Both distributed and channelized systems exist beneath temperate mountain glaciers. Dye tracer tests can show very rapid or very slow output (e.g. Seaberg *et al*, 1988; Fountain, 1993). Boreholes through wet-based ice connect to high-volume water systems only in some instances (e.g. Hodge, 1979; Murray and Clarke, 1995). Deglaciated bedrock can contain potholes worn by streams, but also evidence of sub-glacial cavities and regions of intimate ice–bed contact with carbonate spicules grown in thin water films (e.g. Hallet, 1979; Walder and Hallet, 1979).

One of the most contentious issues in glacier hydrology has been whether basally generated water will collapse into R channels. Walder (1982) showed that a uniform film of water between ice and rock will be unstable against perturbations. However, Walder (1982) also suggested that the interaction of small, but growing, perturbations of water-film thickness with bedrock roughness elements would close the perturbations before they were able to form well-developed channels, assuming water film thicknesses appropriate to most mountain glacier conditions. Calculations following Walder (1982) obtain the same result for most ice-sheet conditions.

For further discussion of film–channel stability, see Weertman (1972), Weertman and Birchfield (1983) and Lliboutry (1979) relative to hard beds, and Boulton and Hindmarsh (1987), Alley (1989) and Walder and Fowler (1994) for deforming beds. The net result of such arguments is probably to strengthen the conclusion of Walder that collapse of a distributed system to a channelized system is not expected under most circumstances. Here the simplifying assumption has been made that because the water pressure in Walder–Fowler 'canals' on soft beds increases with water flux, they can be lumped with the distributed systems. Also note that basal water can form channels if, for example, it is draining a subglacial lake in a jokulhlaup or otherwise has access to much more water than is typically available to a steady-state, basally fed drainage system.

Abundant surface melt near ice-sheet margins clearly feeds channelized basal drainage, as shown, for example, by streams emerging from beneath the modern Greenland ice sheet and by eskers of palaeo-ice sheets. However, the limited data available from beneath interior regions of ice sheets are consistent with the existence of distributed systems, but inconsistent with channelized drainage. The single borehole at Byrd Station, West Antarctica (Gow, 1970; Weertman, 1970; Alley *et al.*, 1987) and the seismic measurements (Blankenship *et al.*, 1987) and nest of boreholes (Engelhardt *et al.*, 1990; Kamb, 1991; Kamb and Engelhardt, 1991) near the Upstream B camp on ice stream B, West Antarctica, encountered water pressures close to ice overburden pressures and water systems with low capacities. The observations are most consistent with low-volume, distributed systems and they cannot rule out the possibility of dominance by Darcian flow in permeable subglacial sediments. However, they do not show the low pressures and high capacity expected of channelized systems (Bindschadler, 1983).

Water pressure in a distributed system

Many basal models assume that the area of ice–bed separation by water increases with increasing water pressure. Increased water pressure will slow cavity closure by creep compared with cavity opening by basal sliding or melting, causing cavities to lengthen (Kamb, 1987). Increased water pressure will cause water films to spread into regions where the normal stress of ice on rock is lower than the increased water pressure (Weertman, 1964; Alley, 1989).

In contrast, the global force balance of a tessellated-bed cavity model by Humphrey (1987) yielded the apparently counterintuitive result that subglacial water pressure is independent of water extent (also see Iken, 1981; Kamb, 1993). The geometry used by Humphrey (1987) is shown in Figure 1. No shear stresses are allowed to act across ice–rock or ice–water interfaces (the usual assumption in basal sliding models). The shear and normal stresses τ_i and σ_i are resolved at the base of the interior flow of the ice (above the level where bedrock bumps affect the flow significantly) and are balanced by the stresses P_w and σ_b on the water and rock. Water occupies X_w and rock occupies $X_b = X - X_w$ out of each X of the bed, and the stoss face of the bedrock is inclined at angle B.

Force balances parallel and normal to the mean bed yield

$$X\tau_i = h(\sigma_b - P_w) \tag{1}$$

$$X\sigma_i = (X - X_w)\sigma_b + X_w P_w. \tag{2}$$

Noting $h = (X - X_w)\tan B$, these yield $P_w = \sigma_i - \tau_i/\tan B$, independent of X_w.

A useful way to visualize this is to note from the balance equations that an increase in P_w with constant bed geometry and interior ice stresses would require an increase in the component of σ_b parallel to the mean bed and a decrease in the component of σ_b normal to the mean bed. However, the requirement that σ_b act normal to the local bed prevents it from rotating, and this fixes the water pressure at a single value regardless of the extent of cavitation X_w.

On a curved bed, σ_b is replaced by some average over the curving faces, $\bar{\sigma}_b$. As shown in Figure 2, rotation of $\bar{\sigma}_b$ can be achieved by a change in X_w flooding or revealing a section of bed with a different slope than the mean over the stoss side of the bedrock. In particular, if the cavity ends in a convex-up region of the bed, increased P_w can be balanced by shrinkage of the cavity to reveal steeper bedrock (Humphrey, 1987); if the cavity ends in a concave-up region of the bed, increased P_w can be balanced by lengthening of the cavity to flood low-angle bedrock ($\partial X_w/\partial P_w < 0$ for convex-up bed; >0 for concave-up).

A second-order correction is needed to account for the effects of the curvature on h. No closed-form solution is yet available, but I have conducted a variety of simulations assuming faceted or sinusoidal beds, with possible assumptions about the distribution of the magnitude of σ_b as a function of position

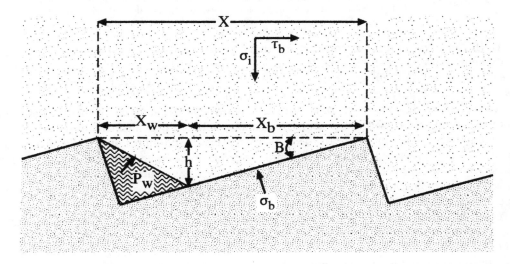

Figure 1. Geometry used in Humphrey (1987) global force balance model, using his notation. Water pressure, P_w, and ice–rock contact stress, σ_b, are constrained to act normal to the local base of the ice, as shown, and their components normal and parallel to the mean bed must balance the pressure σ_i and shear stress τ_i from the interior ice flow

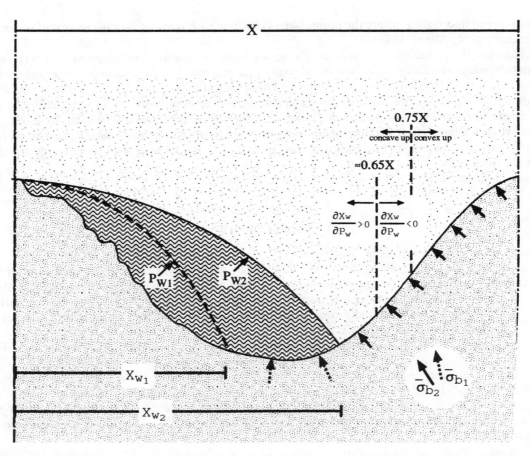

Figure 2. Possible extension of the Humphrey (1987) global force-balance model to a curving bed. The ice–rock stress, $\bar{\sigma}_b$, is an average over the exposed stoss side of the bedrock, and its direction can be changed by changes in cavitation X_w flooding or revealing portions of the bed with different slope than the average. Numerical simulations assuming that the magnitude of σ_b is constant for each interval of ice–rock contact dX_b, or that the magnitude varies sinusoidally, yield $\partial X_w/(\partial P_w)$ as shown to maintain force balance, with only weak dependence on the ratio of amplitude to wavelength of the bedrock roughness

within X_b (for example, that it is constant, or that it varies sinusoidally). I obtain $\partial X_w/\partial P_W < 0$ when the cavity ends in a convex-up or weakly concave-up region of the bed; >0 if strongly concave-up. For example, on a sinusoidal bed with constant magnitude σ_b, cavitation increases with water pressure for $X_w <\approx 0\cdot65X$, only weakly sensitive to the amplitude of the bed relief.

A generalization is that the local solution of cavitation or water film extent increasing with increasing water pressure is consistent with the global force balance solution provided cavitation is not too extensive (cavities end in concave-up regions of the bed). For extensive cavitation (cavities end in convex-up regions), an increase in water pressure leads to a force imbalance that is corrected through changes in τ_i associated with stress transmission in the interior ice flow (Humphrey, 1987). Increased water pressure over a large area of bed (at least many ice thicknesses in lateral extent) with extensive cavitation might rapidly produce a large change in ice flow.

Notice the behaviour of the Weertman (1964) basal model in this context. Weertman assumed cubical bedrock obstacles with faces normal to τ_i and to σ_i. This separates the components of σ_b normal and parallel to the mean bed and allows the bed to support any magnitude of τ_i so long as the obstacles project through the water into the ice. The global force balance solution is always consistent with the local behaviour, and force balance can be maintained even with extensive cavitation in response to high water pressure (Weertman, 1986).

I thus argue that we can use the local solution — cavitation increases with water pressure — unless cavitation is extensive and the bed lacks roughness elements with stoss faces approximately normal to mean ice flow (see also Alley, 1989). The global force balance requires a fixed water pressure or decreasing cavitation with increasing water pressure only if cavitation is extensive and the bed lacks Weertman-type roughness elements (Humphreys, 1987; also see Iken, 1981; Kamb, 1993).

Lubrication of basal sliding

This is probably the least contentious assumption here. Essentially all sliding theories are based on the premise that ice is restrained by interaction with roughness elements in the bed, and that the rate of motion of ice over these roughness elements ('sliding') increases with the stress on them (Weertman, 1964). If some roughness elements that restrain the ice are replaced by water that does not restrain the ice, the stress on roughness elements still in contact with the ice will increase, increasing the sliding velocity. Any water system of non-zero thickness that becomes more extensive in thickness or area will tend to 'drown' some roughness elements and speed sliding, with the efficiency of such 'drowning' a function of the roughness spectrum and the water geometry.

MODEL

A distributed water system beneath a glacier can be characterized by its pressure, P_w, or alternatively by the effective pressure $N \equiv P_i - P_w$, where P_i is the ice pressure (equivalent to σ_i in Humphrey, 1987). In addition, a distributed water system can be characterized by its water storage, its ability to transmit water and the fraction of the bed occupied by water ($f = X_w/X$ in Figure 1). I choose to represent water storage by the equivalent thickness of a uniform sheet, d', and to represent the ability to transmit water by the thickness of a uniform sheet that would transmit an equivalent amount, d. A Weertman film would have $d' = d$; a linked-cavity system would have $d' > d$ or $d' \gg d$. Taking Q as the water flux, we can expect relations of the form $d' \sim 1/N$, $d \sim d'$, $Q \sim d$ and $f \sim 1/N$.

Force balance arguments suggest that the basal shear stress τ_b (equivalent to τ_i of Humphrey, 1987) should increase with increasing effective pressure, giving $\tau_b \sim N$ (e.g. Iken, 1981; Bindschadler, 1983; Humphrey, 1987). The basal sliding velocity u_b should increase with the basal shear stress and with the water storage flooding roughness elements, giving $u_b \sim d'$, $u_b \sim \tau_b$.

For simplicity, I suppose that linear relations exist among the geometric variables

$$f = K_f d \tag{3}$$

$$d' = K_d d. \tag{4}$$

Water flow and its linkage to ice flow can be described by

$$N = K_n \frac{\tau_b}{d'} \tag{5}$$

$$Q = K_w d^a (\nabla \Phi_w)^b \tag{6}$$

$$\Phi_w = \rho_i g(z_s - z_b) + \rho_w g z_b - N \tag{7}$$

$$\nabla \cdot Q = S - \frac{\partial d'}{\partial t} \tag{8}$$

$$u_b = K_s \tau_b^{(c-1)} (1 + K_s' d' \tau_b) \tag{9a}$$

$$u_b = K_s'' \tau_b^c d'. \tag{9b}$$

The force balance equation, Equation (5), is derived from Alley [1989: his equation (22)] using Equations (3) and (4) here. The water flux, Q in Equation (6), has $a = 3$, $b = 1$ and $K_w = 1/(12\nu)$, $\nu = 1\cdot8 \times 10^{-3}$ Pa s for laminar flow between smooth, parallel plates, and $a \approx 0\cdot5$, $b \approx 0\cdot5-0\cdot7$, K_w to be determined from shape and roughness of the conduits for turbulent flow (see review in Weertman, 1972). The water potential, Φ_w, in Equation (7) is the sum of the pressure head and elevation head less the pressure drop associated with ice flow [Equation (5)] and depends on the elevations of the ice surface, z_s, and bed, z_b, the densities of water, ρ_w, and ice, ρ_i, and the acceleration of gravity, g. The source term, S, in the continuity Equation (8), includes any basal melting or freezing, water loss or gain from subglacial aquifers, etc. The sliding velocity, u_b, is modified slightly from Weertman and Birchfield (1982); in their model, d and d' are identical and the power $c = 3$. The approximation in Equation (9b) is valid for d' of the order of 1 mm or larger, following Weertman and Birchfield (1982). The geometrical and physical constants K_d, K_n, K_s, K_s' and K_w must be specified. Notice that if Equations (4) and (5) are used to express d' in terms of N, the sliding relation Equation (9b) includes the inverse dependence on N that is frequently assumed (see review by Bentley, 1987).

Over some sufficiently large area (at least several ice thicknesses in lateral extent; e.g. Kamb and Echelmeyer, 1986; Whillans and Van der Veen, 1993), the basal shear stress, τ_b, must average to the driving stress

$$\bar{\tau}_b = \rho_i g(z_s - z_b)\nabla z_s \tag{10}$$

where the geometrical variables are understood to be averages for the region. The averaging distance is a function of ice flow (the ability of ice to sustain gradients in longitudinal deviatoric stresses) and so requires coupling of basal and ice deformational models for accurate simulations.

MODEL TESTS

Testing parameterizations for ice sheet modelling has proved very difficult. Limited data and numerous free parameters in models almost guarantee the ability to fit modern ice sheets. The short duration of instrumental records compared with the response times of ice sheets makes it difficult to test models against a range of conditions. Mountain glaciers are much more complicated than ice sheets because of the time-varying water input to channelized as well as distributed water systems, but the short response times and easier observations on mountain glaciers and their beds are attractive for model tests.

The onset of springtime surface melt on mountain glaciers fills and pressurizes conduits draining to the bed, which interact with regions of distributed water flow between conduits. The typical result is uplift of the glacier as water spreads across the bed and contacts previously isolated or weakly interconnected regions, increased water flux from the glacier and faster sliding (e.g. Iken *et al.*, 1983; Kamb *et al.*, 1985; Iken and Bindschadler, 1986; Raymond and Malone, 1986; Kamb and Engelhardt, 1987; Collins, 1989; Hooke *et al.*, 1989; Willis *et al.*, 1990).

These data from mountain glaciers further show that individual rain or melt events lasting hours or days affect glacier motion and water flow in a similar manner to the seasonal cycle, especially if the events occur early in the melt season when the channelized drainage system is still rather constricted. Also, there is a general asymmetry of response at a variety of time-scales, with water input and water pressure rise causing rapid flooding of the glacier bed and ice speed-up, followed by less rapid drainage of the bed and ice slowdown.

To learn whether the model sketched here produces such behaviour, I simulate the response of a one-dimensional, horizontal, distributed or film water system connected laterally to a channel. The channel parallels ice flow and the x axis is taken along the bed normal to the channel. Water flow is allowed normal to the channel and to ice flow, with no water flow parallel to the channel in the distributed system. A no-flux boundary condition is specified at some position x_{max}. The physical and geometrical constants K_d, K_n, K_s'' and K_w are assumed independent of position, as are the sliding velocity and the ice pressure. I do not allow for lateral shear of the ice, for the geometrical effect on ice velocity of the opening or closing of cavities (Iken and Bindschadler, 1986), or for bridging stresses (Weertman, 1972; Murray and Clarke, 1995). The basal shear stress is allowed to assume a different value in each cell of the computational grid, but must average to an externally applied value over the x domain. Basal melting or freezing, and water loss or gain to subglacial or englacial aquifers, are assumed small compared with flow to and from the channel, and the source term S is set to zero in Equation (8). Laminar water flow is assumed. Initial values of d are specified and the system of Equations (4)–(9) is forced by specified changes in water pressure at the channel–film interface, $x = 0$.

A change in the channel pressure induces a pressure gradient and water flow in the adjacent film, with the flow-rate increasing with the film thickness as well as with the pressure gradient. Spatial gradients in the water flow cause storage or release of water from the film and change the film thickness and thus the flow rate. Thickness changes in turn cause a redistribution of basal shear stress from thick film, well-lubricated areas to thin film, poorly lubricated areas to balance the average applied shear stress, and the changed patterns of lubrication and redistributed shear stress yield a new sliding velocity. The model is stepped forward in time to simulate the evolution of this system. Stability of the numerical scheme was verified by changing the time step without changing the results.

Results for $d = d'$ are shown in Figure 3a, 3b and 3c. A one-day period of channel pressure variation with a half-amplitude of 1 bar and an average pressure 1·2 bar less than the ice pressure feeds a film extending 500 m to the divide with the next channel. Other constants are listed in the figure captions.

The pressure response is shown in Figure 3a. At the channel ($x = 0$), the variation is specified to be sinusoidal. With increasing distance from the channel, the amplitude is reduced because full adjustment requires longer than allowed by the period of the forcing. In addition, the pressure further from the channel averages higher than in the channel. This is because high pressure in the channel opens up the film near the channel and allows water to flow out easily and increase the pressure farther from the channel, but low pressure in the channel closes down the film nearby and slows drainage back to the channel.

The initial decrease in pressure far from the channel, during 'spin-up' of the model from an assumed initial film of 0·05 mm everywhere, is tied to basal shear stress. As lubrication near the channel increases, the externally applied force that must be balanced by the basal shear stress becomes concentrated on the less-lubricated areas further away, although the water film thickness has not changed there. From Equation (5), this causes a decrease in water pressure. Once the initial transients have died away, the pressure cycle at the flow divide is not damped relative to the next grid point closer to the channel for essentially the same reason.

Figure 3b tells a similar story. The average water-film thickness is greater than expected for the average channel pressure. The easy passage of water away from the channel, but difficult return, increases the thickness, as do interactions with the shear stress through Equation (5).

The sliding velocity is shown in Figure 3b and enlarged in Figure 3c. Again, the average sliding velocity is greater than would be expected for the average channel pressure. The variation in sliding velocity is asymmetrical, with a rapid rise followed by a slow decay. The rise is rapid because the region far from the channel remains relatively well lubricated at all times, so velocity increases as soon as the rising water pressure increases lubrication in the narrow zone near the channel. The asymmetry reflects the difficulty of draining even this narrow zone during falling water pressure.

The modelled characteristics of rapid uplift and speed-up of ice in response to water pressure rise, followed by slow fall, are consistent with observations on glaciers. The biggest difficulty is that modelled perturbations in the channel are transmitted to the divide more rapidly than is realistic — one would be unlikely to obtain the clear seasonal cycle observed in mountain glaciers if they had a response time of around a day. This problem can be solved by allowing parameterized water storage through setting $d' > d$. Such a change is highly reasonable given that glaciers rise centimetres in the springtime owing to cavity filling, but the simulated d values here are only millimetres.

An attempt to simulate a seasonal cycle is shown in Figure 4, with $d' = 10d$, the period of the forcing set to one year, the half-amplitude of the sinusoidal pressure set to 3 bar and physical constants listed in the figure caption. The behaviour is similar to that seen in Figure 3, except that increased storage greatly slows the response. Figure 5 shows the effect of adding the diurnal cycle of Figure 3 (half-amplitude 1 bar, but with effective pressure in no instance less than 0·2 bar) to the seasonal cycle from Figure 4, during the summer quarter of the year. Notice that the diurnal cycle has a significant effect on the ice motion, showing the responsiveness of the system.

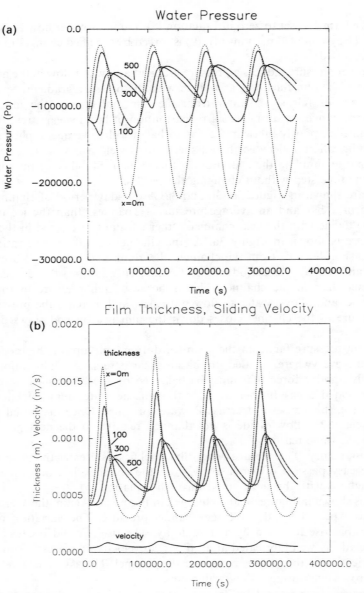

Figure 3. Modelled response of distributed water system to diurnal variation in pressure of adjacent channel, in the absence of linked cavities ($K_d = 1$, so $d' = d$). Simulation assumes laminar water flow [$a = 3$, $b = 1$, $K_w = 1/(12\nu)$, $\nu = 1\cdot8 \times 10^{-3}$ Pa s], $K_n = 5 \times 10^{-4}$ m, $K_s'' = 10^{-11}$ Pa^{-2} s^{-1}, $\bar{\tau}_b = 10^5$ Pa (1 bar) and $c = 2$ in Equation (9b), equivalent to a stress-independent controlling obstacle size in the Weertman sliding model. The channel pressure averages 120 kPa ($1\cdot2$ bar) less than the ice overburden pressure and varies sinusoidally with a half-amplitude of 100 kPa (1 bar) and a period of 86 400 s (1 day). (a) Water pressure in Pa relative to ice pressure assumed zero, at the channel (labeled 0 m) and 100, 300 and 500 m away from the channel, where 500 m is the no-flow boundary of the solution domain. (b) Water film thickness and velocity. Thickness is plotted for same sites as pressure in (a). (c) Ice-flow velocity, enlarged from (b). Because I have not allowed lateral shear of the ice, the velocity is the same for all points and increases as the average lubrication increases

If turbulent flow occurs, the water flux is smaller for a given thickness and pressure gradient than for the laminar-flow calculations shown. Channel pressure variations in turbulent flow are transmitted through the distributed system more slowly than for laminar flow.

The main observations that I am testing against are the existence of annual and diurnal cycles of glacier behaviour in response to water pressure forcing in moulins, in which water pressure increase causes rapid

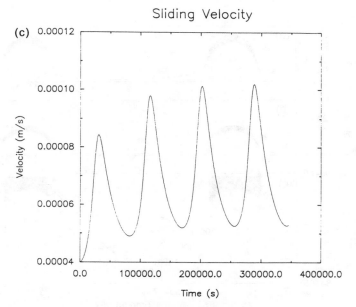

Figure 3 (continued from previous page)

uplift and speed-up of the ice as water accesses more of the glacier bed, followed by slower subsidence and slowdown of the ice as water drains back into the channels. I believe that any model will exhibit these characteristics if increased water pressure is modelled to cause increased water storage and transmission and increased basal sliding lubrication ($d \sim P_{\mathrm{w}}$, $d \sim d'$, $u_{\mathrm{s}} \sim d'$). I have not tested the exact forms of the relations used in Equations (4)–(9), except in the general sense that when taken together, they simulate

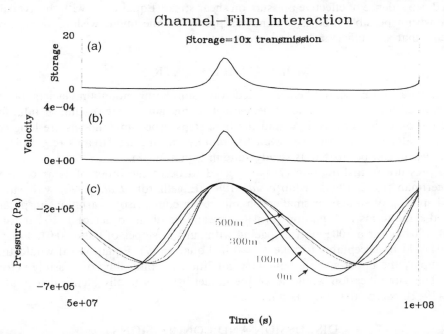

Figure 4. Modelled response of distributed water system to seasonal variation in pressure of adjacent channel, with small linked cavities ($K_{\mathrm{d}} = 10$, so $d' = 10d$). Simulation assumes laminar water flow, $K_{\mathrm{n}} = 5 \times 10^{-3}\,\mathrm{m}$, $K_{\mathrm{s}}'' = 10^{-12}\,\mathrm{Pa}^{-2}\,\mathrm{s}^{-1}$, $\bar{\tau}_{\mathrm{b}} = 10^5\,\mathrm{Pa}$ (1 bar) and $c = 2$. The channel pressure averages 320 kPa (3·2 bar) less than the ice overburden pressure and varies sinusoidally with a half-amplitude of 300 kPa (3 bar) and a period of $3·16 \times 10^7\,\mathrm{s}$ (1 year). (a) Water storage in cubic metres. (b) Ice velocity in m s^{-1}. (c) Water pressure in Pa relative to ice pressure assumed zero, at the channel and 100, 300 and 500 m away

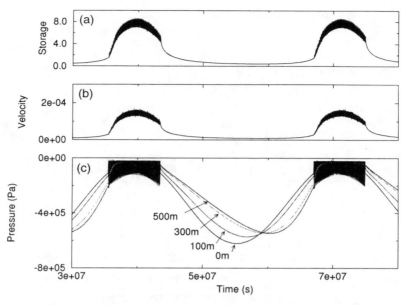

Figure 5. Modelled response of distributed water system to seasonal and diurnal variation in pressure of adjacent channel, with small linked cavities. The one-day, 100 kPa (1 bar) sinusoidal variation of Figure 3 is added to the simulation of Figure 4 during the quarter of the Figure 4 year with highest water pressures, using the same constants as in Figure 4. However, the water pressure is not allowed to rise above 20 kPa (0·2 bar) less than the ice pressure. (a) Water storage in cubic metres. (b) Ice velocity in m s^{-1}. (c) Water pressure in Pa relative to ice pressure assumed zero, at the channel and 100, 300 and 500 m away

appropriate response rates. The forms are based on physical reasoning or on numerical simplicity. I cannot test the assumed dependence of effective pressure on shear stress (Equation 5) with the available data. Thus the data do constrain possible models, but not as tightly as one might wish, and my favoured model contains elements that are sufficient, but not necessary.

MODEL SIMPLIFICATION

All of the model equations presented here are needed for simulating mountain glaciers, but some simplifications are warranted for ice sheets. Except near ice-sheet margins where moulin or tidal forcing may be significant, the response times of ice flow and the time steps of ice-flow models are long compared with the time-scale of water system response, even for rapidly varying ice streams (e.g. Alley et al., 1994). Thus the water system can be treated as steady-state in Equation (8).

The effective pressure N in Equation (7) is expected to be of the order of 1 bar or less for ice-sheet conditions (Weertman, 1972; Blankenship et al., 1987; Engelhardt et al., 1990) with variations over a 10–100 km grid spacing of similar or smaller magnitude. In comparison, variations in potential from the surface- and bed-slope terms in Equation (7) will typically be of the order of 1–10 bar for a 10 km grid spacing and 10–100 bar for a 100 km grid spacing with surface slopes of 0·001–0·01. For some ice-sheet applications it should be acceptable to take $N = 0$ in Equation (7). Equation (5) would then be used to calculate N to supply the upper-surface boundary condition for modelling of possibly deforming subglacial sediments. This simplification would leave the model here essentially identical to that of Weertman and Birchfield (1982), except that they assumed $d = d'$.

DISCUSSION AND CONCLUSIONS

Despite decades of concentrated research, it remains unclear how to treat basal processes in ice-sheet modelling. For example, Bentley (1987) reviewed several models that had been applied to West Antarctic ice streams and showed how new data partially or completely invalidated all of them for a region near the Upstream B camp, despite the fact that some had been fitted to the ice stream. This difficulty arises primarily from our profound ignorance of basal conditions of ice sheets, but also from the problem of properly parameterizing the details of ice–water–rock interactions for models with grid spacing of the order of 10 km or more.

The lack of data on ice-sheet basal processes and conditions has helped maintain the interest of ice-sheet glaciologists in mountain glaciers: mountain glaciers are more complex than ice sheets owing to the inputs of surface water, but the accessibility and short response times of mountain glaciers and their beds offer many avenues of research not available on ice sheets. However, translation of the mountain glacier results to ice-sheet models has been relatively slow, in part because of difficulties with parameterizations. Here, I have tried to use insights from mountain glaciers and ice sheets to constrain the likely behaviour of sub-ice-sheet water, and to test a parameterization based on these insights.

Some theory, and observations in West Antarctica and beneath temperate glaciers, favour the view that basally derived meltwater in steady ice-contact drainage remains in a high-pressure, distributed system unless it encounters a low-pressure channel fed by surface melt (or unless there are strongly nonsteady effects such as jokulhlaups). Observations on mountain glaciers show that in such a distributed system, the water storage, the fraction of the bed occupied by water, the water flux and the lubrication of ice sliding (and thus the ice velocity) all increase with water pressure.

The simplest physically based model of distributed basal water incorporating these elements probably is that of Weertman (1972). To allow the large water storage required by observations such as the spring uplift of temperate glaciers, I have suggested two minor modifications to the model: allowing more direct coupling between water pressure and the other variables [Equation (5)]; and allowing total storage to increase more rapidly than storage in the water transmission elements [Equation (4)], by perhaps 1–10 for smooth, soft tills and 10–100 for bedrock (parameterized cavitation; see also Weertman and Birchfield, 1983; Weertman, 1986).

A qualitative test of this modified Weertman model shows agreement with observations from mountain glaciers. Both model and actual glaciers respond rapidly to overpressurization of drainage conduits caused by water input from the surface. Both show rapid filling of the distributed water system and rapid speed-up of the glacier, followed by less-rapid drainage and slowdown. This appears to be a necessary feature of any drainage system in which water storage, transmission and lubrication of sliding increase with water pressure. During rising water pressures in the channel, the region of the distributed system near the channel is 'opened up', allowing water to move out of the channel easily. During falling water pressures in the channel, the region nearby 'closes down' as it is drained of water first, and the water further away must then pass through a low-volume, hence inefficient drainage system to reach the channel.

Of course, other interactions (e.g. with shear stress) may complicate this picture. Also, I have ignored the effects of local changes in ice pressure caused by opening and closing of the basal water system (Murray and Clarke, 1995). Modelled timing of events could be changed by a switch from laminar to turbulent flow, as well as by adjusting the other physical constants. But I take the general agreement between model and observations as an encouraging sign that the parameterizations advanced here are worth testing in ice-sheet modelling. I urge caution in applying this model to mountain glaciers, despite the fact that I have done so; this clearly is a parameterization of more complicated processes.

ACKNOWLEDGEMENTS

I thank S. Anandakrishnan, N. Humphrey, J. Walder and two anonymous reviewers for helpful suggestions, and the National Science Foundation Office of Polar Programs, the D. and L. Packard Foundation and NASA-EOS for financial support.

REFERENCES

Alley, R. B. 1989. 'Water-pressure coupling of sliding and bed deformation: I. Water system', *J. Glaciol.*, **35**, 108–118.
Alley, R. B., Blankenship, D. D., Rooney, S. T., and Bentley, C. R. 1987. 'Continuous till deformation beneath ice sheets', *IAHS Publ.*, **170**, 81–91.

Alley, R. B., Anandakrishnan, S., Bentley, C. R., and Lord, N. 1994. 'A water-piracy hypothesis for the stagnation of ice stream C', *Ann. Glaciol.*, **20**, 187–194.

Bentley, C. R. 1987. 'Antarctic ice streams: a review', *J. Geophys. Res.*, **92**, 8843–8858.

Bindschadler, R. A. 1983. 'The importance of pressurized subglacial water in separation and sliding at the glacier bed', *J. Glaciol.*, **29**, 3–19.

Blankenship, D. D., Bentley, C. R., Rooney, S. T., and Alley, R. B. 1987. 'Till beneath Ice Stream B. 1. Properties derived from seismic travel times'. *J. Geophys. Res.*, **92**, 8903–8911.

Boulton, G. S. and Hindmarsh, R. C. A. 1987. 'Sediment deformation beneath glaciers: rheology and geological consequences', *J. Geophys. Res.*, **92**, 9059–9082.

Collins, D. N. 1989. 'Seasonal development of subglacial drainage and suspended sediment delivery to melt waters beneath an alpine glacier', *Ann. Glaciol.*, **13**, 45–50.

Engelhardt, H., Humphrey, N., Kamb, B., and Fahnestock, M. 1990. 'Physical conditions at the base of a fast moving Antarctic ice stream', *Science*, **248**, 57–59.

Fountain, A. G. 1993. 'Geometry and flow conditions of subglacial water at South Cascade Glacier, Washington State, U.S.A.; an analysis of tracer injections', *J. Glaciol.*, **39**, 143–156.

Gow, A. J. 1970. 'Preliminary results of studies of ice cores from the 2164 m deep drill hole, Byrd Station, Antarctica', *IAHS Publ.*, **86**, 78–90.

Hallet, B. 1979. 'Subglacial regelation water film', *J. Glaciol.*, **23**, 321–333.

Hodge, S. M. 1979. 'Direct measurement of basal water pressures: progress and problems', *J. Glaciol.*, **23**, 309–319.

Hooke, R. LeB., Calla, P., Holmlund, P., Nilsson, M., and Stroeven, A. 1989. 'A 3 year record of seasonal variations in surface velocity, Storglaciären, Sweden', *J. Glaciol.*, **35**, 235–247.

Humphrey, N. F. 1987. 'Coupling between water pressure and basal sliding in a linked-cavity hydraulic system', *IAHS Publ.*, **170**, 105–119.

Iken, A. 1981. 'The effect of the subglacial water pressure on the sliding velocity of a glacier in an idealized numerical model', *J. Glaciol.*, **27**, 407–421.

Iken, A. and Bindschadler, R. A. 1986. 'Combined measurements of subglacial water pressure and surface velocity of Findelengletscher, Switzerland: conclusions about drainage system and sliding mechanism', *J. Glaciol.*, **32**, 101–119.

Iken, A., Röthlisberger, H., Flotron, A., and Haeberli, W. 1983. 'The uplift of Unteraargletscher at the beginning of the melt season — a consequence of water storage at the bed?' *J. Glaciol.*, **29**, 28–47.

Kamb, B. 1987. 'Glacier surge mechanism based on linked cavity configuration of the basal water conduit system', *J. Geophys. Res.*, **92**, 9083–9100.

Kamb, B. 1991. 'Rheological nonlinearity and flow instability in the deforming-bed mechanism of ice-stream motion', *J. Geophys. Res.*, **96**, 16 585–16 595.

Kamb, B. 1993. 'Glacier flow modeling'. In Stone, D. B. and Runcorn, S. K. (Eds), *Flow and Creep in the Solar System: Observations, Modeling and Theory*. Kluwer Academic, Dordrecht. pp. 417–460.

Kamb, B. and Echelmeyer, K. A. 1986. 'Stress-gradient coupling in glacier flow: I. Longitudinal averaging of the influence of ice thickness and surface slope', *J. Glaciol.*, **32**, 267–284.

Kamb, B. and Engelhardt, H. 1987. 'Waves of accelerated motion in a glacier approaching surge: the mini-surges of Variegated Glacier, Alaska, U.S.A.', *J. Glaciol.*, **33**, 27–46.

Kamb, B. and Engelhardt, H. 1991. 'Antarctic ice stream B: conditions controlling its motions and interactions with the climate system', *IAHS Publ.*, **208**, 145–154.

Kamb, B., Raymond, C. F., Harrison, W. D., Engelhardt, H., Echelmeyer, K. A., Humphrey, N., Brugman, M. M., and Pfeffer, T. 1985. 'Glacier surge mechanism: 1982–1983 surge of Variegated Glacier, Alaska', *Science*, **227**, 469–479.

Lliboutry, L. 1979. 'Local friction laws for glaciers: a critical review and new openings', *J. Glaciol.*, **23**, 67–94.

Murray, T., and Clarke, G. K. C. 1995. 'Black-box modeling of the subglacial water system', *J. Geophys. Res.*, **100**, 10231–10245.

Nye, J. F. 1973. 'Water at the bed of a glacier', *IAHS Publ.*, **95**, 189–194.

Raymond, C. F. and Malone, S. 1986. 'Propagating strain anomalies during mini-surges of Variegated Glacier, Alaska, U.S.A.', *J. Glaciol.*, **32**, 178–191.

Röthlisberger, H. 1972. 'Water pressure in intra- and subglacial channels', *J. Glaciol.*, **11**, 177–203.

Seaberg, S. Z., Seaberg, J. Z., Hooke, R. LeB., and Wiberg, D. W. 1988. 'Character of the englacial and subglacial drainage system in the lower part of the ablation area of Storglaciären, Sweden, as revealed by dye-trace studies', *J. Glaciol.*, **34**, 217–227.

Walder, J. S. 1982. 'Stability of sheet flow of water beneath temperate glaciers and implications for glacier surging', *J. Glaciol.*, **28**, 273–293.

Walder, J. S. 1986. 'Hydraulics of subglacial cavities', *J. Glaciol.*, **32**, 439–445.

Walder, J. S. and Fowler, A. C. 1994. 'Channelised subglacial drainage ovr a deformable bed', *J. Glaciol.*, **40**, 3–15.

Walder, J. S. and Hallet, B. 1979. 'Geometry of former subglacial water channels and cavities', *J. Glaciol.*, **23**, 335–346.

Weertman, J. 1964. 'Glacier sliding', *J. Glaciol.*, **5**, 287–303.

Weertman, J. 1970. 'A method for setting a lower limit on the water layer thickness at the bottom of an ice sheet from the time required for upwelling of water into a borehole', *IAHS Publ.*, **86**, 69–73.

Weertman, J. 1972. 'General theory of water flow at the base of a glacier or ice sheet', *Rev. Geophys. Space Phys.*, **10**, 287–333.

Weertman, J. 1986. 'Basal water and high-pressure basal ice', *J. Glaciol.*, **32**, 455–463.

Weertman, J. and Birchfield, G. E. 1982. 'Subglacial water flow under ice streams and West Antarctic ice-sheet instability', *Ann. Glaciol.*, **3**, 316–320.

Weertman, J. and Birchfield, G. E. 1983. 'Stability of sheet water flow under a glacier', *J. Glaciol.*, **29**, 374–382.

Whjllans, I. M. and Van der Veen, C. J. 1993. 'New and improved determinations of velocity of ice streams B and C, West Antarctica', *J. Glaciol.*, **39**, 483–490.

Willis, I. C., Sharp, M. J., and Richards, K. S. 1990. 'Configuration of the drainage system of Midtdalsbreen, Norway, as indicated by dye-tracing experiments', *J. Glaciol.*, **36**, 89–101.

INDEX

$\delta^{18}O$
    at Austre Okstindbreen 35, 40, 41
    of snow cover 36–7
ablation patterns 269
aufeis 50
Austre Brøggerbreen, Svalbard 65, 68–78
    subglacial drainage 76–8
Austre Okstindbreen 29
    discharge variations 32–6

basal
    conduits 187
    drainage systems 175–88
    processes 329, 338
    sliding 333
    till 188
    water pressures 175, 265
    water velocity 210–14
baseflow 29
boreholes 175–88, 219, 224
    classification 179–82
    connected 180, 243–4
    electrical conductivity profiling 225, 256–8
    seasonal reorganization 239–65
    studies 4–5
    video 191–202
    water chemistry 229–35
    water level 175, 191, 194
    water level and water quality 245–50
Burroughs Glacier, Alaska 16

channelized drainage system 68, 103, 139
chemical weathering in sub-aerial environments 60–2
cold-based glacier, hydrochemistry 47–62
conduit system
    braided 187
    broad anastomosing 187
cross-correlation analyses 250–2
crushing 165–7

delayed flow 68, 139, 155
DEM 275, 305
dissolution experiments 139, 140–1
distributed drainage system 65, 68, 81, 103, 129, 139, 187, 330
    water pressure 330–3
drainage system reorganization 214–16, 239
drilling-produced voids 198–9
dye return curves 106–7
dye tracing experiments 103–17, 129, 272, 276–9

EC stratification 199
electrical conductivity 191, 205, 208–9, 225, 239
electrical conductivity sensors 207–8
englacial channels 194–8, 200–1

field studies 2–3
    modelling 3–4
firn hydrology 21–2
forcing–response plots 252–6

geothermal activity 81–99
geothermal fluid recruitment 81
glacier
    bed 201–2
    drainage systems
      borehole studies 4–5
      surface/subsurface linkages 8–9
    geometry 269, 273–6
    hydrology 31–2, 119–37, 191, 269
      of Austre Okstindbreen 31–2
      of water storage 34
    meltwater
      chemical evolution 41
      quality 7
    motion 119–37
      sediment flushes and 134
      thermal regime/meltwater quality 7
ground survey 272
groundwater flow 81

H_2S concentration 81

Haut Glacier d'Arolla, Switzerland 65, 68–78, 103, 219
    borehole video 191–202
    drainage reconstruction 273–6
    glacier geometry 273–6
    modelling of glacier hydrology 269–97, 299–326
    post-mixing chemical reactions 139–53
    subglacial drainage 67–8, 74–6, 239–65
    suspended sediment 155–72
    water balance 279–81
hot water drilling 221
hydraulic
    conductivity 217
    geometry 103, 104
hydraulics 175
hydrochemistry 7, 30, 81, 139, 147, 219–20, 269, 281–91
    of cold-based glacier 47–62
    dissolution experiments 51
    of subglacial drainage system 65–78
hydrographs 147
hydrological model 269–97
    distributed 293–6
    distributed, physically based model 299–326
    ice-sheet 329–39
    lumped 291–3

ice-bed separation 329
ice-cap 88
ice-cored moraine 49
ice marginal channels 49
ice sheets
    hydrological models 9–10
    simulations 329–39
ice thickness 272
icing 50, 59–60
intraglacial throughflow velocities 96

Jökulá á Sólheimasandi, Iceland, geothermal activity 81–99

laboratory dissolution 139, 140–1,
    155
linked-cavity system 136

meltwater chemistry
    atmospheric influences 6
    bulk 7–8
    lithological influences 6
Midtdalsbreen, Norway 119–37
modelling, integrated approach to
    269–97
Myrdaksjökull 81
    subglacial topography 84, 88–90

naled 50
nearest neighbour analysis 182–5
Norway 29, 119–37

particle size 159–61
post-mixing chemical reactions
    139–53

quickflow 68, 139, 155

radio-echo sounding 272
rainfall 43
R-channel 136
repeated wetting 167–9

Scott TurnerBreen, Svalbard 47–62
sediment exhaustion 131
sediment flushes 119, 120, 131
    glacier motion and 134

sediment supply, rainfall-induced
    variations 128
sliding model 329
Small River Glacier, British
    Columbia 175
snow chemistry 22–4
    solute flushing 24
snow hydrology 16–21
soil-bed drainage 187
solute acquisition 66–7, 91, 98,
    155
South Cascade Glacier 20, 21
stratification 226
subglacial
    channels 103
    formation 263–5
    convective hydrothermal
        circulation 81
    drainage system 9, 119, 134–6,
        228–9, 258–63, 299, 329
    flow conditions 103–17
    geothermal system 92–5
    hydraulic conditions 216
    hydraulic potential 306
    hydrological system 119, 134–6,
        228–9
    hydrology
        beneath sub-Polar glaciers
            67–8
        submodel 299, 302–5
    karst 175
    meltwaters 219–35
    sediment layer 134

sediment velocities 135
water fluxes 265
water pressures 134, 299
water provenance 214–16
water quality 205–17
water velocities 135, 216
surface ablation 279
surface energy balance model 279,
    299, 300
    equations 324–6
surface flow routing model 299,
    300–2
suspended sediment 119–37, 147,
    155–72, 265, 269, 281–91
    concentration 57
    modelling 123–9
    rating curves 125

Trapridge Glacier, Yukon Territory,
    Canada 205
turbidity 205, 208
turbidity sensors 206–7
two-component mixing model 291

velocity–discharge relationships
    103–17
    hysteresis 107, 114
video camera, miniature 191

water balance 269
water quality 91–2, 199–200
    basal 205–17
    modelling of 269–97

There is in easing scientific evidence t... ...ly changing the earth's climate. Greenhou... ...emi... ...ssi... use are altering the atmosphere, creating a... uncertain fu... re of global wa...ing, altered patterns of precipitation, and sea level rise for the generations to come. In an effort to prevent further damage to the fragile atmosphere, and in the belief that action is required now, the scientific community has been prolific in its dissemination of information on climate change. This literature has brought climate change the attention it surely warrants, but it has inundated the public with a deluge of information that cannot be easily absorbed.

Inspired by the Intergovernmental Panel on Climate Change's Second Assessment Report, Jepma and Munasinghe set out to create a concise, practical, and compelling approach to climate change issues. In *Climate Change Policy: Facts, Issues, and Analyses,* they deftly explain the implications of global warming and the risks involved in attempting to mitigate climate change. They look at how and where to initiate action, and what sort of organization is needed to implement the changes.

After reviewing the best scientific evidence available, the authors describe how tools and insights derived from economics and related disciplines can be used to address global climate change. They clearly explain how greenhouse abatement policies are most likely to evolve, focusing particular attention on the position of developing countries. The authors provide a comprehensive overview of the potential damage arising from climate change, the costs of no-action scenarios, and the potential for technical response actions. They also offer a rigorous evaluation of the mitigation strategies and policies currently under consideration.

This book represents a much-needed synopsis of climate change and its real impact on society. It will be an essential text for climate change researchers, policy analysts, university students studying the environment, and anyone else with an interest in climate change issues.